Principles of Engineering Economic Analysis

Principles of Engineering Economic Analysis

Fourth Edition

John A. White
Georgia Tech University

Kenneth E. Case
Oklahoma State University

David B. Pratt
Oklahoma State University

Marvin H. Agee

John Wiley & Sons, Inc.
New York   Chichester   Weinheim   Brisbane   Singapore   Toronto

ACQUISITIONS EDITOR	Wayne Anderson
MARKETING MANAGER	Karen Allman
PRODUCTION EDITOR	Ken Santor
DESIGNER	Kevin Murphy
ILLUSTRATION COORDINATOR	Gene Aiello

This book was set in Times Roman by Bi-Comp, Inc. and printed and bound by the Hamilton Printing Company. The cover was printed by Phoenix Color Corporation.

This book is printed on acid-free paper. ∞

The paper in this book was manufactured by a mill whose forest management programs include sustained yield harvesting of its timberlands. Sustained yield harvesting principles ensure that the numbers of trees cut each year does not exceed the amount of new growth.

Copyright © 1998 John Wiley & Sons, Inc. All rights reserved.

No part of this publication may be reproduced, stored in a retrieval system or transmitted in any form or by any means, electronic, mechanical, photocopying, recording, scanning or otherwise, except as permitted under Sections 107 or 108 of the 1976 United States Copyright Act, without either the prior written permission of the Publisher, or authorization through payment of the appropriate per-copy fee to the Copyright Clearance Center, 222 Rosewood Drive, Danvers, MA 01923, (508) 750-8400, fax (508) 750-4470. Requests to the Publisher for permission should be addressed to the Permissions Department, John Wiley & Sons, Inc., 605 Third Avenue, New York, NY 10158-0012, (212) 850-6011, fax (212) 850-6008, E-Mail: PERMREQ@WILEY.COM.

Library of Congress Cataloging in Publication Data:
White, John A., 1939–
    Principles of engineering economic analysis / John H. White,
Kenneth E. Case, David B. Pratt.—4th ed.
        p.   cm.
    Includes index.
    ISBN 0-471-11027-2 (cloth : alk. paper)
    1. Engineering economy.   I. Case, Kenneth E.   II. Pratt, David B.
III. Title.
TA177.4.W48   1998
658.15'5—dc21   97-20876
                                                            CIP

Printed in the United States of America

10 9 8 7 6 5 4 3 2 1

Preface

This textbook is for those who are interested in learning the principles involved in analyzing economic investment alternatives. Although the material is useful for accounting, economics, finance, and management students, its primary audience will be engineering and engineering technology students. Courses for which the book is intended are variously titled engineering economy, economic analysis, industrial economy, economic decision analysis, engineering decision analysis, and engineering economic analysis.

Over the past 30 years we have taught engineering economic analysis to more than 6,000 persons, including undergraduate and graduate students from engineering and nonengineering academic programs. Additionally, we have taught refresher courses for professional engineering registration examinations and continuing education short courses on the subject. However, in addition to teaching others to perform economic analyses, we have also put into practice the principles and techniques we have taught. Specifically, our consulting and industrial experience required the application of the material presented here. Based on that experience, we perceived a need for a textbook that provided a unified treatment of economic analysis principles. We also felt the need to address the subject from a *cash flow viewpoint*.

The following are some of the important features of this textbook.

1. A cash flow approach is taken throughout.
2. A systematic seven-step approach is given for performing a comparison of investment alternatives.
3. The study period must be specified explicitly instead of unknowingly employing a least common multiple of lives study period.
4. Comprehensive treatments are provided of:
   a. Cost concepts, including cost terminology, cost estimation, accounting principles, nonmonetary considerations, and cost of capital determinations.
   b. Inflation, as it affects the economic comparison of investment alternatives.
   c. MACRS, ACRS, and pre-1987 methods of depreciation.
   d. Income taxes, including capital gains and losses, depreciation recapture, and depletion.
   e. Economic analyses in the public sector, emphasizing benefit-cost analysis procedures and the revenue requirements method.
   f. Break-even, sensitivity, and risk analyses, with the latter including analytical and simulation approaches.
5. Approximately 200 worked-out examples and 400 problems are provided, with answers to even-numbered problems given at the end of the text.

6. Section numbers are provided for material covered by each end-of-chapter problem to facilitate a self-study program.

7. Modeling uniform, gradient, and geometric series, modeling changing interest rates, and modeling equity amounts in loan payments are presented to emphasize the importance of *modeling* cash flow series instead of using more conventional cookbook approaches.

8. Tables of compound interest factors are provided, including both discrete and continuous compounding, discrete and continuous flows, and gradient and geometric cash flow series.

9. Separate chapters on measuring investment worth and the comparison of alternatives are provided to permit a detailed understanding of the measures of investment worth prior to their use in comparing alternatives.

10. Methods of measuring investment worth include present worth, annual worth, future worth, capitalized worth, payback period, savings/investment ratio, internal rate of return, and external rate of return methods.

11. Methods of comparing investment alternatives include both ranking and incremental cash flow approaches.

12. Depreciation is covered in conjunction with the discussion of income taxes, since depreciation is not a cash flow and it influences cash flows through taxation.

13. Replacement analysis is covered in the discussion of comparison of alternatives, since a cash flow analysis of replacement problems is essentially no different than any other type of investment alternative.

14. The data have been packaged in a eight-chapter format in the sequence most natural to the coverage of the material.

15. The text is practice-oriented; only material that is required in the performance of engineering economic analyses is presented, since we believe that including tangential and purely theoretical material has a deleterious effect on the student.

16. A detailed Instructor's Manual has been prepared that includes detailed solutions to all exercises.

The manuscript has been tested in the classroom. The orientation of the material and its organization have been influenced by the recommendations of numerous colleagues and undergraduate and graduate students.

All eight chapters have been developed from our collective experience of teaching the subject to engineering sophomores from all disciplines with the exception of Section 5.7 on performing supplementary analysis. The material in Section 5.7 has been covered in a senior-level elective course. Although elementary calculus is a sufficient prerequisite for the majority of the text, an introductory course in probability theory is suggested as a prerequisite for Section 5.7.

A web site has been established for this text on the Wiley server. It contains a description of the book and a description of the student workbook. Additionally, it provides a sample case study (from the workbook), Excel template files for downloading, and a "Latest Information" section. The latest information will include recent news for students and faculty relative to engineering economics—for example, news about tax law changes, etc. The URL for the Web Site is www.wiley.com/college/whitePEEA.

Many people influenced the development of this textbook. In particular, we have benefited from either taking coursework from or teaching with John R. Canada, Wolter J. Fabrycky, Richard S. Leavenworth, William T. Morris, G. T. Stevens, Jr., and Gerald J. Thuesen, who have authored or coauthored one or more texts on the subject. Our approach to the subject has also been influenced by our associations with Ronald G. Askin, Richard H. Bernhard, William E. Biles, Leland T. Blank, James R. Buck, Lynn E. Bussey, Claus Christiansen, Thomas P. Cullinane, Stuart E. Dreyfus, Carl B. Estes, Gerald A. Fleischer, David R. Freeman, Jorge Haddock, Lynwood A. Johnson, Marilyn S. Jones, Raymond P. Lutz, Leon F. McGinnis, Chan S. Park, Robert S. Kaplan, W. J. Kennedy, Jr., Jack R. Lohmann, Paul A. Nelson, M. Wayne Parker, Gunter P. Sharp, Donald R. Smith, William G. Sullivan, J. M. A. Tanchoco, Suleyman Tufekci, and Donovan B. Young. Therefore, the influence of others will be evident in various places in the book. Because it is impossible to credit the many individual contributions, we simply acknowledge gratefully their influence on our understanding of the subject.

Finally, we thank our wives Mary Lib, Lynn, Jan, and Barbara for their patience and understanding during the project. They constantly demonstrated considerable insight on the subject by the gentle reminder, "If you know so much about investments, why aren't we rich?"

John A. White
Kenneth E. Case
David B. Pratt
Marvin H. Agee

Contents

Chapter 1

Introduction

1.1 BACKGROUND

The subject matter discussed in this book is variously referred to as *economic analysis, engineering, economy, economic justification, capital investment analysis,* and *economic decision analysis,* among other names. Traditionally, the application of economic analysis techniques in the comparison of engineering design alternatives has been referred to as *engineering economy.* However, the emergence of a widespread interest in economic analysis in public sector decision making has brought about greater use of the more general term, *economic analysis.*

Our approach to the subject will be a *cash flow approach.* A cash flow occurs when money actually changes hands from one individual to another or, for that matter, from one organization to another, Thus, money received and money dispersed (spent) constitute cash flows. We also will emphasize the fundamentals of economic analysis without dwelling on some of the philosophical issues associated with the subject; for those interested in pursuing the latter, references are provided.

1.2 THE PROBLEM-SOLVING PROCESS

Economic analyses are typically performed as a part of the overall problem-solving process. In designing a new or improved product, a manufacturing process, or a system to provide a desired service, the engineer is involved in performing the following steps.

1. Formulation of the problem.
2. Analysis of the problem.
3. Search for alternative solutions to the problem.
4. Selection of the preferred solution.
5. Specification of the preferred solution.

The problem-solving or engineering design process begins when a decision maker is dissatisfied with something, or recognizes a need, and then decides to do something about it. The process ends with plans for correcting the dissatisfaction or satisfying the need. The term *problem* should therefore be considered in

1

a general and unlimited context. Examples range from easily defined problems such as high material scrap rates in production, unsafe working conditions, and poor utilization of machinery to somewhat hazy problems such as "something must be done to maintain business survival."

The *formulation of the problem* involves the establishment of the boundaries for the problem. While it is tempting to underscope the problem and focus on what appears to be the right problem, care must be taken that the underlying cause/effect relationships are not missed because of underscoping the problem. Likewise, you must be careful not to develop a scope that is so broad that nothing is accomplished. Finally, care must be taken to avoid being overly influenced by the presence of a present solution; frequently, the existence of a present solution biases the problem-solving process and produces incremental improvements, rather than radically new solutions. Somewhere between a narrowly focused "clean up, fix up, paint up" scope and a "save the world" scope lies the appropriate scope for your problem. We have found this to be the most critical step; it deserves the time and attention of engineers and managers.

The *analysis of the problem* consists of a relatively detailed phrasing of the characteristics of the problem, including restrictions and the criteria to be used in evaluating alternatives. Considerable fact gathering is involved, including the real restrictions on the problem, which must be satisfied. Consequently, any budget, quality, safety, personnel, environmental, and service level constraints that may exist are identified.

In this step of the problem-solving process, the problem definition is judged to be critical since, in most cases, the solution of a problem is directly determined by the problem definition. According to Stoll [13], engineering design is accomplished through the problem-solving process. In the context of designing new products for manufacture, engineering design changes are inherent in the process as more complete design information is obtained (from prototype tests, analysis, customer feedback, etc.). Stoll emphasizes that time and effort spent "up front" on careful and thorough definition of the problem is necessary to achieve a design solution having acceptable quality, cost, and delivery schedule.

The *search for alternative solutions to the problem* involves the use of the engineer's creativity in developing feasible solutions to the design problem.

To stimulate the development of creative alternatives, we recommend using a team. Rarely is one person able to generate the range of creative options generated by a team, particularly a team that is diverse in experiences, education, personalities, and ethnicity. In fact, a number of firms require that their employees undergo personality tests to ensure that teams include individuals from the range of personality types represented by the test. Likewise, many firms insist on having gender and racial diversity on the teams, particularly when the solution will impact men and women of various races.

Note: We structured our discussion of the third step of the design process, generating alternative solutions, around teams for two reasons: Teams have become the preferred problem-solving approach in most organizations and so few engineering students have had much experience working on teams. Our educational systems and society in general tend to reward individuals and individual performance. As a result, it is logical for engineering students to become individual problem solvers. Unfortunately, students who are unprepared for or incapable of working effectively on teams are in for a rough time when they are assigned to design teams.

Brainstorming, as well as other group process techniques, should be used to stimulate the creativity of the team members. Likewise, the basic and advanced tools of *total quality management* can be useful in organizing and analyzing the data produced by the team. [1],[2] Finally, to achieve *breakthrough thinking* in your problem-solving activity, keep in mind the following seven principles developed by Nadler and Hibino [9].

1. ***Uniqueness.*** Recognize that each problem is unique and should be approached as such; do not become preoccupied with the present solution or with solutions by others (competitors) to similar problems.
2. ***Purposes.*** Focus on the underlying purposes to strip away nonessential and distracting aspects of the problem; at each step in formulating the problem, ask what the purpose is for every aspect of the solution; eliminate from further consideration those that either have no purpose or have purposes that are not aligned with the overall purpose of the firm or system in question.
3. ***Solution-after-next.*** Recognize that you are designing for the future; think about the next generation of designs, rather than the generation you are working on currently. Focus on the future, far enough to free up your thinking from short-term influences, but not so far that you have unrealistic expectations of what the requirements will be for the next generation. Recognize your solution to the problem will likely have consequences that will generate new problems; anticipate new problems that will be generated and anticipate new customer expectations and new technologies for solving tomorrow's problems.
4. ***Systems.*** Solutions to problems seldom exist in isolation; there is a bigger system that will be impacted by your solution. How will your solution fit within the broader system? What will be the influence of your solution on the larger system? How should the larger system impact your solution? Is there a representative on the design team with a good perspective of the broader system?
5. ***Limited information collection.*** Engineers tend to be data-driven and err on the side of collecting too much data, too soon. This principle reminds us that there is a time for data collection, and it is not at the initiation of the problem-solving process. This principle is particularly appropriate for the first step of the problem-solving process we described, formulating the problem; this is not the time to become well-schooled in all the hows, whys, wherefores of the problem. Take time to be creative. The time for data collection will come soon enough.
6. ***People design.*** Because people are involved in developing solutions and people will be impacted by the solutions, care must be taken to ensure that sensitivites, politics, and egos are dealt with appropriately. If you want those who will be impacted by the solution to "own" it, then they should be involved in developing the solution.
7. ***Betterment timeline.*** One of the worst axioms ever articulated is "if it isn't broken, don't fix it!" There are many times when we want to scream, "if

it isn't broken, break it!" As advocates of continuous improvement, we believe, "fix it before it breaks!" and "whether it is broken or not, improve it!" Underlying the principle is the concept of continuous improvement. The problem-solving process never ends; as soon as you solve today's problem for tomorrow, begin designing for the solution-after-next. Of course, it makes no sense to constantly change a production system; things could be in constant turmoil and steady state conditions would never exist. The point of the principle is captured in "know when to improve it." In defining the principle, Nadler and Hibino noted, "prepare a schedule for change and improvement of a solution when you are implementing it. Identify the elements of the change you are installing that could be changed later to move it toward the solution-after-next." [9, p. 258]

Although the problem-solving or engineering design process was given as a straightforward, sequential process, it is rare for problems to be solved or products, facilities, or systems to be designed in the linear process presented. Feedback and feedforward occur in the process, such that at intermediate points in the design process the problem formulation is re-visited, additional data are collected and analyzed, and the design process, itself, is sub-divided into stages. For example, one might negotiate the problem-solving process in designing sub-systems and then repeat the process at the system level. Likewise, one might perform a "rough-cut" design, first, and then do a detailed design on the rough-cut design selected.

Rouse [10] presents several design frameworks, depending on what one is designing and the level of detail one wants to provide in the framework. For example, he presents a four-step process:

1. Identify product or system goals, functions, and objectives.
2. Identify specific information and control requirements.
3. Develop conceptual design of product or system.
4. Develop final design of product or system.

He also presents the following five-step simplified system design process:

1. Recognition of need.
2. Problem formulation.
3. Requirements analysis.
4. Synthesis of solution.
5. Fabrication of solution.

In designing support systems, Rouse recommends an eight-step design process:

1. Define tasks.
2. Map to general tasks.
3. Map to limitations.
4. Requirements analysis.
5. Cluster requirements.
6. Map to support concepts.
7. Cluster support concepts.
8. Functional analysis.

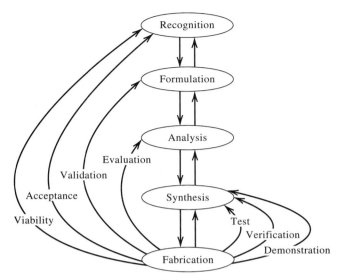

Figure 1.1 Rouse's five-step design framework illustrating the integration of measurement issues and the design process. Reprinted from [11], with permission of the author and publisher.

However, the design framework on which Rouse's book is based is that depicted in Figure 1.1. Advocating the integration of measurement issues with system design, Rouse's framework shows the importance of flexibility in the design process; it also illustrates the importance of information feedforward and feedback.

In this section, we are not attempting to provide a comprehensive treatment of engineering design. Instead, our objective is to establish the importance of economic justification in engineering design.

The *selection of the preferred solution* consists of the measurement of the alternatives, using the appropriate criteria. The alternatives are compared with the constraints, and infeasible alternatives are eliminated. The benefits produced by the feasible alternatives are then compared. Among the criteria considered for choosing among alternatives is the economic performance of each alternative.

The *specification of the preferred solution* consists of a detailed description of the solution to be implemented. Predictions of the performance characteristics of the solution to the problem are included in the specification.

The problem is not solved, however, until the preferred solution is successfully implemented. Upon completion of the problem-solving process discussed before, someone must approve the recommended solution and authorize its implementation. (This is discussed later in Section 1.3.2). Implementation may itself be considered another problem (or project) whereby similar steps are followed to accomplish the solution implementation. Once a preferred solution is accepted and authorized by the appropriate decision maker, implementation must be planned and then managed through to completion. Typically, a project team is assigned to accomplish implementation, with a project manager assigned to be accountable for completion.

EXAMPLE 1.1

As an illustration of the application of the problem-solving approach, consider the case of a locally owned electrical power generating company faced with an air pollution problem that had to be solved. The power plant was using low-grade coal with sulfur content of 1.9% by weight, and the air pollution control board for the state directed the power plant to comply with newly adopted standards for air quality. A consulting engineer was hired to study the problem and recommend the best course of action.

The engineer met with representatives of the air pollution control board, the power company, county and city officials, and suppliers of various fuel alternatives and associated power-generating equipment. An analysis of the problem yielded an agreement to use a 20-year planning horizon, forecasts of power requirements over the planning horizon, and a specification of the criteria to be used in evaluating the alternatives. Specifically, a benefit-cost analysis was to be performed; a cost of $150/ton of sulfur pollutant released into the air was to be employed in the analysis.

A search for alternative solutions to the problem resulted in a consideration of bunker C fuel oil, low-sulfur furnace fuel oil, and natural gas. The methods described in Chapter Seven were employed to compare the alternatives using a benefit-cost analysis; bunker C fuel oil was the recommended energy source.

Based on the report provided by the consulting engineer, the city contracted for the installation of storage tanks for the fuel oil, piping, pumps, and the other equipment necessary for converting the furnaces from coal to fuel oil.[1]

EXAMPLE 1.2

As a second illustration of the problem-solving procedure, a leading manufacturer of automotive bearings was faced with the need to expand its distribution operations. A study team was formed to analyze the problem and develop a number of feasible alternative solutions. After analyzing the problem and projecting future distribution requirements, the following feasible alternatives were determined:

1. Consolidate all distribution activities and expand the existing distribution center, located in Michigan.
2. Consolidate all distribution activities and construct a new distribution center, location to be determined.
3. Decentralize the distribution function and build several new distribution centers geographically dispersed in the United States.

After considering the pros and cons of each alternative, the officers of the company directed the study team to pursue the second alternative.

An extensive plant location study was performed. The location study resulted

---

[1] For additional discussion of the case study on which the above was based, see [5].

in five candidate locations being selected for final consideration. The criteria used to make the final selection included:

1. Land cost and availability.
2. Labor availability and cost.
3. Proximity to supply and distribution points (present and future).
4. Taxes (property and income) and insurance rates.
5. Transportation (access to rail and interstate highways).
6. Community attitudes.
7. Building costs.

Based on site visits to each location, the officers of the company selected a site in Alabama and directed the study team to develop design alternatives for the material-handling system to be used in the distribution center.

Applying the problem-solving procedure, four alternatives were obtained for evaluation. The first alternative involved the use of pallet racks, lift trucks, and flow racks; the second alternative included the use of an automated stacker crane system, lift trucks, conveyors, and flow racks; the third alternative suggested narrow-aisle, guided picking machines, high-rise shelving, conveyors, driverless tractor trains, and lift trucks; the fourth alternative consisted of an automated stacker crane system, a rail-guided picking vehicle system, a sortation and accumulation conveyor system, and high-rise, narrow-aisle lift trucks.

A planning horizon of 10 years was used in performing an economic analysis of each design alternative. The fourth alternative was the most economical and was recommended to top management. Based on the detailed presentation and the economics involved, management approved a budget of $9 million to implement the recommendations of the study team.

As demonstrated by the preceding illustrations, engineers and systems analysts must be prepared to defend their solutions to problems. Economic performance is among the several criteria used to evaluate each alternative. Monetary considerations seldom can be ignored. If economics are not considered in the criteria used in the final evaluation of the alternatives, they are usually involved in an initial screening of the alternatives. In fact, money can be a constraint on the alternatives that can be considered, as well as the basis for the final selection.

This textbook treats in detail the step in the problem-solving process that involves the selection of the preferred solution. The process of measuring cash flows and benefits and the consideration of multiple objectives in selecting the preferred design are also treated.

In comparing alternatives, the *differences* in the alternatives will be emphasized. Consequently, the aspects of the alternatives that are the same normally will not be included in the analysis.

1.3   NONMONETARY CONSIDERATIONS

1.3.1   Multiple Criteria in the Decision Process

The principal emphasis of this textbook is on the use of logical methodology to choose an investment project from among several investment projects available to the decision maker, where the criterion for choice will be some single economic measure of effectiveness. The investment project chosen, however, may be different from the project recommended by the engineer, who based the recommendation solely on an economic criterion. Managers typically have multiple criteria to be considered in reaching a final decision about the alternative to be adopted. Among the factors to be considered are quality, safety, environmental impact, community attitudes, labor-management relationships, cash flow position, risks, system reliability, system availability, system maintainability, system operability, system flexibility, impact on personnel levels, training requirements, comparisons with competitors, impact on the different units within the organization, ego, customers' preferences, capital requirements, and economic justification.

To illustrate a situation in which both economic factors and noneconomic, or nonmonetary, factors are involved, suppose that the management of a firm, because of a significant increase in market demand, has the choice between (1) going to a second shift of production using existing machinery and (2) installing more highly mechanized machinery in order to meet the new production schedules. Clearly, a comparison of the costs for each of these alternatives over a planning horizon is in order, and one objective of management would no doubt be to meet production schedules at the lowest possible annual cost of production.

Although considerable investigation and study are required, many of the factors involved in a decision that may at first seem noneconomic can be expressed in (or reduced to) monetary values. For instance, in the above example, maintenance personnel may have to be trained in the installation and repair of the new machinery. Most likely, these training costs can be determined and charged to the "mechanization" alternative. However, other factors involved are not so easily reduced to monetary values and, indeed, some factors cannot be so reduced. For example, the installation of the mechanized machinery would probably reduce the number of personnel required with the present production system, whereas the second shift alternative would result in additional employees. If persons are laid off or transferred to other jobs within the firm, their salaries could be considered annual savings accruing to the mechanized machinery alternative and, conversely, the salaries of additional employees integral with the second shift alternative would be annual costs. However, layoffs or transfers could have a deleterious effect on the morale of the remaining employees. On the other hand, adding new personnel could have a beneficial effect on morale. Assessing the cost or gain of these expected changes in morale is virtually impossible. Nevertheless, the potential effect on production that morale changes may have must be considered by management.

Factors that affect a decision but cannot be expressed in monetary terms are often called *intangibles* or *irreducibles*. Practically all real-world business decisions involve both monetary and intangible factors. In the above production example, assume that the alternative of a second shift results in a total annual cost of $100,000 for both shifts over a 5-year planning horizon with no persons laid off or transferred or new employees hired. For the mechanized machinery alternative, there would be an estimated annual cost of $85,000 over the 5-year planning horizon with six persons laid off. Thus, this oversimplified decision is reduced to two measures: a monetary annual cost and the single intangible factor of employee morale. If management chooses the second shift alternative, this would imply that management places a higher subjective utility value on "good employee morale" than on the $15,000 difference in annual costs between the two alternatives. That is, the objective of good employee morale is more important than the objective of lowest annual cost (at least for this difference of $15,000). Instances of decision situations in which intangible considerations outweigh monetary ones are frequent, and the engineer should not become distraught over this fact. Within the hierarchy of management responsibilities for a firm, the higher the level of management, the more likely it is that intangible considerations will be given greater subjective weight. Engineering project proposals should therefore reflect a knowledge of intangibles that management will wish to consider.

Miller and Starr made the following observations in regard to the decision-making process.[2]

1. Being unable to satisfactorily describe goals in terms of one objective, *people customarily maintain various objectives.*
2. Multiple objectives are frequently in conflict with each other, and when they are, a *suboptimization* problem exists.
3. At best, we can only optimize as of *that time* when the decision is made. This will frequently produce a suboptimization when viewed in subsequent times.
4. Typically, decision problems are so complex that any attempt to discover *the* set of optimal actions is useless. Instead, people set their goals in terms of outcomes that are *good enough.*
5. Granted all the difficulties, human beings make every effort to be *rational* in resolving their decision problems.

Today's business world is characterized by continual and rapid change. Managers of these businesses are faced with an ever-increasing amount of new information and technology, coupled with ever-increasing global competition. These managers must establish long-term (strategic) goals and objectives for their firms to survive and progress. Example goals might be to (1) shorten the time between the design of a new product or service and customer delivery, (2) improve the quality of the products or services, and (3) reduce the amount of materials inventoried. Associated with each of the goals could be several objectives. By the same token, a single objective may affect each of the goals. For example, the objective of automating a particular area of a manufacturing plant could contribute toward shortening the lead time from design to customer delivery, improving product quality, and lowering the amount of work-in-process inventory. An engineering project to automate the particular area of the plant could generate alternative hardware/software systems to accomplish the objective. The evaluation of each alternative system should therefore include factors that would measure progress toward each of the three goals. These factors would most likely be both economic (initial investment costs and annual operating expenses) and noneconomic (expanded capacity, machine utilization, etc.) in nature.

---

[2] David W. Miller and M. K. Starr, *Executive Decisions and Operations Research,* Second Edition, Prentice-Hall, 1969, pp. 52–53.

1.3.2   Selling the Investment Alternative

Although we will revisit the subject in Chapter Five, it is important to emphasize at the beginning of the book the importance of gaining management acceptance of your investment recommendation. One of the more important ingredients in gaining management approval for an economically and technically justified investment alternative is the sales technique of the individual who presents the alternative to management. It is not unusual for management to adopt a weaker alternative because of the persuasive power of the person who presented the investment alternative.

Remember, *what is said* to support an investment justification and *how it is said* influence the ultimate decision regarding the investment. The strengths and weaknesses of the communication often outweigh the technical aspects of the investment alternative. How well you communicate with the decision makers often will determine whether or not the alternative is funded. To maximize your chance of success, it is important to speak the language of the listeners, to understand what motivates them, to see the world as they see it, and to understand what rules and criteria they use. Your proposal must be structured in such a way that it communicates to *them* the message *you* want to deliver.

Engineers are prone to spend far more time doing the economic analysis and far too little time preparing the presentation to management to sell the preferred investment alternative. Indeed, the expression *paralysis of analysis* comes to mind, as does the saying, "there comes a time in the life of every project when you must shoot the engineers and get on with production." Engineers tend to become enamored of the technical aspects of the solution they have developed or designed and fail to understand that management often speaks financial language, not the language of technology.

Effective managers are concerned with making as good a product or delivering as good a service as possible, in as little time as possible, and for the smallest possible cost. They realize they must deliver value to the owners of the business. If you can convince them that your investment alternative will deliver value beyond that delivered by competing alternatives and beyond the cost of capital to the firm, then it has a very good chance of being accepted and implemented.

In presenting investment recommendations to management, remember many others are being considered; you are competing for limited funds. Even if you do a superb job of analyzing, documenting, and presenting your investment recommendation, the ultimate decision will be influenced by how your investment fits into the firm's overall investment portfolio, how it fits the firm's investment strategy, how it meshes with the firm's goals and objectives, how much risk or uncertainty exists relative to your and other investment alternatives, and how good the competing investments are. Also, the impact of the investment on the various organizational units within the firm will be a factor in the final decision; for example, if you are recommending automating a production department and the maintenance workers are incapable of maintaining the equipment you are recommending, then resistance to your recommendation might arise.

A proposal requiring capital expenditures may require approval at several managerial levels. The level(s) and number of levels required is a function of multiple variables, but particularly the magnitude of the capital request. For example, a departmental manager may have the authority to approve requests up to $10,000, a plant manager may have approval authority up to $100,000, but corporate management must approve requests exceeding $100,000. The decision criteria at each management level may be, and probably are, different. The engineer is therefore reminded again to practice effective communication by using management's rules, words, and decision criteria. In this regard, it should be noted that many senior managers do not have technical degrees or backgrounds. The analyst or engineer making the proposal should therefore explain complex technical topics in a manner which an intelligent nontechnical manager can understand.

Given that a proposed investment project is technically sound and economically justified, the proposal should also clearly specify and emphasize any noneconomic benefits expected to accrue from project implementation. The nature of the noneconomic benefits depends of course on the type of project proposed. However, some example noneconomic benefits might be better utilization of floor space, reduction in computer data entry errors, reduction in material flow congestion and backtracking, safer working conditions, improvements in production scheduling, standardization of spare parts inventory, and so forth. Although managers necessarily have to focus attention on economic benefits, they can and do appreciate that noneconomic benefits directly or indirectly contribute to the long-range progress and success of the business.

It is important to write a proposal that includes an *executive summary*. The executive summary normally reduces the performance characteristics of the recommendation to a few pages. Technical aspects of the recommended solution and the technical details of the economic analysis are usually provided in the main report for the manager who wishes to obtain additional understanding of the information in the executive summary. A face-to-face presentation is also quite important in achieving effective communication.

Communication techniques such as the use of visual aids, voice control, dress, structure, and clarity of expression are necessary to sell the proposed solution. However, it is probably more important to know your audience well and to prepare your proposal using their language. Since the language of managers is often the language of finance, Chapter Eight introduces some fundamental financial terminology used by managers.

1.4   PRESENT ECONOMY EXAMPLES

Most of the methodology presented in this book is based on the fundamental concept that money has a time value. Investment projects that have a life cycle of several years will almost invariably have associated revenues (or savings) and expenses that occur periodically over the life cycle. This series of cash flows constitute a particular cash flow pattern, and alternative projects can have different cash flow patterns over their respective life cycles. In order to compare such alternative projects at a common point in time by an economic measure of effectiveness, the time value of money must be considered.

The time value of money concept simply recognizes that a given dollar amount has different values at different times. In order to adjust these different values to a common reference point on the time scale, an interest rate is used to make the adjustment. This time value of money concept will be developed and illustrated in Chapter Two. Economic measures of effectiveness will also be defined and illustrated in Chapter Four.

There are other situations in which time is not a significant economic factor and thus, the time value of money does not have to be considered. Degarmo [5], one of the early authors on the subject of engineering economy, labeled these types of economic problems as *present economy studies*. He stated that the time value of money did not have to be considered when one or more of the following conditions held: (1) no investment of capital and only out-of-pocket costs were involved, (2) after paying any first cost (capital investment), the long-term costs would be the same for all alternatives or would be proportional to the first cost, and (3) the alternatives would have essentially identical results regardless of the capital investment involved.

In the business and industrial world, present economy studies typically involve the choice among alternative designs, materials, or methods. Eight examples are subsequently presented for illustrative purposes, and the end-of-chapter problems provide additional insight into the nature of such problems. Two of the illustrative examples involve capital investments but the decision question posed can be answered without considering the time value of money.

EXAMPLE 1.3

A particular metal component part can be machined on either an engine lathe or a turret lathe. In either case, the machines are not used exclusively for this particular part. Rather, parts are produced in batches according to a customer's order. If a customer's order is produced on the engine lathe, the following information is relevant:

Set-up time required per order	= negligible
Direct labor and machine time per part	= 10 minutes
Direct labor rate	= $7.50/hour
Machine rate	= $12.00/hour
Cutting tool cost per part	= $0.20/unit

If the part is produced on the turret lathe, a set-up of the machine is required each time a customer's order is processed. The set-up is required only once per order if the order size is less than or equal to 300 units. If the order size is greater than 300 units, then a set-up is required every 300 units. Relevant time and cost data are given below:

Set-up time required per order (both the operator and machine time is charged to the order)	= 2 hours
Direct labor and machine time per part	= 5 minutes
Direct labor rate	= $9.00/hour
Machine rate	= $15.00/hour
Cutting tool cost per part	= $0.10/unit

The material cost per unit is the same regardless of the machine used. Which is the economical machine to use for order sizes of 25, 100, and 500 units?

Let x = the number of units produced per customer order. Then, it will be convenient to develop a cost-per-order formula for each type of machine. That is,
Engine lathe cost per order:

Direct labor cost/minute	= $7.50/60 = $0.125/minute
Direct labor cost/unit	= ($0.125/minute)(10 minute/unit)
	= $1.25/unit
Machine cost/minute	= $12.00/60 = $0.20/minute
Machine cost/unit	= ($0.20/minute)(10 minute/unit)
	= $2.00/unit
Cutting tool cost/unit	= $0.20/unit
Thus, the cost per order is	
$(1.25 + 2.00 + 0.20)/	
unit (x units/order)	= $3.45x

Turret lathe cost per order:

Set-up cost/order	= (n)(2 hour/set-up)
	× ($9.00/hour + $15.00/hour)
	= ($48/set-up)$(n)$
where n	= the number of set-ups
	required per order
Direct labor cost/minute	= $9.00/60 = $0.15/minute
Direct labor cost/unit	= ($0.15/minute)(5 minute/unit)
	= $0.75/unit
Machine cost/minute	= $15.00/60 = $0.25/minute
Machine cost/unit	= ($0.25/minute)(5 minute/unit)
	= $1.25/minute
Cutting tool cost/unit	= $0.10/unit
The cost per order is then	
$48 n + $(0.75 + 1.25 + 0.10)/unit	
(x)	= $48 n + $2.10 x

The comparative costs per order size of 25, 100, and 500 units are calculated as follows:
For an order size of 25 units

Engine lathe cost =	$3.45(25)	= $ 86.25
Turret lathe cost =	$48(1) + $2.10(25)	= $100.50

For an order size of 100 units

Engine lathe cost =	$3.45(100)	= $345.00
Turret lathe cost =	$48(1) + $2.10(100)	= $258.00

For an order size of 500 units

Engine lathe cost =	$3.45(500)	= $1725
Turret lathe cost =	$48(2) + $2.10(500)	= $1146

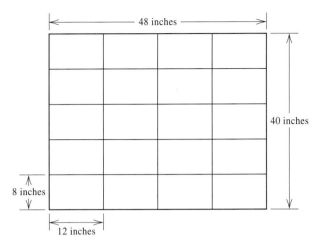

Figure 1.2 Pallet pattern.

EXAMPLE 1.4

The SMA Corporation, a food processing firm, ships cartons of canned food to wholesale distributors via motor freight (tractor and trailer). The cartons are 8 inches wide × 12 inches long × 6 inches high and are stacked in layers on wooden pallets, with the largest surface area of the cartons (8 inches × 12 inches) resting on the pallet. The pallets are 40 inches long × 48 inches wide × 6 inches high and the cartons are arranged on the pallet in the pattern sketched in Figure 1.2. Three layers of cartons are stacked on each pallet and then "strapped" with metal bands to stabilize the load. Each pallet with 60 cartons is then termed a unit load. The total height of the unit load is 24 inches, as shown in Figure 1.3.

The inside dimensions of the trailer (closed van) are 34 feet long × 7 feet 6 inches wide × 6 feet 10 inches high. The freight cost to ship to a particular wholesale distributor is $1000/trailer load. The SMA Corporation therefore wishes to ship the maximum number of unit loads per trip. The company has a choice of two types of wooden pallets. The types are two-way entry and four-way entry, that is, a fork-lift truck (used to place and retrieve the unit loads from the trailer) can insert the forks into the pallet from two directions or from

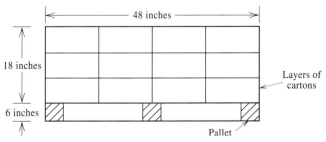

Figure 1.3 Unit load.

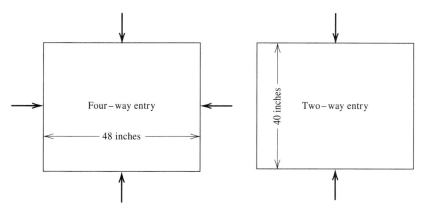

Figure 1.4 Pallet styles.

four directions, as shown in Figure 1.4. The implication of this choice is that, for the two-way entry pallet, the fork-lift truck can only address the pallet perpendicular to the 48-inch dimension of the pallet. The two-way pallets cost $6 each and the four-way pallets cost $8 each. If the unit loads can be stacked only three tiers high inside the trailer (three unit loads per stack), which pallet style will result in the lowest freight cost per unit load? (Assume negligible clearance required between each adjacent stack and between a stack and the walls of the trailer).

The trailer is 7 feet 6 inches (or 90 inches) wide. Since the 40-inch × 48-inch, two-way entry pallet is 48 inches wide, two rows of pallets cannot be stacked in the trailer (in the width direction). The trailer is 34 feet (or 408 inches) long and the number of unit load stacks that can be placed in the trailer length-direction is 408 inches/40 inches = 10.2, or 10 stacks. With 3 unit loads/stack, 10 stacks/row, and 1 row/trailer, 30 unit loads can be shipped per trailer. This results in a unit load cost per load of $1000/30 = $33.33/unit load. The unit cost of the pallets is $6, and the total cost per unit load shipped is $(33.33 + 6) = $39.33.

If the four-way entry pallet is used, two rows of pallets can be stacked in the trailer (in the width direction), with 10 inches of clearance between the rows. That is, 90-inch trailer width − 2(40-inch unit load width) = 10-inch clearance. With a trailer length of 408 inches, the number of load stacks per row is 408 inches/48 inches = 8.5, or 8 stacks/row. Thus, two rows/trailer × 8 stacks/row × 3 unit loads/stack = 48 unit loads/trailer. The freight cost per unit load shipped is then $1000/48 = $20.83. With a unit cost of $8/pallet, the total cost per unit load shipped is $28.33.

The four-way entry pallets should be purchased.

EXAMPLE 1.5

The Single Product Manufacturing Company manufactures a single product but in six different styles. The market area for the product is intrastate only, and the company also owns six retail outlet stores within the state. Because of customer

Table 1.1 Sales Volume Data for Example 1.5

	Sales Volume	
Style	Present Sales Volume (units/month)	Predicted Sales Volume (percent of present volume)
1	860	120
2	5300	140
3	370	200
4	900	100
5	2200	80
6	230	150

preferences, only certain styles are inventoried at each outlet store. The decision at the present time concerns the warehouse space at each outlet store.

Based upon an analysis of past sales records and other current marketing information, an increase in future sales for each store is predicted. An increase in warehouse space at each store location is therefore required. If the company only has $100,000 available for warehouse expansion, how much capital must be borrowed to finance the construction of the needed space?

Relevant data are given in Tables 1.1 and 1.2. Although some of the warehouses have a small amount of unused space currently available, assume they are at capacity. Further, as a gross estimate, assume that the space required is proportional to the volume of product (style) manufactured per month.

The predicted sales for each style are

$$\text{Style 1} = (\ 860)(1.20) = 1032 \text{ units/month}$$
$$\text{Style 2} = (5300)(1.40) = 7420 \text{ units/month}$$
$$\text{Style 3} = (\ 370)(2.00) = \ 740 \text{ units/month}$$
$$\text{Style 4} = (\ 900)(1.00) = \ 900 \text{ units/month}$$
$$\text{Style 5} = (2200)(0.80) = 1760 \text{ units/month}$$
$$\text{Style 6} = (\ 230)(1.50) = \ 345 \text{ units/month}$$

Table 1.2 Space Requirement Data for Example 1.5

	Space Requirements		
Outlet Store	Styles Inventoried	Present Warehouse Space (square feet)	Cost of Additional Space (dollars/square foot)
A	1,2,5	4000	25
B	2,4	2000	35
C	3,5,6	3600	30
D	1,2,4	7400	40
E	1,3,6	700	25
F	3,5,6	4200	30

The present sales indices for each outlet store are

Store A = (860 + 5300 + 2200) = 8360 units/month
Store B = (5300 + 900)   = 6200 units/month
Store C = (370 + 2200 + 230) = 2800 units/month
Store D = (860 + 5300 + 900) = 7060 units/month
Store E = (860 + 370 + 230) = 1460 units/month
Store F = (370 + 2200 + 230) = 2800 units/month

The indices for each outlet store, based on forecast sales, are

Store A = (1032 + 7420 + 1760) = 10212 units/month
Store B = (7420 + 900)   = 8320 units/month
Store C = (740 + 1760 + 345) = 2845 units/month
Store D = (1032 + 7420 + 900) = 9352 units/month
Store E = (1032 + 740 + 345) = 2117 units/month
Store F = (740 + 1760 + 345) = 2845 units/month

An estimate of the total warehouse space needed at each store is

Store A = (10212/8360)(4000 square feet) = 4886 square feet
Store B = (8320/6200)(2000 square feet) = 2684 square feet
Store C = (2845/2800)(3600 square feet) = 3658 square feet
Store D = (9352/7060)(7400 square feet) = 9802 square feet
Store E = (2117/1460)(700 square feet) = 1015 square feet
Store F = (2845/2800)(4200 square feet) = 4268 square feet

The cost of the additional warehouse space needed at each store is

Store A = (4886 − 4000)($25) = $ 22,150
Store B = (2684 − 2000)($35) =   23,940
Store C = (3658 − 3600)($30) =   1,740
Store D = (9802 − 7400)($40) =   96,080
Store E = (1015 − 700)($25) =   7,875
Store F = (4268 − 4200)($30) =   2,040

TOTAL $153,825

Since the $153,825 needed is greater than the $100,000 available, the company would have to borrow $53,825 to finance all the estimated warehouse expansion required.

EXAMPLE 1.6

For one product that a company manufactures, the purchased raw material is 2-inch diameter × 10-feet long extruded aluminum bar stock. This material costs $1.05/pound. Since the company purchases an average of 100,000 pounds/month, it is considering manufacturing the raw material, which would require the purchase of an extrusion machine and a cutoff saw. Raw material for the extruder can be purchased in 4-inch $\times$ 4-inch $\times$ 57-inch-long cast aluminum ingots for $0.50/pound. The company's process engineer estimates the extruding operation would cost $0.30/pound, and the subsequent cutoff operation would cost $0.05/pound. The engineer also estimates there would be a 10% scrap loss of material at the extrusion stage and a 5.7% loss of material at the cutoff operations. (1) If the company manufacturers the extruded rod, what are the estimated annual savings? (2) If the company manufactures the extruded rod, what is the production cost for each rod? (Aluminum density = 0.1 pound/cubic inch).

(1) The present annual cost of the raw material is

$$(100,000 \text{ pounds/month})(\$1.05/\text{pound})(12 \text{ months/year}) = \$1,260,000$$

If the company manufactures the extruded rod from cast ingots, the scrap losses must be considered to arrive at the average quantity of ingots to purchase each month. Thus,

$$\text{let } x = \text{ the pounds of ingots purchased per month}$$

Then, since there is a 10% loss at the extrusion stage, 90% or $0.90x$ will output from this stage. The quantity, $0.9x$, then enters the cutoff stage. Since there is a 5.7% loss at this stage, then 94.3% of the input quantity will output from the cutoff stage, or $(0.90x)(0.943) = 0.8487x$ is the quantity output from the cutoff operation. This quantity must equal 100,000 pounds/month. The monthly purchase quantity should therefore be

$$0.8487x = 100,000 \text{ pounds/month}$$
$$x = 117,827.3 \text{ pounds/month}$$

The monthly input and output quantities for each manufacturing stage are

Extrusion input:	117,827.3 pounds
Extrusion output:	$(117,827.3)(.90) = 106,044.6$ pounds
Cutoff input:	106,044.6 pounds
Cutoff output:	$(106,044.6)(.943) = 100,000$ pounds

The monthly production costs are

Raw material costs:	$(117,827.3 \text{ pounds/month})(\$0.50/\text{pound})$ = $58.913.65/month
Extrusion costs:	$(117,827.3 \text{ pounds/month})(\$0.30/\text{pound})$ = $35,348.19/month
Cutoff costs:	$(106,044.6 \text{ pounds/month})(\$0.05/\text{pound})$ = $5302.23/month

The total costs are

$$(\$58,913.65 + \$35,348.19 + \$5302.23)(12)$$
$$= \$1,194,768.80/\text{year}$$

Thus, the annual savings from manufacturing the extruded rod are

$$\$1,260,000 - \$1,194,768.80 = \$65,231.20$$

(2) The volume of each extruded rod is

$$(\pi)(10)(12) = 377.00 \text{ cubic inches}$$

and the weight of each rod is

$$(377.00 \text{ cubic inch})(0.1 \text{ pound/cubic inch}) = 37.70 \text{ pounds}$$

The number of rods extruded per month is

$$(100{,}000 \text{ pounds/month}) \div (37.70 \text{ pounds/unit}) = 2652.52 \text{ units/month}$$

The cost of an extruded rod is

$$\frac{(\$58{,}913.65 + \$35{,}348.19 + \$5{,}302.23)/\text{month}}{2652.52 \text{ units/month}}$$

of \$37.54/unit.

EXAMPLE 1.7

The National Honor Society at a local high school wants to raise money by selling Christmas trees. However, they can't decide how many trees to order. If they place their order early, they can purchase trees for \$5; they believe they can sell them for \$25. The lot they want to use will cost them \$2,000. Any trees unsold at the end of the selling season are disposed of at no cost or revenue. Based on past experience, they estimate that the demand for trees will have the following probability distribution.

Number Demanded	Probability	Number Demanded	Probability
200	0.10	400	0.20
250	0.10	450	0.15
300	0.15	500	0.10
350	0.20		

Based on the probabilities, if they purchase 400 trees, their cost will be \$2,000 for the trees and \$2,000 for the lot; their expected income will be

$$\$25[200(0.10) + 250(0.10) + 300(0.15) + 350(0.20) + 400(0.20 + 0.15 + 0.10)]$$

or \$8,500. Hence, their expected profit will be \$4,500. Performing a similar calculation for purchase quantities equal to the various demand quantities yields the results in Table 1.3.

From Table 1.3, the optimum purchase quantity is 450 trees. There is a probability of 0.10 of having two few trees and a probability of 0.75 of having too many trees. The expected number of trees requiring disposal after the season is

$$250(0.10) + 200(0.10) + 150(0.15) + 100(0.20) + 50(0.20)$$

or 97.5 trees.

Table 1.3 Number of Trees Sold for Various Demands and Purchase Quantities

SP =	$25						
VC =	$5						
FC =	$2,000						

		Number of Trees Purchased					
Demand	Probability	200	250	300	350	400	450	500
200	0.10	200	200	200	200	200	200	200
250	0.10	200	250	250	250	250	250	250
300	0.15	200	250	300	300	300	300	300
350	0.20	200	250	300	350	350	350	350
400	0.20	200	250	300	350	400	400	400
450	0.15	200	250	300	350	400	450	450
500	0.10	200	250	300	350	400	450	500
exp. units sold =		200	245	285	317.5	340	352.5	357.5
revenue =		$5,000.00	$6,125.00	$7,125.00	$7,937.50	$8,500.00	$8,812.50	$8,937.50
variable cost =		($1,000.00)	($1,250.00)	($1,500.00)	($1,750.00)	($2,000.00)	($2,250.00)	($2,500.00)
fixed cost =		($2,000.00)	($2,000.00)	($2,000.00)	($2,000.00)	($2,000.00)	($2,000.00)	($2,000.00)
profit =		$2,000.00	$2,875.00	$3,625.00	$4,187.50	$4,500.00	$4,562.50	$4,437.50

SP: sales price per trees
VC: variable cost (purchase price) per tree
FC: fixed cost (rental cost for lot)

EXAMPLE 1.8

A local retailer sells a product that has a constant demand rate over the year, averaging 10,000 units sold per year. Each unit of product costs the retailer $10. Each time an order is placed, it costs $150 to process and receive the order. The retailer estimates that it costs $0.25 per year for each dollar tied up in inventory; the estimated inventory carrying cost includes insurance, taxes, storage, and interest costs on money tied up. The retailer wants to determine the optimum size of a replenishment order for the product.

Letting D represent the annual demand, C represent the unit cost for the product, I represent the inventory carrying cost rate, A represent the cost of placing and processing an order, and Q represent the purchase quantity for the product, the total annual cost (TC) for the product can be represented as

$$TC = CD + AD/Q + ICQ/2$$

The first term (CD) represents the amount spent annually in purchasing the units; the second term (AD/Q) represents the amount spent on placing and processing orders annually, since D/Q orders will be placed each year; and the last term ($ICQ/2$) represents the inventory carrying cost, based on an average inventory level of $Q/2$ during the year.

To determine the value of Q that will minimize the total annual cost, take the derivative of TC with respect to Q, set the result equal to zero, and solve for Q. (To verify that the value obtained is, indeed, the optimum value, take the second derivative of TC with respect to Q; since, the second derivative is positive for all values of Q, the cost function is convex and the value obtained is the minimum value.) The result obtained for Q is

$$Q = [2AD/IC]^{1/2}$$

For the example in question, $A = \$150$, $C = \$10$, $I = 25\%$, $D = 10,000/\text{yr}$, and

$$Q = [2(150)(10,000)/(0.25)(10)]^{1/2}$$
$$Q = 1,095 \text{ units/order}$$

EXAMPLE 1.9

The Ajax Machining Company has a number of identical automatic machines that are manually loaded and unloaded; the time required to load a machine is 2 minutes, and the time to unload the machine is 1.5 minutes. During the time the operator is loading or unloading a machine, no production can occur on the machine. However, while the machine is operating during its 30-minute automatic machining cycle, the operator can perform certain independent activities, such as inspecting parts, packaging parts, completing paperwork, and walking between machines; a total of 5 minutes is spent on independent activities per machine cycle. Each operator costs $15.00 per hour, and each machine costs $145.00 per hour. The objective is to assign a number of machines to each operator such that costs per unit produced is minimized.

From the perspective of a machine, its cycle time is $2 + 30 + 1.5$, or 33.5 minutes; from the perspective of the operator, a cycle totals $1.5 + 2 + 5$, or 8.5 minutes. Therefore, under ideal conditions, an operator should be able to handle 33.5/8.5 or 3.9412 machines. Assuming operators do not have shared responsibility for machines, an integer number of machines should be assigned to an operator. If three are assigned, then there will be labor idle time; if four are assigned, then machines will be idle. Which is the most economic?

If a three-machine assignment is made, the cycle time will be equal to a machine cycle, or 33.5 minutes. Assuming one unit of product is produced per cycle, 3 units will be produced every 33.5 minutes. The cost incurred during the 33.5 minute cycle will be [\$15 + 3(\$145)](33.5/60), or \$251.25. Hence, the cost per unit produced will be \$83.75.

If a four-machine assignment is made, the cycle time will be equal to an operator cycle, or 4(8.5), which equals 34 minutes. The cost per unit produced will be [\$15 + 4(\$145)](8.5/60), or \$84.29. Hence, to minimize the cost per unit produced, three machines should be assigned to an operator, as depicted in Figure 1.5.

Now, suppose the firm owns 20 machines, how should they be assigned to operators? Based on the preference of three machines per operator, it appears that the best assignment would be six three-machine assignments and one two-machine assignment. However, it is possible that the optimum assignment is some other combination, such as four three-machine assignments and two four-machine assignments. Further, it might be possible to further reduce the cost by having operators share machines. (Notice, what seemed like a fairly simple decision problem has become complex. Hence, we advocate the use of economic

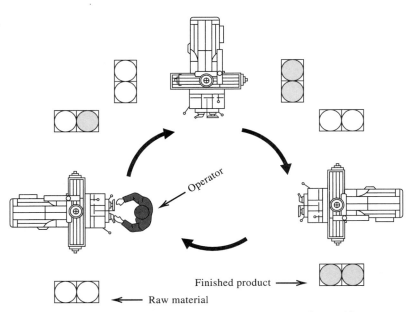

Figure 1.5 Assignment of one operator to three automatic machines.

analyses, rather than relying solely on intuition. For a fuller treatment of the assignment of multiple machines to a single operator, see pp. 94–97 in [6].)

1.5   OVERVIEW OF THE TEXT

The material in this book has been organized in a manner consistent with the logical sequence of steps followed in performing an economic justification of investment alternatives.

Chapter Two presents important fundamental concepts involving the *time value of money*. In fact, Chapter Two provides the foundation for the remainder of the book; therefore the reader must understand the time value of money operations presented in this chapter.

Chapter Four builds upon these fundamental concepts by adding application examples and introducing reality issues that are likely to be encountered in solving time value of money problems. The application examples include evaluation of loans and bonds. The reality issues include consideration of inflation and handling changing interest rates.

Chapter Four presents various *economic measures for determining a project's worth.* Included are the measures of annual worth, future worth, present worth, rates of return, savings/investment ratios, and payback period.

The economic measures for determining project worth are used in Chapter Five for *comparing investment alternatives.* The subjects of sensitivity analysis and replacing capital assets are also emphasized in this chapter.

Chapter Six addresses the issues of *depreciation allowances* and *income taxes* and their incorporation in economic analyses; Chapter Seven presents *benefit-cost analysis* which is often used for the economic analysis of projects in the public sector.

The text concludes with Chapter Eight, which provides a discussion of cost concepts and includes discussions of elements of *costs:* future costs, sunk costs, fixed costs, variable costs, average costs, and an introduction to project cost estimation. The basic concepts of ratio analyses, and activity-based costing are also introduced.

BIBLIOGRAPHY

1. Brassard, Michael, *The Memory Jogger Plus +: Featuring the Seven Management and Planning Tools,* GOAL/QPC, Methuen, MA, 1989.
2. Brassard, M. and D. Ritter, *The Memory Jogger: A Pocket Guide of Tools for Continuous Improvement,* GOAL/QPC, Methuen, MA, 1988.
3. Canada, John R., and Sullivan, William G., *Economic and Multiattribute Evaluation of Advanced Manufacturing Systems,* Prentice Hall, Englewood Cliffs, NJ, 1989.
4. Cox, John L., and Dimsdale, Parks B., "Selling New Ideas Properly Aids Their Acceptance Within an Organization," *Industrial Engineering,* December 1986, pp. 30–41.
5. Degarmo, E. Paul, Canada, John R., and Sullivan, William G., *Engineering Economy,* 7th edition, Macmillan, Inc., New York, 1984.

6. Francis, Richard L., McGinnis, Leon F. Jr., and White, John A., *Facility Layout and Location: An Analytical Approach*, 2nd edition, Prentice Hall, Upper Saddle River, NJ, 1992.

7. Frazelle, Edward H., "Suggested Techniques Enable Multi-Criteria Evaluation of Material Handling Alternatives," *Industrial Engineering*, February, 1985, pp. 42–48.

8. Gessner, Robert A., *Manufacturing Information Systems: Implementation Planning*, John Wiley & Sons, New York, 1984.

9. Nadler, Gerald, and Shozo Hibino, *Breakthrough Thinking: Why We Must Change the Way We Solve Problems, and the Seven Principles to Achieve This,* Prima Publishing and Communications, Rocklin, CA, 1990.

10. O'Guin, Michael C., "Information Age Calls for New Methods of Financial Analysis in Implementing Manufacturing Technologies," *Industrial Engineering*, November 1987, pp. 36–40.

11. Rouse, William B., *Design for Success: A Human-Centered Approach to Designing Successful Products and Systems,* John Wiley & Sons, New York, 1991.

12. Starr, Martin K., *Product Design and Decision Theory*, Prentice Hall, Englewood Cliffs, NJ, 1963.

13. Stoll, Henry W., "Design for Manufacture," Chapter 13 in *Tool and Manufacturing Engineers Handbook,* vol. 5—*Manufacturing Management,* Society of Manufacturing Engineers, 1988.

14. Sullivan, William G., "Models IE's Can Use to Include Strategic, Non-Monetary Factors in Automation Design," *Industrial Engineering,* March 1986, pp. 42–50.

15. Tompkins, James A., White, John A., Bozer, Yavuz A., Frazelle, Edward H., Tanchoco, J. M. A., and Trevino, Jaime, *Facilities Planning,* 2nd edition, John Wiley & Sons, New York, 1996.

PROBLEMS

1. A particular product can be manufactured by either of two machining sequences, S_1 or S_2. Sequence S_1 consists of processing through five machines: M1, M2, M3, M4, and M5. For Sequence S_2, processing is by three machines: MA, MB, and MC.

   The scrap rate for each of these machines is as follows: M1 = 3%, M2 = 5%, M3 = 5%, M4 = 4%, M5 = 4%, MA = 5%, MB = 3%, and MC = 3%. Assume a batch size of 1000 units is to be processed through each sequence. The raw material cost is $1.00 per unit. The total processing cost through sequence S_1 is $5500 and through sequence S_2 is $6200. Which sequence of processing results in the lowest unit cost of final product? (1.4)*

2. A frequently occurring maintenance task is currently performed by a work crew of three persons. This crew typically accomplishes the task in 6 hours. However, when the three-person crew performs the task, only one person is busy 100% of the time. The second person is busy 80% of the time, and the third person is busy only 50% of the time. The task can be performed by either one, two, or three persons.

   The regular wage rate for each of the three persons is $12.00/hour. The overtime (greater than 8 hours) wage rate is $18.00/hour. Whenever the task is performed, tools and equipment are required, and these are valued at $15.00/hour.

   Based on the total cost (labor and equipment) for the task, what is the most economical

---

* Indicates the relevant section in the chapter.

crew size? *Hint:* First determine the total work content (hours) of the task. Then determine the total cost for each crew size. (1.4)

3. A study made in the assembly department of a small electronics manufacturing firm revealed that workers produce from 100 to 120 assemblies in an 8-hour working day. For simplicity, assume the production rates are 100, 110, or 120 units per day. If 100 units are produced, an average of 3% are rejected. When 110 units are produced, an average of 5% are rejected, and 8% are rejected at the 120 units/day rate. The total material cost per unit is $7.50.

   If workers are paid $1.20 per acceptable assembly, what is the preferred production rate from the worker's point of view? Which production rate results in the lowest unit cost for the company? (1.4)

4. Barbara Smith is considering the purchase of a new car for business purposes. She has narrowed her choice to three cars; the cost details are given below:

Car Type	A	B	C
Fixed cost/month (dollars)	$450	$350	$750
Miles per gallon of fuel	25	18	28
Estimated cost of maintenance (dollars/mile)	$ 0.15	$ 0.18	$ 0.12

Ms. Smith estimates she will travel about 2500 miles/month. The cost of fuel is $1.25/gallon.
   a. Based on a total cost/month comparison, determine the most economical car to purchase.
   b. If Car Type C were purchased, what would be the average cost per mile of travel? (1.4)

5. A small grocery store is considering the purchase of a personal computer (PC) system to be able to keep more complete inventory records. There is a choice of three alternative local suppliers. Specifications and cost details for each PC system are given below:

PC Type	A	B	C
Time required to perform all the desired activities (hours/day)	6	8	5
Hourly wage of the operator (dollars/hour)	$10	$10	$10
Fixed cost for system (dollars/day)	$16	$10	$30
Cost of power (dollars/hour)	$ 0.30	$ 0.30	$ 0.60

Assume that the operator of the PC system can do other productive work at the store when not using the microcomputer. Based on a total cost per day comparison.
   a. Determine the most economical system.
   b. Would your purchase recommendation change if the hourly wage were $15? *Note:* Total cost per day equals the sum of the fixed costs, labor cost, and power costs. (1.4)

6. The relationship between cutting tool life (T) in minutes, and cutting speed (V) in feet/minute, is expressed by Taylor's equation:

$$VT^n = K$$

where K is a constant. From the above equation, it is possible to mathematically derive an equation to find the tool life resulting in minimum unit cost. The equation is:

$$T_c = \left(t_c + \frac{c_c}{c_o} \right) \left(\frac{1}{n} - 1 \right)$$

where T_c = tool life for minimum unit cost, minutes
t_c = tool change time, minutes
c_c = cost per cutting edge, dollars/edge
c_o = cost of labor and overhead, dollars/minute
n = Taylor's tool-life exponent

A metal machining company is evaluating three different types of cutting tool inserts for one of their high-volume NC turning operations. The specifications and cost information available for the inserts are as follows:

	Insert Type		
	Tungsten Carbide	Coated Carbide	Ceramic
n	0.22	0.27	0.38
t_c (minutes)	2	2	2
c_c (dollars/edge)	$ 0.20	$ 0.90	$ 1.50
c_o (dollars/minute)	$ 0.5	$ 0.5	$ 0.5
K	150	250	550

For each insert type, determine the tool life for minimum unit cost and the corresponding cutting speed. Which insert type is preferred to maximize the metal removal rate (proportional to cutting speed) and hence the production rate? (1.4)

7. A group of 10 university students is planning a 7-day trip to Daytona Beach, Florida. This destination is 950 miles away from their campus. The group has decided to share automobile expenses equally and is considering three alternative car rentals. For simplicity, assume only the following cost data are relevant. The cost for Car A is $16.00/day plus $0.15/mile traveled, and the capacity of Car A is four people. The cost for Car B is $27.50/day plus $0.10 per mile traveled, with a capacity of four people. The cost for Car C is $35.00/day plus $0.12 per mile traveled, with a capacity of six people. The group estimates that local travel in Daytona will be about 350 miles. (1.4)
   a. What is the most economic alternative travel plan for the group of students?
   b. Given the answer from (*a*), determine the per person cost for travel.

8. The process engineer in a medium-sized machine shop has the choice of machining a particular part on either of two machines. Orders for this part are received regularly, but on a recurring basis. The order size varies. When the order is processed on Machine A, four setup operations of 1 hour each are required. Once set up, the per unit machining time on Machine A is 0.5 hours.

   When the order is processed on Machine B, eight setup operations of 0.75 hour each are required. Once set up, the per unit machining time on Machine B is 0.55 hours. The hourly wage rates for the machine operators are $10 and $9 for Machine A and Machine B, respectively. The hourly overhead rates (including setup time) are $15 and $13.50 for Machine A and Machine B, respectively. (1.4)
   (a) Which machine is preferred if the order size is 100 units? 500 units?
   (b) What is the break-even order size?

9. The American Grinding Company is faced with a decision to select one of the following two types of grinding wheels for a rough grinding operation. The criterion of selection is to minimize the unit cost per cubic inch of metal removed. The unit cost is given by the following expression:

$$\text{Unit cost} = \text{cost of machining} + \text{cost of wheel}$$

$$= \frac{c_o}{R} + \frac{c_w}{G_r}$$

where c_o = Cost of labor and overhead, dollars/hour
R = Material removal rate, cubic inch/minute
c_w = Cost of wheel, dollars/cubic inch
G_r = Grinding ratio, cubic inch of metal/cubic inch of wheel

Also, the grinding ratio and the material removal rate are related by

$$G_r R^n = K$$

where n and K = constants.
The alternatives being considered are:

	Grinding Wheel Type	
	A	B
c_w	0.1	0.09
c_o	10	10
n	1	1
K	4400	2600

For each wheel type, develop a unit cost equation as a function of the variable R. Compare the equations and recommend which grinding wheel to purchase. (1.4)

10. In Example 1.6, assume that the company has decided to manufacture extruded rods and has two alternative sources of aluminum ingots. One source provides 4-inch × 4-inch × 57-inch ingots for $0.75/pound and the other source provides 5-inch × 6-inch × 100-inch ingots for $0.60/pound. The process engineer estimates a 8% scrap loss for the 57-inch-long ingot and a 12% scrap loss in the 100-inch-long ingot at the extrusion stage. The cutoff loss is estimated at 4% for each of the ingot types. Assume the monthly consumption of the extruded bar stock is 100,000 pounds. Which size ingot should the company purchase? (1.4)

11. For Example 1.3, determine the order size at which it will be economical to switch over from the engine lathe to the turret lathe. (1.4) *Hint:* Define turret lathe cost per order equations over the production ranges, $1 \le x < 300$ and $300 \le x < 600$. Then compare engine lathe and turret lathe production costs using the boundary values for each range. Draw conclusions from these comparisons.

12. In Example 1.4, assume that the alternative pallets available to the company are 32 inches long × 32 inches wide × 6 inches high, and 40 inches long × 48 inches wide × 6 inches high, both of two-way entry type. The pallets cost $6 and $10, respectively. Find which pallet style will result in the lowest freight cost per carton. (1.4)

13. In Example 1.6, let there be a fixed overhead cost of $750/month connected with the extruder machine. For simplicity, it is assumed there is a zero fixed cost associated with

the cutoff saw. Using a total monthly production cost comparison (purchase extruded stock vs. produce extruded stock), find the volume of production at which it will be economical for the company to start manufacturing the extruded rods. *Hint:* Define y = the required production volume in pounds/month, which is the output quantity from the cutoff operation, and define x = the required purchase volume in pounds/month, which is the quantity required as input to the extruder. Express total monthly production costs as a function of the variable y; set the two comparison equations equal to each other and solve for y. (1.4)

14. In Example 1.7, suppose each left over tree must be disposed of at a cost of $1 and the selling price averages only $20 per tree. How many should be ordered?

15. A foundry produces castings to order. An order for 20 special castings has been received, at a price of $1,250 per casting. Unfortunately, the casting process is highly variable, and not all castings produced are good. The cost of producing each casting is $550; the additional cost of finishing those castings that are good is $125. If a casting is not good, it is recycled and is judged to have a value of $75; likewise, if a good casting is produced in excess of the customer's order quantity, it is recycled. If fewer than 20 good castings are produced, the customer will accept the good castings produced so long as the number of good castings produced is at least equal to 15. If fewer than 15 castings are produced, none are purchased by the customer. Probability distributions for the number of good castings produced in a batch of varying sizes are given below. How many castings should be scheduled for the production order in order to maximize expected profit?

#Good Castings	Number of Castings Scheduled										
	20	21	22	23	24	25	26	27	28	29	30
10	0.05	0.00	0.00	0.00	0.00	0.00	0.00	0.00	0.00	0.00	0.00
11	0.05	0.05	0.00	0.00	0.00	0.00	0.00	0.00	0.00	0.00	0.00
12	0.05	0.05	0.05	0.00	0.00	0.00	0.00	0.00	0.00	0.00	0.00
13	0.05	0.05	0.05	0.05	0.00	0.00	0.00	0.00	0.00	0.00	0.00
14	0.05	0.05	0.05	0.05	0.05	0.00	0.00	0.00	0.00	0.00	0.00
15	0.05	0.05	0.05	0.05	0.05	0.05	0.00	0.00	0.00	0.00	0.00
16	0.10	0.05	0.05	0.05	0.05	0.05	0.05	0.00	0.00	0.00	0.00
17	0.10	0.10	0.05	0.05	0.05	0.05	0.05	0.05	0.00	0.00	0.00
18	0.15	0.10	0.10	0.05	0.05	0.05	0.05	0.05	0.05	0.00	0.00
19	0.15	0.15	0.10	0.10	0.05	0.05	0.05	0.05	0.05	0.05	0.00
20	0.20	0.15	0.15	0.10	0.10	0.05	0.05	0.05	0.05	0.05	0.05
>20	0.00	0.20	0.35	0.50	0.60	0.70	0.75	0.80	0.85	0.90	0.95

16. A neighborhood market is planning on placing a single order for turkeys in preparation for Thanksgiving. Each turkey will sell for $15. By placing a bulk order early, turkeys can be purchased for $5, each. Any turkeys not sold for $15 will be donated to a homeless shelter; although a tax deduction of $2 can result from the donation, the owner does not choose to claim such deductions. The following estimates have been made concerning the number of turkeys that can be sold by the market.

Turkeys Sold	Probability
10	0.10
15	0.15
20	0.20
25	0.25
30	0.15
35	0.10
40	0.05

(a) How many turkeys should be ordered in order to maximize the expected profit?

(b) If 25 turkeys are ordered, what is the probability of realizing a net profit from the transaction?

(c) If 30 turkeys are ordered, what is the probability of a net loss occuring?

(d) If the owner of the market decides to take a tax deduction for turkeys donated to the homeless shelter, what number ordered will maximize the expected profit?

17. Annual demand for a particular model of facsimile machines is anticipated to be 10,000. Each facsimile machine costs $45. An annual carrying cost rate of 20 percent applies to the facsimile machines. The cost of placing an order for, receiving, and storing a new shipment of facsimile machines is $500. How many facsimile machines should be ordered if total cost is to be minimized?

18. In Example 1.8, suppose the cost of placing and processing an order increases by 20% and the cost of each individual item increases by 20%. What will be the economic order quantity? What if demand also increases by 20%?

19. In Example 1.9, how should the 20 machines be staffed? Do not consider teaming options.

20. In Example 1.9, suppose the load time, unload time, and independent operator activity double in value. What is the economic assignment of machines to operators? Assuming load time, unload time, and independent operator activity increase proportionately, how much can they increase without affecting the optimum assignment of machines to operators?

Chapter 2

Time Value of Money Operations

2.1 INTRODUCTION

Design alternatives are normally compared by using a host of different criteria, including system performance and economic performance. Among the system performance characteristics that are of concern, quality, safety, and customer service considerations are of primary importance. Among the economic performance characteristics normally considered are initial investment requirements, return on investment, and the cash flow (CF) profile. Since the cash flow profiles are usually quite different among the several design alternatives, in order to compare the economic performances of the alternatives, one must compensate for the differences in the timing of cash flows. A fundamental concept underlies much of the material covered in the text: *money has a time value*. By this, we mean that the *value* of a given sum of money depends not only on the amount of money but *when* the money is received.

This chapter introduces the mathematics and basic operations needed to perform economic analysis incorporating time value of money concepts. Sections 2.1 through 2.3 provide motivation and background and help provide a rationale for the mathematical treatment. Sections 2.4 and 2.5 are critical to the remainder of the text. A solid understanding of the mathematics and concepts contained in these sections is essential. Section 2.6 generalizes the basic material in Sections 2.4 and 2.5 for the case of non-annual compounding; a case frequently encountered in practice. Section 2.7 introduces continuous compounding. While important in many areas of engineering practice, this section can be skipped without loss of continuity in the text. Section 2.8 clarifies the definitions of two terms, equivalence and indifference, which are implicitly defined in the earlier sections. Experience has shown that students benefit from explicit treatment and reinforcement of these terms, but the material can be omitted without loss of content. Section 2.9 introduces, by way of an example, the notion of project measures of worth. This section provides a satisfying conclusion to the basic time value of money concepts by illustrating their use in project evaluation. Alternatively, this section can be used as a lead-in to Chapter 4, "Measuring the Worth of

Investments." Either way, the material contained in Section 2.9 must be understood prior to undertaking the material in Chapter 4.

EXAMPLE 2.1

To illustrate the concept of the time value of money, suppose a wealthy individual approaches you and says, "Because of your outstanding ability to manage money, I am prepared to present you with a tax-free gift of $1000. However, if you prefer, I will postpone the presentation for 1 year, at which time I will guarantee that you will receive a tax-free gift of $X." Would you choose to receive the $1000 now or the $X 1 year from now if $X equaled (1) $1000, (2) $1100, (3) $2000, (4) $10,000?

In presenting this situation to numerous students, no students preferred receiving $X in case (1), very few students preferred $X in case (2), most students preferred receiving $2000 a year from now, and all students indicated a preference for $10,000 a year from now. The point is that the value of $1000 1 year from now was perceived to be less than the value of $1000 at present. For most students, the value of $1100 1 year from now was believed to be less than the value of $1000 at present. Only a few students felt that $2000 a year from now was less valuable than $1000 at present. All students believed $10,000 a year from now was more valuable than $1000 at present. Thus, for each individual student, some value (or range of values) of $X exists for which one would be indifferent (i.e., have no preference) between receiving $1000 now versus receiving $X a year from now. If, for example, one is indifferent for $X equal to $1250, then we would conclude that $1250 occurring one year from now has a *present value of* $1000 *for that particular individual.*

Some would argue that the reason the students prefer receiving $X today rather than receiving $X 1 year later is due to the effects of *inflation*. They would contend that the purchasing power of $X decreases over time during periods of inflation. In fact, most people have experienced the impact of inflation and cost of living increases.

While it is true that the *value, economic worth,* or *purchasing power of money* changes over time during periods of inflation and deflation, it is also true that people will prefer to receive $X today rather than 1 year later even when there is no inflation or deflation. One reason for this preference is the *opportunity cost of money*.

The term *opportunity cost* will be developed more fully in Chapter Eight, but, for now, it can be taken to mean the cost of forgoing the opportunity to earn interest, or a return, on investment funds. The $X received today can be invested at some interest rate so that 1 year later the investment will be worth more than $X.

Because people favor current consumption to postponed consumption, it is necessary that postponed consumption be compensated or rewarded with "interest" or a "return on investment." Without such compensation, consumption would not be postponed and capital would not be available for investment.

The concept of inflation and its consideration in evaluating investment alternatives is extremely important. However, before addressing the subject of inflation, it is essential that a number of fundamental concepts be addressed related to the time value of money in the comparison of the economic performances of design alternatives. In the next chapter we consider inflation effects. In this chapter we examine a number of mathematical operations that are based on the time value of money, with an emphasis on modeling cash flow profiles.

The following examples are intended to illustrate the time value of money in the absence of inflation or deflation.

EXAMPLE 2.2

To continue our consideration of different cash flow situations, examine closely the two cash flow profiles given in Table 2.1. Both alternatives involve an investment of $100,000 in ventures that last for 4 years. Alternative A involves an investment in a minicomputer by a consulting engineer who is planning on providing computerized design capability for clients. Since the engineer anticipates that competition will develop very quickly if the plan proves to be successful, a declining revenue profile is anticipated.

Alternative B involves an investment in a land development venture by a group of individuals. Different parcels of land are to be sold over a 4-year period. The land is anticipated to increase in value. There will be differences in the sizes of parcels to be sold. Consequently, an increasing revenue profile is anticipated.

The consulting engineer has available funds sufficient to undertake either investment, but not both. The cash flows shown are cash flows after taxes and other expenses have been deducted. Both investments result in $160,000 being received over the 4-year period; hence, a net cash flow of $60,000 occurs in both cases.

Which would you prefer? If you prefer Alternative B, then you are not acting in a manner consistent with the concept that money has a time value. The $60,000 difference at the end of the first year is worth more than the $60,000 difference at the end of the fourth year. Likewise, the $20,000 difference at the end of the

Table 2.1 Cash Flow Profiles for Two
Investment Alternatives

End of Year (EOY)	CF		
	A	B	(A − B) Difference
0	−$100,000	−$100,000	$0
1	+ 70,000	+ 10,000	+60,000
2	+ 50,000	+ 30,000	+20,000
3	+ 30,000	+ 50,000	−20,000
4	+ 10,000	+ 70,000	−60,000

second year is worth more than the $20,000 difference at the end of the third year. One of the goals of this book is to allow a person to evaluate alternatives such as these on the basis of sound time value of money principles and techniques rather than intuitive arguments.

EXAMPLE 2.3

As another illustration of the impact of the time value of money on the preference between investment alternatives, consider investment alternatives C and D, having the cash flow profiles depicted in Figure 2.1. The cash flow diagrams indicate that the positive cash flows for Alternative C are identical to those for Alternative D, except that the former occur 1 year sooner; both alternatives require an investment of $6000. *If exactly one of the alternatives must be selected,* then Alternative C would be preferred to Alternative D, based on the time value of money.

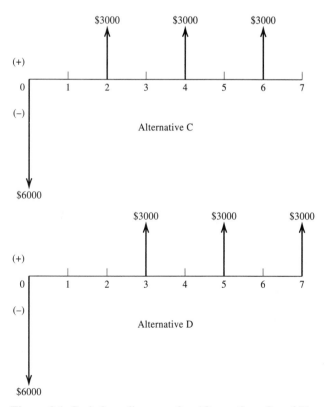

Figure 2.1 Cash flow diagrams for Alternatives C and D.

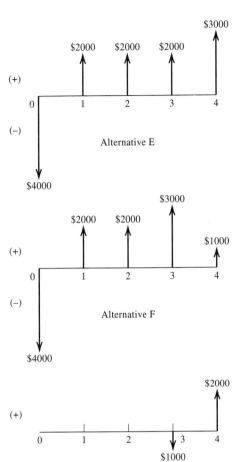

Figure 2.2 Cash flow diagrams for Alternatives E and F.

EXAMPLE 2.4

A third illustration of the effect of the time value of money on the selection of the preferred investment alternative is presented in Figure 2.2. Either Alternative E or Alternative F must be selected; the only differences in the performance characteristics of the two alternatives are economic differences. As shown in Figure 2.2, the economic differences reduce to a situation in which the receipt of $1000 is delayed in order to receive $2000 a year later. For this illustration, we would conclude that most of the students polled earlier would prefer Alternative E to Alternative F, since most of the students preferred X when it equaled $2000.

In order to provide motivation and to provide a familiar scenario, the treatment of time value of money operations will be presented frequently in the context of personal finance. However, each of the concepts examined in this chapter can and does occur in the business world.

In this chapter we emphasize end-of-period cash flows and end-of-period compounding.[1] Depending on the financial institution involved, in personal finance transactions, savings accounts might not pay interest on deposits made in "the middle of a compounding period." Consequently, answers obtained using the methods we describe should not be expected to be exactly the same as those provided by the financial institution.

The beginning-of-period cash flows can be handled very easily by noting that the end of period t is the beginning of period $t + 1$. To illustrate, rental payments might be made at the beginning of each month. However, one can think of the payment made at the beginning of, say, March as having been made at the end of February.

In this and subsequent chapters, end-of-year cash flows are assumed unless otherwise noted. It is realized that in many cases monetary transactions take place during a calendar year, but it is convenient to ignore any compounding effects within a year and deal directly with end-of-year cash flows.

2.2   CASH FLOW DIAGRAMS

In the previous two examples, cash flow diagrams were used to illustrate the timing and magnitude of the cash flows associated with an investment alternative. Cash flow diagramming is a powerful descriptive technique that is used extensively in evaluating economic alternatives. A cash flow diagram is constructed using a segmented horizontal line as a time scale and vertical arrows to indicate cash flows. An upward arrow indicates a cash inflow, and a downward arrow indicates a cash outflow. The arrows are placed along the time scale to correspond with the problem specified timing of the cash flows. The length of the cash flow arrows can be used to suggest the relative magnitude of the corresponding cash flows, but, in most cases, little is gained by precise scaling of the arrows. A rate of interest can be written at the end of the time scale to indicate the interest rate applicable to the analysis. An example of a cash flow diagram for a loan transaction is illustrated in Figure 2.3. The loan illustrated involves a borrowed amount of $1000 at $t = 0$ (today) followed by five annual loan payments of $231 at the end of each of the next five years, $t = 1, 2, 3, 4,$ and 5. The applicable interest rate is 5%/year.

It should be noted that the cash flow diagram illustrated in Figure 2.3 is drawn from the borrower's perspective. This is apparent based on the fact that the loan amount ($1000) is shown as a cash inflow while the payments ($231) are shown as cash outflows. A cash flow diagram for this transaction drawn from the lender's perspective could easily be developed. The direction of all cash flow arrows drawn from the lender's perspective would be the reverse of those shown in Figure 2.3. In essence, a cash inflow to the borrower is a cash outflow from the

---

[1] The exception to this will be the treatment of continuous cash flows.

Figure 2.3 Cash flow diagram for a uniform series.

lender, and a cash outflow from the borrow is a cash inflow to the lender. When drawing cash flow diagrams it is important to ensure that only a single perspective is reflected throughout the drawing.

Drawing cash flow diagrams for economic transactions is important for at least two reasons. First and foremost, cash flow diagrams are powerful communication tools. A cash flow diagram presents a clear, concise, and unambiguous description of the amount and timing of all cash flows associated with an economic analysis. Usually, a well-drawn cash flow diagram can be readily understood by all parties to an economic transaction regardless of whether they have had any formal training in economic analysis.

A second important reason for drawing cash flow diagrams is that they frequently aid in the identification of significant cash flow patterns which might exist within an economic transaction. One such pattern, illustrated in Figure 2.3, is a *uniform* series. In the figure, the loan payments represent a uniform series of length 5 and magnitude 231. In later sections of this chapter, the significance and usefulness of recognizing this pattern and other patterns will become more apparent.

2.3 INTEREST CALCULATIONS

In considering the time value of money, it is convenient to represent mathematically the relationship between the current or *present* value of a single sum of money and its *future* value. Letting time be measured in years, if a single sum of money has a current or *present* value of P, its value in n years would be equal to

$$F_n = P + I_n$$

where F_n is the accumulated value of P over n years, or the *future* value of P, and I_n is the increase in the value of P over n years. I_n is referred to as the accumulated *interest* in borrowing and lending transactions and is a function of P, n, and the *annual interest rate, i*. The annual interest rate is defined as the change in value for $1 over a 1-year period.

Over the years, two approaches have emerged for computing the value of I_n. The first approach considers I_n to be a linear function of time. Since i is the rate of change over a 1-year period, it is argued that P changes in value by an amount Pi each year. Hence, it is concluded that I_n is the product of P, i, and n, or,

$$I_n = Pin$$

and

$$F_n = P(1 + in)$$

This is called the *simple interest* approach.

The second approach used to compute the value of I_n is to interpret i as *the rate of change in the accumulated value of money*. Hence, it is argued that the following relation holds:

$$I_n = \sum_{t=1}^{n} iF_{t-1}$$

where t increments the years from 1 to n and $F_0 = P$ and

$$F_n = F_{n-1}(1 + i)$$

This approach is referred to as the *compound interest* approach.

The approach to be used in any particular situation depends on how the interest rate is defined. Since practically all monetary transactions are based currently on compound interest rates instead of on simple interest rates, we will assume compounding occurs unless otherwise stated.

A convenient method of representing the time value of money is to visualize positive and negative cash flows as though they were generated by a borrower and a lender. In particular, suppose you loaned $1000 for 1 year to an individual who agreed to pay you interest at a rate of 10%/year. The $1000 is referred to as the *principal* amount. At the end of 1 year, you would receive $1100 from the borrower. Thus, we might say that $1100 1 year from now is worth $1000 today based on a 10% interest rate or, conversely, we might say that $1000 today has a value of $1100 1 year from now based on a 10% interest rate.

If the individual borrowed the $1100 for an additional year, then you would be owed $1210, since the interest on $1100 for 1 year equals $(0.10) \times (\$1100)$, or $110. Equivalently, borrowing $1000 for 2 years at 10% interest yields $1210 owed if interest is *compounded* annually. Compound interest involves the computation of interest charges during a time period based on the unpaid principal amount plus any accumulated interest charges up to the beginning of the time period. The interest amount due for a given interest period converts to principal for the purpose of calculating the interest amount due in the subsequent interest period. The relationship between the principal amount and the compounding of interest is given in Table 2.2.

Recall that whenever the interest charge for any time period is based only on the unpaid principal amount and not on any accumulated interest charges, simple interest calculations apply. The interest due for a given interest period does not convert to principal for the purpose of calculating the interest amount due in the subsequent interest period. In the preceding illustration, the amount owed at the end of 2 years would be the $1000 principal plus the interest charges of $(0.10)(\$1000) = \$100$ each year of $1200 total. Thus, in this example, the $1210 for the compounding case compares with the $1200 for the simple interest case. (In some cases, the difference is much more dramatic!)

Table 2.2 Illustrating the Effect of Compound Interest

End of Period	(A) Amount Owed	(B) Interest for Next Period	(C) = (A) + (B) Amount Owed for Next Period*	
0	P	Pi	$P + Pi$	$= P(1 + i)$
1	$P(1 + i)$	$P(1 + i)i$	$P(1 + i) + P(1 + i)i$	$= P(1 + i)^2$
2	$P(1 + i)^2$	$P(1 + i)^2 i$	$P(1 + i)^2 + P(1 + i)^2 i$	$= P(1 + i)^3$
3	$P(1 + i)^3$	$P(1 + i)^3 i$	$P(1 + i)^3 + P(1 + i)^3 i$	$= P(1 + i)^4$
⋮	⋮	⋮	⋮	
$n - 1$	$P(1 + i)^{n-1}$	$P(1 + i)^{n-1}i$	$P(1 + i)^{n-1} + P(1 + i)^{n-1}i = P(1 + i)^n$	
n	$P(1 + i)^n$			

*Notice, the value in column (C) for the end of period $(n - 1)$ provides the value in column (A) for the end of period n.

EXAMPLE 2.5

Person A borrows $4000 from Person B and agrees to pay $1000 plus accrued interest at the end of the first year and $3000 plus the accrued interest at the end of the fourth year. What are the amounts for the two payments if 8% annual simple interest applies? For the first year, the payment is $1000 + (0.08)($4000) = $1320. For the fourth year, the payment is $3000 + (0.08)($4000 − $1000)(3) = $3720.

2.4 SINGLE SUMS OF MONEY

To illustrate the mathematical operations involved in modeling cash flow profiles using compound interest, first consider the investment of a single sum of money, P, in a savings account for n interest periods. Let the interest rate per interest period be denoted by i, and let the accumulated total in the fund n periods in the future be denoted by F. (This F is the same as F_n from the previous section. Therefore, the n subscript will be dropped.) As shown in Table 2.2, assuming no monies are withdrawn during the interim, the amount in the fund after n periods equals $P(1 + i)^n$. As a convenience in computing values of F (the future worth) when given values of P (the present worth), the quantity $(1 + i)^n$ is tabulated in Appendix A for various values of i and n. The quantity $(1 + i)^n$ is referred to as the *single sum, future worth factor* and is denoted $(F|P\ i,n)$. The expression $(F|P\ i,n)$ is read as "the F, given P factor at $i\%$ for n periods." The above discussion is summarized as follows.

Let P = the equivalent value of an amount of money at time zero, or present worth
F = the equivalent value of an amount of money at time n, or future worth
i = the interest rate per interest period
n = the number of interest periods

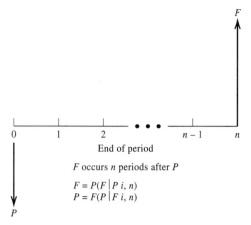

Figure 2.4 Cash flow diagram of the time relationship between P and F in a savings account.

Thus, the future worth is related to the present worth as follows:

$$F = P(1 + i)^n \qquad (2.1)$$

where i *is expressed as a decimal amount.* Equivalently,

$$F = P(F|P\ i,n) \qquad (2.2)$$

where i is expressed as a percentage amount.[1] A cash flow diagram depicting the relationship between F and P for the savings account example is given in Figure 2.4. Remember that F occurs n periods after P.

EXAMPLE 2.6

An individual borrows $1000 at 12% compounded annually. The loan is paid back after 5 years. How much should be repaid?

Using the compound interest tables in Appendix A for 12% and five periods, the value of the $(F|P\ 12,5)$ factor is found to be 1.7623. Thus,

$$F = P(F|P\ 12,5)$$
$$= \$1000(1 + 0.12)^5$$
$$= \$1000(1.7623)$$
$$= \$1762.30$$

The amount to be repaid equals $1762.30.

---

[1] A comma may sometimes be placed after the factor identifier and before the i. Thus, for example, the representations $(F/P\ i,n)$ and $(F/P,i,n)$ are equivalent.

Since we are able to determine conveniently values of F when given values of P, i, and n, it is a simple matter to determine values of P when given values of F, i, and n. In particular, since

$$F = P(1 + i)^n$$

on dividing both sides by $(1 + i)^n$, we find that the present worth and future worth have the relation

$$\boxed{P = F(1 + i)^{-n}} \tag{2.3}$$

or

$$\boxed{P = F(P|F\,i,n)} \tag{2.4}$$

where $(1 + i)^{-n}$ and $(P|F\,i,n)$ are referred to as the *single sum, present worth factor.*

EXAMPLE 2.7

To illustrate the computation of P given F, i, and n, suppose you wish to accumulate \$10,000 in a savings account 4 years from now and the account pays interest at a rate of 9% compounded annually. How much must be deposited today?

$$P = F(P|F\,9,4)$$
$$= \$10,000(0.7084)$$
$$= \$7084$$

2.5   SERIES OF CASH FLOWS

Having considered the transformation of a single sum of money to a future worth equivalent when given a present worth amount and vice versa, we generalize that discussion to consider the conversion of a series of cash flows to present worth and future worth equivalents. In particular, let A_t denote the magnitude of a cash flow (receipt or disbursement) at the end of time period t. Using discrete compounding, the present worth equivalent for the cash flow series is equal to the sum of the present worth equivalents for the individual cash flows. Consequently,

$$P = A_1(1 + i)^{-1} + A_2(1 + i)^{-2} + \cdots + A_{n-1}(1 + i)^{-(n-1)} + A_n(1 + i)^{-n} \tag{2.5}$$

or, using the summation notation,

$$P = \sum_{t=1}^{n} A_t(1 + i)^{-t} \tag{2.6}$$

or, equivalently,

$$\boxed{P = \sum_{t=1}^{n} A_t(P|F\,i,t)} \tag{2.7}$$

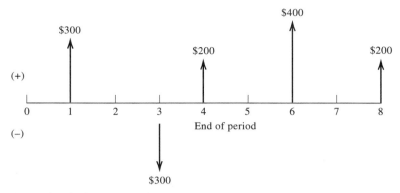

Figure 2.5 Series of cash flows.

EXAMPLE 2.8

Consider the series of cash flows depicted by the cash flow diagram given in Figure 2.5. Using an interest rate of 6%/interest period, the present worth equivalent is given by

$$P = \$300(P|F\,6,1) - \$300(P|F\,6,3) + \$200(P|F\,6,4)$$
$$+ \$400(P|F\,6,6) + \$200(P|F\,6,8)$$
$$= \$300(0.9434) - \$300(0.8396) + \$200(0.7921)$$
$$+ \$400(0.7050) + \$200(0.6274)$$
$$= \$597.04$$

The future worth equivalent is equal to the sum of the future worth equivalents for the individual cash flows. Thus,

$$F = A_1(1 + i)^{n-1} + A_2(1 + i)^{n-2} + \cdots + A_{n-1}(1 + i) + A_n \qquad \textbf{(2.8)}$$

or, using the summation notation,

$$F = \sum_{t=1}^{n} A_t(1 + i)^{n-t} \qquad \textbf{(2.9)}$$

Notice in Equations 2.8 and 2.9 that the exponent of the interest factor counts the number of periods between the cash flow and the time period where F is located.

$$F = \sum_{t=1}^{n} A_t(F|P\,i, n - t) \qquad \textbf{(2.10)}$$

Alternately, since we know the value of future worth is given by

$$F = P(1 + i)^n \qquad \textbf{(2.11)}$$

substituting Equation 2.6 into Equation 2.11 yields

$$F = (1 + i)^n \sum_{t=1}^{n} A_t(1 + i)^{-t}$$

Hence,

$$F = \sum_{t=1}^{n} A_t (1 + i)^{n-t} \tag{2.12}$$

EXAMPLE 2.9

Given the series of cash flows in Figure 2.4, determine the future worth at the end of the eighth period using an interest rate of 6%/interest period.

$$F = \$300(F|P\,6,7) - \$300(F|P\,6,5) + \$200(F|P\,6,4)$$
$$+ \$400(F|P\,6,2) + \$200$$
$$= \$300(1.5036) - \$300(1.3382) + \$200(1.2625)$$
$$+ \$400(1.1236) + \$200$$
$$= \$951.56$$

Alternately, since we know the present worth is equal to \$597.04,

$$F = P(F|P\,6,8)$$
$$= \$597.04(1.5938)$$
$$= \$951.56$$

Note that in computing the future worth of a series of cash flows, *the future worth amount obtained occurs at the end of time period* n. Thus, if a cash flow occurs at the end of time period n, it earns no interest in computing the future worth amount at the end of the nth period.

Obtaining the present worth and future worth equivalents of cash flow series by summing the individual present worths and future worths, respectively, can be quite time-consuming if many cash flows are included in the series. However, with the development and widespread use of pocket calculators (some of which have the capability of performing compound interest calculations automatically), personal computers, minicomputers, and time-sharing systems, it is not uncommon to treat all series in the manner described above.

When such hardware is not available, it is possible to use more efficient solution procedures if the cash flow series have one of the following forms.

Uniform Series of Cash Flows

$$A_t = A \qquad\qquad t = 1, \ldots, n$$

Gradient Series of Cash Flows

$$A_t = \begin{cases} 0 & t = 1 \\ A_{t-1} + G & t = 2, \ldots, n \end{cases}$$

Geometric Series of Cash Flows

$$A_t = \begin{cases} A & t = 1 \\ A_{t-1}(1 + j) & t = 2, \ldots, n \end{cases}$$

2.5.1   Uniform Series of Cash Flows

A uniform series of cash flows exists when all of the cash flows in a series are equal. In the case of a uniform series the present worth equivalent is given by

$$P = \sum_{t=1}^{n} A(1 + i)^{-t} \tag{2.13}$$

where A is the magnitude of an individual cash flow in the series.

Letting $X = (1 + i)^{-1}$ and bringing A outside the summation yields

$$P = A \sum_{t=1}^{n} X^t$$

$$= AX \sum_{t=1}^{n} X^{t-1}$$

Letting $h = t - 1$ gives the geometric series

$$P = AX \sum_{h=0}^{n-1} X^h \tag{2.14}$$

Since the summation in Equation 2.14 represents the first n terms of a geometric series, the closed form value for the summation is given by

$$\sum_{h=0}^{n-1} X^h = \frac{1 - X^n}{1 - X} \tag{2.15}$$

Hence, on substituting Equation 2.15 into Equation 2.14, we obtain

$$P = AX \left(\frac{1 - X^n}{1 - X} \right)$$

Replacing X with $(1 + i)^{-1}$ yields the following relationship between P and A.

$$P = A \left[\frac{(1 + i)^n - 1}{i(1 + i)^n} \right] \tag{2.16}$$

more commonly expressed as

$$P = A(P|A \; i,n) \tag{2.17}$$

where $(P|A, i,n)$ is referred to as the *uniform series, present worth factor* and is tabulated in Appendix A for various values of i and n.

EXAMPLE 2.10

An individual wishes to deposit a single sum of money in a savings account so that five equal annual withdrawals of $2000 can be made before depleting the fund. If the first withdrawal is to occur 1 year after the deposit and the fund pays interest at a rate of 12% compounded annually, how much should be deposited?

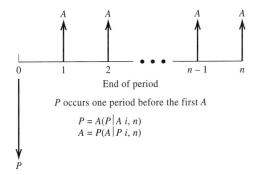

Figure 2.6 Cash flow diagram of the relationship between P and A in a savings account.

Because of the relationship of P and A, as depicted in Figure 2.6, in which P occurs one period before the first A, we see that

$$P = A(P|A\ 12,5)$$
$$= \$2000(3.6048)$$
$$= \$7209.60$$

Thus, if $7209.60 is deposited in a fund paying 12% compounded annually, then five equal annual withdrawals of $2000 can be made. After the fifth withdrawal, the fund will be depleted.

EXAMPLE 2.11

In Example 2.10, suppose that the first withdrawal will not occur until 3 years after the deposit.

As depicted in Figure 2.7, the value of P to be determined occurs at $t = 0$, whereas a straightforward application of the $(P|A\ 12,5)$ factor will yield a single sum equivalent at $t = 2$. Consequently, the value obtained at $t = 2$ must be moved backward in time to $t = 0$. The latter operation is easily performed using the $(P|F\ 12,2)$ factor. Therefore,

$$P = A(P|A\ 12,5)(P|F\ 12,2)$$
$$= \$2000(3.6048)(0.7972)$$
$$= \$5747.49$$

Deferring the first withdrawal for 2 years reduces the amount of the deposit by $7209.60 - $5747.49 = $1462.11.

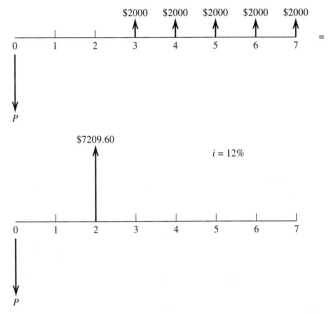

Figure 2.7 Equivalent cash flow diagrams.

The reciprocal relationship between P and A can be expressed as

$$A = P\left[\frac{i(1 + i)^n}{(1 + i)^n - 1}\right] \qquad \textbf{(2.18)}$$

or as

$$A = P(A|P\ i,n) \qquad \textbf{(2.19)}$$

The expression $(A|P\ i,n)$ is called the *capital recovery factor* for reasons that will become clear in Chapter Four. The $(A|P\ i,n)$ factor is used frequently in both personal financing and in comparing economic investment alternatives.

EXAMPLE 2.12

Suppose $10,000 is deposited into an account that pays interest at a rate of 15% compounded annually. If 10 equal, annual withdrawals are made from the account, with the first withdrawal occurring 1 year after the deposit, how much can be withdrawn each year in order to deplete the fund with the last withdrawal? Since we know that A and P are related by

$$A = P(A|P\ i,n)$$

then

$$A = P(A|P\ 15,10)$$
$$= \$10,000(0.1993)$$
$$= \$1993.00$$

EXAMPLE 2.13

Suppose that in Example 2.12 the first withdrawal is delayed for 2 years, as depicted in Figure 2.8. How much can be withdrawn each of the 10 years?

The amount in the fund at $t = 2$ equals

$$V_2 = P(F|P\ 15,2)$$
$$= \$10,000(1.3225)$$
$$= \$13,225$$

Therefore, the size of the equal annual withdrawals will be

$$A = V_2(A|P\ 15,10)$$
$$= \$13,225(0.1993)$$
$$= \$2635.74$$

Thus, delaying the first withdrawal for 2 years increases the size of each withdrawal by $642.74.

The future worth of a uniform series is obtained by recalling that

$$F = P(1 + i)^n \qquad \textbf{(2.20)}$$

Substituting Equation 2.16 into Equation 2.20 for P and reducing yields

$$\boxed{F = A \left[\frac{(1 + i)^n - 1}{i} \right]} \qquad \textbf{(2.21)}$$

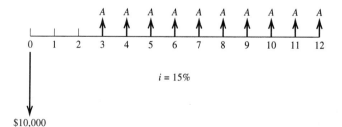

Figure 2.8 Cash flow diagram of deferred payment example.

or, equivalently,

$$F = A(F|A, i,n)$$ **(2.22)**

where $(F|A\ i,n)$ is referred to as the *uniform series, future worth factor*.

EXAMPLE 2.14

If annual deposits of $1000 are made into a savings account for 30 years, how much will be in the fund immediately after the last deposit if the fund pays interest at a rate of 8% compounded annually?

$$F = A(F|A\ 8,30)$$
$$= \$1000(113.2831)$$
$$= \$113,283.10$$

The reciprocal relationship between A and F is easily obtained from Equation 2.21. Specifically, we find that

$$A = F \left[\frac{i}{(1 + i)^n - 1} \right]$$ **(2.23)**

or, equivalently,

$$A = F(A|F\ i,n)$$ **(2.24)**

The expression $(A|F\ i,n)$ is referred to as the *sinking fund factor*, since the factor is used to determine the size of a deposit one should place (sink) in a fund in order to accumulate a desired future amount. As depicted in Figure 2.9, *F occurs at the same time as the last A*. Thus, the last A or deposit earns no interest.

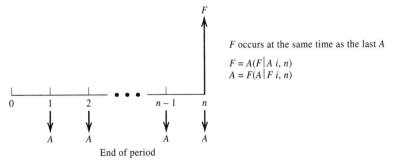

Figure 2.9 Cash flow diagram of the relationship between A and F in a savings account.

EXAMPLE 2.15

If $150,000 is to be accumulated in 35 years, how much must be deposited annually in a fund paying 8% compounded annually in order to accumulate the desired amount immediately after the last deposit?

$$A = F(A|F\,8,35)$$
$$= \$150,000(0.0058)$$
$$= \$870.00$$

2.5.2   Gradient Series of Cash Flows

A gradient series of cash flows occurs when the value of a given cash flow is greater than the value of the previous cash flow by a constant amount, G. Consider the series of cash flows depicted in Figure 2.10. The series can be represented by the sum of a uniform series and a gradient series. By convention, the gradient series is defined to have the first positive cash flow occur at the end of the second time period. The size of the cash flow in the gradient series occurring at the end of period t is given by

$$A_t = (t - 1)G \qquad t = 1, \ldots, n \qquad (2.25)$$

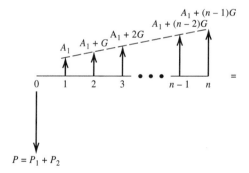

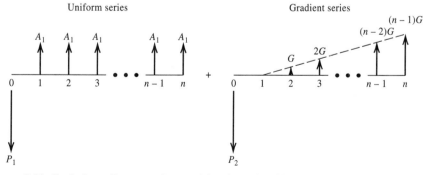

Figure 2.10 Cash flow diagram of a combination of uniform and gradient series in a savings account.

The gradient series arises when the value of an individual cash flow differs from the preceding cash flow by a constant, G. As an illustration, if an individual receives an annual bonus and the size of the bonus increases by $100 each year, then the series is a gradient series. Also, operating and maintenance costs tend to increase over time because of both inflation effects and a gradual deterioration of equipment; such costs are often approximated by a gradient series.

The present worth equivalent of a gradient series is obtained by recalling

$$P = \sum_{t=1}^{n} A_t(1 + i)^{-t} \tag{2.26}$$

Substituting Equation 2.25 into Equation 2.26 gives

$$P = \sum_{t=1}^{n} (t - 1)G(1 + i)^{-t} \tag{2.27}$$

or, equivalently,

$$P = G \sum_{t=1}^{n} (t - 1)(1 + i)^{-t} \tag{2.28}$$

As an exercise, you may wish to show that the summation reduces to

$$P = G \left[\frac{1 - (1 + ni)(1 + i)^{-n}}{i^2} \right] \tag{2.29}$$

Expressing the term in brackets in terms of interest factors already treated yields

$$P = G \left[\frac{(P|A\ i,n) - n(P|F\ i,n)}{i} \right] \tag{2.30}$$

or, equivalently,

$$P = G(P|G\ i,n) \tag{2.31}$$

where $(P|G\ i,n)$ is the *gradient series, present worth factor* and is tabulated in Appendix A.

A uniform series equivalent to the gradient series is obtained by multiplying the value of the gradient series present worth factor by the value of the $(A|P\ i,n)$ factor to obtain

$$A = G \left[\frac{1}{i} - \frac{n}{i}(A|F\ i,n) \right]$$

or, equivalently,

$$A = G(A|G\ i,n) \tag{2.32}$$

where the factor $(A|G\ i,n)$ is referred to as the *gradient-to-uniform series conversion factor* and is tabulated in Appendix A. To obtain the future worth equivalent of a gradient series at time n, multiply the value of the $(A|G\ i,n)$ factor by the value of the $(F|A\ i,n)$ factor.

It is not uncommon to encounter a cash flow series that is the sum or difference of a uniform series and a gradient series. To determine present worth and future worth equivalents of such a composite, one can deal with each special type of series separately.

EXAMPLE 2.16

Maintenance costs for a particular production machine increase by $1000/year over the 5-year life of the equipment. The initial maintenance cost is $3000. Using an interest rate of 8% compounded annually, determine the present worth equivalent for the maintenance costs.

A cash flow diagram for this example is given in Figure 2.11. Note that the cash flow series consists of the sum of a uniform series of $3000 and a gradient series with G equal to $1000. Converting the uniform series to a present worth amount gives

$$P_1 = \$3000(P|A\ 8,5)$$
$$= \$3000(3.9927)$$
$$= \$11,978.10$$

Converting the gradient series to a present worth amount gives

$$P_2 = G(P|G\ 8,5)$$
$$= \$1000(7.3724)$$
$$= \$7372.40$$

(*Notice that* n *equals 5 even though only four positive cash flows are present in the gradient series.*)

Hence, the present worth equivalent of the maintenance costs will be

$$P = P_1 + P_2$$
$$= \$11,978.10 + \$7372.40$$
$$= \$19,350.50$$

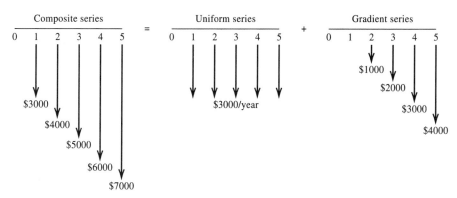

Figure 2.11 Cash flow diagram for a gradient series example.

The uniform series equivalent of the cash flow series given in Figure 2.11 can be determined by summing the base series of $3000 and the uniform series equivalent of the gradient series. Hence,

$$A = \$3000 + G(A|G\,8,5)$$
$$= \$3000 + \$1000(1.8465)$$
$$= \$4846.50$$

Converting the maintenance costs to a future worth amount can be performed either by converting the present worth amount to a future worth amount,

$$F = \$19,350.50(F|P\,8,5)$$
$$= \$19,350.50(1.4693)$$
$$= \$28,431.69,$$

or by converting the uniform series amount to a future worth amount,

$$F = \$4846.50(F|A\,8,5)$$
$$= \$4846.50(5.8666)$$
$$= \$28,432.48,$$

with the $0.79 difference due to round-off error in the interest tables.

EXAMPLE 2.17

Five annual deposits are made into a fund that pays interest at a rate of 8% compounded annually. The first deposit equals $800; the second deposit equals $700; the third deposit equals $600; the fourth deposit equals $500; and the fifth deposit equals $400. Determine the amount in the fund immediately after the fifth deposit.

As depicted by the cash flow diagrams in Figure 2.12, the cash flow series can be represented by the difference in a uniform series of $800 and a gradient series of $100. The uniform series equivalent of the gradient series is given by

$$A = G(A|G\,8,5)$$
$$= \$100(1.8465)$$
$$= \$184.65$$

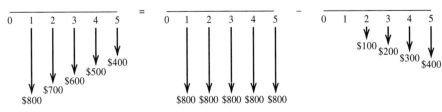

Figure 2.12 Cash flow diagrams for the decreasing gradient series example.

Therefore, a uniform series having cash flows equal to $800 − $184.65, or $615.35, is equivalent to the original cash flow series. The future worth equivalent is found to be

$$F = A(F|A\ 8,5)$$
$$= \$615.35(5.8666)$$
$$= \$3610.01$$

2.5.3   Geometric Series of Cash Flows

The geometric cash flow series, as depicted in Figure 2.13, occurs when the size of a cash flow increases (decreases) by a fixed percent from one time period to the next. If j denotes the percent change in the size of a cash flow from one period to the next, the size of the tth flow can be given by

$$A_t = A_{t-1}(1 + j) \qquad t = 2, \ldots, n$$

or, more conveniently,

$$A_t = A_1(1 + j)^{t-1} \qquad t = 1, \ldots, n \qquad \textbf{(2.33)}$$

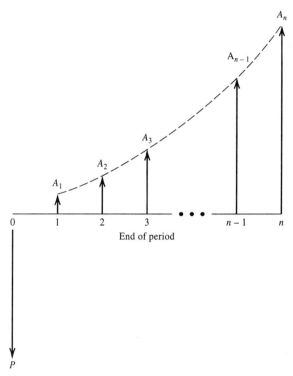

Figure 2.13 Cash flow diagram of the geometric series in a savings account.

The geometric series is used to represent the growth (positive j) or decay (negative j) of costs and revenues undergoing annual percentage changes. As an illustration, if labor costs increase by 10% a year, then the resulting series representation of labor costs will be a geometric series.

The present worth equivalent of the cash flow series is obtained by substituting Equation 2.33 into Equation 2.6 to obtain

$$P = \sum_{t=1}^{n} A_1(1+j)^{t-1}(1+i)^{-t} \tag{2.34}$$

or

$$P = A_1(1+j)^{-1} \sum_{t=1}^{n} \left(\frac{1+j}{1+i}\right)^t \tag{2.35}$$

As an exercise, the reader may wish to show that the following relationship results:

$$P = \begin{cases} A_1 \left[\dfrac{1-(1+j)^n(1+i)^{-n}}{i-j} \right] & i \neq j \\[3ex] \dfrac{nA_1}{1+i} & i = j \end{cases} \tag{2.36}$$

or

$$P = A_1(P|A_1\ i,j,n) \tag{2.37}$$

where $(P|A_1\ i,j,n)$ is the *geometric series, present worth factor* and is tabulated in Appendix A for various values of i, j, and n.

For the case of $j \geq 0$ and $i \neq j$, the relationship between P and A can be conveniently expressed in terms of compound interest factors previously considered

$$P = A_1 \left[\frac{1-(F|Pj,n)(P|Fi,n)}{i-j} \right] \qquad i \neq j, \qquad j \geq 0 \tag{2.38}$$

EXAMPLE 2.18

A company is considering purchasing a new machine tool. In addition to the initial purchase and installation costs, company management is concerned about the machine's maintenance costs. The maintenance costs of the machine tool are expected to be $1000 at the end of the first year of the machine's life and increase 8% per year thereafter. The expected life of the machine tool is 15 years. Company management would like to endow a maintenance fund for expected costs. If the endowment account earns 10% per year compounded annually, how much money must be initially deposited in the account?

$$P = A_1(P|A_1, 10, 8, 15)$$
$$= \$1,000(12.0304)$$
$$= \$12,030.40$$

The future worth equivalent of the geometric series is obtained by multiplying the value of the geometric series present worth factor and the $(F|P\ i,n)$ factor to obtain

$$F = \begin{cases} A_1 \left[\dfrac{(1+i)^n - (1+j)^n}{i-j} \right] & i \neq j \\ nA_1(1+i)^{n-1} & i = j \end{cases} \qquad (2.39)$$

or

$$F = A_1(F|A_1\ i,j,n) \qquad (2.40)$$

where $(F|A_1\ i,n)$ is the *geometric series, future worth factor* and is tabulated in Appendix A. From Equation 2.39, notice that $(F|A_1\ i,j,n) = (F|A_1\ j,i,n)$.

EXAMPLE 2.19

An individual receives an annual bonus and deposits it in a savings account that pays 8% compounded annually. The size of the bonus increases by 10% each year; the initial deposit was $500. Determine how much will be in the fund immediately after the tenth deposit.

In this case, $A_1 = \$500$, $i = 8\%$, $j = 10\%$, and $n = 10$. Thus, the value of F is given by

$$F = A_1(F|A_1\ 8,10,10) = A_1(F|A_1\ 10,8,10)$$
$$= \$500(21.7409)$$
$$= \$10,870.45$$

The interest factors developed to this point are summarized in Table 2.3. Values of the factors given in Table 2.3 are provided in Appendix A.

Table 2.3 Summary of Discrete Compounding Interest Factors

To Find	Given	Factor	Symbol	Name
P	F	$(1 + i)^{-n}$	$(P\|F,i,n)$	Single payment, present worth factor
F	P	$(1 + i)^n$	$(F\|P,i,n)$	Single payment, compound amount factor
P	A	$\dfrac{(1 + i)^n - 1}{i(1 + i)^n}$	$(P\|A,i,n)$	Uniform series, present worth factor
A	P	$\dfrac{i(1 + i)^n}{(1 + i)^n - 1}$	$(A\|P,i,n)$	Uniform series, capital recovery factor
F	A	$\dfrac{(1 + i)^n - 1}{i}$	$(F\|A,i,n)$	Uniform series, compound amount factor
A	F	$\dfrac{i}{(1 + i)^n - 1}$	$(A\|F,i,n)$	Uniform series, sinking fund factor
P	G	$\dfrac{[1 - (1 + ni)(1 + i)^{-n}]}{i^2}$	$(P\|G,i,n)$	Gradient series, present worth factor
A	G	$\dfrac{(1 + i)^n - (1 + ni)}{i[(1 + i)^n - 1]}$	$(A\|G,i,n)$	Gradient series, uniform series factor
P	A_1, j	$\left(\dfrac{1 - (1 + j)^n(1 + i)^{-n}}{i - j}\right)$ for $i \neq j$	$(P\|A_1,i,j,n)$	Geometric series, present worth factor
F	A_1, j	$\dfrac{(1 + i)^n - (1 + j)^n}{i - j}$ for $i \neq j$	$(F\|A_1,i,j,n)$	Geometric series, future worth factor

2.6 MULTIPLE COMPOUNDING PERIODS IN A YEAR

Not all interest rates are stipulated as annual compounding rates. In the business world it is common practice to express a nonannual interest rate as follows:

12% per year compounded monthly or 12% per year per month

When expressed in this form, 12% per year per month is known as the *nominal annual* interest rate, r[1]. Unfortunately, the techniques we have learned up to now cannot be used directly to solve an economic analysis problem of this type because the interest period (per year) and compounding period (monthly) are not the same. To solve this type of problem, one of two approaches must be employed; the period interest rate approach or the effective interest rate approach.

---

[1] In this text we will consider nominal interest rates that are annual and compounding periods that are either annual or more frequent than the annual. The approach presented in this section can easily be generalized to other cases.

To utilize the period interest rate approach, we must define a new term, the *period* interest rate:

$$\text{Period interest rate} = \frac{\text{Nominal annual interest rate}}{\text{Number of interest periods per year}} \quad\quad \text{(2.41)}$$

Using the example above,

$$\text{Period interest rate} = \frac{12\% \text{ per year per month}}{12 \text{ month per year}} = 1\% \text{ per month/month}$$

Since the interest period and the compounding period are now the same (monthly), the factors in the back of the book can be applied directly. *Note, however, that the number of interest periods (n) must be adjusted to match the new frequency.*

EXAMPLE 2.20

$2000 is invested in an account which pays 12% per year compounded monthly. What is the balance in the account after 3 years?

$$\text{Nominal annual interest rate} = 12\%/\text{year/month}$$

$$\text{Period interest rate} = \frac{12\%/\text{year/month}}{12 \text{ month/year}} = 1\%/\text{month/month}$$

$$\text{Number of interest periods} = 3 \text{ years} \times 12 \text{ month/year}$$
$$= 36 \text{ interest periods (months)}$$
$$F = P(F|P, i, n) = \$2000 \, (F|P, 1, 36) = \$2000 \, (1.4308) = \$2861.60$$

In a similar manner:
12% per year compounded semiannually:

$$\text{Nominal annual interest rate} = 12\%/\text{year/6-month}$$

$$\text{Period interest rate} = \frac{12\%/\text{year/6-month}}{2 \text{ 6-month/year}} = 6\%/\text{6-month/6-month}$$

Number of interest periods = 3 years × 2 6-month/year = 6 interest periods
$$F = P(F|P, i, n) = \$2000 \, (F|P, 6, 6) = \$2000 \, (1.4185) = \$2837.00$$

12% per year compounded quarterly:

$$\text{Nominal annual interest rate} = 12\%/\text{year/quarter}$$

$$\text{Period interest rate} = \frac{12\%/\text{year/quarter}}{4 \text{ quarter/year}} = 3\%/\text{quarter/quarter}$$

Number of interest periods = 3 years × 4 quarter/year = 12 interest periods
$$F = P(F|P, i, n) = \$2000 \, (F|P, 3, 12) = \$2000 \, (1.4258) = \$2851.60$$

12% per year compounded daily:

$$\text{Nominal annual interest rate} = 12\%/\text{year/day}$$

$$\text{Period interest rate} = \frac{12\%/\text{year/day}}{365 \text{ days/year}} = 0.0329\%/\text{day/day}$$

Number of interest periods = 3 years × 365 days/year = 1095 interest periods

$F = P(F|P, i, n) = \$2000\ (F|P, 0.0329, 1095) = \$2000\ (1.000329)^{1095}$

$\quad = \$2000\ (1.4336) = \$2867.22$

EXAMPLE 2.21

What are the monthly payments on a 5-year car loan of $12,500 at 6% per year compounded monthly.

Nominal annual interest rate = 6%/year/month

$$\text{Period interest rate} = \frac{6\%/\text{year/month}}{12\ \text{month/year}} = 0.5\%/\text{month/month}$$

Number of interest periods = 5 years × 12 month/year = 60 interest periods

$A = P(A|P, i, n) = \$12,500\ (A|P, 0.5, 60) = \$12,500\ (0.0193) = \$241.25$

The second approach to solving problems where the compounding is not annual is the effective interest rate approach. The *effective annual* interest rate is the annual interest rate that would be equivalent to the period interest rate as previously calculated.

For example: If the nominal interest rate is 12%/year/quarter, then period interest rate is 3%/quarter/quarter. One dollar invested for 1 year at 3%/quarter/quarter would have a future worth of:

$\quad F = P(F|P, i, n) = \$1\ (F|P, 3, 4) = \$1\ (1.03)^4 = \$1\ (1.1255) = \$1.1255$

To get this same value in one year would require an annual rate of 12.55%/year/year. This value is called the effective annual interest rate. The effective annual interest rate is given by $(1.03)^4 - 1 = 0.1255$ or 12.55%.

The general equation for the effective annual interest rate is:

$$\text{Effective annual interest rate} = i_{\text{eff}} = (1 + r/m)^m - 1 \qquad \textbf{(2.42)}$$

where r = nominal annual interest rate

$\quad m$ = number of interest periods per year

EXAMPLE 2.22

Calculate the effective annual interest rate for each of the following cases:

12% per year compounded semiannually:

    nominal annual interest rate = 12%/year/6-month

    period interest rate = 6%/6-month/6-month

    effective annual interest rate = $(1 + 0.12/2)^2 - 1 = 0.1236$ or 12.36%

12% per year compounded monthly:

    nominal annual interest rate = 12%/year/month

    period interest rate = 1%/month/month

    effective annual interest rate = $(1 + 0.12/12)^{12} - 1 = 0.1268$ or 12.68%

12% per year compounded daily:

Nominal annual interest rate = 12%/year/day

Period interest rate = 0.0329%/day/day

Effective annual interest rate = $(1 + 0.12/365)^{365} - 1 = 0.1275$ or 12.75%

EXAMPLE 2.23

An individual borrowed $1000 and paid off the loan with interest after 4.5 years. The amount paid was $1500. What was the effective annual interest rate for this transaction? Letting the interest period be a 6-month period, it is seen that the payment of $1500 and the debt of $1000 are related by the expression

$$F = P(F \mid P \; i,n)$$

Thus,

$$\$1500 = \$1000(F \mid P \; i,9)$$

or

$$\$1500 = \$1000(1 + i)^9$$

Dividing both sides by $1000 gives

$$1.50 = (1 + i)^9$$

Taking the logarithm of both sides yields

$$\log 1.50 = 9 \log (1 + i)$$

Dividing both sides by 9 and taking the antilog of the result provides the relation

$$(1 + i) = 1.046$$
$$i = 0.046$$

Thus, the 6-month interest rate is approximately 4.6%. Computing the effective annual interest rate yields

$$i_{eff} = (1 + i)^2 - 1$$
$$= (1 + 0.046)^2 - 1$$
$$= 0.0943$$

The effective annual interest rate for the loan transaction was approximately 9.43%.

Instead of using logarithms, we could have searched the interest tables for a value of i that yielded a value of 1.50 for the $(F \mid P \; i,9)$ factor. With interpolation, *i* would be found to equal approximately 4.6%. The computation of the effective annual interest rate would follow the same procedure used to determine the effective annual interest rate for 9.2% compounded semiannually.

In the United States, the federal government requires that anyone borrowing money be informed of the total amount of interest paid and the *annual percentage rate* associated with the transaction. Depending on how the lender computes the annual percentage rate, the percentage quoted can be either the nominal annual interest rate, the effective annual interest rate, or something else entirely. To illustrate, many revolving charge accounts charge an interest rate of 1.5%/month, which is equivalent to 18% compounded monthly or 19.56% compounded annually. In many cases, the annual percentage rate is stated to be 18%.

Since, in personal financing, the annual percentage rate might be computed differently among the alternative sources of funds, the effective annual interest rate is a very good basis for comparing alternative financing plans.

2.7   CONTINUOUS COMPOUNDING

In the discussion of the effective interest rate in the previous section, you should note that as the frequency of compounding in a year increases, the effective interest rate increases. Since monetary transactions occur daily or hourly in most businesses, and money is normally "put to work" for the business as soon as it is received, compounding is occurring quite frequently. If one wishes to account explicitly for such rapid compounding, then *continuous compounding relations should be used.* Continuous compounding means that each year is divided into an infinite number of interest periods. Mathematically, the single payment compound amount factor under continuous compounding is given by

$$\lim_{m \to \infty} \left(1 + \frac{r}{m}\right)^{mn} = e^{rn}$$

where *n* is the number of years, *m* is the number of interest periods per year, and *r* is the nominal annual interest rate. Given *P*, *r*, and *n*, the value of *F* can be computed using continuous compounding as follows.

$$F = P\, e^{rn} \tag{2.43}$$

or

$$F = P(F|P\ r, n)_\infty \tag{2.44}$$

where $(F|P\ r, n)_\infty$ denotes the *continuous compounding, single sum, future worth factor.* The subscript ∞ is provided to denote that continuous compounding is being used. The interest tables for continuous compounding are given in Appendix B.

EXAMPLE 2.24

If $2000 is invested in a fund that pays interest at a rate of 12% compounded continuously, after 5 years the cumulative amount in the fund will total

$$F = P(F|P\,12\%,5)_\infty$$
$$= \$2000(1.8221)$$
$$= \$3644.20$$

Thus, a withdrawal of $3644.20 will deplete the fund after 5 years.

The effective interest rate under continuous compounding is easily obtained using the relation

$$\boxed{i_{\text{eff}} = e^r - 1} \tag{2.45}$$

or

$$\boxed{i_{\text{eff}} = (F|P\,r,1)_\infty - 1} \tag{2.46}$$

To illustrate, if interest is 12% compounded continuously, then the effective interest rate is given by

$$i_{\text{eff}} = (F|P\,12,1)_\infty - 1$$
$$= 0.1275$$

Thus, 12.75% compounded annually is equivalent to 12% compounded continuously.

The inverse relationship between F and P indicates that

$$\boxed{P = F\,e^{-rn}} \tag{2.47}$$

or

$$\boxed{P = F(P|F\,r,n)_\infty} \tag{2.48}$$

where $(P|F\,r,n)_\infty$ is called the *continuous compounding, single sum, present worth factor.*

2.7.1 Discrete Flows

If it is assumed that cash flows are discretely spaced over time, then the continuous compounding relations for the uniform, gradient, and geometric series can be obtained. Substituting e^{-rn} for $(1 + i)^{-n}$, $e^r - 1$ for i, and e^{rn} for $(1 + i)^n$ in the remaining discrete compounding formulas yields the continuous compounding interest factors summarized in Table 2.4. Values for these factors are provided in Appendix B.

Table 2.4 Summary of Continuous Compounding Interest
Factors for Discrete Flows

To Find	Given	Factor	Symbol
P	F	e^{-rn}	$(P\|F\ r,n)_\infty$
F	P	e^{rn}	$(F\|P\ r,n)_\infty$
F	A	$\dfrac{e^{rn}-1}{e^r-1}$	$(F\|A\ r,n)_\infty$
A	F	$\dfrac{e^r-1}{e^{rn}-1}$	$(A\|F\ r,n)_\infty$
P	A	$\dfrac{e^{rn}-1}{e^{rn}(e^r-1)}$	$(P\|A\ r,n)_\infty$
A	P	$\dfrac{e^{rn}(e^r-1)}{e^{rn}-1}$	$(A\|P\ r,n)_\infty$
P	G	$\dfrac{e^{rn}-1-n(e^r-1)}{e^{rn}(e^r-1)^2}$	$(P\|G\ r,n)_\infty$
A	G	$\dfrac{1}{e^r-1}-\dfrac{n}{e^{rn}-1}$	$(A\|G\ r,n)_\infty$
P	A_1, c	$\dfrac{1-e^{(c-r)n}}{e^r-e^c}$	$(P\|A_1\ r,c,n)_\infty^*$
F	A_1, c	$\dfrac{e^{rn}-e^{cn}}{e^r-e^c}$	$(F\|A_1\ r,c,n)_\infty^*$

* $r \neq c$.

EXAMPLE 2.25

To illustrate the use of the continuous compounding interest factors, suppose
$1000 is deposited each year into an account that pays interest at a rate of
12% compounded continuously. Determine both the amount in the account
immediately after the tenth deposit and the present worth equivalent for 10 deposits.

The amount in the fund immediately after the tenth deposit is given by the
relation

$$F = \$1000(F\|A\ 12,10)_\infty$$
$$= \$1000(18.1974)$$
$$= \$18,197.40$$

The present worth equivalent for 10 deposits is obtained using the relation

$$P = \$1000(P\|A\ 12,10)_\infty$$
$$= \$1000(5.4810)$$
$$= \$5481$$

In the case of the geometric series the size of the tth cash flow will be assumed to be given by

$$A_t = A_{t-1}e^c \qquad t = 2, \ldots, n \qquad\qquad (2.49)$$

or, equivalently,

$$A_t = A_1 e^{(t-1)c} \qquad t = 1, \ldots, n \qquad\qquad (2.50)$$

where c is the nominal compound rate of increase in the size of the cash flow. The resulting expressions for the *continuous compounding, geometric series present worth factor* and the *continuous compounding, geometric series future worth factor* are given, respectively, by $(P|A_1\ r,c,n)_\infty$ and $(F|A_1\ r,c,n)_\infty$, as given in Table 2.4 for the case of $r \ne c$. As an exercise you may wish to derive the appropriate expressions when $r = c$.

EXAMPLE 2.26

An individual receives an annual bonus and deposits it in a savings account that pays 8% compounded continuously. The size of the bonus increases each year at a rate of 10% compounded continuously; the initial deposit was $500. Determine how much will be in the fund immediately after the tenth deposit.

In this case, $A_1 = \$500$, $r = 8\%$, $c = 10\%$, and $n = 10$. Thus, the value of F is given by

$$F = A_1(F|A_1\ 8,10,10)_\infty$$
$$= \$500(22.5162)$$
$$= \$11,258.10$$

Comparing the result with that obtained in Example 2.19, the effect of continuous compounding increases the amount in the fund by $\$11,258.10 - \$10,870.45 = \$387.65$.

Since the differences in discrete and continuous compounding are not great in most cases, it is not uncommon to see discrete compounding used when continuous compounding is more appropriate. The arguments given for this are that errors in estimating the cash flows will probably offset any attempts to be very precise by using continuous compounding and that the interest rate used in discrete compounding is actually the effective interest rate resulting from continuous compounding.

2.7.2   Continuous Flow

Thus far only discrete cash flows have been assumed. It was assumed that cash flows occurred at, say, the end of the year. In some cases money is expended throughout the year on a somewhat uniform basis. Costs of labor, carrying inventory, and operating and maintaining equipment are typical examples. Others include capital improvement projects that conserve energy, water, or process steam.

Consequently, as a mathematical convenience, instead of assuming that money flows in discrete increments at the end of monthly, weekly, daily, or hourly time periods, it is assumed that money flows continuously during the time period at a uniform rate. Instead of having a uniform series of discrete cash flows of magnitude A, it assumed that a *total* of $\overline{A}$ dollars flows uniformly and continuously throughout a given time period. Such an approach to modeling cash flows is referred to as the *continuous flow* approach.

To illustrate the continuous flow concept, suppose you are to divide $1000 into k equal amounts to be deposited at equally spaced points in time during a year. The interest rate per period is defined to be r/k, where r is the nominal rate. Thus, the present worth of the series of k equal amounts is

$$P = \frac{\$1000}{k}\left(P|A\,\frac{r}{k},k\right)$$

or

$$P = \frac{\$1000}{k}\left[\frac{(1 + (r/k))^k - 1}{(r/k)(1 + (r/k))^k}\right]$$

which reduces to

$$P = \frac{\$1000}{k}\left[\frac{1}{r} - \frac{1}{r(1 + (r/k))^k}\right]$$

Taking the limit of P as k approaches infinity gives

$$\lim_{k\to\infty} P = \$1000\left(\frac{1}{r} - \frac{1}{re^r}\right)$$

or

$$P = \$1000\left(\frac{e^r - 1}{re^r}\right)$$

In general, for n years,

$$P = \overline{A}\left(\frac{e^{rn} - 1}{re^{rn}}\right) \tag{2.51}$$

or

$$P = \overline{A}(P|\overline{A}\ r,n) \tag{2.52}$$

where $(P|\overline{A}\ r,n)$ is referred to as the *continuous flow, continuous compounding uniform series present worth factor* and is tabulated in Appendix B.

The remaining continuous flow, continuous compound interest factors are summarized in Table 2.5. Values of the factors are given in Appendix B for

Table 2.5 Summary of Continuous Compounding Interest Factors for Continuous Flows

To Find	Given	Factor	Symbol
P	$\overline{A}$	$\dfrac{e^{rn} - 1}{re^{rn}}$	$(P|\overline{A}\ r,n)$
$\overline{A}$	P	$\dfrac{re^{rn}}{e^{rn} - 1}$	$(\overline{A}|P\ r,n)$
F	$\overline{A}$	$\dfrac{e^{rn} - 1}{r}$	$(F|\overline{A}\ r,n)$
$\overline{A}$	F	$\dfrac{r}{e^{rn} - 1}$	$(\overline{A}|F\ r,n)$

various values of r and n. With continuous flow and continuous compounding, the continuous annual cash flow, $\overline{A}$, is equivalent to $Ar/(e^r - 1)$, when discrete flow and continuous compounding is used. Hence, the discrete flow equivalent of $\overline{A}$ is given by

$$A = \frac{\overline{A}(e^r - 1)}{r}$$

or

$$A = \overline{A}(F|\overline{A}\ r,1)$$

<hr>

EXAMPLE 2.27

What are the present worth and future worth equivalents of a uniform series of continuous cash flows totalling \$10,000/year for 10 years when interest is compounded continuously at a rate of 20%/year?
The present worth equivalent is given by

$$P = \$10,000(P|\overline{A}\ 20,10)$$
$$= \$10,000(4.3233)$$
$$= \$43,233$$

The future worth equivalent is given by

$$F = \$10,000(F|\overline{A}\ 20,10)$$
$$= \$10,000(31.9453)$$
$$= \$319,453$$

Continuous flow and continuous compounding are concepts that are used to represent more closely the realities of business transaction. In fact, several of the computer programs developed by industries and governmental agencies and used to perform economic analyses incorporate both concepts. We have presented both concepts since each has experienced increased usage during the past decade.

2.8 EQUIVALENCE AND INDIFFERENCE

Throughout the preceding sections, we have used the term *equivalence* without defining precisely what the term meant. This was done intentionally in order to introduce subtly the notion that two cash flow series or profiles are equivalent at some specified interest rate $k\%$, if their present worths are equal using an interest rate of $k\%$. We will now define equivalence (and a related concept, indifference) more formally and discuss their importance in economic analysis.

2.8.1 Equivalence

In economic analysis, "equivalence" means "the state of being equal in value." The concept is primarily applied in the comparison of two or more cash flow profiles. Specifically, two (or more) cash flow profiles are *equivalent* if their time-value-of-money worths at a common point in time are equal.

Question: Are the following two cash flows equivalent at 15%/yr?
   Cash Flow 1: Receive $1322.50 two years from today
   Cash Flow 2: Receive $1000.00 today
Analysis Approach 1: Compare worths at $t = 0$ (present worth)
   $PW(1) = 1322.50 \times (P|F, 15, 2) = 1322.50 \times 0.756147 = 1000$
   $PW(2) = 1000$
Answer: Cash Flow 1 and Cash Flow 2 are equivalent
Analysis Approach 2: Compare worths at $t = 2$ (future worth)
   $FW(1) = 1322.50$
   $FW(2) = 1000 \times (F|P, 15, 2) = 1000 \times 1.3225 = 1322.50$
Answer: Cash Flow 1 and Cash Flow 2 are equivalent
Analysis Approach 3: Compare worths at $t = 1$
   $W_1(1) = 1322.50 \times (P|F, 15, 1) = 1322.50 \times 0.869565 = 1150.00$
   $W_1(2) = 1000 \times (F|P, 15, 1) = 1000 \times 1.15 = 1150.00$
Answer: Cash Flow 1 and Cash Flow 2 are equivalent
Analysis Approach 4: Compare worths at $t = 6$
   $W_6(1) = 1322.50 \times (F|P, 15, 4) = 1322.50 \times 1.749006 = 2,313.06$
   $W_6(2) = 1000 \times (F|P, 15, 6) = 1000 \times 2.313061 = 2,313.06$
Answer: Cash Flow 1 and Cash Flow 2 are equivalent

Note that the selection of the point in time, t, at which to make the comparison is completely arbitrary. Clearly, however, some choices are more intuitively appealing than others ($t = 0$ and $t = 2$ in the above example).

EXAMPLE 2.28

What single sum of money at $t = 6$ is equivalent to the cash flow profile shown in Figure 2.14 if $i = 10\%$?

The present worth of the cash flow profile is given by

$$P = -\$400(P|F\,10,1) + \$100(P|A\,10,3)(P|F\,10,1)$$
$$+\$100(P|A\,10,3)(P|F\,10,5)$$
$$= -\$400(0.9091) + \$100(2.4869)(0.9091) + \$100(2.4869)(0.6209)$$
$$= \$16.85$$

Moving $16.85 forward in time to $t = 6$ gives

$$F = \$16.85(F|P\,10,6)$$
$$= \$16.85(1.7716)$$
$$= \$29.86$$

Thus, at 10% a positive cash flow of $29.86 at $t = 6$ is equivalent to the cash flow profile shown in Figure 2.14.

EXAMPLE 2.29

Using an 8% discount rate, what uniform series over five periods, [1, 5], is equivalent to the cash flow profile given in Figure 2.15?

The cash flow profile in Figure 2.15a consists of the difference in a uniform series of $500 and a gradient series, with $G = \$100$. A uniform series equivalent of the cash flow profile can be obtained for the interval [2, 6] as follows:

$$A = \$500 - \$100(A|G\,8,5)$$
$$= \$500 - \$100(1.8465)$$
$$= \$315.35$$

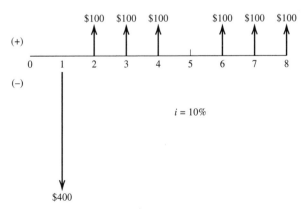

Figure 2.14 Cash flow diagram for the equivalence example.

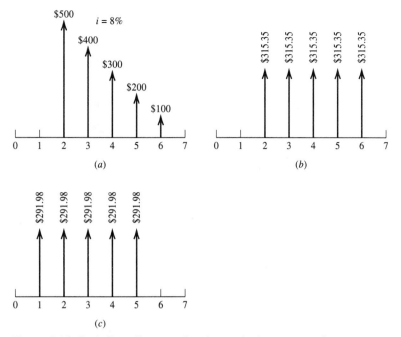

Figure 2.15 Cash flow diagrams for the equivalence example.

The uniform series of $315.35 over the interval [2, 6] must be converted to a uniform series over the interval [1, 5]. Thus, *each* of the five cash flows must be moved back in time one period. The discounted value of $315.35 over one time period using an 8% discount rate is

$$P = \$315.35(P\,|\,F\,8,1)$$
$$= \$315.35(0.9259)$$
$$= \$291.98$$

Consequently, a uniform series of $291.98 over the interval [1, 5] is equivalent to the cash flow profile given in Figure 2.15*a*. If you have doubts concerning the equivalence, compare their present worths using an 8% interest rate.

EXAMPLE 2.30

Determine the value of X that makes the two cash flows, given in Figure 2.16, equivalent when a discount rate of 15% is used.

Equating the future worths of the two cash flow profiles at $t = 4$ gives

$$\$200(F\,|\,A\ 15,4) + \$100(F\,|\,A\ 15,3) + \$100 = [\$200 + X(A\,|\,G\ 15,4)](F\,|\,A\ 15,4)$$

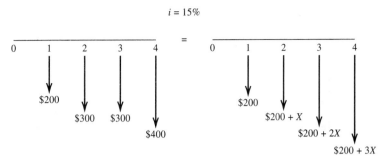

Figure 2.16 Cash flow diagrams for the equivalence example.

Canceling $200(F|A\ 15,4)$ on both sides yields

$$\$100(3.4725) + \$100 = X(1.3263)(4.9934)$$

Solving for X gives a value of $67.53.

EXAMPLE 2.31

For what interest (discount) rate are the two cash flow profiles shown in Figure 2.17 equivalent?

Converting each cash flow profile to a uniform series over the interval [1, 5] gives

$$-\$4000(A|P\ i,5) + \$1500 = -\$7000(A|P\ i,5) + \$1500 + \$500(A|G\ i,5)$$

or

$$\$3000(A|P\ i,5) = \$500(A|G\ i,5)$$

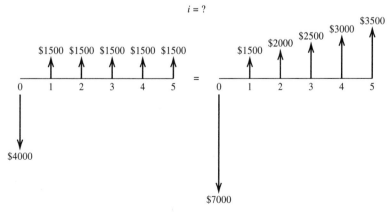

Figure 2.17 Cash flow diagrams for the equivalence example.

which reduces to

$$(A|G\ i,5) = 6(A|P\ i,5)$$

On searching through the interest tables at $n = 5$, it is found that the $(A|G\ i,5)$ factor is six times the value of the $(A|P\ i,5)$ factor for an interest rate between 12% and 15%. Specifically, with a 12% interest rate,

$$(A|G\ 12,5) - 6(A|P\ 12,5) = 1.7746 - 6(0.2774) = 0.1102$$

and, using a 15% interest rate,

$$(A|G\ 15,5) - 6(A|P\ 15,5) = 1.7228 - 6(0.2983) = -0.0670$$

Interpolating for i gives

$$i = 0.12 + \frac{(0.15 - 0.12)(0.1102)}{(0.1102 + 0.0670)}$$

or

$$i = 0.1386$$

Therefore, using a discount rate of approximately 13.86% will establish an equivalence relationship between the cash flow profiles given in Figure 2.17.

2.8.2   Indifference

A concept that is closely related to equivalence is the concept of indifference. In economic analysis, "indifference" means "to have no preference." The concept is primarily applied in the comparison of two or more cash flow profiles. Specifically, a potential investor is *indifferent* between two (or more) cash flow profiles if they are equivalent.

Question: Given the following two cash flows at 15%/year, which do you prefer?
    Cash Flow 1: Receive $1322.50 two years from today
    Cash Flow 2: Receive $1000.00 today
Answer: Based on the equivalence calculations above, given these two choices, I am indifferent.

2.8.3   Applying the Concepts of Equivalence and Indifference

The concept of equivalence can be used to break a large, complex problem into a series of smaller, more manageable ones. This is done by taking advantage of the fact that, in calculating the economic worth of a cash flow profile, any part of the profile can be replaced by an *equivalent* representation without altering the worth of the profile at an arbitrary point in time.

Question: You are given a choice between (1) receiving P dollars today or (2) receiving $2000 per year for 5 years starting at $t = 3$. What must the value of P be for you to be indifferent between the two choices if $i = 10\%$/yr?

Analysis Approach: To be indifferent between the choices, P must have a value such that the two alternatives are equivalent at 10%/year. If we select $t = 0$ as the common point in time upon which to base the analysis (present worth approach), then the analysis proceeds as follows.

$PW(\text{Alt 1}) = P$

$PW(\text{Alt 2})$

  Step 1—Replace the uniform series ($t = 3$ to 7) with an equivalent single sum, V_2, at $t = 2$ (why $t = 2$?):

  $V_2 = 2000*(P|A, 10, 5) = 2000 \times 3.6048 = 7209.60$

  Note that since the A's were positive, V_2 must also be positive for equivalence to hold.

  Step 2—Replace the single sum V_2, with an equivalent value V_0 at $t = 0$:

  $PW(\text{Alt 2}) = V_0 = V_2*(P|F, 10,2) = 7209.60 \times 0.7972 = 5747.49$

Answer: To be indifferent between the two alternatives, they must be equivalent at $t = 0$. To be equivalent, P must have a value of $5747.49

2.9   INTRODUCTION TO MEASURES OF WORTH

Most frequently the time value of money concepts and factors presented in this chapter are used to calculate measures of worth of investment alternatives. Several important measures of worth will now be introduced by way of the following example. These measures of worth, as well as several others, will be formally treated in Chapter Four.

EXAMPLE 2.32

A firm purchases a machine for $30,000, keeps it for 5 years, and sells it for $6000. During the time the machine was owned by the company, operating and maintenance costs totaled $8000 the first year, $9000 the second year, $10,000 the third year, $11,000 the fourth year, and $12,000 the fifth year. The firm uses a 15% interest rate in performing economic analyses. Determine the single sum of money occurring at (1) $t = 0$ and (2) $t = 5$, which is equivalent to the cash flow history for the machine. Also, determine the uniform series occurring over the interval [1, 5] that is equivalent to the cash flow profile for the machine.

   In the economic analysis literature, the single sum equivalent at time zero for a cash flow profile is called the *present worth* or *present value* for the cash flow profile. We will use the term present worth and denote it as PW. For the example,

$$PW = -\$30,000 - \$8000(P|A\ 15,5) - \$1000(P|G\ 15,5)$$
$$+\$6000(P|F\ 15,5)$$
$$= -\$59,609.57$$

Hence, a single expenditure of $59,609.57 at time zero would have been equivalent to the cash flows experienced during the ownership of the machine.

The single sum equivalent at the end of the life of a project is termed the *future worth* or *future value* for the project. We denote the future worth by *FW* and compute its value as follows:

$$FW = -\$30{,}000(F|P\ 15{,}5) - [\$8000 + 1000(A|G\ 15{,}5)](F|A\ 15{,}5)$$
$$+\$6000$$
$$= -\$119{,}897$$

Alternatively, the future worth can be obtained from the present worth:

$$FW = PW(F|P\ i{,}n)$$
$$= -\$59{,}609.57(F|P\ 15{,}5)$$
$$= -\$119{,}898.69$$

(The difference of $1.69 is due to round-off error in the tables). Hence, a single expenditure of $119,897 at time 5 would have been equivalent to the cash flows associated with the machine.

A uniform series equivalent for a series of yearly cash flows is referred to as the *annual worth* or *equivalent uniform annual cost* for the project. The latter expression is most appropriate for the type of example under consideration, since the resulting uniform series is a cost, not an income, series. The annual worth designation is *AW*; *EUAC* denotes the equivalent uniform annual cost. We will use both designations throughout the text.

For the example problem, the *EUAC* determination is performed as follows:

$$EUAC = \$30{,}000(A|P\ 15{,}5) + [\$8000 + \$1000(A|G\ 15{,}5)]$$
$$-\$6000(A|F\ 15{,}5)$$
$$= \$17{,}782$$

Hence an annual expenditure of $17,782/year for 5 years is equivalent to the cash flow profile associated with the machine investment. Alternately, the equivalent uniform cost can be obtained from either the present worth or the future worth.

$$EUAC = -PW(A|P\ i{,}n)$$
$$EUAC = -FW(A|F\ i{,}n)$$

Present worth, future worth, and annual worth computations are used often in comparing economic investment alternatives having different cash flow profiles. Consequently, we will have need for *PW, FW,* and *AW* calculations for the discussion of alternative comparisons in Chapter Four.

2.10 SUMMARY

In this chapter, we developed the time value of money concept and defined a number of mathematical operations consistent with that concept. In subsequent chapters we apply the concepts developed in this chapter to personal financing and to the study of investment alternatives from the viewpoint of an ongoing enterprise.

BIBLIOGRAPHY

1. Altany, D. R., "Running Out of Room to Grow," *Industry Week,* 236, (6), March 21, 1988, pp. 19–20.
2. Blank, L. T., and Tarquin, A. T. *Engineering Economy,* 3rd ed., McGraw-Hill, 1989.
3. Canada J. R., Sullivan, W. G., and White, J. A. *Capital Investment Decision Analysis for Management and Engineering,* Prentice-Hall, 1996.
4. DeGarmo, E. P., Sullivan, W. G., and Boutadelli, J. A. *Engineering Economy,* 9th ed., Macmillan, 1993.
5. Grant, E. L., Ireson, W. G., and Leavenworth, R. S. *Principles of Engineering Economy,* 7th ed., Wiley, 1982.
6. Lutz, R. P., "Discounted Cash Flow Techniques," Chapter 9.3, *Handbook of Industrial Engineering,* (G. Salvendy, ed.), John Wiley, 1982.
7. Newman, D. G., and Johnson, R. *Engineering Economic Analysis,* 5th ed., Engineering Press, 1995.
8. Riggs, J. L., Bedworth, D. D., and Randhawa, S. U. *Engineering Economics,* 4th ed., McGraw-Hill, 1996.
9. Thuesen, G. J., and Fabrycky, W. J. *Engineering Economy,* 7th ed., Prentice Hall, 1989.
10. Thuesen, G. J., "Economic Analysis," Chapter 6.2, *Production Handbook,* 4th ed., (J. A. White, ed), Wiley, 1987.

PROBLEMS

1. Pat sold Mike a used automobile for $3000. Mike paid $500 as a down payment and gave Pat three personal notes for the remainder due. The principal of each note was $1000, $750, and $750. The $1000 note was due 1 year after the purchase, the second note was due 2 years after the purchase, and the third note was due 3 years after the purchase. The annual simple interest agreed to by Pat and Mike was 10% for the first note, 12% for the second note, and 15% for the third note. How much total interest was paid to Pat by Mike? (2.3)
2. A debt of $2500 is incurred at $t = 2$. An annual simple interest rate of 8% on the unpaid balance is agreed on. Three equal payments are made at $t = 4, 5,$ and 6 to pay off the debt, including interest payments. Determine the size of each payment. (2.3)
3. How long does it take a deposit in a 4% compound interest fund to double in value? (2.4)
4. How much money today is equivalent to $10,000 in 12 years, with interest at 10% compounded annually? (2.4)
5. If $5000 is deposited into a fund paying 8% compounded annually, what sum will be accumulated at the end of 10 years? What would be the sum accumulated at the end of 5 years if the fund paid 16% compounded annually? What is suggested regarding doubling the interest rate and halving the length of the time period? If you had $5000 available for investment and the two options were available, which would you choose if you had to choose one of them? Justify your choice. If you chose the shorter duration investment, what will you do with your accumulated monies over the next 5-year period? Should the answer to this question influence your choice? (2.4)
6. If a fund pays 12% compounded annually, what single deposit now will accumulate $12,000 at the end of the tenth year? If the fund pays 6% compounded annually, what single deposit is required now in order to accumulate $6000 at the end of the tenth year? (2.4)

7. Maria deposits $1200, $500, and $2000 at $t = 1$, 2, and 3, respectively. If the fund pays 8% compounded per period, what sum will be accumulated in the fund at (a) $t = 3$ and (b) $t = 6$? (2.5)

8. Suppose you wanted to become a millionaire at retirement. If an annual compound interest rate of 8% could be sustained over a 40-year period, how much would have to be deposited yearly in the fund in order to accumulate $1 million? What if the interest rate is 10%?, 12%? (2.5)

9. Juan deposits $1000 in a savings account that pays 8% compounded annually. Exactly 2 years later he deposits $3000; 2 years later he deposits $4000; and 4 years later he withdraws all of the interest earned to date and transfers it to a fund that pays 10% compounded annually. How much money will be in each fund 4 years after the transfer? (2.5)

10. A debt of $1000 is incurred at $t = 0$. What is the amount of four equal payments at $t = 1$, 2, 3, and 4 that will repay the debt if money is worth 10% compounded per period? (2.5.1)

11. Five deposits of $500 each are made at $t = 1$, 2, 3, 4, and 5 into a fund paying 6% compounded per period. How much will be accumulated in the fund at (a) $t = 5$, and (b) $t = 10$? (2.5.1)

12. What equal annual deposits must be made at $t = 2$, 3, 4, 5, and 6 in order to accumulate $25,000 at $t = 8$ if money is worth 10% compounded annually? (2.5.1)

13. If annual deposits of $1000 are made in a fund paying 12% interest compounded annually, how much money will be in the fund immediately after the fifth deposit? (2.5.1)

14. Kim deposits $1000 in a savings account; 4 years after the deposit, half of the account balance is withdrawn. $2000 is deposited annually for an 8-year period, with the first deposit occurring 2 years after the withdrawal. The total balance is withdrawn 15 years after the initial deposit. If the account earned interest of 8% compounded annually over the 15-year period, how much was withdrawn at each withdrawal point? (2.5.1)

15. John borrows $15,000 at 18% compounded annually; he pays off the loan over a 5-year period with annual payments. Each successive payment is $700 greater than the previous payment. How much was the first payment? (2.5.2)

16. Solve Problem 15 for the case in which each successive payment is $700 less than the previous payment. (2.5.2)

17. Land is purchased for $75,000. It is agreed that land will be paid for over a 5-year period with annual payments and using a 12% annual compound interest rate. Each payment is to be $3,000 greater than the previous payment. Determine the size of the last payment. (2.5.2)

18. Solve Problem 17 for the case in which each successive payment is $3000 less than the previous payment. (2.5.2)

19. Solve Problem 15 for the case in which each successive payment is to be 10% greater than the previous payment. (2.5.3)

20. Solve Problem 15 for the case in which each successive payment is to be 10% less than the previous payment. (2.5.3)

21. Solve Problem 17 for the case in which each successive payment is to be 12% greater than the previous payment. (2.5.3)

22. Solve Problem 17 for the case in which each successive payment is to be 20% less than the previous payment. (2.5.3)

23. In a new, highly automated factory, labor costs are expected to decrease at an annual compound rate of 5%; material costs are expected to increase at an annual compound

rate of 6%; and energy costs are expected to increase at an annual compound rate of 3%. The labor, material and energy costs the first year are $3 million, $2 million, and $1,500,000. (2.5.3)

a. What will be the value of each cost during the fifth year?

b. Using an interest rate of 10% compounded annually, what uniform annual costs over a 5-year period would be equivalent to the cumulative labor, material, and energy costs?

24. Mary works for a company that pays an annual bonus, the size of which is based on experience with the company. At the end of her first year, the bonus equals $1500. The size of the bonus compounds at an annual rate of 6%. Mary decides to place two-thirds of her annual bonus in a fund that pays 10% compounded annually. How much will be in the fund immediately after the thirtieth deposit? (2.5.3)

25. Using a compound interest rate of $i\%$ per period, determine the present worth equivalent for a series of cash flows having the following form:

$$A_t = [A + G(t - 1)](1 + j)^{t-1} \qquad \text{for } t = 1, \ldots, n$$

Note:

$$\sum_{t=1}^{n} t = n(n + 1)/2$$

$$\sum_{t=1}^{n} x^t = \frac{x(1 - x^n)}{1 - x} \qquad \text{for } |x| \neq 1$$

$$\sum_{t=1}^{n} tx^t = \frac{x - (1 - x)(n + 1)x^{n+1} - x^{n+2}}{(1 - x)^2} \qquad \text{for } |x| \neq 1$$

(2.5.3)

26. Yavuz wishes to make a single deposit P at $t = 0$ into a fund paying 15% compounded quarterly such that $1000 payments are received at $t = 1, 2, 3$, and 4 (periods are 3-month intervals), and a single payment of $7500 is received at $t = 12$. What single deposit is required? (2.6)

27. Dr. Shieh deposits $3000 in a money market fund. The fund pays interest at a rate of 12% compounded annually. Just 3 years after making the single deposit, he withdraws one third the accumulated money in his account. Then, 5 years after the initial deposit, he withdraws all of the accumulated money remaining in the account. How much does he withdraw 5 years after his initial deposit? (2.6)

28. Lynn borrows $5000 at 15%/year compounded monthly. She wishes to repay the loan with 12 end-of-month payments. She wishes to make her first payment 3 months after receiving the $5000. She also wishes that, after the first payment, the size of each payment be 10% greater than the previous payment. What is the size of her sixth payment? (2.6)

29. Solve Problem 28 for the case in which the size of each payment is $60 greater than the previous payment. (2.6)

30. David borrows $25,000 at 8% compounded quarterly. He wishes to repay the money with 10 equal semiannual installments. What must be the size of the payment if the first payment is made 1 year after obtaining the $25,000? (2.6)

31. Kristin deposits $200/month into an account paying 8% compounded quarterly. Twelve monthly deposits are made. Determine how much will be accumulated in the account 1 year after the last deposit. (2.6)

32. Barbara makes four consecutive annual deposits of $2000 in a savings account that pays interest at a rate of 10% compounded semiannually. How much money will be in the account 2 years after the last deposit? (2.6)

33. Lynn makes monthly deposits of $500 in a savings account that pays interest at a rate equivalent to 8% compounded quarterly. How much money should be in the account immediately after the sixtieth deposit? If no interest is earned on money deposited during a quarter and the first deposit coincides with the beginning of a quarter, what will be the account balance immediately after the sixtieth deposit? (2.6)

34. Stephen makes semiannual deposits of $1500 into an account that pays interest equivalent to 12% compounded monthly. Determine the account balance immediately *before* the eighth deposit. (2.6)

35. Mary Lib purchases a house for $250,000; a down payment of $20,000 is made at the time of purchase; and the balance is financed at 12% compounded monthly, with monthly payments made over a 10-year period. (2.6)
   a. What is the size of the monthly payments?
   b. If the loan period had been 20 years, what would have been the size of the monthly payments?

36. What is the effective interest rate of 18% compounded monthly? (2.6)

37. Marvin borrows $2500 from the Shady Dealings Loan Company. He is told the interest rate is merely 2%/month and his payment is computed as follows:

Loan period	= 30 months
Interest = 30(0.02)($2500)	= $1500
Credit investigation and insurance	= $ 50
Total amount owned = $2500 + $1500 + $50	= $4050
Size of payment = $4050/30	= $135/month

What effective interest rate is paid for the transaction? (2.6)

38. Lynne borrows $15,000 at 1.5%/month. She desires to repay the money using equal monthly payments over 36 months. Lynne makes four such payments and decides to pay off the remaining debt with one lump sum payment at the time for the fifth payment. What should be the size of the payment if interest is truly compounded at a rate of 1.5%/month? (2.6)

39. Approximately how long will it take a deposit to triple in value if money is worth 12% compounded monthly? (2.6)

40. Ken borrows $5000 and pays the loan off, with interest, after 3 years. He pays back $8000. What is the effective interest rate for this transaction? (2.6)

41. An automobile agency advertises lease arrangements available for a new model car. The lease period is 48 months. A total of $949.11 will be due at the beginning of the lease, including a refundable $225 security deposit, a $500 down payment, and taxes and fees. The size of the monthly payments is advertised to be $224.10. The 4-year lease allows 60,000 miles to be driven; for each additional mile driven, there is a penalty of $0.06 per mile. To purchase the car will cost $9876.50 initially; after four years' use, its resale value will be approximately $3000. Assuming no more than 60,000 miles are driven during the 4-year period and the operating and maintenance costs will be the same under leasing and purchasing, what is the effective interest rate for the lease? (2.6)

42. Ludo deposits $1000 now, $2000 2 years from now, and $5000 5 years from now into a fund paying 12% compounded semiannually. (2.6)

a. What sum of money will be accumulated at the end of the sixth year?

b. What equal deposits of size A, made every 6 months (with the first deposit at $t = 0$ and the last deposit at the end of the sixth year), are equivalent to the three deposits?

43. Semiannual deposits of $500 are made into a fund paying 10% compounded continuously. What is the accumulated value in the fund after 12 such deposits? (2.7.1)

44. Four equal, quarterly deposits of $1000 each are made at $t = 0, 1, 2$, and 3 (time periods are 3-month intervals) into a fund that pays 8% compounded *continuously*. Then, at $t = 8$ and $t = 10$, withdrawals of size $A are made so that the fund is depleted at $t = 10$. What is the size of the withdrawals? (2.7.1)

45. Jaime receives an annual bonus from his employer. He wishes to deposit the bonus in a fund that pays interest at a rate of 6% compounded continuously. His first bonus is $1000. The size of his bonus is expected to increase at a rate of 4%/year. How much money will be in the fund immediately after the twelfth deposit? (2.7.1)

46. Operating and maintenance costs for year k are given as $C_k = (e^{0.10})C_{k-1}$, $k = 2$, $3, \ldots, 15$, with $C_1 = 1500. Determine the equivalent uniform annual operating and maintenance cost based on continuous compounding with a nominal interest rate of (a) 10% and (b) 20%. (2.7.1)

47. A person borrows $10,000 and wishes to pay it back with 9 equal annual payments. What will the size of the payments be if the interest chart is 12% compounded (a) annually, (b) semiannually, and (c) continuously? (2.7.1)

48. For the cash flow series given below, determine (a) the present worth equivalent, (b) the future worth equivalent, and (c) the uniform annual series equivalent using a 10% continuous compound interest rate. (2.7.1)

EOY	Cash Flow	EOY	Cash Flow
0	$-$10,000	6	$5,000e^{0.10}$
1	1,000	7	$5,000e^{0.20}$
2	2,000	8	$5,000e^{0.30}$
3	3,000	9	$5,000e^{0.40}$
4	4,000	10	$5,000e^{0.50}$
5	5,000		

49. Labor costs occur continuously during the year. At the end of each year a new labor contract becomes effective for the following year. Let $\overline{A}_j$ denote the cumulative labor cost occurring uniformly during year j, where $\overline{A}_j = 1.05 \overline{A}_{j-1}$. If money is worth 10% compounded continuously and $\overline{A}_1 = $25,000$, determine the present worth equivalent for 5 years of labor costs. (2.72)

50. Operating and maintenance costs for a production machine occur continuously during the year. If the total annual operating and maintenance cost is $24,000 and money is worth 20% compounded continuously, what single sum of money at the present is equivalent to 5 years of operating and maintenance costs? (2.7.2)

51. Consumption of fuel oil occurs continuously in a manufacturing plant. The total monthly cost in January is $25,000. The demand for fuel oil increases at a monthly continuous compound rate of 1.5%. Determine the equivalent end-of-year (future worth) cost for one year's consumption using a monthly nominal continuous compound rate of 2.0%. (2.7.2)

52. For the continuous cash flows shown below, determine (a) the present worth equivalent and (b) the future worth equivalent using a continuous compound interest rate of 15%. (2.7.2)

Year	Continuous Cash Flow, $\overline{A}$
1	$10,000
2	12,000
3	14,000
4	16,000
5	18,000

53. Solve Problem 52 for the following cash flow series (2.7.2):

Year	Continuous Cash Flow, $\overline{A}$
1	$10,000
2	$10,000e^{0.20}$
3	$10,000e^{0.40}$
4	$10,000e^{0.60}$
5	$10,000e^{0.80}$

54. A firm buys a new computer that costs $100,000. It may either pay cash now or pay $20,000 down and $30,000/year for 3 years. If the firm can earn 15% on investments, which would you suggest? (2.8)

55. Kareem invests $10,000 in a venture that returns him $500(0.80)^{t-1}$ at the end of year t for $t = 1, \ldots, 20$. With an interest rate of 12%, what is the equivalent uniform annual profit (cost) for the venture? (2.8)

56. Hugh is considering two investment alternatives. Alternative A requires an initial investment of $10,000; it will yield incomes of $3000, $3500, $4000, and $4500 over its 4-year life. Alternative B requires an initial investment of $12,000; it is anticipated that the revenue received will increase at a compound rate of 10%/year. Based on an interest rate of 12% compounded annually, what must be the revenue the first year for B in order for A and B to be equivalent? (2.8)

57. It is desired to determine the size of the uniform series over the time period [2, 5] that is equivalent to the cash flow profile shown below using an interest rate of 10%. (2.8)

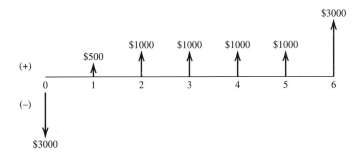

58. Given the following cash flows, what single sum at $t = 4$ is equivalent to the given data. Assume $i = 15\%$. (2.8)

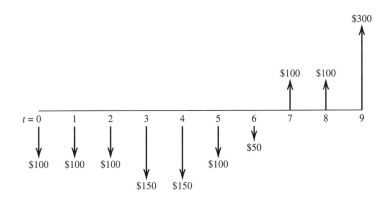

59. A machine is purchased at $t = 0$ for \$40,000 (including installation costs). Net annual revenues resulting from operating the machine are \$15,000. The machine is sold at the end of 8 years, $t = 8$, for \$5000. The cash flow series for the machine is equivalent to what single sum, \$X, at (a) $t = 6$ and (b) $t = 8$ if money is worth 15\% compounded continuously? (2.8)

60. Assume the following two investment plans.

  a. Purchase for \$4000 (a negative cash flow) and receive (1) \$400 at the end of each 6 months for 4 years ($t = 1, 2, \ldots, 8$), and (2) a single payment for \$2000 at the end of the fourth year.

  b. Purchase for \$3850 and receive (1) \$900 at the *beginning* of each year for the 4-year period, and (2) a single payment of \$X at the end of the fourth year.

  If money is worth 8\% compounded semiannually, what is the value of X so that the two investments are equivalent? (2.8)

61. Gerardo invests \$5000 and receives \$600 each year for 12 years, at which time he sells out for \$1000. With a 10\% interest rate, what equal annual cost (profit) is equivalent to the venture? (2.8)

62. Consider the following cash flow series.

EOY	CF
0	$-$\$10,000
1	6,500
2	6,000
3	5,500
4	5,000
5	4,500
6	4,000
7	3,500
8	3,000

At 10% annual compound interest, what uniform annual cash flow is equivalent to the above cash flow series? (2.8)

63. Susan borrows $1000 from a bank at $t = 0$ at 10% simple interest for 3 years. She pays the total interest due for the 3-year period at $t = 0$ and thus receives $700 $t = 0$. If she pays back $1000 at $t = 3$, Susan is, in effect, paying an interest rate of $X\%$ compounded annually. Solve for X. (2.8)

64. Assume payments of $2000, $3000, and $5000 are received at $t = 3, 4$, and 5, respectively. What five equal payments occurring at $t = 1, 2, 3, 4$, and 5, respectively, are equivalent if $i = 12\%$ compounded per period? (2.8)

65. What single deposit of size X into a fund paying 10% compounded annually is required at $t = 0$ in order to make withdrawals of $700 each at $t = 4, 5, 6$, and 7 and a single withdrawal of $1000 at $t = 10$? (2.8)

66. If the withdrawals in Problem 65 are immediately placed into another fund pay 12% compounded annually, what amount will be accumulated in this fund at $t = 10\%$ (2.8).

67. Ms. Torro-Ramos borrows $7000 and repays the loan with four quarterly payments of $600 during the first year and four quarterly payments of $1500 during the second year after receiving the $6000 loan. Determine the effective interest rate of the loan transaction. (2.8)

68. Quarterly deposits of $1000 are made at $t = 1, 2, 3, 4, 5, 6$, and 7. Then withdrawals of size A are made at $t = 12, 13, 14$, and 15, and the fund is depleted with the last withdrawal. If the fund pays 8% compounded quarterly, what is the value of A? (2.8).

69. Given the cash flow diagram shown below and a nominal interest rate of 12% compounded continuously, solve for the value of an equivalent amount at (a) $t = 5$, (b) $t = 10$, (c) $t = 15$. (*Note:* Upward arrows represent cash inflows and downward arrows represent cash outflows.) (2.8)

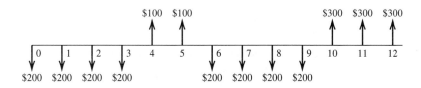

70. What single sum of money at $t = 4$ is equivalent to the cash flow profile shown below? Use a 10% interest rate in your analysis. (2.8)

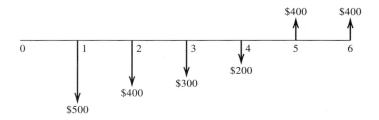

71. Determine the value of X so that the following cash flow series are equivalent at 8% interest. (2.8)

EOY	CF(A)	CF(B)
0	$-\$8,000$	$-\$15,000$
1	6,000	4,000
2	5,000	$3,000 + X$
3	4,000	$2,000 + 2X$
4	5,000	$3,000 + 3X$
5	6,000	$4,000 + 4X$
6	5,000	$3,000 + 5X$

72. Given the cash flow profiles shown below, determine the value of X so that the two cash flow profiles are equivalent at a 10% interest rate. (2.8)

EOY	CF(A)	CF(B)
1	$-\$12,000$	$-\$10,000$
2	1,000	7,000
3	3,000	$6,000 + 0.5X$
4	5,000	$5,000 + 1.0X$
5	7,000	$4,000 + 1.5X$
6	9,000	$3,000 + 2.0X$

73. Given the two cash flow profiles shown below, for what value of X are the two series equivalent using an interest rate of 10%? (2.8)

EOY	CF(A)	CF(B)
0	$-\$200,000$	$-\$140,000$
1	24,000	16,000
2	32,000	16,000
3	40,000	$16,000 + 2X$
4	48,000	$16,000 + 4X$
5	56,000	$16,000 + 6X$
6	64,000	$16,000 + 8X$
7	72,000	$16,000 + 10X$
8	80,000	$16,000 + 12X$

74. Given the cash flow profiles shown below, determine the value of X such that the two cash flow profiles are equivalent at 20% compounded annually. (2.8)

EOY	CF(A)	CF(B)
1	−$12,000	$ −X
2	1,000	7,000
3	4,000	9,000
4	6,000	10,000
5	7,000	10,000
6	5,000	7,000

75. Given the two cash flow profiles shown below, for what value of X are the two series equivalent using an interest rate of 10%? (2.8)

EOY	CF(A)	CF(B)
0	−$100,000	−$70,000
1	12,000	8,000
2	16,000	8,000 +
3	20,000	8,000 +
4	24,000	8,000 + X
5	28,000	8,000 + 2X
6	32,000	8,000 + 3X
7	36,000	8,000 + 4X
8	40,000	8,000 + 5X

76. Given the three cash flow profiles shown below, determine the values of X and Y so that all three cash flow profiles are equivalent at an annual interest rate of 15%. (2.8)

EOY	CF(A)	CF(B)	CF(C)
0	−$1,000	−$2,500	$ Y
1	X	3,000	Y
2	1.5X	2,500	Y
3	2.0X	2,000	2Y
4	2.5X	1,500	2Y
5	3.0X	1,000	2Y

77. Given the cash flow profiles shown below, determine the value of X such that the two cash flow profiles are equivalent at 20% *compounded continuously*. (2.8)

EOY	CF(A)	CF(B)
1988	$-\$75,000.00$	$\$ -X$
1989	10,000.00	5,000
1990	12,500.00	10,000
1991	15,625.00	15,000
1992	19,531.25	20,000
1993	24,414.06	25,000
1994	30,517.58	30,000

78. Given the two cash flow profiles shown below, for what value of X are the two series equivalent using an interest rate of 10%? (2.8)

EOY	CF(A)	CF(B)
0	$-\$60,000$	$-\$80,000$
1	15,000	15,000
2	$15,000(1 + x)$	20,000
3	$15,000(1 + x)^2$	25,000
4	$15,000(1 + x)^3$	30,000
5	$15,000(1 + x)^4$	35,000

79. The Ajax Chemical Company sells a particular chemical to the A. C. Works. A 5-year contract has been negotiated with the following terms. A constant volume, V_t, measured in cubic feet will be supplied uniformly and continuously during year t, with $V_t = 1.10V_{t-1}$. The price per cubic foot supplied during year t will be p_t, where $p_t = 1.08p_{t-1}$. The Ajax Chemical Company wishes to determine the equivalent *present value* of the contract using a continuous compounding rate of 10%/year. Express the equivalent present value in terms of p_1 and V_1. (2.8)

80. If a machine costs $10,000 and lasts for 15 years, at which time it is sold for $2000, what equal annual cost over its life is equivalent to these two cash flows with a 20% annual interest rate? (2.8)

81. Given the following profile of continuous flows, determine the equivalent series of annual discrete cash flows using a continuous compounding rate of 15%/year. (2.8)

EOY	$\overline{A}$
1	$10,000
2	12,000
3	14,000
4	16,000
5	18,000

Chapter 3

Reality Issues and Practical Applications

3.1 INTRODUCTION

In Chapter Two, basic time value of money concepts and mathematics were introduced. This chapter extends these basic concepts with issues frequently encountered in real world analyses and with applications commonly associated with personal financial decision making. The chapter concludes with consideration of special cases and relationships among the interest factors.

Two important issues that were not addressed in Chapter Two are frequently encountered when economic analysis problems are addressed in the real world. These issues are economic analysis subject to changing interest rates and economic analysis under inflation. Dealing with changing interest rates is a straight-forward embellishment of the techniques present in Chapter Two. Most students readily absorb this material. Contrarily, inflation is a complex and challenging subject. Many students find the concepts and techniques for dealing with inflation difficult to grasp (at least initially). This chapter presents an overview of inflation in broad terms and introduces two approaches to economic analysis under inflation. More detailed consideration of analysis under inflation can be found in [1,2].

Time value of money problems occur regularly and repeatedly in the business world. The majority of examples and problems in this text consider issues that are based on business scenarios. The factors and techniques presented are equally applicable to personal financial decision making. To demonstrate this applicability, two broad classes of problems that students frequently encountered in personal financial planning are introduced in this chapter. They are principal and interest payments on loans and bond calculations.

Many times, when first learning time value of money factors, students focus almost exclusively on formulas and calculations. While this is appropriate at the introductory level, it is also valuable to consider relationships that exist between the factors, limiting cases of the factors, and certain special cases of factor values. All of these can extend and solidify the student's understanding of the underlying concepts and mathematics. This chapter closes with a brief discussion of these topics.

3.2   CHANGING INTEREST RATES

The preceding chapter assumed that the interest rate did not change during the time period of concern. Recent experience indicates that such a situation is not likely if the time period of interest extends over several years (i.e., more than one interest rate may be applicable). Considering a single sum of money and discrete compounding, if i_t denotes the interest rate appropriate during time period t, the future worth equivalent for a single sum of money can be expressed as

$$F = P(1 + i_1)(1 + i_2) \cdot \cdot \cdot (1 + i_{n-1})(1 + i_n) \tag{3.1}$$

and the inverse relation

$$P = F(1 + i_n)^{-1}(1 + i_{n-1})^{-1} \cdot \cdot \cdot (1 + i_2)^{-1}(1 + i_1)^{-1} \tag{3.2}$$

EXAMPLE 3.1

Consider the situation depicted in Figure 3.1 in which an individual deposited $1000 in a savings account that paid interest at an annual compounding rate of 8% for the first 3 years, 10% for the next 4 years, and 12% for the next 2 years. Note how the cash flow diagram has been modified to incorporate the notation of changing interest rates. How much was in the fund at the end of the ninth year?

Letting V_t denote the value of the account at the end of time period t, we see that

$$V_3 = \$1000(F|P\ 8,3)$$
$$= \$1000(1.2597)$$
$$= \$1259.70$$

Likewise,

$$V_7 = \$1259.7(F|P\ 10,4)$$
$$= \$1259.7(1.4641)$$
$$= \$1844.33$$

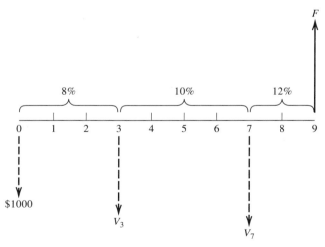

Figure 3.1 Cash flow diagram for a changing interest rate example.

Similarly,

$$F = V_9 = \$1844.33(F|P\,12,2)$$
$$= \$1844.33(1.2544)$$
$$= \$2313.53$$

Alternately, the amount in the account at the end of 9 years is given by

$$F = \$1000(1.08)(1.08)(1.08)(1.10)(1.10)(1.10)(1.10)(1.12)(1.12)$$
$$= \$2313.55$$

(The difference in answers is due to cumulative round-off errors in the interest tables.)

Extending the consideration of changing interest rates to series of cash flows, the present worth of a series of cash flows can be represented as

$$P = A_1(1 + i_1)^{-1} + A_2(1 + i_1)^{-1}(1 + i_2)^{-1} + \cdots \qquad \textbf{(3.3)}$$
$$+ A_n(1 + i_1)^{-1}(1 + i_2)^{-1} \cdots (1 + i_n)^{-1}$$

The future worth of a series of cash flows can be given by

$$F = A_n + A_{n-1}(1 + i_n) + A_{n-2}(1 + i_{n-1})(1 + i_n) + \cdots \qquad \textbf{(3.4)}$$
$$+ A_1(1 + i_2)(1 + i_3) \cdots (1 + i_{n-1})(1 + i_n)$$

EXAMPLE 3.2

Consider the cash flow diagram given in Figure 3.2 with the appropriate interest rates indicated. Determine the present worth, future worth, and uniform series equivalents for the cash flow series.

Computing the present worth gives

$$P = \$200(P|F\,10,1) - \$200(P|F\,10,1)(P|F\,10,1)$$
$$+ \$300(P|F\,8,1)(P|F\,10,1)(P|F\,10,1)$$
$$+ \$200(P|F\,12,1)(P|F\,8,1)(P|F\,8,1)(P|F\,10,1)(P|F\,10,1)$$
$$= \$200(P|F\,10,1) - \$200(P|F\,10,2)$$
$$+ \$300(P|F\,10,2)(P|F\,8,1) + \$200(P|F\,10,2)(P|F\,8,2)(P|F\,12,1)$$
$$= \$372.62$$

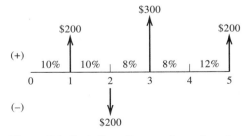

Figure 3.2 Cash flow diagram for a changing interest rate example.

The future worth is given by

$$F = \$200 + \$300(F|P\,8,1)(F|P\,12,1) - (\$200(F|P\,8,2)(F|P\,12,1)$$
$$+ \$200(F|P\,10,1)(F|P\,8,2)(F|P\,12,1)$$
$$= \$589.01$$

The uniform series equivalent is obtained as follows.

$$P = A(P|F\,10,1) + A(P|F\,10,2) + A(P|F\,8,1)(P|F\,10,2)$$
$$+ A(P|F\,8,2)(P|F\,10,2) + A(P|F\,12,1)(P|F\,8,2)(P|F\,10,2)$$
$$\$372.62 = A[(0.9091) + (0.8264) + (0.9259)(0.8264) + (0.8573)(0.8264)$$
$$+ (0.8929)(0.8573)(0.8264)]$$
$$\$372.62 = 3.842\,A$$
$$A = \$96.99$$

Thus, $96.99/time period for five time periods is equivalent to the original cash flow series.

3.3   CONSIDERATION OF INFLATION

Inflation is characterized by a decrease in the purchasing power of money caused by an increase in general price levels of goods and services without an accompanying increase in the value of the goods and services. Inflationary pressure is created when more dollars are put into an economy without an accompanying increase in goods and services. In other words, printing more money without an increase in economic output generates inflation. A complete treatment of inflation is beyond the scope of this chapter. A good summary can be found in Sullivan and Bontadelli [1980].

Any consideration of cash flows in today's economy must include a consideration of inflation. Although it is important to consider the impact of inflation on investments made within one country, it is especially important to do so in multinational investment situations. The inflation rates in Argentina, Brazil, Canada, France, Great Britain, Italy, Japan, Mexico, and the United States, for example, can be dramatically different. Hence, a firm that is faced with making decisions concerning the investment of capital in various nations must give strong consideration to the inflationary economies in the countries in question.

Inflation affects adversely the purchasing power of money. To illustrate what we mean by the term, *purchasing power of money,* suppose a firm purchases 1 million pounds of material each year, and the price of the material increases by 10% per year. Obviously, the quantity of material the firm can purchase with a fixed amount of money, that is, the purchasing power of the firm's money, decreases over time. The only way the firm can afford to continue purchasing the material is to decrease its usage rate or to increase its own source of funds. In the latter case, the firm might increase the price of the products it produces; if so, then the purchasing power of its customers' money will be decreased. In these situations, the continuing spiral of price increases does not contribute real increases in the firm's profits; instead, it results in an inflated representation of the firm's profits. The overall process is referred to as *inflation*.

Over the past three decades, we have experienced price level increases at both single-digit and double-digit rates. In some countries, triple-digit rates describe the price increases they have experienced. It has become widely accepted to refer to price level increases as inflation and to refer to price level decreases as deflation. (Under deflationary conditions, the deflation rate can be incorporated in economic analyses as a negative inflation rate.)

Although we have become accustomed to living in inflationary times, we are often very uncertain about how inflation should be treated, if at all, in an economic analysis. This section will present some background material on inflation, followed by two equivalent approaches for performing economic analyses under inflation.

3.3.1 Background on Inflation

One of the most difficult aspects of inflation is measuring it. To illustrate our claim, consider the following types of rates: *inflation rates, real interest rates, combined interest rates,* and *commodity escalation rates.*

As noted above, in recent years it has been apparent that the costs of goods and services are often affected by inflation. As a result, it is appropriate for interest rates and rates of return to reflect both the time value of money and the effects of inflation. For example, your decision to invest in a fund paying 12% interest would be quite different under negligible inflation conditions and 10% inflation conditions. We call the time value of money (or desired return) *in the absence of inflation* the *real interest rate* and denote it by d. The *combined interest rate* is denoted by i and the *inflation rate* is denoted by j. The three rates are related as follows:

$$1 + i = (1 + d)(1 + j)$$

or

$$i = d + j + dj \tag{3.5}$$

EXAMPLE 3.3

If the inflation rate (j) is 3%/year and the real time value of money (d) is 15%/year, what is the combined interest rate (i)?

$$1 + i = (1 + j) \times (1 + d)$$
$$1 + i = (1 + 0.03) \times (1 + 0.15)$$
$$1 + i = 1.1845$$
$$i = 1.1845 - 1 = 0.1845 = 18.45\%$$

EXAMPLE 3.4

If the combined interest rate is 26%/year and the real-time value of money is 20%/year, what is the inflation rate?

$$1 + i = (1 + j) \times (1 + d)$$
$$1 + 0.26 = (1 + j) \times (1 + 0.20)$$
$$1 + j = 1.26/1.20$$
$$j = 1.05 - 1 = 0.05 = 5.0\%$$

To complicate matters, it is not uncommon to find that prices and expenses for various goods and services increase (and decrease) at different rates, due to different external forces. For example, the price paid for a dozen eggs has scarcely changed over the past decade or more; whereas, the cost of new housing has increased dramatically in some parts of the United States. Oil prices change at different rates than the price of bread; changes in aluminum and steel prices are not closely correlated with changes in textile prices or the prices for microcomputers. In this case, we would consider the rate at which unit prices changed for each commodity to be an individual *commodity escalation rate*. (It is important to note that the commodity escalation rate for an individual commodity can be greater or less than the general inflation rate.)

In practice, when it comes time to measure the inflation rate, what we observe are the values of commodity escalation rates and the combined interest rates. The value of the inflation rate is not as evident. As a result, in estimating the value of the underlying inflation rate, a surrogate measure is generally used. One popular surrogate measure of the inflation rate in the United States is the yearly change in the *Consumer Price Index* (CPI); another is the *Producer Price Index* (PPI), also known as the *Wholesale Price Index* (WPI). A number of composite indexes are used by firms to forecast the inflation rate; for example both *Business Week* and *Forbes* publish their own indexes.

The challenge is not in obtaining values of indexes. Rather, it is in knowing how to convert the values into forecasts of future inflation rates. As an example, *Industry Week* regularly publishes monthly updates on a dozen indexes; it also provides monthly updates on approximately twenty "business barometers," including aluminum output (in thousands of metric tons), appliance shipments (in millions of units), auto output (in thousands of units), electric power output (in billions of kilowatt-hours), machine tool orders (in millions of dollars), new construction (in billions of dollars), prime interest rate, and retail sales (in billions of dollars).

Table 3.1 presents CPI values for the years since the base year 1967. From annual changes in the CPI, a value can be assigned to the overall annual inflation rate.

Some do not agree with the use of changes in the CPI as estimates of inflation. They argue that the inflation rate cannot be an exact representation of the purchasing power of *your* money without *your* purchases being accurately reflected by the makeup of commodities included in the computation of the CPI.

Table 3.1 CPI as an Indicator of U.S. Inflation[a]

Year	Consumer Price Index 1967 = 100.0	CPI Rate of Inflation Percent Change in CPI
1967	100.0	
1968	104.2	4.2
1969	109.8	5.4
1970	116.3	5.9
1971	121.3	4.3
1972	125.3	3.3
1973	133.1	6.2
1974	147.7	11.0
1975	161.2	9.1
1976	170.5	5.8
1977	181.5	6.5
1978	195.4	7.7
1979	217.4	11.3
1980	246.8	13.5
1981	272.4	10.4
1982	289.1	6.1
1983	298.4	3.2
1984	311.1	4.3
1985	322.2	3.6
1986	328.4	1.9
1987	340.4	3.7
1988	354.3	4.1
1989	371.3	4.8
1990	391.4	5.4
1991	408.0	4.2
1992	420.3	3.0
1993	432.7	3.0
1994	444.0	2.6
1995	456.5	2.8
1996	469.9	2.9

[a] Source: Bureau of Labor Statistics

As an example, if you never intend to buy or sell a house, then it would seem that the change of prices in houses will not have an impact on the purchasing power of your money. Likewise, if you are a vegetarian, then the change in prices for beef, pork, and chicken would not seem to affect you directly. However, although changes in some prices might not affect you directly, they will affect others—and those people, in turn, will increase the prices of their goods and services in order to recover the increased amount of money spent on items you do not buy.

Because the overall economy is so interdependent, it is difficult to determine the inflation rate for a person or an individual firm. Hence, changes in composite indexes, such as the CPI and PPI (or WPI), are used by individuals and organizations to estimate the impact of inflation on their economic investments. Since the rates of change in CPI and PPI (or WPI) reflect the changes in prices of a mixture of commodities, then any estimate of inflation derived from these rates will be a weighted average of the annual rates of change in the costs of a wide range of commodities. For this reason, the resulting estimate of the inflation rate is referred to as a "market basket" rate.

In addition to consumer prices, producer prices, and wholesale prices, the Commerce Department and the Labor Department, among others, publish a wide range of economic data for a variety of industry groups. Depending on the industry, inflation estimates might be more accurately developed using one or more such sources.

When consideration of inflation is introduced into economic analysis, future cash flows can be stated in terms of either constant-worth dollars or then-current dollars. *Then-current* cash flows are expressed in terms of the *face amount* of dollars (actual number of dollars) that will change hands when the cash flow occurs. Alternatively, *constant-worth* cash flows are expressed in terms of the *purchasing power* of dollars relative to a fixed point in time known as the base period. Then-current cash flows explicitly incorporate inflation into the cash flows; constant-worth cash flows do not.

EXAMPLE 3.5

For the next four years, a family anticipates buying $1000 worth of groceries each year. If inflation is expected to be 3%/year what are the then-current cash flows required to purchase the groceries.

To buy the groceries, the family will need to take the following face amount of dollars to the store. We will somewhat artificially assume that the family only shops once per year, buys the same set of items each year, and that the first trip to the store will be one year from today.

Year 1: dollars required $1000.00 \times (1.03) = $1030.00
Year 2: dollars required $1030.00 \times (1.03) = $1060.90
Year 3: dollars required $1060.90 \times (1.03) = $1092.73
Year 4: dollars required $1092.73 \times (1.03) = $1125.51

What are the constant-worth cash flows, if today's dollars are used as the base year?

The constant-worth dollars are inflation free dollars; therefore, the $1000 of groceries costs $1000 each year.

Year 1: $1000.00
Year 2: $1000.00
Year 3: $1000.00
Year 4: $1000.00

When constant-worth cash flows can be represented as a uniform series, the equivalent then-current cash flows will be represented by a geometric series. Let year zero be used to establish the base for constant-worth cash flows and C_k denote the constant-worth amount of a cash flow occurring at the end of year k. If

$$C_k = T_0 \qquad \text{for } k = 1, \ldots, n \tag{3.6}$$

then the then-current equivalent cash flow will be

$$
\begin{aligned}
T_k &= T_0(1+j)^k \\
&= T_0(F|P\,j,k)
\end{aligned}
\qquad \text{for } k = 1, \ldots, n \tag{3.7}
$$

EXAMPLE 3.6

The need for maintenance occurs at an increasing rate of 8%/year; the maintenance cost this year totals $1000. Furthermore, the labor and parts required to maintain the equipment increase due to inflation at a rate of 10%/year. The real interest rate for the firm is 12%/year. It is desired to determine the present worth equivalent of the then-current and constant-worth maintenance cost in the fifth year. In this case, $v = 0.08$, $j = 0.10$, $T_0 = \$1000$, and $d = 0.12$.

The commodity escalation rate for maintenance costs is made up of an 8% real escalation rate and a 10% inflation rate. Hence, the commodity escalation rate will be equal to the sum of 8%, 10%, and their product, or 18.8%.

The *then-current maintenance cost* will total $1000(F|P\,18.8,5)$ or $2366.31 in the fifth year; the *constant-worth maintenance cost* in the fifth year will be $1000(F|P\,8,5)$ or $1469.30. To obtain the present worth equivalent of the maintenance cost, the real interests rate of 12% is applied to the constant-worth amount, whereas a combined interest rate of $0.12 + 0.10 + 0.12(0.10)$ or 0.232 is applied to the then-current amount. Thus,

$$
\begin{aligned}
P &= \$1469.30(P|F\,12,5) \\
&= \$833.68
\end{aligned}
$$

or

$$
\begin{aligned}
P &= \$2366.31(P|F\,23.2,5) \\
&= \$2366.31(1.232)^{-5} \\
&= \$833.72
\end{aligned}
$$

(The $0.04 difference in the two answers is due to round-off error in the value of the $(P|F\,12,5)$ interest factor.)

Table 3.2 Approaches to Economic Analysis under Inflation

	Then-current Analysis	Constant worth analysis
Cash flows	Then-current dollars Inflation explicitly accounted for in cash flows	Constant-worth dollars Cash flows are inflation free
Discount rate for time Value of money factors	Combined interest rate	Real interest rate

3.3.2 Economic Analysis Under Inflation

The key to proper economic analysis under inflation is to base the value of the discount rate on the types of cash flows. If the cash flows contain inflation, then the value of the discount rate should also be adjusted for inflation. Alternatively, if the cash flows do not contain inflation, then the value of the discount rate should be inflation free.

If the cash flows of a project are stated in terms of then-current dollars, the appropriate value of the discount rate is the combined rate. Analysis done in this way is referred to as *then-current analysis*. If the cash flows of a project are stated in terms of constant-worth dollars, the appropriate value of the discount rate is the real rate. Analysis done in this way is referred to as then *constant worth analysis*. Table 3.2 summarizes these two approaches.

EXAMPLE 3.7

Using the cash flows of Examples 3.5 and interest rates of Example 3.3, determine the present worth of the grocery purchases using a constant worth analysis.

Constant worth analysis requires constant worth cash flows and the real interest rate.

$$PW = 1000 \times (P|A, 15,4) = 1000 \times (2.8550) = \$2855.00$$

EXAMPLE 3.8

Using the cash flows of Examples 3.5 and interest rates of Example 3.3, determine the present worth of the grocery purchases using a then current analysis.

Then current analysis requires then current cash flows and the combined interest rate

$$PW = 1030.00 \times (P|F, 18.45,1) + 1060.90 \times (P|F, 18.45,2) + 1092.73 \times (P|F, 18.45,3) + 1125.51 \times (P|F, 18.45, 4)$$
$$PW = 1030.00 \times (0.8442) + 1060.90 \times (0.7127) + 1092.73 \times (0.6017) + 1125.51 \times (0.5080)$$
$$PW = 869.53 + 756.10 + 657.50 + 571.76 = 2854.89$$

The notable result of Examples 3.7 and 3.8 is that the present worths determined by the constant worth approach ($2855.00) and the then-current approach ($2854.89) are *equal* (the $0.11 difference is due to factor rounding). This result is often unexpected but mathematically sound. The important conclusion is that if care is taken to appropriately match the cash flows and value of the discount rate, the level of general price inflation is *not* a determining factor in the acceptability of projects. To make this important result hold, inflation must either (1) be included in both the cash flows and the discount rate (the then-current approach) or (2) be included in neither the cash flows nor the discount rate (the constant worth approach). In subsequent sections of this book, we implicitly assume that one of the above two valid approaches has been followed in determining cash flows and discount rates.

Inflation is a much-discussed subject in the area of economic investment analysis. Some argue that inflation effects can be ignored, since inflation will affect all investments in roughly the same way. Thus, it is argued, the relative differences in the alternatives will be approximately the same with or without inflation considered. Others argue that the inflation rate during the past few decades has been so dynamic that an accurate prediction of the true inflation rate and its impact on future cash flows is not possible. Another argument for ignoring explicitly the effects of inflation is that it is accounted for implicitly, since cash flow estimates for the future are made by individuals conditioned by an inflationary economy. Thus, it is argued, any estimates of future cash flows probably incorporate implicitly inflationary effects. A final argument for ignoring inflation in comparing investment alternatives involving only negative cash flows is that an alternative that is preferred by ignoring inflation effects will be even more attractive when effects of inflation are incorporated in the analysis.

The above arguments are certainly valid in some instances. However, counterarguments can be given for each. Actually, it is not difficult to explicitly account for inflation. Use of either of the two approaches illustrated is recommended.

3.4 PRINCIPAL AMOUNT AND INTEREST AMOUNT IN LOAN PAYMENTS

With both personal and corporate investments that are financed from borrowed funds, income taxes might be affected by the amount of interest paid. Hence, it is quite important to know how much of each payment is interest and how much is reducing the principal amount borrowed initially. To illustrate this situation, suppose you borrowed $10,000 and paid it back using four equal annual installments, with interest computed at 10% compounded annually. The payment size is computed to be

$$A = P(A|P\ 10,4)$$
$$= \$10,000(0.3155)$$
$$= \$3155$$

The interest accumulation the first year is 0.10(10,000), or $1000. Therefore, the first payment consists of a $1000 interest payment and a $2155 principal payment. The unpaid balance at the beginning of the second year is $10,000 − $2155, or $7845; consequently, the interest charge the second year is 0.10($7845), or

Table 3.3 Loan Repayment (Amortization) Schedule

Year	Beginning Balance	Loan Payment	Interest Component	Principal Component	Ending Balance
1	$10,000.00	$3155.00	$1,000.00	$2155.00	$7845.00
2	$7845.00	$3155.00	$784.50	$2370.50	$5474.50
3	$5474.50	$3155.00	$547.45	$2607.55	$2866.95
4	$2866.95	$3155.00	$286.70	$2868.30	−$1.35

$784.50. Thus, the second payment consists of a $784.50 interest payment and a principal payment of $3155 − $784.50, or $2370.50. The unpaid balance at the beginning of the third year is $7845 − $2370.50, or $5474.50, so the interest charge the third year is 0.10($5474.50), or $547.45. Thus, the third payment consists of a $547.45 interest payment and a principal payment of $3155 − $547.45, or $2607.55. The unpaid balance at the beginning of the fourth year is $5474.50 − $2607.55, or $2866.95; the interest charge the fourth year is 0.10($2866.95), or $286.70. Thus, the fourth payment consists of a $286.70 interest payment and a principal payment of $3155 − $286.70, or $2868.30. (The principal payment in the fourth payment should equal the unpaid balance of $2866.95 at the beginning of the fourth year. The difference of $1.35 is due to round-off errors in computing the payment size.) These results are summarized in Table 3.3.

The results above can be generalized in the following way. If P is borrowed (at $t = 0$) with a period interest rate $i\%$/period for n periods and is to be repaid in n equal end-of-period payments (starting at $t = 1$), then the amount of the payments, A, can be determined by the relationship $A = P(A|P, i, n)$. For any given payment, part of the payment goes to pay interest and the remaining part reduces the principal amount (unpaid balance). While the payments remain constant at A, the portions of each payment that go to pay interest and to reduce the principal amount vary over time.

The amount of principal remaining to be repaid immediately after making payment at time t can be found by stripping off the interest on the remaining $(n - t)$ payments. Letting U_t denote the unpaid principal after making t payments, we have

$$U_t = A(P|A, i, n - t) \tag{3.8}$$

where A, n, and i are defined as noted above.

A related quantity is the payoff quantity. Payoff$_t$ is the total amount required to pay off the loan at t including both the current payment and the unpaid balance.

$$\text{Payoff}_t = A + U_t \tag{3.9}$$
$$\text{Payoff}_t = A + A(P|A, i, n - t)$$
$$\text{Payoff}_t = A[1 + (P|A, i, n - t)] \tag{3.10}$$

The interest accrued during any payment period is given by the unpaid balance at the beginning of the period times the period interest rate. Letting I_t denote the portion of payment t that goes to pay interest, we have

$$I_t = iU_{t-1} \tag{3.11}$$

Substituting the results of Equation 3.8 into Equation 3.11, the following relationship can be derived.

$$I_t = A\,[1 - (P|F, i, n - t + 1)] \tag{3.12}$$

The portion of the payment which does not pay accrued interest goes to principal reduction. Letting E_t denote the portion of payment t that goes to reduce principal, we have

$$E_t = A - I_t \tag{3.13}$$

or

$$E_t = A(P|F, i, n - t + 1) \tag{3.14}$$

EXAMPLE 3.9

In order to pay for the new car you just purchased, you borrow $12,000 at 12%/year/month for four years. You agree to repay the loan with 48 equal end of month payments.

(a) What are the payments?
(b) What is the remaining balance immediately after the 24th payment?
(c) If you choose to pay off the loan at $t = 36$, how much must you pay?
(d) What portion of the 12th payment is interest?
(e) What portion of the 12th payment is principal repayment?

Part a

Approach: $A = P(A|P, i, n)$

Period interest rate $= \dfrac{12\%/\text{year/month}}{12\ \text{month/year}} = 1\%/\text{month/month}$

$n = 48$ months

$A = P(A|P, 1, 48) = 12{,}000\,(0.0263) = \$315.60/\text{month}$

Part b

Approach: $U_t = A(P|A, i, n - t)$

$n = 48$

$t = 24$

$n - t = 48 - 24 = 24$

$U_{24} = A(P|A, 1, 24) = 315.60\,(21.2434) = \$6704.42$

Part c

Approach: $\text{Payoff}_t = A + A(P|A, i, n - t)$

$n = 48$

$t = 36$

$n - t = 48 - 36 = 12$

$\text{Payoff}_{36} = A + A(P|A, 1, 12) = 315.60 + 315.60\,(11.2551) = \$3867.71$

Part d

Approach: $I_t = iU_{t-1}$

$n = 48$

$t = 12$

$t - 1 = 12 - 1 = 11$

$U_{11} = A(P|A, 1, 37) = 315.60\,(30.7995) = \$9720.32$

$I_{12} = iU_{11} = 0.01\,(9720.32) = \$97.20$

Part e

Approach: $E_t = A - I_t$

$t = 12$

$E_{12} = A - I_{12} = 315.60 - 97.20 = \$218.40$

EXAMPLE 3.10

To illustrate the use of the formulas for computing the values of E_t and I_t, suppose \$10,000 is borrowed at 12% annual interest and repaid with five equal annual payments. The payment size is found to be

$$A = \$10,000(A|P\,12,5)$$
$$A = \$10,000(0.2774)$$
$$= \$2774$$

Thus

$$E_1 = \$2774(P|F\,12,5) = \$1574, \qquad I_1 = \$1200$$
$$E_2 = \$2774(P|F\,12,4) = \$1763, \qquad I_2 = \$1011$$
$$E_3 = \$2774(P|F\,12,3) = \$1975, \qquad I_3 = \$799$$
$$E_4 = \$2774(P|F\,12,2) = \$2211, \qquad I_4 = \$563$$
$$E_5 = \$2774(P|F\,12,1) = \$2477, \qquad I_5 = \$297$$

EXAMPLE 3.11

An individual purchases a $130,000 house and makes a down payment of $30,000. The remaining $100,000 is financed over a 25-year period at 12% compounded monthly. The monthly house payment is computed to be

$$A = \$100,000(A|P\,1,300)$$
$$= \$1053$$

The individual keeps the house for 5 years and decides to sell it. How much equity is there in the house?

The amount of principal remaining to be paid can be determined by computing the present worth of the remaining 240 monthly payments using a 1% interest rate, or

$$P = \$1053(P|A\,1,240)$$
$$= \$95,633$$

Thus, the individual's equity equals ($100,000 − $95,633) or $4367, plus the value of the down payment, or $34,367. It is interesting (or perhaps depressing if you are a home owner) to determine that over the 5-year period the individual made payments totaling $63,180, of which $58,813 was for interest; that is, $58,813 = $63,180 − $4367.

3.5 BOND PROBLEMS

Although bonds are important financial instruments in the business world as investment opportunities, there are additional reasons for considering bond problems in an economic analysis text, as well as at this particular point in the text.

1. Bond problems illustrate the notion of equivalence covered previously in Chapter Two. That is, the purchase price of a bond is equivalent to the returns from the bond at an appropriate compound interest rate. Hence, they provide a useful mechanism for demonstrating a number of the time value of money calculations developed earlier.
2. Bond problems are convenient investment opportunities and will be used to motivate our presentation in Chapter Four of various measures of the economic worths of investment alternatives.
3. The issuance and sale of bonds is a mechanism by which capital may be raised to finance engineering projects; as such, they contribute to the firm's overall cost of capital, as described in Chapter Five.
4. Bonds represent investment opportunities, which many individual investors choose to include in their personal financial plans.

An organizational unit desiring to raise capital may issue bonds totaling, say, $1 million, $5 million, $25 million, or more. A financial brokerage firm usually handles the issue on a commission basis and sells smaller amounts to other organizational units or individual investors. Individual bonds are normally issued in even denominations such as $500, $1000, or $5000. The stated value on the individual bond is termed the *face* or *par value*. The par value is to be repaid by the issuing organization at the end of a specified period of time, say 5, 10, 15, 20, or even 50 years. Thus, the issuing unit is obligated to *redeem* the bond at par value at *maturity*. Furthermore, the issuing unit is obligated to pay a stipulated *bond rate* on the face value during the interim between date of issuance and date of redemption. This might be 10% payable quarterly, 9½% payable semiannually, 11% payable annually, and so forth. For the purpose of the problems to follow, it is emphasized that the bond rate applies to the par value of the bond.

EXAMPLE 3.12

A person purchases a $5000, 5-year bond on the date of issuance for $5000. The stated bond rate is 10% semiannually, and the interest payments are received on schedule until the bond is redeemed at maturity for $5000. The bond rate per interest period is 10%/2 = 5%. Thus, the bond holder receives (0.10/2) ($5000) = $250 payments every 6 months. A cash flow diagram for the duration of the investment is given in Figure 3.3, where time periods are 6-month intervals.

It is noted from Figure 3.3 that the $5000 expenditure at $t = 0$ yields $250 each interest period for 10 periods and a $5000 redemption value at $t = 10$. Thus, the $5000 investment at $t = 0$ yielded the revenues from $t = 1$ through $t = 10$. What annual interest rate or *return on investment* did the $5000 investment yield?

One might intuitively answer the above question by stating that the $5000 investment was exactly returned (no loss or gain in capital) at redemption and, since during the interim an interest rate of 10% semiannual was received, then the yield *must* be 10% semiannual. Accepting this argument for the moment, one might further pose certain hypotheses concerning the transaction. For instance, it is hypothesized that the $5000 at $t = 0$ is equivalent (has the same present worth) to the revenue cash flows if the time value of money is 10% compounded

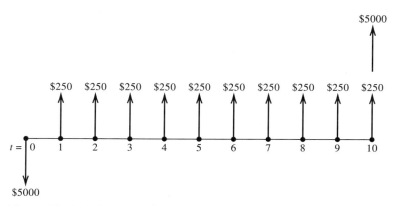

Figure 3.3 Cash flow for a $5000 bond.

semiannually. That this hypothesis is true is shown by the relation

$$P = A(P|A\ 5,10) + F(P|F\ 5,10)$$

or

$$\$5000 = (0.05)(\$5,000)(7.7217) + \$5000(0.6139),$$

and

$$\$5000 = \$5000$$

A second hypothesis concerning the transaction is that if more than the par value is paid for the bond at $t = 0$ and all revenue figures remain the same, then a return on investment less than the bond rate of 10% semiannual will be received on the investment. For example, if $5500 were paid for the bond at $t = 0$, the present worth of the $550 outgo is not equal (equivalent) to the present worth of the revenues at a 10% semiannual yield. This is shown by

$$\$5500 \neq (0.05)(\$5000)(P|A\ 5,10) + \$5000(P|F\ 5,10)$$
$$\$5500 \neq \$250(7.7217) + \$5000(0.6139)$$
$$\$5500 \neq \$5000$$

This result now raises the pertinent question, what is the rate of return on the investment if $5500 is paid for the bond and the revenues remain the same? Intuition may suggest that the yield will be less than 10% semiannual because the purchase value of $5500 decreases (a loss of investment capital) to $5000 on redemption. Furthermore, the semiannual payments of $250 are not equal to 5% of the $5500 purchase price. In order to answer the question more precisely, let us answer the alternate question of "what interest rate per period (or yield per period) will make the future worth of all cash flows equal to zero?" That is, what interest rate satisfies the following equation?

$$-\$5500(F|P\ i,10) + (.05)(\$5000)(F|A\ i,10) + \$5000 = \$0$$

The solution to the above equation gives the answer to the original question, and the task is to solve for the positive roots of the polynomial. However, an approximate solution to avoid such tedium is by trial and error. An interative procedure follows.

For $i = 3\%$ (6% semiannual)

$$-\$5500(F|P\ 3,10) + (.05)(\$5000)(F|A\ 3,10) + \$5000 \neq \$0$$
$$-\$5500(1.3439) + (\$250)(11.4639) + \$5000 \neq \$0$$
$$\$474.53 \neq \$0$$

For $i = 4\%$ (8% semiannual)

$$-\$5500(F|P\ 4,10) + (.05)(\$5000)(F|A\ 4,10) + \$5000 \neq \$0$$
$$-\$5500(1.4802) + (\$250)(12.0061) + \$5000 \neq \$0$$
$$-\$139.58 \neq \$0$$

Table 3.4 Bond Yield Interpolation

For $i =$	0.03	X	0.04
FW of revenues $=$	$474.53	$0.00	$-$139.58

From these two trials (for $i = 3\%$ and $i = 4\%$), the future worth of $0 is bracketed, as shown in Table 3.4. Using the data of Table 3.4, we can solve for X by linear interpolation, or

$$\frac{0.04 - 0.03}{-\$139.58 - \$474.53} = \frac{0.04 - X}{-139.58 - \$0.00}$$

or

$$X = 0.3772 \quad \text{or} \quad 3.772\%$$

Thus, the equivalent yield on the $5500 investment is approximately 3.772%/period, or $(2)(3.772) = 7.544\%$ compounded semiannually, or an effective annual yield of $[(1 + 0.03772)^2 - 1](100\%) = 7.686\%$. The figures support the *a priori* intuition of a return on investment less than 10% semiannual, and the second hypothesis is accepted.

Bond problems arise in economic analysis because many bonds trade daily through financial markets such as the New York Stock Exchange. Thus, bonds may be purchased for less than, greater than, or equal to par value, depending on the economic environment. They may also be sold for less than, greater than, or equal to par value. Furthermore, once purchased, bonds may be kept for a variable number of interest periods before being sold. A variety of situations can occur, but only three basic types of bond problems can occur. These will be presented after formalizing the discussion thus far. We now employ the following notation.

$P =$ the purchase price of a bond
$F =$ the sales price (or redemption value) of a bond
$V =$ the par or face value of a bond
$r =$ the bond rate per interest period
$i =$ the yield rate (return on investment or rate of return) per interest period
$n =$ the number of interest payments received by the bondholder
$A = Vr =$ the interest or coupon payment received

The general expression relating these is

$$P = Vr(P|A\ i, n) + F(P|F\ i, n) \tag{3.15}$$

Now, the three types of bond problems follow:

1. Given P, r, n, V, and a desired i, find the sales price F.
2. Given F, r, n, V, and a desired i, find the purchase price P.
3. Given P, F, r, n, and V, find the yield rate i that has been carried on the investment.

Each of these cases is illustrated in the following examples.

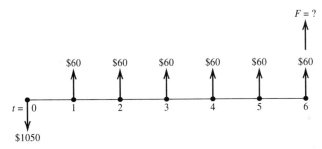

Figure 3.4 Cash flow for a $1000 bond—determine sales price.

EXAMPLE 3.13

A $1000, 12% semiannual bond is purchased for $1050 at an arbitrary $t = 0$. If the bond is sold at the end of 3 years and six interest payments, what must be the selling price if this bond is to be a good investment compared to investing at 5% per 6-month period? The cash flow for this example is given by Figure 3.4.

The present worth of the cash flow depicted in Figure 3.4 follows.

$$P = Vr(P|A \; 5,6) + F(P|F \; 5,6)$$

or

$$\$1050 = (\$1000)(0.06)(5.0757) + F(0.7462)$$

Solving the above equation for F yields a value of $999.01. As long as the selling price at the end of three years is at least $999.01 the bond will yield a return of at least 10% compounded semiannually.

EXAMPLE 3.14

If a $1000, 12% semiannual bond is purchased at an arbitrary $t = 0$, held for 3 years and six interest payments, and redeemed at par value, what must the purchase price have been in order for the bond to be preferred over investing at 14% compounded semiannually?

From the cash flow given in Figure 3.5, the present worth of future cash flows is

$$P = Vr(P|A \; 7,6) + F(P|F \; 7,6)$$

or

$$P = (\$1000)(0.06)(4.7665) + \$1000(0.6663)$$

and

$$P = \$952.29$$

As long as the purchase price was less than $952.29, the bond was better than investing at 14% compounded semiannually.

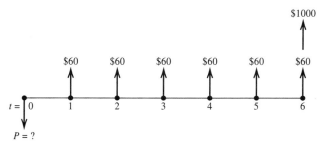

Figure 3.5 Cash flow for a $1000 bond—determine purchase price.

EXAMPLE 3.15

If a $1000, 12% quarterly bond is purchased at $t = 0$ for $1020 and sold 3 years later for $950, (a) what was the quarterly yield on the investment, and (b) what was the effective annual return? From the cash flow diagram in Figure 3.6, and setting the present worth equal to zero, we have

$$-\$1020 + (\$1000)(0.03)(P|A\ i,12) + \$950(P|F\ i,12) = \$0$$

It is then necessary to solve for the unknown i by trial and error as follows: For $i = 2\%$,

$$-\$1020 + \$30(10.5753) + \$950(0.7885) \neq \$0$$
$$-\$46.33 \neq \$0$$

For $i = 2\frac{1}{2}\%$,

$$-\$1020 + \$30(10.2578) + \$950(0.7436) \neq \$0$$
$$\$5.85 \neq \$0$$

Then, interpolating gives

$$\frac{0.025 - 0.020}{\$5.85 - (-\$46.33)} = \frac{0.025 - X}{\$5.85 - \$0}$$

$$X = 0.02444 \qquad \text{or} \qquad 2.444\% \text{ per quarter}$$

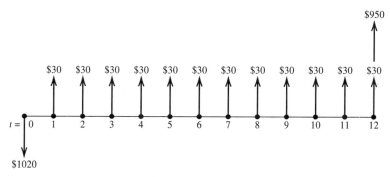

Figure 3.6 Cash flow for a $1000 bond—determine yield.

The effective annual return is

$$[(1 + 0.02444)^4 - 1]100\% = 10.140\%$$

3.6   SPECIAL AND LIMITING CASES OF TIME VALUE OF MONEY FACTORS

It is instructive to give consideration to certain special and limiting cases of the time value of money factors. Several of the relationships presented below have been discussed previously in the text and are summarized here for convenience. Others (particularly the limiting cases) are new and may require some reflective thought on the part of the student. Derivation and/or proofs of the relationships are left as exercises for the student.

3.6.1   Mathematical Relationships

The equations below represent mathematical relationships that hold among the time value of money factors. While proofs of each can be derived mathematically, it is perhaps more instructive to intuitively rationalize the relationships using simple examples and cash flow diagrams.

Inverse Relationships

$$(F|P, i, n) = \frac{1}{(P|F, i, n)} \tag{3.16}$$

$$(A|P, i, n) = \frac{1}{(P|A, i, n)} \tag{3.17}$$

$$(A|F, i, n) = \frac{1}{(F|A, i, n)} \tag{3.18}$$

Series to Series and Series to Single Sum Relationships

$$(A|P, i, n) = (A|F, i, n) + i \tag{3.19}$$

$$(P|A, i, n) = \sum_{t=1}^{n} (P|F, i, t) \tag{3.20}$$

$$\frac{1}{(A|P, i, n)} = \sum_{t=1}^{n} \frac{1}{(F|P, i, t)} \tag{3.21}$$

Single Sum Timing Relationships

$$(P|F, i, n + m) = (P|F, i, n) \times (P|F, i, m) \tag{3.22}$$

$$(P|F, i, n - m) = \frac{(P|F, i, n)}{(P|F, i, m)} \tag{3.23}$$

$$(P|F, i, n \times m) = (P|F, i, n)^m = (P|F, i, m)^n \tag{3.24}$$

Table 3.5 Limiting Cases

Expression	Variable and Limit	Result
$F = P\,(F|P, i, n)$	$n \to \infty$	$F \to \infty$
$F = P\,(F|P, i, n)$	$i \to 0$	$F \to P$
$P = F\,(P|F, i, n)$	$n \to \infty$	$P \to 0$
$P = F\,(P|F, i, n)$	$i \to 0$	$P \to F$
$A = P\,(A|P, i, n)$	$n \to \infty$	$A \to Pi$
$A = P\,(A|P, i, n)$	$i \to 0$	$A \to P/n$
$P = A\,(P|A, i, n)$	$n \to \infty$	$P \to A/i$
$P = A\,(P|A, i, n)$	$i \to 0$	$P \to nA$
$F = A\,(F|A, i, n)$	$n \to \infty$	$F \to \infty$
$F = A\,(F|A, i, n)$	$i \to 0$	$F \to nA$
$A = F\,(A|F, i, n)$	$n \to \infty$	$A \to 0$
$A = F\,(A|F, i, n)$	$i \to 0$	$A \to F/n$

3.6.2 Limiting Cases

Table 3.5 presents limiting cases of time value of money relationships. Each row is meant to be considered independently. In each case the expression in the left column approaches the result in the right column as the limiting condition in the middle column is approached. In each case, all variables except the limiting condition variable are assumed to hold constant.

3.6.3 Special Cases for $n = 1$

The equations presented below represent special cases that hold when the value of n is equal to 1.

$$(P|G, i, 1) = 0 \tag{3.25}$$
$$(A|G, i, 1) = 0 \tag{3.26}$$
$$(A|P, i, 1) = (F|P, i, 1) = 1 + i \tag{3.27}$$
$$(A|F, i, 1) = (F|A, i, 1) = 1 \tag{3.28}$$

3.7 SUMMARY

In this chapter, we have extended basic time value of money concepts to address the issues of changing interest rates and inflation. These two situations are frequently encountered in real-world analyses. To motivate the application of the interest factors, we have presented loan and bond calculations as they might be encountered in personal financial decision making. Finally, we have presented several special cases and relationships among the interest factors.

BIBLIOGRAPHY

1. Jones, B. W., *Inflation in Engineering Economy Analysis*, John Wiley, 1982.
2. Sullivan, W. G., and Bontadelli, J. A., "The Industrial Engineer and Inflation," *Industrial Engineering*, 12 (3), pp. 24–33.
3. Thuesen, G. J. and Fabrycky, W. J., *Engineering Economy*, 6th ed., Prentice-Hall, 1984.
4. Thuesen, G. J., "Economic Analysis," Chapter 6.2, *Production Handbook*, 4th ed., (J. A. White, editor), Wiley, 1987.
5. White, J. A., "Engineering Economy," *Handbook of Engineering Fundamentals*, 4th ed., (M. Souders and O. W. Eshbach, editors), Wiley, 1990.

PROBLEMS

1. Maria deposits $2000 in a savings account that pays 7% compounded annually; 2 years after the deposit, the interest rate increases to 8% compounded annually. A second deposit of $3000 is made immediately after the interest rate changes to 8%. How much will be in the fund 7 years after the second deposit? (3.2)
2. Based on continuous cash flows and continuous compounding, compute the present worth for the following situation: (3.2)

Year	$\overline{A}$	Nominal Interest Rate During Year
1	$2,000e^{0.05}$	0.25
2	$2,000e^{0.10}$	0.20
3	$2,000e^{0.15}$	0.15
4	$2,000e^{0.20}$	0.10
5	$2,000e^{0.25}$	0.05

3. Labor costs over a 4-year period have been forecast in then-current dollars as follows: $10,000, $12,000, $15,000, and $17,500. The general inflation rate for the 4 years is forecast to be 5%. Determine the constant worth dollar amounts of labor over each of the 4 years. (3.3)
4. Yearly labor costs of a highway maintenance group are currently $420,000/year. If labor rates escalate at a 12% rate and general inflation increases at 9%, determine for each of the next 6 years the labor cost in then-current and constant worth dollars. (3.3)
5. If you desire a real return of 8% on your money, excluding inflation, and inflation is running at 3%, what combined discount rate should you be seeking? (3.3)
6. Mellin Transformers Co. uses a required return of 15% in all alternative evaluations. Inflation is running at 5%. What real discount rate, exclusive of inflation, are they implicitly using? (3.3)

7. The following material costs are anticipated over a 5-year period: $9000, $11,000, $14,000, $18,000, and $23,000. It is estimated that a 4% inflation rate will apply over the time period in question. The material costs given above are expressed in then-current dollars. The time value of money, excluding inflation, is estimated to be 7%. Determine the present worth equivalent for material cost using the following. (3.3)

   a. Then-current costs

   b. Constant worth costs

8. A landfill has a first cost of $270,000. Annual operating and maintenance costs for the first year will be $40,000. These costs will increase at 11%/year. Income for dumping rights at the landfill will be held fixed at $120,000/year. The landfill will be in operation for 10 years. Inflation will average 8%, and a real return of 3.6% is desired. (3.3)

   a. Determine the present worth of this project using then-current dollars.

   b. Determine the present worth of this project using constant worth dollars.

9. The inflation rates for 4 years are forecast to be 3%, 3%, 4%, and 5%. The interest rate exclusive of inflation is anticipated to be 6%, 5%, 4%, and 5% over this same period. If labor is projected to be $1000, $15000, $2000, and $1000 in then-current dollars, during those years, determine the present worth equivalent for labor cost. (3.3)

10. The unit price of personal computers, measured in constant dollars, is expected to decrease at an annual rate of 10%. However, the number of microcomputers purchased by the company is expected to increase over a 5-year period at an annual rate of 30%. The unit price of a personal computer is currently $2000. This year the company will purchase 100 microcomputers at the current price. Using a real interest rate of 8% and an inflation rate of 6%, determine the following:

   a. the yearly expenditures for personal computers over the 5-year period, measured in constant worth dollars.

   b. the yearly expenditures for personal computers over the 5-year period, measured in then-current dollars.

   c. the single sum equivalent at the end of the current year of the expenditures on microcomputers over the 5-year period, measured in current dollars.

   d. the future worth equivalent of the expenditures on microcomputers at the end of the 5-year period, measured in then-current dollars. (3.3)

11. $90,000 is invested in a program to reduce the material requirements in a production process. As a result of the investment, the annual material requirement is reduced by 10,000 pounds. The present unit cost of the material is $2 per pound. The price of a pound of the material is expected to increase at an annual rate of 8%, due to inflation. Determine the combined interest rate that equates the present worth of the savings to the present worth of the investment over a 5-year period; based on the result obtained, determine the real interest rate that equates the two present worths. (3.3)

12. At $t = 0$, Martin borrows $5000 at 6%/period. Twelve equal payments are used to repay the loan at $t = 1, \ldots, 12$. Determine the amount of interest included in the fourth payment. (3.4)

13. Chris and Debbie borrow $10,000 at 7% compounded monthly. The loan is to be paid off with 48 equal monthly payments. One month after making the thirtieth payment, they decide to pay off the unpaid balance on the note. How much should be repaid? (3.4)

14. Tom and Dale purchase a boat for $150,000; the down payment is $15,000; the balance is financed over a 10-year period. Equal monthly payments are made. Determine the amount of interest paid the first month if the monthly interest rate is 1%. (3.4)

15. Dale borrows $8000 at 6% compounded annually. Four equal annual payments are used to repay the loan, with the first payment occurring 2 years after receiving the $8000. Determine the amount of principal included in each payment. (3.4)

16. Dr. Schultz is considering purchasing a bond having a face value of $2500 and a bond rate of 10% payable semiannually. The bond has a remaining life of 8 years. How much should she pay for the bond in order to earn a return on investment of 14% compounded semiannually? Assume the bond will be redeemed for face value. (3.5)

17. A 10,000, 10-year, 12% semiannual bond is purchased at $t = 0$ by Dr. Rich for par value. After receiving the twelfth dividend, Dr. Rich sells the bond to Dr. Ruth at a price sufficient to yield a return on the original purchase price equivalent to 9% compounded semiannually. What was Dr. Rich's selling price? If Dr. Ruth keeps the bond until maturity and redeems it for $5000, what approximate annual effective yield rate will Dr. Ruth receive? (3.5)

18. Dr. Ramirez wishes to purchase a bond having a face value of $10,000 and a bond rate of 15% payable annually. The bond has a remaining life of 8 years. In order to earn a 20% rate of return on the investment, what amount should be paid for the bond? (3.5)

19. Elizabeth wishes to sell a bond that has a face value of $1000. The bond has a bond rate of 9% with bond premiums paid annually. Four years ago, she paid $900 for the bond. At least a 15% return on investment is desired. What must be the minimum selling price for the bond in order to earn the desired return on investment? (3.5)

20. Dr. Schultz purchases a $10,000 face value bond for $10,500. It matures in 15 years after paying $1200 at the end of each year. What yield rate was earned on the investment? (3.5)

21. Kristin buys a $2000 bond for $2100. The bond has a bond rate of 12% with bond premiums paid annually. If the bond is kept for 8 years and sold for par value, determine the equivalent annual yield rate (rate of return) for Kristin's bond investment. (3.5)

22. Dr. Bozer is considering purchasing a bond having a face value of $2500 and a bond rate of 8% payable semiannually. The bond has a remaining life of 10 years. How much should Dr. Bozer pay for the bond in order to yield a return of 10% compounded semiannually? Assume the bond will be redeemed for face value. (3.5)

23. Develop a mathematical proof of Equation 3.19. (3.6)

24. Develop a mathematical proof of Equation 3.21. (3.6)

25. Explain why Equation 3.22 holds. Develop a mathematical proof. (3.6)

26. Explain why A approaches Pi in the expression $A = P(A|P, i, n)$ as n approaches infinity. How is this expression related to the concept of an endowment? (3.6)

27. Explain why A approaches P/n in the expression $A = P(A|P, i, n)$ as i approaches zero. (3.6)

28. Explain why Equation 3.25 holds. (3.6)

Chapter 4

Measuring the Worth of Investments

4.1 INTRODUCTION

Chapter One recommended that engineers and systems analysts solve problems by formulating and analyzing the problem, generating a number of feasible solutions (alternatives) to the problem, comparing the investment alternatives, selecting the preferred solution, and implementing the solution. The process of evaluating the investment alternatives and selecting the preferred solution is the subject of the remainder of this book.

In this chapter we develop a number of methods and criteria for measuring the worth of investments. This material forms the basis for techniques used in Chapter Five to compare alternative investments and select those preferred from the economic point of view. The methods and criteria of this chapter must be understood and properly applied if valid economic analyses are to be assured.

Before proceeding with a treatment of the various methods of measuring the economic worth of investment alternatives, it is important to note that all but one of the measures of worth of an investment we treat involve the use of a compound interest rate. The interest rate used is variously referred to as a *minimum attractive rate of return* (MARR), *hurdle rate, required rate of return, return on investment,* and *discount rate.* (Recall, in Chapter Three we referred to the interest rate as the *yield rate* for a bond.) Where compound interest calculations are used, the measures of investment worth are referred to as *discounted cash flow (DCF) measures;* the methods used to compute the values of investment worth are called *discounted cash flow methods.*

Various methods exist for determining the value of the interest rate to use. However, as explained subsequently in Chapter Five, we will consider the interest rate to represent a minimum attractive rate of return (*MARR*) on investment capital. If an investment proves to be profitable after considering the *MARR*, then it will be a viable candidate for investment. Furthermore, if money is not invested in the alternative in question, then we assume it will be invested elsewhere and earn a return at least equal to the *MARR*.

4.2   METHODS OF MEASURING INVESTMENT WORTH

As noted in Chapter Two in discussing equivalence and measures of worth, the economic worth of an investment can be measured in a number of ways. Present worth (PW) and annual worth (AW) methods are two commonly used approaches. Among the several methods of measuring investment worth are

1. Present worth method (PW).
2. Annual worth method (AW).
3. Future worth method (FW).
4. Internal rate of return method (IRR).
5. External rate of return method (ERR).
6. Savings/investment ratio method (SIR).
7. Payback period method (PBP).
8. Capitalized worth method (CW).

Each of the above measures of merit or measures of effectiveness has been used numerous times in evaluating real-world investments. They may be described briefly as follows:

1. Present worth method converts all cash flows to a single sum equivalent at time zero using $i = MARR$.
2. Annual worth method converts all cash flows to an equivalent uniform annual series of cash flows over the planning horizon using $i = MARR$.
3. Future worth method converts all cash flows to a single sum equivalent at the end of the planning horizon using $i = MARR$.
4. Internal rate of return method determines the interest rate that yields a future worth (or present worth or annual worth) of zero.
5. External rate of return method determines the interest rate that yields a future worth of zero, explicitly assuming reinvestment of recovered funds at the $MARR$.
6. Savings/investment ratio method determines the ratio of the present worth of savings to the present worth of the investment.
7. Payback period method determines how long it will take to recover the initial investment.
8. Capitalized worth method determines the single sum at time zero that is equivalent at $i = MARR$ to a cash flow pattern that continues indefinitely.

With the exception of the payback period and capitalized worth methods, all of the measures listed are equivalent methods of measuring investment worth. Hence, applying each of the first six measures of merit to the same investment alternative will yield the same recommendation.

Since the present worth, annual worth, future worth, internal rate of return, external rate of return, and savings/investment ratio methods are equivalent, why do more than one method exist? The primary reason for having different, but equivalent, measures of effectiveness for economic alternatives appear to be the differences in preferences among managers. Some individuals (and firms) prefer to express the net economic worth of an investment alternative as a single sum amount; hence, either the present worth method or the future worth method is used. Other individuals prefer to see the net economic worth spread out uniformly over the planning horizon, so the annual worth method is used by them. Yet another group of individuals wishes to express the net economic worth as a rate or percentage; consequently, one of the rate-of-return methods would be preferred. Finally, some individuals prefer to see the net economic worth expressed as a percentage of the investment required; the savings/investment ratio is one method of providing such information.

Since many organizations have established procedures for performing economic analyses, it seems worthwhile to consider in this chapter the more popular measures of merit that are used. Among those listed, it appears that the present worth, rate-of-return, and payback period methods are currently the most popular. However, a number of governmental agencies have adopted some version of the savings/investment (or benefit/cost) ratio method for purposes of comparing investment alternatives; hence, it is popular in some sectors.

4.2.1   Present Worth Method

The present worth of investment alternative j can be represented as

$$PW_j(i) = \sum_{t=0}^{n} A_{jt}(1 + i)^{-t} \tag{4.1}$$

with $PW_j(i)$ = present worth of Alternative j using $MARR$ of $i\%$
n = planning horizon
A_{jt} = net cash flow for Alternative j at the end of period t
i = $MARR$

The *present worth method* is the most popular measure of merit available. When it is used in Chapter Five to compare alternatives, the one having the greatest present worth is the alternative recommended.

EXAMPLE 4.1

A pressure vessel was purchased for $16,000, kept for 5 years, and sold for $3000. Annual operating and maintenance costs were $4000. Using a 12% minimum attractive rate of return, what was the present worth for the investment?

$$PW_1(12\%) = -\$16,000 - \$4000(P|A\ 12,5) + \$3000(P|F\ 12,5)$$
$$= -\$16,000 - \$4000(3.6048) + \$3000(.5674)$$
$$= -\$28,717$$

We can see that a single expenditure of $28,717 at time zero is equivalent to the cash flow profile for the pressure vessel.

Example 4.1 deals with an investment where no returns other than the salvage value are shown, resulting in a negative present worth. Example 4.2 considers an investment having a series of positive net cash flows.

EXAMPLE 4.2

Improved tooling for numerical control machinery will cost $10,000, last 6 years, and have no salvage value at that time. Due to this investment, net income will increase by $2525 during each of the first 3 years and by $3840 during each of the remaining 3 years. Using a 15% *MARR*, what is the present worth for the investment?

$$PW_1(15\%) = -\$10,000 + \$2525(P|A\ 15,3) + \$3840(P|A\ 15,3)(P|F\ 15,3)$$
$$= -\$10,000 + \$2525(2.2832) + \$3840(2.2832)(.6575)$$
$$= \$1529.70$$

A single sum of $1529.70 at time zero is equivalent to the cash flow profile for this investment.

4.2.2   Annual Worth Method

The annual worth of investment alternative j can be computed as

$$AW_j(i) = \left[\sum_{t=0}^{n} A_{jt}(P|F\ i,t)\right](A|P\ i,n) \tag{4.2}$$

or

$$AW_j(i) = PW_j(i)(A|P\ i,n) \tag{4.3}$$

where $AW_j(i)$ denotes the annual worth of alternative j using $i = MARR$. The *annual worth method* and the *present worth method* are related by a constant multiplier, $(A|P\ i,n)$, and are therefore equivalent measures of investment worth. The investment alternative having the greatest annual worth will be preferred when the annual worth method is used to compare alternatives in Chapter Five.

EXAMPLE 4.3

Let us determine the annual worth for the pressure vessel of Example 4.1 in two different ways. Recall that $P = \$16,000$, $n = 5$, $F = \$3000$, annual operating and maintenance were $4000, and $i = 12\%$.

$$AW_1(12\%) = -\$16,000(A|P\ 12,5) - \$4000 + \$3000(A|F\ 12,5)$$
$$= -\$16,000(.2774) - \$4000 + \$3000(.1574)$$
$$= -7966.20/\text{year}$$

Or, using Equation 4.3, and the answer to Example 4.1,

$$AW_1(12\%) = (-\$28{,}717)(A|P\,12{,}5)$$
$$= (-\$28{,}717)(.2774)$$
$$= -7966.10/\text{year}$$

The minor difference in answers is due to round-off error. An expenditure of $7966.20 at the end of each of 5 years is equivalent to the cash flow profile for the pressure vessel.

EXAMPLE 4.4

We want to determine the annual worth for the improved numerical control tooling of Example 4.2. Recall that the tooling lasts 6 years and costs $10,000, but net incomes increase by $2525 during each of the first 3 years and by $3840 thereafter. *MARR* is 15%.

$$AW_1(15\%) = -\$10{,}000(A|P\,15{,}6) + \$2525 + \$1315(F|A\,15{,}3)(A|F\,15{,}6)$$
$$= -\$10{,}000(.2642) + \$2525 + \$1315(3.4725)(.1142)$$
$$= \$404.48/\text{year}$$

or

$$AW_1(15\%) = PW_1(15\%)(A|P\,15{,}6)$$
$$= \$1529.70(.2642)$$
$$= \$404.15/\text{year}$$

Note that the difference between $404.48/year and $404.15/year is due to round-off errors.

4.2.3   Future Worth Method

The future worth of investment alternative j can be determined using the relationship

$$\boxed{FW_j(i) = \sum_{t=0}^{n} A_{jt}(1 + i)^{n-t}} \qquad (4.4)$$

or

$$FW_j(i) = PW_j(i)(F|P\,i,n) \qquad (4.5)$$

or

$$FW_j(i) = AW_j(i)(F|A\,i,n) \qquad (4.6)$$

where $FW_j(i)$ is defined as the future worth of investment j using a *MARR* of $i\%$. The *future worth method* is equivalent to the *present worth method* and the *annual worth method*, since the ratio of $FW_j(i)$ and $PW_j(i)$ equals a constant, $(F|P\ i,n)$, and the ratio of $FW_j(i)$ and $AW_j(i)$ equals a constant, $(F|A\ i,n)$. The alternative having the greatest future worth will be preferred when the future worth method is used in Chapter Five to compare alternatives.

EXAMPLE 4.5

For the previous two examples, the future worth is given by

$$FW_1(15\%) = -\$10{,}000(F|P\ 15{,}6) + \$2525(F|A\ 15{,}6) + \$1315(F|A\ 15{,}3)$$
$$= -\$10{,}000(2.3131) + \$2525(8.7537) + \$1315(3.4725)$$
$$= \$3538.43$$

or

$$FW_1(15\%) = PW_1(15\%)(F|P\ 15{,}6)$$
$$= \$1529.70(2.3131)$$
$$= \$3538.35$$

or

$$FW_1(15\%) = AW_1(15\%)(F|A\ 15{,}6)$$
$$= \$404.48(8.7537)$$
$$= \$3540.70$$

Again, differences in answers are due to round-off errors.

4.2.4 Internal Rate of Return Method

The rate of return for an alternative can be defined as the interest rate that equates the future worth to zero. Letting i_j^* denote the rate of return for alternative j, Equation 4.4 becomes

$$0 = \sum_{t=0}^{n} A_{jt}(1 + i_j^*)^{n-t} \tag{4.7}$$

This method of defining the rate of return is referred to in the economic analysis literature as the discounted *cash flow rate of return, internal rate of return,* and the *true rate of return.* We prefer the term *internal rate of return (IRR).*

Note that the present worth and the annual worth can be obtained by multiplying both sides of Equation 4.4 by the appropriate interest factor [i.e., $(P|F\ i_j^*,n)$ and $(A|F\ i_j^*,n)$]. Hence, the internal rate of return can also be defined to be the interest rate that yields either a present worth or an annual worth of zero. Depending on the form of a particular cash flow profile, it might be more convenient to use a present worth or an annual worth formulation to determine the internal rate of return for an alternative.

It is important to understand the definition of rate of return inherent in the use of the *IRR* method. In particular, the internal rate of return on an investment is defined as *the rate of interest earned on the unrecovered balance of an investment.* This concept was demonstrated in Chapter Three in discussing the amount of a loan payment that was principal. It is illustrated again in Table 4.1, where $10,000 is invested to obtain the receipts shown over a 6-year period. A_t denotes the cash flow at the *end* of period t, B_t represents the unrecovered balance at the *beginning* of period t, E_t is the unrecovered balance at the *end* of period t, and I_t is defined as the interest on the unrecovered balance *during* period t. The following relationships exist.

$$E_0 = A_0$$
$$B_t = E_{t-1} \qquad t = 1, \ldots, n$$
$$I_t = B_t i \qquad t = 1, \ldots, n$$
$$E_t = A_t + B_t + I_t \qquad t = 1, \ldots, n$$

If i is the internal rate of return, then E_n will equal zero. As indicated in Table 4.1, if i is 20%, E_n is approximately zero. Consequently, i^* is approximately 20%. The equivalence of E_n being zero and the future worth being zero is easily understood by recognizing that E_n is actually the future worth of the cash flow profile. To see why this is true, notice that

$$E_n = A_n + B_n + I_n$$

Employing the definition of I_n,

$$E_n = A_n + B_n(1 + i)$$

By the relationship between B_n and E_{n-1}, it is seen that

$$E_n = A_n + E_{n-1}(1 + i)$$

Since a similar relationship exists between E_{n-1} and E_{n-2}, we note that

$$E_n = A_n + A_{n-1}(1 + i) + E_{n-2}(1 + i)^2$$

Generalizing, the recursive relationship between E_t and E_{t-1} gives

$$E_n = A_n + A_{n-1}(1 + i) + A_{n-2}(1 + i)^2 + \cdots + A_0(1 + i)^n$$

Table 4.1 Data Illustrating the Meaning of the Internal Rate of Return

t	A_t	B_t	I_t^a	E_t
0	−10,000			−10,000
1	2,525	−10,000	−2,000	− 9,475
2	2,525	− 9,475	−1,895	− 8,845
3	2,525	− 8,845	−1,769	− 8,089
4	3,840	− 8,089	−1,618	− 5,867
5	3,840	− 5,867	−1,173	− 3,200
6	3,840	− 3,200	− 640	0

[a] Based on $i = 0.20$.

Hence, we see that

$$E_n = FW(i\%)$$

as anticipated.

The above example illustrates that the time value of money operations involved in the *IRR* method are equivalent to assuming that all monies received are reinvested and earn interest at a rate equal to the internal rate of return. This can also be seen mathematically by letting

$$A_{jt} = \text{net cash flow for Investment } j \text{ in period } t$$

$$R_{jt} = \begin{cases} A_{jt}, \text{ if } A_{jt} \geq 0 \\ 0, \text{ otherwise} \end{cases}$$

$$C_{jt} = \begin{cases} -A_{jt}, \text{ if } A_{jt} < 0 \\ 0, \text{ otherwise} \end{cases}$$

$r_t = $ reinvestment rate for positive cash flows occurring in period t
$i' = $ the rate of return for negative cash flows

Then the following relationship can be defined.

$$\sum_{t=0}^{n} R_{jt}(1 + r_t)^{n-1} = \sum_{t=0}^{n} C_{jt}(1 + i')^{n-t} \qquad (4.8)$$

Note that the future worth of reinvested monies received must equal the future worth of investments.

If r_t equals i', Equation 4.8 becomes

$$0 = \sum_{t=0}^{n} (R_{jt} - C_{jt})(1 + i')^{n-t} \qquad (4.9)$$

Letting A_{jt} equal $R_{jt} - C_{jt}$ defines the *IRR* method given by Equation 4.7. Hence, we see that the rate of return obtained using the *IRR* method can be interpreted as the reinvestment rate for all recovered funds.

It should be noted that it is not essential that recovered funds be reinvested at the solving rate of return in order for the initial investment to have yielded the internal rate of return. For example, suppose an investment of $100,000 yields $10,000 annually for a 4-year period, and at the end of the fifth period the investment is terminated, and $110,000 is recovered. Clearly, the investment yielded an internal rate of return of 10%, regardless of what was done with the annual recovery of $10,000. The recovered amounts could have been invested in, say, the stock market and resulted in positive or negative results. However, the original investment of $100,000 did, in fact, yield a 10% return. [26]

Since the determination of the rate of return involves solving Equation 4.7 for i_j^*, it is seen that (for a given investment j) it is necessary to determine the values of x that satisfy the n degree polynomial $0 = A_0 x^n + A_1 x^{n-1} + \cdots + A_{n-1}x + A_n$ where $x = (1 + i_j^*)$. In general, there can exist n distinct roots (values of x) for an n-degree polynomial; however, most cash flow profiles encountered in practice will have a unique root (internal rate of return).

Descartes' rule of signs indicates the n-degree polynomial will have a single positive real root if there is a single sign change in the sequence of cash flows, $A_0, A_1, \ldots, A_{n-1}, A_n$. If there are two sign changes, then there will be either two or no positve real roots; if there are three sign changes, there will be either three or one positive real roots; if there are four sign changes, there will be either four, two, or no positive real roots; if there are five sign changes, there will be either five, three, or one positive real roots; and so forth. Since the typical cash flow pattern begins with a negative cash flow followed by a positive cash flow pattern begins with a negative cash flow followed by a positive cash flow, a unique root normally exists. (Note, since $x = (1 + i_j^*)$, a positive real root can occur even when i_j^* is negative; e.g., x will be positive when $i_j^* > -1$.)

EXAMPLE 4.6

As an illustration of a cash flow profile having multiple roots, consider the data given in Table 4.2. The future worth of the cash flow series will be zero using a 20, 40, or 50% interest rate.

$$FW_1(20\%) = -\$1000(1.2)^3 + \$4100(1.2)^2 - \$5580(1.2) + \$2520 = 0$$
$$FW_1(40\%) = -\$1000(1.4)^3 + \$4100(1.4)^2 - \$5580(1.4) + \$2520 = 0$$
$$FW_1(50\%) = -\$1000(1.5)^3 + \$4100(1.5)^2 - \$5580(1.5) + \$2520 = 0$$

A plot of the future worth for this example is given in Figure 4.1. The future worth polynomial is a third-degree polynomial, and there are three changes of sign in the ordered sequence of cash flows $(-, +, -, +)$. In this case there are three unique positive real roots, corresponding to $i = .20$, $i = .40$, and $i = .50$. This is seen by factoring the future worth polynomial, which can be written as

$$FW_1(i) = \$1000(1.2 - x)(1.4 - x)(1.5 - x)$$

where $x = (1 + i)$.

Multiple rates of return as seen in Example 4.6 are difficult to interpret properly without other information such as the future worth, present worth, or annual worth. For example, the pattern of cash flows shown in Table 4.2 is preferred

Table 4.2 Cash Flow Profile

EOY	CF
0	−$1,000
1	4,100
2	− 5,580
3	2,520

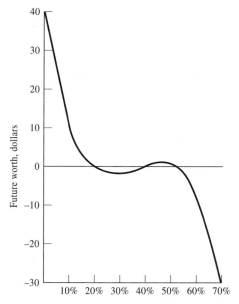

40

30

20

10

0

−10

−20

−30

Future worth, dollars

10%   20%   30%   40%   50%   60%   70%

Figure 4.1 Plot of future worth for Example 4.6.

to investing at the *MARR* if 0% ≤ *MARR* < 20% or 40% < *MARR* < 50%. This is easily seen by looking at Figure 4.1, but is not immediately obvious from the three roots of the future worth equation.

The difficulty of interpreting the polynomial roots in complex cash flow profiles is a severe drawback to the use of the *IRR* approach. For additional discussion of the subject of multiple roots in internal rate of return calculations, as well as their interpretation, see Bernhard [7–10] and Bussey [11].

4.2.5   External Rate of Return Method

The possibility of multiple roots occurring in the internal rate of return calculation has led to the development of an alternative rate of return method, called the *external rate of return method*. The external rate of return (*ERR*) method consists of the determination of the value of i' that satisfies Equation 4.8, which is repeated here for convenience.

$$\sum_{t=0}^{n} R_{jt}(1 + r_t)^{n-t} = \sum_{t=0}^{n} C_{jt}(1 + i')^{n-t}$$ **(4.10)**

In this equation, i' is now known as the external rate of return. As defined previously, R_{jt} and C_{jt} are positive and negative net cash flows for investment j during period t, respectively, and r_t is the reinvestment rate for net positive cash flows occurring in period t. Normally, r_t equals the minimum attractive rate of return, since the *MARR* reflects the opportunity cost for money available for investment. That is, we assume that positive net cash flows can always be reinvested at the *MARR*.

The external rate of return method is one of several rate of return approaches that consider the explicit reinvestment of positive cash flows, positive net cash flows, accumulated cash flows, and so forth. The *ERR* method offers an advantage over the *IRR* approach, since a unique value of i' satifies Equation 4.10. That is, there is only one external rate of return for a given cash flow profile. If i' exceeds the *MARR*, the investment is preferred over investing at the *MARR*.

As an exercise, you might want to show that the value of the external rate of return will always be between the values of the interal rate of return and the *MARR*. Because of this property, if you use either the *ERR* or the *IRR* method, your economic choice will be the same.

EXAMPLE 4.7

Let us reconsider the cash flows of Example 4.6 in Table 4.2. The value of the *ERR* may be determined by solving the equation

$$4100(1 + r_t)^2 + 2520 = 1000(1 + i')^3 + 5580(1 + i')$$

for i' where $r_t = MARR$. This results in the following series of values for a wide range of the *MARR*.

MARR Value (r_t)	External Rate of Return (i')
.10	.101302
.15	.150458
.20	.200000
.25	.249817
.30	.299812
.35	.349898
.40	.400000
.45	.450053
.50	.500000
.55	.549795
.70	.697893

There is not much difference between the value of r_t and its corresponding value i'. This is simply a function of the particular cash flow profile and is not important (note that the future worths in Figure 4.1 did not deviate far from zero either). What *is* important is that the *ERR* exceeds the *MARR* in the range $0 \leq MARR < 20\%$ and $40\% < MARR < 50\%$. These are precisely the same ranges for which the future worths were positive in Figure 4.1.

This example illustrates the consistency of the ERR method with the future worth, present worth, and annual worth methods when evaluating a single investment, even when that investment has a complex cash flow pattern.

EXAMPLE 4.8

For the data provided in Table 4.1, suppose money received from the initial investment is reinvested and earns 15% interest. At the end of the sixth year the reinvested funds total

$$\$2525(F|A\ 15,6) + \$1315(F|A\ 15,3)$$
$$= \$2525(8.7537) + \$1315(3.4725)$$
$$= \$26,669.43$$

Consequently, the *external rate of return* is defined as the interest rate such that the future worth of the $10,000 investment equals $26,669.43. Thus,

$$\$10,000(1 + i')^6 = \$26,669.43$$

Taking the logarithm and solving for i yields a value of 17.76% as the external rate of return.

EXAMPLE 4.9

Now, rework Example 4.8 assuming recovered funds are reinvested at 20%. The future worth of the reinvested funds will be

$$\$2525(F|A\ 20,6) + \$1315(F|A\ 20,3)$$
$$= \$2525(9.9299) + \$1315(3.6400)$$
$$= \$29,859.60$$

Setting the future worth of the $10,000 investment equal to $29,859.60 yields

$$\$10,000(1 + i')^6 = \$29,859.60$$

Solving for i' gives a value of 20% as the external rate of return, as anticipated.

4.2.6   Savings/Investment Ratio Method

The *savings/investment ratio method* can be defined in many ways; however, we will employ the following formulations,

$$SIR_j(i) = \frac{\sum_{t=0}^{n} R_{jt}(1 + i)^{-t}}{\sum_{t=0}^{n} C_{jt}(1 + i)^{-t}} \tag{4.11}$$

where $SIR_j(i)$ is the savings/investment ratio for investment alternative j based on a *MARR* of $i\%$. We again define A_{jt}, R_{jt}, and C_{jt} as follows:

$$A_{jt} = \text{net cash flow for investment } j \text{ in periods } t$$

$$R_{jt} = \begin{cases} A_{jt}, \text{ if } A_{jt} \geq 0 \\ 0, \text{ otherwise} \end{cases}$$

$$C_{jt} = \begin{cases} -A_{jt}, \text{ if } A_{jt} < 0 \\ 0, \text{ otherwise} \end{cases}$$

Note that this *SIR* formulation is the present worth of net positive cash flows divided by the present worth of net negative cash flows. Therefore, for a project to be preferred over investing at the *MARR*, the ratio must be greater than one.

EXAMPLE 4.10

Consider the cash flows given in Table 4.1 and let the *MARR* be 15%. The present worth of the net positive cash flows is

$$\sum_{t=0}^{n} R_{1t}(1 + i)^{-1} = \$2525(P|A\,15,3) + \$3840(P|A\,15,3)(P|F\,15,3)$$

$$= \$2525(2.2832) + \$3840(2.2832)(.6575)$$
$$= \$11,529.70$$

The present worth of the net negative cash flows is

$$\sum_{t=0}^{n} C_{1t}(1 + i)^{-t} = -\$10,000$$

The savings/investment ratio using Equation 4.11 is given by

$$SIR_1(15\%) = \frac{\$11,529.70}{\$10,000} = 1.15297$$

Note that the ratio is greater than one, indicating that an investment resulting in the cash flow profile of Table 4.1 is preferable to investing at the *MARR*.

An alternative label for the savings/investment ratio is the *benefit-cost ratio.* The savings are interpreted as benefits derived from a venture, and the investment corresponds to the cost of providing the benefits. The benefit-cost ratio is used extensively in government and public sector economic analyses. Like the savings / investment ratio, the benefit-cost ratio has many possible formulations. We explore benefit-cost analysis more fully in Chapter Seven.

4.2.7 Payback Period Method

Each of the preceding methods for measuring project worth is equivalent or consistent. That is, each will provide the same decision regarding the desirability of investing in a project versus investing at the *MARR*. The *payback period method,* unfortunately, is not an equivalent method. Yet, it is used frequently in economic analyses; therefore, we should know about its good points as well as its pitfalls.

The payback period method used most often involves the determination of the length of time required to recover the initial investment based on a zero interest rate. Letting C_{j0} denote the initial cost of investment alternative j, and R_{jt} denote the net revenue received from investment j during period t, if we assume no other negative net cash flows occur, then the smallest value of m_j such that

$$\boxed{\sum_{t=1}^{m_j} R_{jt} \geq C_{j0}} \qquad\qquad (4.12)$$

define the payback period for investment j. The investment alternative having the smallest payback period is the preferred alternative using the payback period method.

EXAMPLE 4.11

Based on the data given in Table 4.1,

$$\cdot \sum_{t=1}^{3} R_{1t} = \$2525 + \$2525 + \$2525 = \$7575 < C_{10} = \$10{,}000$$

and

$$\sum_{t=1}^{4} R_{1t} = \$2525 + \$2525 + \$2525 = \$3840 = \$11{,}415 > C_{10} = \$10{,}000$$

Consequently, m equals 4, indicating that 4 years are required to pay back the original investment.

A number of variations of the payback or payout method have been used by different organizations. For example, if negative net cash flows occur after year zero, the payback period might be defined as the smallest value of m_j in which the sum of the net cash flows $\sum_{t=0}^{m_j} A_{jt}$ first exceeds zero. Also, the payback period may be noninteger, assuming cash flows are uniformly distributed throughout the year. However, the basic deficiencies of the payback period method are present in most variations; the *timing* of cash flows and the *duration* of the project are ignored.

To illustrate the deficiencies of the payback period, consider the cash flow profiles depicted in Figure 4.2a. Using the payback period, profile B would be preferred, even though we would most likely choose profile A. Likewise, in Figure 4.2b, profiles C and D are equally preferred using the payback period, even though profile D is clearly better.

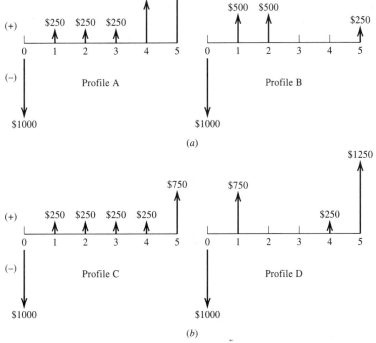

Figure 4.2 Four different cash flow profiles for illustrating deficiencies of payback period.

Despite its obvious deficiencies, the payback period method continues to be one of the most popular methods of judging the desirability of investing in a project. The reasons for its popularity include the following:

1. Does not require interest rate calculations.
2. Does not require a decision concerning the discount rate (*MARR*) to use.
3. Is explained and understood easily.
4. Reflects a manager's attitudes when investment capital is limited.
5. Hedges against uncertainty of future cash flows.
6. Provides a rough measure of the liquidity of an investment.

To determine how long it takes to fully recover an investment without ignoring the time value of money, an alternate version of the payback period method, called the *discounted payback period method*, can be used. The discounted payback period is determined by adding cash flows year-by-year to the present worth calculation until the present worth is nonnegative. At such time as the present worth is nonnegative, one can claim that the investment "has paid off."

Expressed mathematically, the discounted payback period is the smallest value of *m* for which the following relationship holds:

$$\sum_{t=0}^{m} A_{jt}(1 + i)^{-j} \geq 0 \tag{4.13}$$

where, as previously, A_{jt} denotes the net cash flow for Alternative j at the end of period t.

EXAMPLE 4.12

To illustrate the discounted payback period method, consider an example involving an initial investment of $10,000, followed by five annual cash flows of $4000. As depicted in the spreadsheet given shown in Table 4.3, using a MARR of 15% the present worth through the third year is $-$867.10$, whereas the present worth through the fourth year is $1419.91. Since PW(15%) through 4 years is positive, the initial investment has paid off and the payback period is 4 years. (Assuming the cash flow in year 4 occurs uniformly over the year and using linear interpolation, a payback period of 3.38 years can be estimated.)

Although the discounted payback period method considers the time value of money, it ignores cash flows occurring after the payback period. To understand the implication of ignoring future cash flows, note that the investment considered in Example 4.12 is preferred to an investment of $10,000 that returns $3500 annually for 5 years and is fully recoverable at the end of the fifth year, i.e., has a $10,000 salvage value. (As an exercise, verify that the latter investment has a greater present worth than the investment given in Example 4.12 for any MARR < 73.41%.)

Regardless of the version of the payback period method used, we recommended that it be a secondary method of measuring investment worth. In cases where multiple investments have approximately the same present worth, for example, the payback period might be used to choose among them. We do not recommend the use of the payback period as the sole criterion for choosing among investment alternatives.

Table 4.3 Discounted Payback Period Calculation for Example 4.12

End of Year	Net Cash Flow	Cumulative Present Worth
0	($10,000)	($10,000.00)
1	$4000	($6521.74)
2	$4000	($3497.16)
3	$4000	($867.10)
4	$4000	$1419.91
5	$4000	$3408.62

4.2.8 Perpetuities and Capitalized Worth Method

A specialized type of cash flow series is a *perpetuity*, a uniform series of cash flows which continues indefinitely. This is a special case, since an infinite series of cash flows would rarely be encountered in the business world; rather, a finite series of cash flows is the general rule. However, for such very long-term investment projects as bridges, highways, forest harvesting, or the establishment of endowment funds where the estimated life is 50 years or more, an infinite cash flow series may be appropriate.

If a present value P is deposited into a fund at interest rate i per period so that a payment of size A may be withdrawn each and every period forever, then the following relation holds between P, A, and i.

$$Pi = A \qquad (4.14)$$

Thus, as depicted in Figure 4.3, P is a present value that will pay out equal payments of size A indefinitely if the interest rate per period is i. The present value P is termed the *capitalized worth* of A, the size of each of the perpetual payments.

EXAMPLE 4.13

What deposit at $t = 0$ into a fund paying $9\frac{1}{2}\%$ annually is required in order to pay out \$5000 each year forever? The solution is straightforward from

$$P = \frac{A}{i} = \$5000/.095 = \$52{,}631.58$$

By means of the subsequent example, let us now broaden the meaning of capitalized worth to be that present value that would pay for the first cost of some project and provide for its perpetual maintenance at an interest rate i.

EXAMPLE 4.14

Project ABC consists of the following requirements:

1. A \$50,000 first cost at $t = 0$.
2. A \$5000 expense every year.

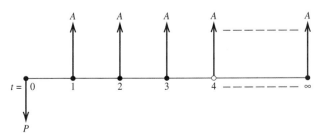

Figure 4.3 An infinite cash flow series.

3. A $25,000 expense every third year forever, with the first expense occurring at $t = 3$.

What is the capitalized cost of Project ABC if $i = 15\%$ annually?

It will be instructive to determine the capitalized cost of each requirement separately and then sum the results. First, the capitalized cost of "25,000 every third year forever" may be determined from any of three points of view. One view is that a value P is required at the beginning of a 3-year period such that, with interest compounded at 15% annually, a sum of $P + \$25,000$ will accrue at the end of the 3-year period. Thus, $25,000 would be withdrawn, thereby leaving the value P to repeat the cycle indefinitely each 3-year period. This logic is illustrated in Figure 4.4.

Thus,

$$P(F|P\ 15,3) = P + \$25,000$$

or

$$1.5209P = P + \$25,000$$

and the capitalized worth (CW) is

$$CW = P = \$47{,}993.86$$

which is the capitalized cost of the requirement.

A second view is to reason that, for each 3-year period, three equal deposits of size A are required that will amount to $25,000 at the end of the third year. Then, if these payments of size A occur every year, $25,000 will be available every third year forever. The present value P that yields the required payment A is the solution, as shown in Figure 4.5.

The required payment A is given by

$$A = F(A|F\ 15,3) = \$25{,}000(0.2880)$$
$$= \$7200$$

Then

$$CW = \frac{A}{i} = \frac{\$7200}{0.15} = \$48{,}000.00$$

which is approximately the same answer as from the first approach. The difference of $6.14 is due to rounding in the interest factor calculations.

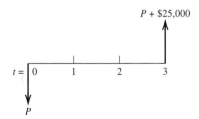

Figure 4.4 Capitalized cost of $25,000 every third year—first view.

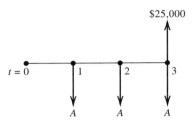

Figure 4.5 Capitalized cost of $25,000 every third year—second view.

A third point of view for the $25,000 requirement is to consider the infinite series depicted in Figure 4.6, such that $P = A/i$ would yield the desired result. However, the value of i required is the effective interest rate for a 3-year period. Thus,

$$i = [(1 + 0.15)^3 - 1] = 0.5209$$

and

$$CW = \frac{\$25,000}{0.5209} = \$47,996.16$$

The capitalized cost of Project ABC is computed as follows:

1. Capitalized cost of $50,000 first cost   = $ 50,000
2. Capitalized cost of $5000 every year = $5000/0.15 = $ 33,333
3. Capitalized cost of $25,000 every third year   = $ 48,000

TOTAL CAPITALIZED COST $131,333

The capitalized worth of $131,333 would provide approximately $A = Pi = \$131,333(0.15) = \$19,700$ every year forever.

The capitalized worth may be looked upon as a present worth of some cash flow pattern that repeats indefinitely.

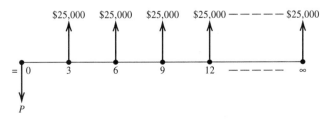

Figure 4.6 Capitalized cost of $25,000 every third year—third view.

EXAMPLE 4.15

Recall that the annual worth of the cash flows in Example 4.4 was calculated to be $404.48/year (these were the same cash flows later shown in Table 4.1). If this cash flow pattern were to repeat every 6 years forever, the capitalized worth would be

$$CW = \frac{A}{i} = \frac{\$404.48}{.15} = \$2699.20$$

Note that this is not drastically higher than the present worth of $1529.70 calculated in Example 4.2 for the first 6 years. This is because the distant cash flows are substantially discounted at $i = 5\%$ in the capitalized worth calculation.

4.3   CAPITAL RECOVERY FORMULA

In engineering economic analyses, it is common to refer to the *capital recovery cost* of an asset. Figure 4.7 illustrates an investment of $P in an asset having a life of n years and disposed of for a salvage value of $F. The capital recovery cost is a uniform annual amount defined as

$$CR = P(A|P\,i,n) - F(A|F\,i,n) \tag{4.15}$$

The salvage value at year n is an income, and therefore a negative cost.

EXAMPLE 4.16

A process computer is purchased for $P = \$82,000$, has a service life of $n = 7$ years, and a salvage value of $F = \$5000$. At an interest rate of 15%, the capital recovery cost is

$$CR = \$82,000(A|P\,15,7) - \$5000(A|F\,15,7)$$
$$= \$82,000(.2404) - \$5000(.0904)$$
$$= \$19,260.80/\text{year}$$

From Chapter Two and the interest tables we know that

$$(A|P\,i,n) = (A|F\,i,n) + i \tag{4.15}$$

Figure 4.7 Cash flow diagram of investment in an asset.

and similarly

$$(A|F\,i,n) = (A|P\,i,n) - i \qquad\qquad (4.16)$$

Substituting Equations 4.15 and 4.16 into Equation 4.14 leads to two more capital recovery cost formulas. They are, respectively,

$$CR = (P\text{-}F)(A|F\,i,n) + Pi \qquad\qquad (4.17)$$

and

$$CR = (P\text{-}F)(A|P\,i,n) + Fi \qquad\qquad (4.18)$$

Among the three methods of computing capital recovery costs, Equation 4.18 appears to be the most popular. We tend to use Equation 4.14 since it is a direct application of the interest factors to the actual cash flows. The relationship between capital recovery cost and depreciation or cost recovery is explained in Chapter Six.

EXAMPLE 4.17

We can apply Equations 4.17 and 4.18 to the previous example problem having $P = \$82{,}000$, $n = 7$, $F = \$5000$, and $i = 15\%$. Using Equation 4.17, we have

$$CR = (\$82{,}000 - \$5000)(A|F\,15{,}7) + \$82{,}000(.15)$$
$$= \$77{,}000(.0904) + \$82{,}000(.15)$$
$$= \$19{,}260.80/\text{year}$$

Using Equation 4.18 we have

$$CR = (\$82{,}000 - \$5000)(A|P\,15{,}7) + \$5{,}000(.15)$$
$$= \$77{,}000(.2404) + \$5{,}000(.15)$$
$$= \$19{,}260.80/\text{year}$$

4.4 USING SPREADSHEETS TO MEASURE INVESTMENT WORTH

In this and previous chapters, we developed and used a variety of compound interest formulas to facilitate the measurement of investment worth. In general, the formulas were derived to facilitate calculations that would be laborious in the absence of computational hardware and software. However, given the widespread use of spreadsheets and sophisticated calculators and hand-held computers, there is less reliance today on compound interest formulas and their tabulated values than in the past. Given the existence of sophisticated software to perform financial analyses, why did we include compound interest formulas and tabulations of their values? The principal reasons for doing so are the differences in assumptions underlying the algorithms used in financial calculators and computer software packages and the importance of understanding how the formulas are obtained. We have found the latter to be particularly valuable; it is important for the user to understand what is behind the results obtained from the software.

However, we would be remiss if we did not acknowledge the valuable role software packages play in measuring investment worth. No contemporary treatment of the subject would be complete without including a consideration of the use of spreadsheets in economic analyses.

A variety of software packages are available to facilitate the use of spreadsheets in performing economic justifications, including Excel®, Lotus 1-2-3®, and Quattro® Pro. Our purpose in this chapter is not to endorse a particular spreadsheet; instead, we want to demonstrate how spreadsheets can be used to perform financial calculations and sensitivity analyses.

To construct you own spreadsheet, notice that most of what we have presented can be replicated by using the formula, $V_2 = V_1(1 + i)$, where V_t denotes the value of a single sum of money at time t and i denotes the time value of money. Due to the computer's speed, there is no need to take advantage of the structure of series of cash flows; instead, each cash flow can be treated as a single sum of money. To move money forward in time by one time unit, multiply by $(1 + i)$; to move money backward in time by one time unit, divided by $(1 + i)$.

4.4.1   Using Spreadsheets to Compute Present Worths

To illustrate the computation of the present worth for an investment, consider the spreadsheet given in Table 4.4. The columns are identified alphabetically, and the rows are identified numerically. The interest rate is entered, using decimal format, in cell B1. The magnitudes of cash flows are entered in column B, beginning with row 4. The value of the discount factor for a particular time period is obtained by dividing the discount factor for the previous period by the

Table 4.4 Sample Spreadsheet Format

Row/Column	A	B	C	D
1	Discount rate =	insert value for i in %		
2				
3	Time	Cash flow	Discount factor	Present value
4	0	A_0	1.0000	=B4*C4
5	1	A_1	=C4/(1.0 + B1)	=B5*C5
6	2	A_2	=C5/(1.0 + B1)	=B6*C6
7	3	A_3	=C6/(1.0 + B1)	=B7*C7
8	4	A_4	=C7/(1.0 + B1)	=B8*C8
9	5	A_5	=C8/(1.0 + B1)	=B9*C9
10				= SUM(D4:D9)

sum of one and the interest rate. The present value of an individual cash flow is obtained by multiplying the value of the cash flow and the discount rate. The present value is obtained by summing the individual present values, as shown in cell D10.

EXAMPLE 4.18

To illustrate the use of a spreadsheet, consider an investment of $10,000 with $3000 returned annually for 5 years. Using a 15% minimum attractive rate of return, as shown in Table 4.5, a present value of $56.47 results.

When using Excel 5.0,[1] the net present value (NPV) can be obtained directly by entering, =NPV(0.15,B5:B9)+B4, in a cell; for the example, a net present value of $56.47 results. It is important to note that the NPV function computes the present value of a series of cash flows as of one period prior to the first value in the series. Using the NPV calculation shown, the net present value is determined for the cash flows entered in cells B5 through B9; hence, to obtain the present value for the entire cash flow series it is necessary to add the entry in cell B4 to the value obtained by using the NPV function over the range of cells from B5 through B9.

EXAMPLE 4.19

Once the spreadsheet is set up, it is easy to examine the impact of changes in the parameters of the investment. For example, if the annual receipts are reduced

Table 4.5 Example of the Use of a Spreadsheet

R/C	A	B	C	D
1	MARR=	0.15		
2				
3	Time	Cash flow	Discount factor	Present value
4	0	($10,000)	1.000000	($10,000.00)
5	1	$3,000	0.869565	$2,608.70
6	2	$3,000	0.756144	$2,268.43
7	3	$3,000	0.657516	$1,972.55
8	4	$3,000	0.571753	$1,715.26
9	5	$3,000	0.497177	$1,491.53
10			NPV =	$56.47

[1] Microsoft®Excel, Version 5.0, Microsoft Corporation, 1993–1994.

Table 4.6 Using a Spreadsheet to Analyze Changes in Cash Flows

R\C	A	B	C	D
1	MARR =	0.15		
2				
3	Time	Cash flow	Discount factor	Present value
4	0	($10,000)	1.000000	($10,000.00)
5	1	$2,500	0.869565	$2,173.91
6	2	$2,500	0.756144	$1,890.36
7	3	$2,500	0.657516	$1,643.79
8	4	$2,500	0.571753	$1,429.38
9	5	$2,500	0.497177	$1,242.94
10			NPV =	($1,619.61)

to $2,500, then, as shown in Table 4.6, the resulting NPV value will be equal to
−$1619.61. Likewise, as shown in Table 4.7, if the interest rate in Table 4.5 is
changed to 16%, the net present value changes to −$177.12.

4.4.2   Using Spreadsheets to Compute Annual Worths and Future Worths

Excel 5.0 does not include an equivalent *NAV* (*net annual value*) or *NFV* (*net
future value*) function. However, Excel's PMT function converts a present value
to a uniform series of payments; the FV function converts a uniform series of

Table 4.7 Using Spreadsheets to Analyze Changes in the Discount Rate

R/C	A	B	C	D
1	MARR =	0.16		
2				
3	Time	Cash flow	Discount factor	Present value
4	0	($10,000)	1.000000	($10,000.00)
5	1	$3000	0.862069	$2586.21
6	2	$3000	0.743163	$2229.49
7	3	$3000	0.640658	$1921.97
8	4	$3000	0.552291	$1656.87
9	5	$3000	0.476113	$1428.34
10			NPV =	($177.12)

payments to its future worth. The AW can be determined from the PW value by using **=pmt(MARR,N,−PW)**; to obtain the FW value, enter **=fv(MARR, N,−AW)**. (Note the use of a minus sign in each, because the functions are derived for loan transactions.)

EXAMPLE 4.20

As shown in Table 4.8, to determine the AW for a 5-year investment having a present worth given in cell B10 and a MARR given in cell B1, enter **=pmt(B1,5,−B10)** in cell B11 to obtain $16.84. To determine the FW, given the AW calculation in cell B11, enter **=fv(B1,5,−B11)** in cell B12 to obtain $113.57.

4.4.3   Using Spreadsheets to Compute Internal Rates of Return and External Rates of Return

When using a spreadsheet of the type given in Table 4.4, the IRR can be determined by varying the entry in cell B1 until the net present value equals zero. Alternatively, when using Excel, a function is provided for determining the internal rate of return for an investment. For the cash flow series in the spreadsheet shown in Table 4.9, entering **=IRR(B3:B8)** in a cell in the spreadsheet

Table 4.8 Using Excel to Compute Annual Worth and Future Worth Values

R/C	A	B
1	MARR =	0.15
2		
3	Time	Cash flow
4	0	($10,000)
5	1	$3000
6	2	$3000
7	3	$3000
8	4	$3000
9	5	$3000
10	PW =	$56.47
11	AW =	$16.84
12	FW =	$113.57

Table 4.9 Using Excel to Compute the Internal Rate of Return and External
Rate of Return

R/C	A	B	C	D
1	MARR =	0.15		
2	Time	Cash flow	Cash flow	Cash flow
3	0	($10,000)	($10,000)	($10,000)
4	1	$3000	$2500	$1000
5	2	$3000	$2500	$2000
6	3	$3000	$2500	$3000
7	4	$3000	$2500	$4000
8	5	$3000	$2500	$5000
9	IRR =	15.24%	7.93%	12.01%
10	ERR =	15.13%	11.01%	12.92%

will yield the value of the discount rate that results in a net present value of
zero for the cash flows B3 through B8.

EXAMPLE 4.21

From Table 4.9, the internal rate of return for an investment of $10,000 that
returns $3,000 a year for 5 years is 15.24%; for the case of $2,500 per year, the
internal rate of return is 7.93%; and for the case of the gradient series shown,
the internal rate of return is 12.01%.

To determine values of the external rate of return, Excel's MIRR function is
closely related to the ERR. Used to compute a modified internal rate of return
based on reinvestment of recovered capital, the MIRR function allows both the
specification of the reinvestment rate, **rrate**, and an interest rate used to finance
the investment, **frate**. The MIRR function is defined by **MIRR(series,frate,rrate)**.
By letting **frate** = 0 and letting **rrate** = MARR, the ERR can be computed when
the cash flow profile satisfies the periodicity requirements of the MIRR function.

EXAMPLE 4.22

As shown in Table 4.9, the ERR values corresponding to the IRR values cited
above are 15.13%, 11.01%, and 12.92%, respectively. (Recall, the relationships
among IRR, ERR, and MARR are demonstrated through this example:
MARR>ERR>IRR or MARR=ERR=IRR or MARR<ERR<IRR.)

In using financial functions in software packages, recognize that even though they may have the same names we use, they can take on very different meanings. The following additional Excel financial functions might prove useful: FVSCHEDULE, IPMT, NPER, PPMT, PV, and RATE.

4.5   SUMMARY

In this chapter, we considered the subject of measuring the economic worth of capital investments. In so doing, we considered briefly the selection of the minimum attractive rate of return, defined eight measures of investment worth, provided alternative methods of computing capital recovery cost, and considered the use of spreadsheets in measuring investment worth. In the next chapter, we describe a systematic procedure that can be used to perform economic analyses; the procedure we described depends critically on the material presented in this chapter. Consequently, it is important for you to master the material in this chapter before attempting to understand the material presented in Chapter Five.

BIBLIOGRAPHY

1. Adler, M., "The True Rate of Return and the Reinvestment Rate," *The Engineering Economist, 15* (3), 1970, pp. 185–87.
2. Au, T., and Au, T. P., *Engineering Economics for Capital Investment Analysis,* Allyn and Bacon, Boston, MA, 1983.
3. Beaves, R. G., "Net Present Value and Rate of Return: Implicit and Explicit Reinvestment Assumptions," *The Engineering Economist, 33* (4), 1988, pp. 275–302.
4. Beenhakker, H. L., "Discounting Indices Proposed for Capital Investment Evaluation: A Further Examination," *The Engineering Economist, 18* (3), 1973, pp. 149–68.
5. Beranek, W., "The Cost of Capital, Capital Budgeting, and the Maximization of Shareholder Wealth," *Journal of Financial and Quantitative Analysis, 10* (1), 1975, pp. 1–18.
6. Bernhard, R. H., "Discount Methods for Expenditure Evaluation—A Clarification of Their Assumptions," *Journal of Industrial Engineering, 18* (1), 1962, pp. 19–27.
7. Bernhard, R. H., "On the Inconsistency of the Soper and Sturm-Kaplan Conditions for Uniqueness of the Internal Rate of Return," *The Journal of Industrial Engineering, 18* (8), 1967, pp. 498–500.
8. Bernhard, R. H., "A Comprehensive Comparison and Critique of Discounting Indices Proposed for Capital Investment Evaluation," *The Engineering Economist, 16* (3), 1971, pp. 157–186.
9. Bernhard, R. H., "Unrecovered Investment, Uniqueness of the Internal Rate of Return and the Question of Project Acceptability," *Journal of Financial and Quantitative Analysis, 12* (1), 1977, pp. 33–38.
10. Bernhard, R. H., "'Modified' Rates of Return for Investment Project Evaluation—A Comparison and Critique," *The Engineering Economists, 24* (3), 1979, pp. 161–168.
11. Bussey, L. E., *The Economic Analysis of Industrial Projects,* Prentice-Hall, 1978.
12. Canada, J. R., "Rate of Return: A Comparison Between the Discounted Cash Flow Model and a Model Which Assumes an Explicit Reinvestment Rate for the Uniform Income Flow Case," *The Engineering Economist, 9* (3), 1964, pp. 1–15.
13. Canada, J. R., Sullivan, W. G., and White, J. A., *Capital Investment Decision Analysis for Management and Engineering,* 2nd ed., Prentice-Hall, Englewood Cliffs, NJ, 1996.

14. de Faro, C., "On the Internal Rate of Return Criterion," *The Engineering Economist, 19* (3), 1974, pp. 165–94.

15. DeGarmo, E. P., Sullivan, W. G., and Canada, J. R., *Engineering Economy*, 7th ed. Macmillan, New York, 1984.

16. Fleischer, G. A., *Engineering Economy, Capital Allocation Theory*, Brooks/Cole, Monterey, CA, 1984.

17. Freidenfelds, J., and Kennedy, M., "Price Inflation and Long-Term Present-Worth Studies," *The Engineering Economist, 24* (3), 1979, pp. 143–160.

18. Grant, E. L., "Reinvestment of Cash Flow Controversy—A Review of: FINANCIAL ANALYSIS IN CAPITAL BUDGETING by Pearson Hunt," *The Engineering Economists, 11* (3) 1966, pp. 23–29.

19. Jeynes, P. H., "Minimum Acceptable Return," *The Engineering Economist, 9* (4), 1964, pp. 9–25.

20. Jeynes, P. H., "Comment on Ralph O. Swalm's Letter," *The Engineering Economist, 10* (3), 1965, pp. 38–42.

21. Jeynes, P. H., "The Significance of Reinvestment Rate," *The Engineering Economists, 11* (1), 1966, pp. 1–9.

22. Kaplan, S., "A Note on a Method for Precisely Determining the Uniqueness or Nonuniqueness of the Internal Rate of Return for a Proposed Investment," *Journal of Industrial Engineering, 26* (1), 1965, pp. 70–71.

23. Kaplan, S., "Computer Algorithms for Finding Exact Rates of Return," *Journal of Business, 40* (4), 1967, pp. 389–392.

24. Kirshenbaum, P. S., "A Resolution of the Multiple Rate-of-Return Paradox," *The Engineering Economist, 10* (1), 1965, pp. 11–16.

25. Lin, S. A., "The Modified Internal Rate of Return and Investment Criterion," *The Engineering Economist, 21* (4), 1976, pp. 237–248.

26. Lohmann, J. R., "The IRR, NPV and the Fallacy of the Reinvestment Rate Assumptions," *The Engineering Economist, 33* (4), 1988, pp. 303–330.

27. Oakford, R. V., Bhimjee, S. A., and Jucker, J. V., "The Internal Rate of Return, the Pseudo Internal Rate of Return, and NPV and Their Use in Financial Decision Making," *The Engineering Economists, 22* (3), 1977, pp. 187–202.

28. Park, C. S., *Modern Engineering Economic Analysis,* Addison-Wesley, Reading, MA, 1990.

29. Park, C. S., and Sharp-Bette, G. P., Advanced Engineering Economics, John Wiley & Sons, New York, 1990.

30. Sharp, G. P., and Guzman-Garza, A., "Borrowing Interest Rate as a Function of the Debt-Equity Ratio in Capital Budgeting Models," *The Engineering Economist, 26* (4), 1981, pp. 293–315.

31. Sullivan, W. G., and Bontadelli, J. A., "The Industrial Engineer and Inflation," *Industrial Engineering, 12* (3), March 1980, pp. 24–33.

32. Swalm, R. O., "On Calculating the Rate of Return on an Investment," *The Journal of Industrial Engineering, 9* (2), 1958, pp. 99–103.

33. Swalm, R. O., "Comment on 'Minimum Acceptable Return' by Paul H. Jeynes," *The Engineering Economist, 10* (3), 1965, pp. 35–38.

34. Teichroew, D., Robicheck, A. A., and Montalbano, M., "Mathematical Analysis of Rates of Return Under Certainty," *Management Science, 11* (3), 1965, pp. 395–403.

35. Thuesen, G. J., and Fabrycky, W. J., *Engineering Economy,* 7th ed., Prentice-Hall, Englewood Cliffs, NJ, 1989.

36. Solomon, E., "The Arithmetic of Capital-Budgeting Decisions," *Journal of Business, 29* (2), 1956, pp. 124–129.

37. Waters, R. C., and Bullock, R. L., "Inflation and Replacement Decisions," *The Engineering Economists, 21* (4), 1976, pp. 249–258.

38. Weaver, J. B., "False and Multiple Solutions by the Discounted Cash Flow Method for Determining Interest Rate of Return," *The Engineering Economist, 3* (4), 1958, pp. 1–31.

39. White, J. A., Case, K. E., and Agee, M. H., "Rate of Return: An Explicit Reinvestment Rate Approach," *Proceedings of the 1976 AIIE Conference,* American Institute of Industrial Engineers, Norcross, GA., 1976.

40. Wohl, M., "A New Ordering Procedure and Set of Decision Rules for the Internal Rate of Return Method," *The Engineering Economist, 30* (40), 1985, pp. 363–386.

41. Young, D., *Modern Engineering Economy,* John Wiley & Sons, New York, 1993.

PROBLEMS

1. Brock Associates invested $80,000 in a business venture with the following cash flow results:

EOY	CF	EOY	CF	EOY	CF
0	−$80,000	3	$22,000	6	$22,000
1	10,000	4	28,000	7	16,000
2	16,000	5	28,000	8	10,000

If *MARR* is 12%, determine the following.
a. Present worth. (4.2.1)
b. Annual worth. (4.2.2)
c. Future worth. (4.2.3)
d. Internal rate of return. (4.2.4)
e. External rate of return. (4.2.5)
f. Savings/investment ratio. (4.2.6)
g. Payback period with and without considering the time value of money. (4.2.7)

2. Shrewd Endeavors, Inc. invested $70,000 in a business venture with the following cash flow results:

EOY	CF	EOY	CF	EOY	CF
0	−$70,000	7	$14,000	14	$7,000
1	20,000	8	13,000	15	6,000
2	19,000	9	12,000	16	5,000
3	18,000	10	11,000	17	4,000
4	17,000	11	10,000	18	3,000
5	16,000	12	9,000	19	2,000
6	15,000	13	8,000	20	1,000

Assuming *MARR* to be 10%, determine the following.
a. Present worth. (4.2.1)
b. Annual worth. (4.2.2)
c. Future worth. (4.2.3)
d. Internal rate of return. (4.2.4)

   e. External rate of return. (4.2.5)
   f. Savings/investment ratio. (4.2.6)
   g. Payback period with and without considering the time value of money. (4.2.7)
3. An investment of $20,000 is to be made on a computer that will last for 5 years and have a zero salvage value at that time. Operating, maintenance, and software costs are projected to be $15,000 the first 3 years and $20,000 the last 3 years. The minimum attractive rate of return is specified to be 12%. Determine for this investment the following.
   a. Present worth. (4.2.1)
   b. Annual worth. (4.2.2)
   c. Future worth. (4.2.3)
4. Tulsa Precision, Inc. borrows $100,000 to purchase a new numerically controlled milling machine and pays the loan back over a 4-year period with equal payments. Interest on the loan is 15% compounded annually. The machine is estimated to have annual operating and maintenance costs of $24,000/year and have a life of 10 years. Salvage value is estimated to be $25,000. The firm has a *MARR* of 12%. Determine for this investment the following.
   a. Present worth. (4.2.1)
   b. Annual worth. (4.2.2)
   c. Future worth. (4.2.3)
5. A utility vehicle is purchased for $30,000, kept for 4 years, and sold for $7500. Annual operating and maintenance costs were $5000. Using a 10% minimum attractive rate of return, determine the following.
   a. Present worth. (4.2.1)
   b. Annual worth. (4.2.2)
   c. Future worth. (4.2.3)
6. An investment of $20,000 for a new condenser is being considered. Estimated salvage value of the condenser is $5000 at the end of an estimated life of 6 years. Annual income each year for the 6 years is $8500. Annual operating expenses are $2300. Assume money is worth 15% compounded annually. Determine the following measures of investment worth; for each, state whether or not your results indicate the new condenser should be purchased.
   a. Present worth. (4.2.1)
   b. Annual worth. (4.2.2)
   c. Future worth. (4.2.3)
   d. Internal rate of return. (4.2.4)
   e. External rate of return. (4.2.5)
   f. Savings/investment ratio. (4.2.6)
7. Smith Investors places $50,000 in an investment fund. One year after making the investment, Smith receives $7500 and continues to receive $7500 annually until 10 such amounts are received. Smith receives nothing further until 15 years after the initial investment, at which time $50,000 is received. Over the 15-year period, what are each of the following if *MARR* = 10%?
   a. Present worth. (4.2.1)
   b. Annual worth. (4.2.2)
   c. Future worth. (4.2.3)
   d. Internal rate of return. (4.2.4)
   e. External rate of return. (4.2.5)

    f. Savings/investment ratio. (4.2.6)

    g. Payback period with and without considering the time value of money. (4.2.7)

8. Many Chemicals Unlimited purchases a computer-controlled filter for $100,000. Half of the purchase price is borrowed from a bank at 15% compounded annually. The loan is to be paid back with equal annual payments over a 5-year period. The filter is expected to last 10 years, at which time it will have a salvage value of $10,000. Over the 10-year period the operating and maintenance costs are anticipated to equal $20,000/year; however, by making the investment, annual fines of $50,000 for pollution will be avoided. The firm expects to earn 12% on its investments. Determine each of the following measures of investment worth and state whether or not the filter purchase was economically sound.

    a. Present worth. (4.2.1)

    b. Annual worth. (4.2.2)

    c. Future worth. (4.2.3)

    d. Internal rate of return. (4.2.4)

    e. External rate of return. (4.2.5)

    f. Savings/investment ratio. (4.2.6)

9. Owners of an economy motel chain are considering building a new 200-unit motel. The present worth cost of building the motel is $8,000,000; the firm estimates furnishings for the motel will cost an additional $800,000 and will require replacement every 5 years. Annual operating and maintenance costs for the facility are estimated to be $800,000. The average rate for a unit is anticipated to be $60/day. A 15-year planning horizon is used by the firm in evaluating new ventures of this type; a terminal salvage value of 15% of the original building cost is anticipated; furnishings are estimated to have no salvage value at the end of each 5-year replacement interval. Assuming average daily occupancy percentages of 50%, 60%, 70%, 80% for years 1 through 4, respectively, and 90% for the fifth and each remaining year, *MARR* of 12%, 365 operating days/year, and ignoring the cost of the land, should the motel be built? Base your decision upon the following values.

    a. Present worth. (4.2.1)

    b. Annual worth. (4.2.2)

    c. Future worth. (4.2.3)

    d. External rate of return. (4.2.5)

    e. Savings/investment ratio. (4.2.6)

10. Growth Fertilizer Company purchases a gravity settling tank by borrowing the $30,000 purchase price. The loan is to be repaid with four equal annual payments at an annual compound rate of 12%. It is anticipated that the tank will be used for 9 years and then be sold for $2000. Annual operating and maintenance expenses are estimated to be $9000/year. A savings of $15,000/year is realized over the present filtration system. The firm uses a *MARR* of 15% for its economic analyses. Determine the following.

    a. Present worth. (4.2.1)

    b. Annual worth. (4.2.2)

    c. Future worth. (4.2.3)

    d. Internal rate of return. (4.2.4)

    e. External rate of return. (4.2.5)

    f. Savings/investment ratio. (4.2.6)

    g. Payback period with and without considering the time value of money. (4.2.7)

11. A firm purchases a computer for $15,000; the computer is used for 4 years and is then sold for $5,000. Annual disbursement for operating, maintenance, and software costs equals $5000/year over the 4-year period. The computer replaced an older manual system that cost $12,000/year. On the basis of a 10% *MARR*, determine if the decision to buy the analog computer was economically sound. Use the following measures of investment worth.

a. Present worth. (4.2.1)
b. Annual worth. (4.2.2)
c. Future worth. (4.2.3)
d. Internal rate of return. (4.2.4)
e. External rate of return. (4.2.5)
f. Savings/investment ratio. (4.2.6)

12. If a fund pays 12% compounded annually, what deposit is required today such that $1000 can be withdrawn every year forever? (4.2.8)

13. If a fund pays 12% compounded annually, what deposit is required today such that $5000 can be withdrawn every 4 years forever? (4.2.8)

14. A scholarship fund pays $12,000 annually. If the scholarship fund earns 12% compounded quarterly, how much money must be in the fund for this series of payments to continue forever? (4.2.8)

15. Maintenance on a reservoir is cyclic with the following *costs* occurring over a 5-year period: $5000, $2000, $7000, $0, and $2000. It is anticipated that the sequence will repeat itself every 5 years forever. Determine the capitalized cost of the maintenance costs based on a time value of money of 10%. (4.2.8)

16. A flood control project has a construction cost at $t = 0$ of $2,000,000, an annual maintenance cost of $50,000, and a major repair at 5-year intervals projected to cost $250,000 with the first such repair occurring at $t = 5$. If interest is 8% annually, determine the amount of money needed at $t = 0$ to provide for construction and perpetual upkeep. (4.2.8)

17. A firm can invest in a venture which costs $200,000 and returns $75,000/year at the end of each of 4 years. This investment can be renewed perpetually every 4 years. If the firm's *MARR* is 10%, determine the capitalized worth of an infinite series of these investments. (4.2.8)

18. By investing $100,000 in special-purpose production equipment, the required number of units can be manufactured in 1 year for a one-time order and generate net income of $1 million, including the salvage value for the equipment. Alternatively, by using the present production methods, the net income will be $500,000 per year for 2 years. Hence, a $100,000 investment will generate $500,000 in extra income the first year, but will cause $500,000 in income to be lost the second year. Determine the internal rate(s) of return for the incremental cash flows. (4.2.4)

19. In Problem 18 determine the external rate of return when the MARR is (a) 10% and (b) 30%. (4.2.5)

20. An individual is faced with two investment possibilities. In the first case, an investment of $210,000 can be made in a limited partner agreement involving land speculation and development. One year following the investment it is expected $90,000 in income will be obtained; incomes for each of the next four years are estimated to be $90,000, $510,000, $0, and $120,000 as various parcels are developed and sold. As an alternative, an investment of $200,000 can be made; two years following the investment, an income of $400,000 is expected; two years later another $400,000 is expected. The cash flow profile for the difference in the two investments is given below.

End of Year	Difference in Cash Flows
0	−$ 10,000
1	90,000
2	−310,000
3	510,000
4	400,000
5	120,000

Determine the internal rate(s) of return for the difference in cash flows. (Hint: Solve Problem 21 first.) (4.2.4)

21. In Problem 20, determine the external rate of return when the MARR is (a) 0%, (b) 100%, and (c) 200%. (4.2.5)

22. The Acme Manufacturing Company purchased an automatic transfer machine for $25,000 installed. At the end of 10 years the company paid $4000 to have the machine removed and junked. The operating and maintenance-cost history was as follows.

EOY	O&M Cost	EOY	O&M Cost
1	$2000	6	3500
2	2000	7	3500
3	2000	8	3500
4	2000	9	5000
5	3500	10	5000

The Acme Manufacturing Company has a 12% MARR. Supply the values of U, V, W, X, and Y in the following equation for computing the equivalent uniform annual cost for the machine. (4.3)

$$(\$25,000 - U)(A \,|\, F\, 12\%,10) + V(0.12) + W + \$1500[(F \,|\, A\, 12\%,6) + (F \,|\, A\, 12\%,X)](Y, 12\%,10)$$

23. A distillation column is purchased for $300,000. Operating and maintenance costs for the first year are $30,000. Thereafter, operating and maintenance costs increase by 10%/year over the previous year's costs. At the end of 8 years the column is sold for $50,000. During the life of the investment, revenue was produced that could be related directly to the investment in the column. The revenue the first year was $75,000. Thereafter, revenue increased by $10,000 over the previous year's revenue. Using a *MARR* of 12%, determine the equivalent annual worth for the investment. (4.3)

24. Determine the capital recovery cost of a bulldozer costing $200,000, having a life of 10 years, and a salvage value of $50,000. Use an interest rate of 10%. Solve using each of three different capital recovery formulas. (4.3)

25. Determine the capital recovery cost for chemical processing equipment costing $100,000, having a life of 5 years, and a salvage value of −$25,000. Use an interest rate of 12%. (4.3)

26. Determine the capital recovery cost of a ditching machine having a first cost of $35,000, a salvage value of $5000, and a life of 5 years. Use an interest rate of 10%. Solve using each of the three different capital recovery formulas. (4.3)

27. Crush Autosmashers can purchase a new electromagnet for moving cars at a cost of $20,000. When scrapped out, the electromagnet will be worth $1000. If Crush's *MARR*

is 12%, how many years must the unit last so that its capital recovery cost will be $3000/ yr. or less? (4.3)

28. What should be entered in cells B10 through B13 in the spreadsheet to obtain the values for PW, AW, FW, and IRR using Excel? Determine the values for PW, AW, FW, and IRR.

R/C	A	B
1	MARR =	0.12
2		
3	Time	Cash flow
4	0	−$15,000.00
5	1	$2000.00
6	2	$5000.00
7	3	$5000.00
8	4	$5000.00
9	5	$10,000.00
10	PW =	
11	AW =	
12	FW =	
13	IRR =	

PW: present worth
AW: annual worth
FW: future worth
IRR: internal rate of return

29. What should be entered in cells B11 through B15 to obtain the values for PW, AW, FW, IRR, and SIR using Excel? Determine the values for PW, AW, FW, IRR, and SIR.

R/C	A	B
1	MARR =	0.10
2		
3	Time	Cash flow
4	0	−$25,000.00
5	1	− $5000.00
6	2	$8000.00
7	3	$8000.00
8	4	$8000.00
9	5	$8000.00
10	6	$18,000.00
11	PW =	
12	AW =	
13	FW =	
14	IRR =	
15	SIR =	

PW: present worth
AW: annual worth
FW: future worth
IRR: internal rate of return
SIR: savings/investment ratio

30. What should be entered in cells B10 through B13 to obtain the values for PW, FW, IRR, and ERR using Excel? Determine the values for PW, FW, IRR, and ERR.

R/C	A	B
1	MARR =	0.15
2		
3	Time	Cash flow
4	0	−$18,000.00
5	1	− $5000.00
6	2	$8000.00
7	3	$9000.00
8	4	$10,000.00
9	5	$11,000.00
10	PW =	
11	FW =	
12	IRR =	
13	ERR =	

PW: present worth
FW: future worth
IRR: internal rate of return
ERR: external rate of return

Chapter **5**

Comparison of Alternatives

5.1 INTRODUTION

In this chapter, we apply time value of money concepts and measures of investment worth to the comparison of engineering investment alternatives. Although multiple objectives are often involved in performing a comparison of alternatives, for now we concentrate on the comparison of engineering investment alternatives on the basis of monetary considerations alone. We will assume that our set of investment alternatives is collectively exhaustive and mutually exclusive. By collectively exhaustive, we mean that we did not leave out any alternatives; by mutually exclusive, we mean it is an either-or situation; no more than one alternative can be chosen. The adage of "not being able to have one's cake and eat it too" illustrates the notion of mutual exclusivity.

A systematic approach that can be used in comparing the economic worths of engineering investment alternatives is summarized as follows:

1. Define the set of feasible, mutually exclusive alternatives to be compared.
2. Define the planning horizon to be used in the comparison.
3. Develop the cash flow profiles for each alternative.
4. Specify the MARR to be used.
5. Compare the alternatives using a specified measure of worth.
6. Perform supplementary analyses.
7. Select the preferred alternative.

The procedures for comparing engineering investment alternatives outlined in this chapter are intended to aid in making better measurements of the *quantitative aspects* of capital investment alternatives. Following development of the 7-step approach, more detail is provided on comparing alternatives having unequal lives. Two special classes of alternatives are examined further. These include alternatives having no positive cash flows and those involving the possible replacement of existing assets.

The treatment of the 7-step approach in this chapter is broad and comprehensive. As such, this chapter is quite lengthy. There are several ways that the

143

material can be pursued to allow more rapid progress through the chapter. Sections 5.2 through 5.8 present the steps of the approach in sequence. To fully appreciate each of the steps, it is necessary to cover each of these sections. However, several of these sections contain subsections that can be skipped, or treated lightly, if necessary. Section 5.6 addresses the fifth step of the approach, the comparison of alternatives, using eight different measures of worth. The first six of these measures (Sections 5.6.1 through 5.6.6) are entirely consistent if properly applied: i.e., they yield the same preferred alternative. It is therefore possible to skip, or treat lightly, some of these measures without loss of content. Based on our collective experience, we believe that the minimal set to be covered from among these six measures consists of present worth (Section 5.6.1), annual worth (Section 5.6.2), and internal rate of return (Section 5.6.4)[1]. Subsection 5.6.7 deals with the payback period method. This method is used frequently and often inappropriately in practice. Due to this common misapplication, we believe covering this subsection is important. Subsection 5.6.8 deals with a special measure of worth for projects whose cash flows repeat indefinitely. It can be skipped, or treated lightly, unless this type of project is of prime importance.

Section 5.7, Supplementary Analysis, is another section whose subsections can be abbreviated if needed. The introductory material provides an overview of three techniques for supplementary analysis, break-even, sensitivity, and risk. The three subsections which follow elaborate on these topics. We believe that the first two of these, break-even analysis (Section 5.7.1) and sensitivity analysis (Section 5.7.2) are sufficiently important to be covered in all cases. However, Subsection 5.7.3 addresses risk analysis and requires an understanding of elementary probability theory, random variables, and simulation. This topic is frequently a fundamental topic in advanced courses in engineering economy. The material may be skipped here without jeopardizing understanding of the remainder of the text.

Section 5.8 concludes the presentation of the 7-step approach. The next three sections in the chapter (Sections 5.9 through 5.11) introduce special cases and issues associated with applying the approach. While each of these considerations is important when circumstances warrant, they can be skipped, or treated lightly, here without loss of generality with respect to the 7-step approach. Alternatively, this material can be deferred for later coverage as time permits.

The comparison of engineering investment alternatives involves the consideration of many factors, not the least of which are technological and monetary aspects of the engineering design. Quality, safety, performance, delivery, risk, flexibility, appearance, marketing, environmental, employment, engineering and construction capability, and competitive pressures, for example, might dominate the final selection of the design. However, for now, the focus will be on comparing engineering alternatives on the basis of their economic impact on the organization.

Regardless of the type of alternatives being considered, it cannot be too strongly emphasized that *no economic evaluation can replace the sound judgment*

---

[1] Savings investment ratio (Section 5.6.6) is of particular importance in public sector projects. This subject is treated in depth in Chapter 7.

of experienced managers concerned with both the quantitative and nonquantitative aspects of investment alternatives.

5.2 DEFINING INVESTMENT ALTERNATIVES

An individual investment alternative selected from a set of alternatives can be made up of several *investment proposals*. Investment proposals are distinguished from investment alternatives by noting that investment alternatives are decision options; investment proposals are single projects or undertakings that are being considered as investment possibilities.

EXAMPLE 5.1

As an illustration of the distinction between investment proposals and investment alternatives, consider a distribution center that receives pallet loads of product, stores the product, and ships pallet loads of product to various customer locations. A new distribution center is to be constructed, and the following proposals have been made:

1. Proposed methods of moving materials from receiving to storage and from storage to shipping include:
   a. Conventional lift trucks for operating in 12-foot aisles.
   b. Narrow-aisle life trucks for operating in 5-foot aisles.
   c. Automated guided vehicle system.
   d. Towline conveyor system.
   e. Pallet conveyor system.
2. Proposed methods of placing materials in and removing materials from storage include:
   a. Conventional lift trucks for operating in 12-foot aisles.
   b. Very-narrow-aisle lift trucks for operating in 5-foot aisles.
   c. Very-narrow-aisle, operator driven, rail-guided storage/retrieval vehicle.
   d. Very-narrow-aisle, automated, rail-guided storage/retrieval vehicle.
3. Proposed methods of storing materials include:
   a. Block stacking pallet loads of material (8 feet high, 12-foot aisles).
   b. Conventional pallet rack (20 feet high, 12-foot aisles).
   c. Flow rack (20 feet high, 12-foot aisles).
   d. Very-narrow-aisle, pallet rack (20 feet high, 5-foot aisles).
   e. Very-narrow-aisle, flow rack (20 feet high, 5-foot aisles).
   f. Very-narrow-aisle, medium height, pallet rack (35 feet high, 5-foot aisles).
   g. Very-narrow-aisle, high rise, pallet rack (70 feet high, 5-foot aisles).

Given the set of proposals, alternative designs for the material handling system can be obtained by combining a proposed method of moving materials from receiving to storage, a proposed method of placing materials in storage, a proposed method of storage, a proposed method of removing materials from storage, and a proposed method of transporting materials from storage to shipping. Some of the combinations of proposals will be eliminated because of their incompatibility. For example, lift trucks requiring 12-foot aisles cannot be used to place materials in and remove materials from storage when 5-foot aisles are used. Other combinations might be eliminated because of budget limitations; a desire to minimize the variation in types of equipment due to maintainability, availability, reliability, flexibility, and operability considerations; ceiling height limitations; physical characteristics of the product (crushable product might require the use of storage racks); and a host of other considerations. Characteristically, experience and judgment are used to trim the list of possible combinations to a manageable number.

EXAMPLE 5.2

To illustrate the formation of investment alternatives from a set of investment proposals, consider a situation involving m investment proposals. Let x_j be defined to be 0 if proposal j is not included in an alternative and let x_j be defined to be 1 if proposal j is included in an alternative. Using the binary variable x_j we can form 2^m alternatives. Thus, if there are three investment proposals, we can form eight investment alternatives, as depicted in Table 5.1.

Among the alternatives formed, some might not be feasible, depending on the restrictions or constraints placed on the problem. To illustrate, there might be a budget limitation that precludes the possibility of combining all three proposals; thus, Alternative 8 would be eliminated. Additionally, some of the proposals might be *mutually exclusive proposals*. For example, Proposals A and B might be alternative computer designs, and only one is to be selected; in this case Alternative 7 would be eliminated from consideration. Other proposals might be *contingent proposals* so that one proposal cannot be selected unless another proposal is also selected. As an illustration of a contingent proposal, Proposal C might involve the procurement of computer terminals, which require the

Table 5.1 Developing Investment Alternatives from Investment Proposals

Alternative	Proposals			Explanation
	x_A	x_B	x_C	
1	0	0	0	Do nothing (proposals A, B, and C not included)
2	0	0	1	Accept proposal C only
3	0	1	0	Accept proposal B only
4	0	1	1	Accept proposals B and C only
5	1	0	0	Accept proposal A only
6	1	0	1	Accept proposals A and C only
7	1	1	0	Accept proposals A and B only
8	1	1	1	Accept all three proposals

selection of the computer design associated with Proposal B. In such a situation, Alternatives 2 and 6 would be infeasible. Thus, depending on the restrictions present, the number of feasible alternatives that result can be considerably less than 2^m.

In many organizations there is a rather formalized hierarchy for determining how the organization will invest its funds. Typically, the entry point in this hierarchy involves an individual analyst or engineer who is given an assignment to solve a problem; the problem may be one requiring the design of a new product, the improvement of an existing manufacturing process, or the development of an improved system for performing a service. The individual performs the steps involved in the problem-solving procedure and recommends the preferred solution to the problem. In arriving at the preferred solution, a number of alternative solutions are normally compared.

The preferred solution is usually forwarded to the next level of the hierarchy for approval. In fact, one would expect many preferred solutions to various problems to be forwarded to the second level of the hierarchy for approval. Each preferred solution becomes an investment *proposal*, the resulting set of mutually exclusive investment *alternatives* are formed, and the process of comparing economic investment alternatives is repeated. This sequence of operations is usually performed in various forms at each level of the hierarchy until, ultimately, the preferred solution by the individual analyst or engineer is accepted or rejected. In this textbook we concentrate on the process of comparing investment alternatives at the first level of the hierarchy; however, the need for such comparisons at many levels of the organization should be kept in mind.

One danger inherent in the hierarchical approval approach is the use of *size gates*. In particular, companies often delegate approval authority over investment alternatives based on the level of funding required. Depending on the size of the investment, a different *approval gate* might be required. For example, a plant manager might have the authority to make decisions on investments involving less than $100,000 in capital; for those investments requiring from $100,000 to $500,000 in capital, division-level approval might be required; and for those requiring in excess of $500,000 in capital, corporate-level approval is required.

Due to the use of size gates, piecemeal investment strategies are often used. Namely, major investments are subdivided into smaller pieces in order to obtain approval at lower levels of the hierarchy. Rather than seek approval to undertake the entire project in one year, a multiyear strategy is developed—simply because of the "gamesmanship" used to cope with the size gates! The adage of "eating the elephant one bite at a time" becomes the strategy for obtaining approval for major system changes.

Unfortunately, the benefits of an integrated system are postponed for years, due to the incremental approach. Further, it is often the case that one of the "bites" is not approved, simply because of the number of competing proposals that particular year, or because of the difficulty of quantifying the benefits of some critical element of the overall system. As a result, small incremental changes are made, rather than major system-wide improvements. (For further consideration of the impact of approval hierarchies on the generation of investment alternatives, see Kaplan [34].)

Another aspect of integrated systems merits consideration. Namely, in justifying computer intergrated manufacturing (CIM) systems, it is usually difficult to quantify the *synergistic benefits* of systems integration. Aristotle is credited with having said that "the whole is greater than the sum of its parts." Whether or not Aristotle made such a claim is not the issue; instead, while it is easy to accept in principle that two plus two is greater than four in an integrated system, the application of incremental justification techniques often belies the claim. As a result, many firms have found that the best strategy is to "design the whole, sell or justify the whole, and then implement the pieces." In particular, the total system is justified and sold to management; where capital is limited, a phased implementation plan is used and individual components of the total system are implemented over time.

Obviously, the alternatives to be considered must be defined carefully so as not to prematurely eliminate viable candidates. At the same time, care must be taken to include only those alternatives that appear to satisfy the noneconomic considerations that will influence the ultimate selection.

Before concluding our consideration of the generation of engineering investment alternatives, a further word is required concerning the use of the "do nothing" alternative. As in Example 5.2, it is frequently the case that the do-nothing alternative will be included in the set of feasible investment alternatives to be compared. Such an alternative is intended to represent business as usual or maintain the status quo; however, it is rarely the case that business conditions stand still. Doing nothing does not mean that nothing will be done; rather, it might mean that management has opted to pass up the opportunity to influence future events. It is often used as a baseline, against which other investment alternatives are compared.

In this chapter, as a matter of convenience, we often associate zero incremental costs to doing nothing. In practice, when the do-nothing alternative is feasible, extreme care must be taken not to underestimate the cost of doing nothing. For many firms, business as usual is their most expensive alternative; "standing pat" or maintaining the status quo for too long can prove to be a disasterous course for many businesses. For, *while the firm is doing nothing, its competition is generally doing something!*

Over a period of time, a series of decisions to not make capital investments or to do nothing can result in a firm's losing market share due to obsolete products and/or processes. Considered individually, each decision might seem to be the most economic at the time. However, collectively, the decisions might prove to be disasterous to the long-term well-being of the firm, due to their being based on too narrow a perspective. In the defense of the decisions to do nothing, they are made in the absence of perfect knowledge of the future.

Ironically, it is because management cannot accurately predict the economic outcome of change that the absence of change occurs and doing nothing is preferred. Yet, there is also uncertainty concerning the economic impact of doing nothing—of failing to change. One must guard against taking too narrow a view of the economic impact of choosing to follow the path of doing nothing. In considering the long-term impact of doing nothing, managers must decide if they want to be "change masters" or to be mastered by change.

In the case of manufacturing and distribution modernizations, a common reason engineers fail to view the do-nothing alternative from a broad perspective is the propensity to restrict their thinking to the costs that arise *inside* the manufacturing plant or distribution facility. It is seldom the case that an engineer looks beyond the walls of the facility and assesses the impact of a do-nothing decision on, say, the marketplace. However, when the engineer enlists the assistance of persons from the marketing organization in analyzing the impact of doing nothing, it is frequently the case that automated manufacturing and distribution systems are justified and installed successfully.

5.3   DEFINING THE PLANNING HORIZON

In comparing engineering investment alternatives, it is important to compare them over a common period of time. In Chapter Four we define that period of time to be the *planning horizon*. In the case of investments in, say, equipment to perform a required service, the period of time over which the service is required might be used as the planning horizon. Likewise, in one-shot investment alternatives, the period of time over which receipts continue to occur might define the planning horizon.

In a sense, the planning horizon defines the width of the "window" through which the economic performance of each alternative will be viewed. In making the selection, consider the person who looks only a foot ahead when traveling along an unfamiliar path; that person might encounter dead ends and might be required to backtrack many times. Likewise, the person who looks a mile ahead and fails to see the short-term obstacles is very likely to stumble and fall along the journey. Adopting too short a planning horizon might result in the elimination of those alternatives that require radical change, but will produce dramatic payoffs in the long run; however, adopting too long a planning horizon might result in the firm going out of business before realizing the promised benefits.

It is important to distinguish between the length of the planning horizon, the working life of equipment, and the depreciable life of equipment. The working life is the actual period of time the equipment is capable of being used (e.g., 20 years); the depreciable life of the equipment is the allowable period of time for depreciating the asset. The planning horizon might have no relationship to the other two periods of time; it is simply the time frame to be used in comparing the alternatives and should realistically represent the period of time over which reasonably accurate cash flow estimates can be provided. Also, to make an objective evaluation, *the same window must be used in viewing each alternative.*

In some cases the planning horizon is easily determined; in other cases the duration of one or more subjects is sufficiently uncertain to cause concern over the time period to use. Some commonly used methods for determining the planning horizon to use in economy studies include:

1. Least common multiple of lives for the set of feasible alternatives, denoted $\hat{T}$.
2. Shortest life among alternatives, denoted T_s.
3. Longest life among alternatives, denoted T_l.
4. Some other period of time.

The *least common multiple of lives* is the most common method of selecting the planning horizon. In this method, each alternative's cash flow profile is assumed to repeat in the future until a time is reached when all alternatives under consideration conclude at the same time. Using such a procedure when three alternatives are being considered and the individual lives are 3 years, 6 years, and 5 years yields a planning horizon of $\hat{T} = 30$ years. If the lives had been 3 years, 6 years, and 6 years, $\hat{T} = 6$ years. Clearly, strict reliance on $\hat{T}$ as the planning horizon is not advisable. Unfortunately, selection of this method is usually made implicitly, not explicitly, when one simply calculates and compares the annual worths of unequal-lived alternatives. This practice is covered in more detail in Section 5.10.

In using the *shortest life among alternatives* to define the planning horizon, estimates are required for the values of the unused portions of the lives of the remaining alternatives. Thus, for the situation considered above, with $T_s = 3$ years, the salvage or residual values at the end of 3 years' use must be assessed for the other two alternatives.

If the *longest life among alternatives, T_l,* is used in determining the planning horizon, some difficult decisions must be made concerning the period of time between T_s and T_l. If the alternative selected is to provide a necessary service, that service must continue throughout the planning horizon, regardless of the alternative selected. Consequently, when the shortest life alternative reaches the end of its project life, it must be replaced with some other asset capable of performing the required service. However, since technological developments will probably take place during the period of time T_s, new and improved candidates will be available for selection at time T_s. Thus, the specification of the cash flows for the shortest life alternative during the period of time T_s to T_l is a difficult undertaking. As a result, T_l is seldom used as the planning horizon.

Many organizations have adopted a *standard planning horizon* for all economic alternatives in order to ensure some consistency in economic analyses. Letting T denote the planning horizon specified by the organization, different approaches are recommended, depending on whether $T < T_s$, $T_s < T < T_l$, or $T > T_l$. If $T < T_s$, each alternative's cash flow profile must be truncated with salvage or residual values estimated at time T. If $T_s < T < T_l$, alternatives having a life less than T require that plans and cash flows be specified up through time T; those having lives longer than T must be truncated. If $T > T_l$, all alternatives require that plans and cash flows be specified through the lesser of $\hat{T}$ and T. That is, when the planning horizon is greater than or equal to the least common multiple of lives, it is recommended that the economic analysis be based on a period of time equal to the least common multiple of lives. The reason is that at time $\hat{T}$ a new economic analysis can be performed based on the alternatives available at that time. After $\hat{T}$ years new alternatives might be available; furthermore, we can more accurately estimate the values of cash flows occurring after $\hat{T}$ if we wait until nearer time $\hat{T}$ to make the estimates.

EXAMPLE 5.3

To illustrate the difficulties associated with the selection of the planning horizon, consider the three mutually exclusive alternatives whose cash flow profiles are given in Table 5.2. Alternatives 1, 2, and 3 have anticipated lives of 3, 6, and 5 years, respectively. Each has revenues and costs which are used to determine

Table 5.2 Cash Flow Profiles for Three Mutually Exclusive Investment
Alternatives Having Unequal Lives

End of Year, t	Revenues, R_t	Costs, C_t	Net Cash Flows, $R_t - C_t$	Salvage Value if Sold at Time t
Alternative 1				
0	—	—	—	$ 0
1-3	$27,500	$23,000	$ 4,500	0
Alternative 2				
0	—	$50,000	−$50,000	$50,000
1	$30,000	10,000	20,000	35,000
2	30,000	10,000	20,000	25,000
3	30,000	10,000	20,000	15,000
4	30,000	10,000	20,000	5,000
5	30,000	10,000	20,000	0
6	30,000	10,000	20,000	0
Alternative 3				
0	—	$75,000	−$75,000	$75,000
1	$27,500	7,500	20,000	55,000
2	32,500	7,500	25,000	40,000
3	37,500	7,500	30,000	25,000
4	42,500	7,500	35,000	10,000
5	47,500	7,500	40,000	0

the net cash flows. Alternative 1 involves used equipment have a useful life of
3 years. Much of this older equipment is readily available at virtually no cost
other than for operating and maintenance. Alternatives 2 and 3 involve the
purchase and operation of new equipment which will last for up to 6 and 5 years,
respectively; estimated salvage values for the equipment are shown as a function
of time.

Using a least common multiple of lives approach, a planning horizon of $\hat{T} = $
30 years would be used. Using a 30-year planning horizon requires answers to
the following questions. What cash flows are anticipated for years 4 to 30 if
Alternative 1 is selected? What will be the cash flows for Alternative 2 for years
7 to 30? What about Alternative 3 for years 6 to 30?

As shown in Table 5.3, the traditional approach is to assume that Alternative
1 will be repeated 10 times, Alternative 2 will be repeated 5 times, Alternative
3 will be repeated 6 times, and that identical cash flows occur during these
repeating life cycles. Inflation effects, as well as technological improvements and
consumer demand changes tend to invalidate such assumptions.

If the shortest life approach is used, a planning horizon of 3 years would be
used. In such a case, the estimated salvage values of Alternatives 2 and 3 should
be indicated at the end of year 3, as denoted in Table 5.3 by an asterisk.

Using the longest life approach yields a 6-year planning horizon. In this instance
a decision must be made concerning the cash flows of Alternative 1 in years 4,

Table 5.3 Cash Flow Profiles for Various Planning Horizons

End of Year, t	Net Cash Flows For Alternatives		
	A_{1t}	A_{2t}	A_{3t}
$T = \hat{T} = 30$ years			
0	—	−$50,000	−$75,000
1	$4,500	20,000	20,000
2	4,500	20,000	25,000
3	4,500	20,000	30,000
4	4,500	20,000	35,000
5	4,500	20,000	−75,000 + 40,000
6	4,500	−50,000 + 20,000	20,000
7	4,500	20,000	25,000
8	4,500	20,000	30,000
.	.	.	.
.	.	.	.
.	.	.	.
29	4,500	20,000	35,000
30	4,500	20,000	40,000
$T = T_s = 3$ years			
0	—	−$50,000	−$75,000
1	$4,500	20,000	20,000
2	4,500	20,000	25,000
3	4,500	20,000 + 15,000[a]	30,000 + 25,000[a]
$T = T_l = 6$ years			
0	—	−$50,000	−$75,000
1	$4,500	20,000	20,000
2	4,500	20,000	25,000
3	4,500	20,000	30,000
4	4,500	20,000	35,000
5	4,500	20,000	40,000
6	4,500	20,000	4,500
$T = 2$ years $< T_s$			
0	—	−$50,000	−$75,000
1	$4,500	20,000	20,000
2	4,500	20,000 + 25,000[a]	25,000 + 40,000[a]
$T_s < T = 5$ years $< T_l$			
0	—	−$50,000	−$75,000
1	$4,500	20,000	20,000
2	4,500	20,000	25,000
3	4,500	20,000	30,000
4	4,500	20,000	35,000
5	4,500	20,000 + 0[a]	40,000
$T = 8$ years $> T_l$			
0	—	−$50,000	−$75,000
1	$4,500	20,000	20,000
2	4,500	20,000	25,000

Table 5.3 (*Continued*)

End of Year, t	Net Cash Flows For Alternatives		
	A_{1t}	A_{2t}	A_{3t}
3	4,500	20,000	30,000
4	4,500	20,000	35,000
5	4,500	20,000	40,000 − 75,000
6	4,500	20,000 − 50,000	20,000
7	4,500	20,000	25,000
8	4,500	20,000 + 25,000[a]	30,000 + 25,000[a]

[a] Denotes salvage value.

5, and 6 and Alternative 3 in year 6. In the case of Alternative 1, we assume that another piece of used equipment will be employed, resulting in resumption of the $4500 net cash flows during years 4, 5, and 6. For Alternative 3, we *could* assume that another new unit is purchased for $75,000 at the end of year 5. This would provide service and net income of $20,000 during year 6, and then have a salvage value of $55,000. This 1-year purchase is obviously not economically sound. Therefore, we assume that a piece of readily available used equipment is employed for year 6, earning $4500.

Suppose a standard planning horizon of 2, 5, or 8 years is used. The same questions that arise under planning horizons of $\hat{T}$, T_s, or T_l apply when one of these other horizons is used. As shown in Table 5.3, Alternatives 1, 2, and 3 are either truncated at, or repeated until, the end of the planning horizon at which time estimates of terminal salvage values are provided.

Although the use of a standard planning horizon has the benefit of a consistent approach in comparing investment alternatives, there are also some dangers that should be recognized. In some cases the major benefits associated with an alternative might occur in the later stages of its project life. If the planning horizon is less than the project life, such alternatives would seldom be accepted. Just such a practice caused one major textile firm to lose its strong position in the industry. A major modernization of the processing departments had been proposed, but its benefits would not be realized until the bugs had been worked out of the new system, all personnel were trained under the new system, and the marketing people had regained the lost customers. Unfortunately, the planning horizon specified by the firm was too short in duration and the modernization plan was not approved.

EXAMPLE 5.4

As a second illustration of the planning horizon selection process, consider the two cash flow diagrams given in Figure 5.1. The two alternatives are *one-shot* investments. We are unable to predict what investment alternatives will be avail-

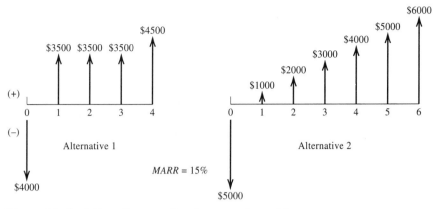

Figure 5.1 Cash flow diagrams for the example problem.

able in the future, but we do anticipate that recovered capital can be reinvested and earn a 15% return.

For this type of situation, a 6-year planning horizon is suggested, with zero cash flows occurring in years 5 and 6 with Alternative 1. At the end of 6 years the net future worths for the two alternatives will be

$$FW_1(15\%) = \$4500(F|P\ 15,2) + \$3500(P|A\ 15,3)(F|P\ 15,6)$$
$$- \$4000(F|P\ 15,6)$$
$$= \$4500(1.3225) + \$3500(2.2832)(2.3131)$$
$$- \$4000(2.3131)$$
$$= \$15,183.29$$
$$FW_2(15\%) = \$1000(F|A\ 15,6) + \$1000(A|G\ 15,6)(F|A\ 15,6)$$
$$- \$5000(F|P\ 15,6)$$
$$= \$1000(8.7537) + \$1000(2.0972)(8.7537)$$
$$- \$5000(2.3131)$$
$$= \$15,546.46$$

Thus, we would recommend Alternative 2.

If we did not give careful thought to the situation involved and simply calculated the annual worth over 4- and 6-year horizons, respectively, annual values of $3000.15 and $1776.20 would result in favor of Alternative 1. This would be comparable to blindly assuming a least common multiple of of lives planning horizon of 12 years, with identical cash flows in repeating life cycles. Hence, it is important to consider the particular situation involved and specify the planning horizon instead of employing a rule of thumb for establishing planning horizons that does not consider the nature of the investments.

It appears that the preferred approach would be to have a "flexible" standard planning horizon. All of the "routine" economic analyses would be based on the standard planning horizon of, say, 5 to 10 years; nonroutine economic analyses would be based on a planning horizon that was appropriate for the situation.

5.4   DEVELOPING CASH FLOW PROFILES

Once the set of investment alternatives has been specified and the planning horizon decision has been made, the cash flow profiles can be developed for the alternatives. As has been emphasized, the cash flow profiles should be developed by giving careful consideration to *future* conditions instead of relying completely on *past* cash flows. The cash flows for an investment alternative are obtained by aggregating the cash flows for all investment proposals included in the investment alternative.

EXAMPLE 5.5

To illustrate the approach to be taken, suppose a planning horizon of 5 years is used and there are three investment proposals. Cash flow profiles for the proposals are given in Table 5.4. A budget limitation of $50,000 is available for investment among the proposals. Proposal B is contingent on Proposal A, and Proposals A and C are mutually exclusive. Based on the restrictions associated with the combinations of proposals, only four investment alternatives are to be considered. These are developed in Table 5.5. Alternative 0 is the "do-nothing" alternative; Alternative 1 involves Proposal C alone; Alternative 2 involves Proposal A alone; and Alternative 3 involves a combination of Proposals A and B. The cash flow profiles for the four alternatives are given in Table 5.6.

In developing the cash flow profiles for each alternative, best estimates are needed of the annual costs and incomes anticipated over the planning horizon, including the salvage values of all assets at the end of the planning horizon. Again, care must be taken to assess the impact of each alternative on the economic performance of the firm, including the economic impact of the do-nothing alternative, if it is a feasible candidate.

In most, if not all, economic evaluations, it is not necessary to develop detailed forecasts of all items of cost, revenue, and investment associated with an alternative. Costs and revenues that will be the same regardless of the alternative selected can be omitted. If a cost reduction alternative will not affect sales revenues, no forecast of such revenues need be developed. Attention is focused

Table 5.4 Cash Flow Profiles for Three Investment Proposals

End of Year, t	New Cash Flows for Proposals		
	CF_{At}	CF_{Bt}	CF_{Ct}
0	−$20,000	−$30,000	−$50,000
1	− 4,000	4,000	− 5,000
2	2,000	6,000	10,000
3	8,000	8,000	25,000
4	14,000	10,000	40,000
5	25,000	20,000	10,000

Table 5.5 Developing Investment Alternatives

Feasible Alternatives	Proposals			Investment Required
	X_A	X_B	X_C	
0	0	0	0	$ 0
1	0	0	1	50,000
~~	~~0~~	~~1~~	~~0~~	30,000
~~	~~0~~	~~1~~	~~1~~	80,000
2	1	0	0	20,000
~~	~~1~~	~~0~~	~~1~~	70,000
3	1	1	0	50,000
~~	~~1~~	~~1~~	~~1~~	100,000

on the items of cost and revenue that will be affected by the alternative selected. However, as noted previously, before concluding that, say, revenues will not be affected by the decision, a careful assessment should be made—including consideration of the competition.

Too frequently, incremental decisions appear to be made in a vacuum. Considerations of the competition and the market environment are seldom incorporated in engineering economic analyses, but they should be! Often, it is blindly assumed that the only cash flows associated with the replacement of existing equipment with state-of-the-art equipment are the costs associated with each alternative. The fact that installing the newest equipment might influence customers to select your product rather than the competition's product is often overlooked.

In determining the costs associated with each alternative, one should not blindly accept cost reduction estimates from a cost accounting system. The fundamental and historical purpose of accounting is to maintain a consistent *historical* record of the financial results of the operations of the organization. Accounting figures are based on definitions derived consistent with this objective. Accounting methods are not designed to determine the economic worth of *future* alternative courses of action. Furthermore, certain accounting practices can create obstacles to capital investment.

Table 5.6 Cash Flow Profiles for Four Investment Alternatives

End of year, t	Net Cash Flows for Alternatives			
	A_{0t}	A_{1t}	A_{2t}	A_{3t}
0	0	−$50,000	−$20,000	−$50,000
1	0	− 5,000	− 4,000	0
2	0	10,000	2,000	8,000
3	0	25,000	8,000	16,000
4	0	40,000	14,000	24,000
5	0	10,000	25,000	45,000

For example, cost accounting systems typically include a proration of overhead on the basis of direct labor. As a result, claimed cost reductions through the elimination of direct labor should be examined carefully to ensure that credit is not being given for reductions in overhead, when, in fact, overhead might be increased by the reduction in direct labor.

Another problem that frequently arises from the allocation of overhead on the basis of direct labor is illustrated by investments in automation. If the use of automation reduces direct labor costs and/or increased indirect labor, the overhead percentage for a plant will increase. Hence, if corporate management places limits on the overhead percentage for a given plant, then the plant might be prevented from investing in automation.

Again, in using a cash flow approach, the actual costs and savings to be realized should be included in the cash flow for each period. Also, the cash flow estimates should reflect the impact of the alternative of the entire system, rather than just the impact on an individual department.

Finally, accurate estimates of cash flows should be used. Too frequently, management suspects that engineers use whatever numbers are needed to satisfy the approval criteria established. As an example, in a conversation with the treasurer of a major U.S. corporation, we learned that he did not believe he could rely on the cash flow estimates obtained from their engineers. Past experience indicated that few of the approved investments came close to delivering the economic performances promised. He noted that for the most important investments made by the corporation, top management depended more on their intuition than the cash flow estimates provided. In many companies, a credibility problem exists regarding the use of discounted cash flow methods. Some managers are arguing for the elimination of the use of DCF methods.

However, the situation that exists is akin to "shooting the messenger because you don't like the message!" The problem is not with the DCF methods, but with their use. DCF methods are GIGO (garbage in, garbage out) methods; the result obtained is no better than the inputs provided.

A primary area of abuse in using DCF methods is in estimating the cash flows associated with an investment. One major firm audited a dozen automation investments and found that 75% of them failed to yield the promised return on investment. In each case, the difficulties of implementing the new equipment had been underestimated. The engineers had shown a single investment occurring in year zero followed by uniform annual savings. Due to late delivery and start-up, the investment occurred over a 2-year period. Furthermore, it took some time for the operators to become familiar with the equipment and to produce quality products at the required quantities. As a result, the promised savings were slower to materialize than had been promised. Due to the poor economic performance of the automation investments, management of the firm became reluctant to approve similar investments.

Depending on the organization, specially designed forms are often provided for aiding the analyst in developing cash flow profiles and conducting the analysis. Sample forms used by one major industrial organization are provided in Figures 5.2 to 5.4. This particular company compares investment alternatives using the internal rate of return method; they refer to it as the discounted cash flow rate of return method.

NET CASH FLOW SCHEDULE

ALTERNATE $\underset{\text{CIRCLE ONE}}{A \text{ OR } B}$ TITLE (PROJECT) _____

TITLE (THIS ALTERNATE) _____

PROJECT NO. _____   DATE _____

LINE	EXPENSES*	PERIOD									
		0	1	2	3	4	5	6	7	8	Total
1	NET BOOK VALUE (BEGINNING OF YEAR)										
2	DEPRECIATION (___ YEARS LIFE)										
3	RENTAL / LEASE COST										
4	IN-PLANT LABOR WITH FRINGE BENEFITS ___%										
5	PURCHASED LABOR										
6	REWORK / SCRAP										
7	TOOLING / ANCILLARY EQUIPMENT										
8	MAINTENANCE & REPAIR										
9	POWER & FUEL / UTILITIES EXPENSE										
10	MATERIALS & SUPPLIES										
11	PRODUCT SHIPPING COSTS										
12	PERSONAL PROPERTY TAX										
13	REAL ESTATE TAX										
14	IMPLEMENTING COST										
15											
16											
17	TOTAL CASH EXPENSES (LINES 2 THRU 16)										
18	AFTER TAX ANNUAL COST** (50% AT LINE 17)										
19	INVESTMENT / EXISTING ASSET VALUE										
20	INVESTMENT TAX CREDIT / REBATE										
21	SALVAGE										
22											
23	NET CASH FLOW ALGEBRAIC SUM OF LINES 18 THRU 22 AND ADD BACK DEPRECIATION										

Figure 5.2

A particularly difficult class of benefits to estimate are those that are categorized as *intangible benefits*. Among the so-called intangible benefits are reduced inventory, increased quality, reduced space, increased flexibility, reduced lead time, increased throughput, reduced scrap, improved tracking of material, and increased safety. We believe that every effort should be made to quantify the benefits of such intangibles, recognizing that precise measurements might be difficult, if not impossible. In those cases where explicit consideration of intangible benefits is not allowed by the firm, then (as noted in Section 5.7) they certainly should be included in the supplementary analyses to be performed.

5.5   SPECIFYING THE *MARR*

The comparison of alternatives on the basis of the time value of money requires an explicit demonstration of the time value to be used. As noted in Chapter Four, the value of money used in discounted cash flow methods is variously referred to as a minimum attractive rate of return, a hurdle rate, a required rate of return, a return on investment, and a discount rate. Regardless of its label, a specified interest rate is needed to convert the mix of cash flows occurring throughout the planning horizon to a meaningful economic measure that can be used rationally to compare engineering alternatives.

DISCOUNTED CASH FLOW SUMMARY

PERIOD	"A" NCF	"B" NCF	"A"–"B" NCF 0% INTEREST	CUMULATIVE CASH FLOW BACK AMOUNT	CUMULATIVE CASH FLOW BACK %	15% INTEREST FACTOR	15% INTEREST PW	25% INTEREST FACTOR	25% INTEREST PW	40% INTEREST FACTOR	40% INTEREST PW	60% INTEREST FACTOR	60% INTEREST PW
-1						1.150		1.250		1.400		1.600	
0						1.000		1.000		1.000		1.000	
TOTAL "X" (DISBURSEMENTS)													
1						.870		.800		.714		.625	
2						.756		.640		.510		.391	
3						.658		.512		.364		.244	
4						.572		.409		.260		.153	
5						.497		.328		.186		.095	
6						.432		.262		.133		.060	
7						.376		.210		.095		.037	
8						.327		.168		.068		.023	
9						.284		.134		.048		.015	
10						.247		.107		.035		.009	
11						.215		.086		.025		.006	
12						.187		.069		.018		.004	

TOTAL "Y" (RECEIPTS)
RATIO "X"/"Y"

PROPOSED INVESTMENT _____

DCF / ROR _____ %

PAYOUT @ 0% _____ YRS.

DIVISION _____

PROJECT _____

_____

_____

_____

_____

PROJECT NO. _____

PREPARED BY _____

APPROVED BY _____

PROJECT NO. _____

PREPARED BY _____

APPROVED BY _____

PAYOUT CHART

CUMULATIVE PERCENT RETURN OF INVESTMENT

YEARS TO PAY OUT

INTERPOLATION CHART

DCF / ROR

RATIO X / Y

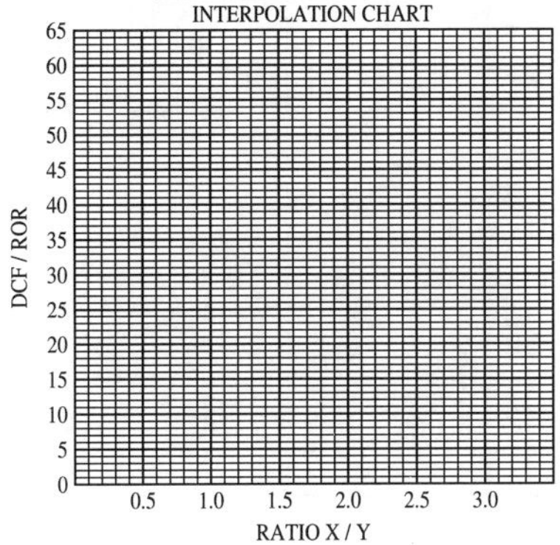

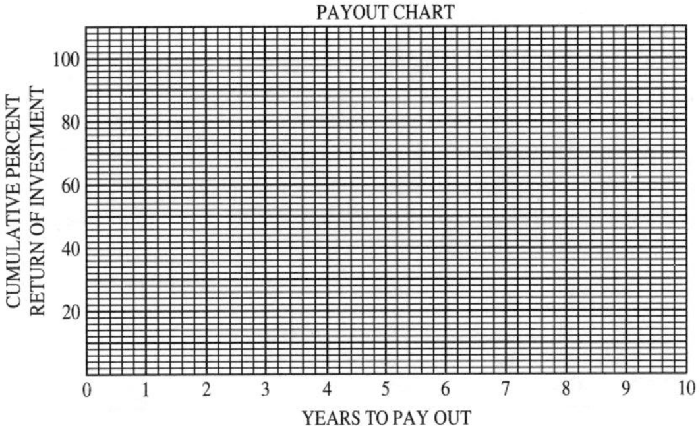

Figure 5.3

159

DCF / ROR DATA SHEET

PROJECT TITLE_____ PROJECT NO._____ DATE_____

PROJECT LIFE IS ___YEARS, DETERMINED BY_____

_____

_____

OTHER ALTERNATIVES CONSIDERED AND REASONS FOR REJECTION:_____

_____

_____

_____

EXPLANATIONS AND CALCULATIONS	REFERENCE LINE	ALTERNATIVE A TITLE_____ METHOD DESCRIPTION_____ _____ _____	REFERENCE LINE	ALTERNATIVE B TITLE_____ METHOD DESCRIPTION_____ _____ _____
DEPRECIATION DESCRIBE EXISTING/PROPOSED EQUIPMENT AND EXPLAIN DEPRECIATION	2		2	
EXPENSES (LINES 3–16) NOTE ZERO YEAR EXPENSES SUCH AS IMPLEMENTATION AND TRAINING. MAINTENANCE/ MAJOR OVERHAUL, OR ANY USUAL EXPENSES. NOTE IF ONLY DELTA COSTS OF ONE ALTERNATIVE ARE BEING SHOWN.				
INVESTMENT SHOWN DERVATION OF EXISTING ASSET VALUE AND PROPOSED CAPITAL INVESTMENT. IDENTIFY ANY GOVT. EQUIP. STATE IF OLD EQUIP TO BE KEPT.	19		19	
INVESTMENT TAX CREDIT OR REBATE SHOW PERCENT USED	20		20	
SALVAGE CALCULATE RESIDUAL AFTER TAX VALUE OF ASSETS	21		21	
OTHER CONSIDERATIONS INCLUDES INTANGIBLES, RISK AND LIKELIHOOD, OTHER ANALYSES AND RECOMMENDATION.				

Figure 5.4

A number of common factors influence the selection of the interest rate to be used. Although some firms treat the *MARR* as a parameter whose value depends on the type of investment in question, others use a fixed value for the *MARR*. Regardless of the basis used for determining the value of the *MARR*, its value should be greater than the *cost of securing capital;* likewise, it should represent the *opportunity cost* associated with investing in the candidate alternative as opposed to other available alternatives. Furthermore, care should be taken to ensure that arbitrarily specified values for the *MARR* do not counteract management's desire to maintain state-of-the-art competency.

5.5.1   The Cost of Capital

Except where other intangible benefits are involved, the discount rate should be greater than the cost of securing additional capital. Indeed it should be greater than the cost of capital by an amount that will cover the unprofitable investments that a firm must make for nonmonetary reasons. Examples of the latter would include investments in environmental compliance equipment, safety devices, and recreational facilities for employees. The discount rate that is specified establishes the firm's minimum attractive rate of return (*MARR*) in order for the investment to be justified.

It costs money to obtain money for investment. Funds for financing projects may consist of *debt capital* and *equity* (or ownership) *capital*. Most often, a major project is financed by a mix of both. In trying to establish the time value of money, it is important that we understand why money has a cost, even when it is our own money that we use.

Debt capital such as that derived from issuing bonds or securing loans results in a specific obligation to pay back both interest and principal. A short-term loan may be obtained from a bank or an insurance company. The interest paid for the use of the money is a cost of capital. Fortunately, not all of the interest is lost. Interest is deductible from income for tax purposes.[1] For example, even though we pay $I in interest during a year, that $I may be subtracted from taxable income, resulting in a tax savings of $I \times T$ where T is our tax rate. That is, the effective interest payment is $I(1 - T)$. Therefore, for a simple loan, the effective after-tax interest rate (k_l) is the product of the effective interest rate and $(1 - T)$. It is given by

$$k_l = \left[\left(1 + \frac{r}{m}\right)^m - 1\right](1 - T) \tag{5.1}$$

where r is the loan's nominal annual interest rate and m is the number of payment periods per year.

---

[1] Tax effects will be only briefly considered here. Income taxes are treated in considerable detail in Chapter Six.

EXAMPLE 5.6

A company has taken out a short-term loan with the bank at a rate of 18%/year with quarterly payments. Assuming the tax rate is 0.36, what is the cost of capital for this debt funding?

Answer:

$$k_l = \left[\left(1 + \frac{r}{m}\right)^m - 1 \right] (1 - T)$$

$$= \left[\left(1 + \frac{.18}{4}\right)^4 - 1 \right] (1 - 0.36) = 0.123$$

or 12.3%.

Debt capital can also be raised through the sale of bonds. As noted in Chapter Three, common debenture bonds have a stated face value, annual rate of interest, payment interval, and maturation period (e.g., $1000, 12%, semiannual, and 15 years). When a bond is sold, not necessarily for its face value, the firm receives money for immediate investment. In return, it gives a promise to pay interest on the face value m times per year at an annual rate r. In addition, the firm promises to repay the face value of the bond at the end of a specified period of time. Much like a loan, bond interest paid by a firm is deductible from income for tax purposes. If the amount paid for the bond equals the face value, and if any costs of preparing and selling the bond are neglected, the effective after-tax interest rate (k_b) is

$$k_b = \left[\left(1 + \frac{r}{m}\right)^m - 1 \right] (1 - T) \tag{5.2}$$

Equity capital is sometimes called *ownership capital*. One source of equity capital is the sale of stock. Money raised through a stock issue is available to the company for investment. However, there is no guarantee of a return to an investor in stock. Of course, the investor expects to receive periodic dividends and also hopes that the trading price of the stock will increase, resulting in a gain when the stock is sold to someone else in the future. Company management usually decides upon the amount of dividends to declare for stockholders after the earnings of the firm are known for the year. Management desires to seek a suitable balance between the amount of dividends paid to keep stockholders happy and the amount of earning retained in the company for investment in the future. Perhaps the most simplistic estimate of the cost of equity capital involving stock (k_s) is

$$k_s = \frac{D}{P_s} \tag{5.3}$$

where D is the dividend per share of stock and P_s is the present trading price or current value of the stock. Since dividends are paid out of after-tax earnings,

k_s is already an after-tax rate. This estimate assumes that dividends will remain constant into the future.

EXAMPLE 5.7

A company issues dividends of $6/share on stock selling at $50 per share on the financial market. What is the cost of capital for this equity funding?
   Answer:

$$k_s = \frac{D}{P_s}$$

$$= \frac{\$6}{\$50} = .12$$

or 12.0%.

Another source of equity capital is the earnings retained within the company and not declared as dividends. Even though this money appears free, there is a cost of capital. Remember, stockholders are owners of the company and therefore the retained earnings are theirs. The cost of using that money for investment carries an opportunity cost, and that is the cost of dividends forgone by the stockholders from their own earnings. For this reason, one estimate of the cost of capital for retained earnings (k_r) is

$$k_r = \frac{D}{P_s} = k_s \qquad (5.4)$$

The overall weighted average cost of capital (k_a) is simply the individual source costs of capital weighted by the fraction of total funding from that source. That is,

$$k_a = k_l p_l + k_b p_b + k_s p_s + k_r p_r \qquad (5.5)$$

where p_l, p_b, p_s, and p_r are the fractions of funding from loans, bonds, stocks, and retained earnings, respectively.

EXAMPLE 5.8

The capital structure of our firm is made up as follows:

	Funding	**Fraction**
Loans	$ 1,000,000	1/13
Bonds	1,800,000	1.8/13
Stock	8,000,000	8/13
Retained earnings	2,200,000	2.2/13
	$13,000,000	13/13

The costs of capital for these various sources are 10.4, 7.2, 12.0, and 12.0%, respectively. What is the overall weighted average cost of capital for the company?

Answer:

$$k_a = .104(1/13) + .072(1.8/13) + .120(8/13) + .120(2.2/13)$$
$$= .1121$$

or 11.21%.

The foregoing analysis implicitly assumed either zero inflation or an inflation-based cost of capital. Kaplan [34] noted that studies of returns to investors in equity and fixed-income markets during the period from 1926 to 1984 indicated that the average total return from a diversified portfolio of common stock was 11.7%/year; fixed-income securities averaged nominal before-tax returns of less than 5%/year. However, both figures included the effect of inflation over the same period. Factoring out the impact of inflation, the returns were approximately 8.5% and 1.5%, respectively. Kaplan concluded that "a mixture of debt and equity financing produces a total real cost of capital of less than 8%." He contended that one of the primary reasons American industry has tended to underinvest in capital improvements, such as automation and computer integrated manufacturing, is the use of an overly large discount rate in discounting the future cash flows anticipated from such investments.

It is important for the cost of capital to reflect the true cost of capital to the firm during the future investment period. For this reason, we prefer to determine the cost of capital *absent* inflation effects.

We briefly introduced tax and inflation considerations in order to describe more realistically the cost of capital. Further discussion of taxes will be deferred until Chapter Six where much more complete coverage is given. Likewise, in Chapter Six we build on the discussion of inflation given in Chapter Three.

5.5.2   The Minimum Attractive Rate of Return

Some firms establish a standard discount rate or minimum attractive rate of return to be used in all economy studies; others maintain a flexible posture. As one firm described it,

The XYZ Company return on investment (ratio of earnings to gross invest-ment) has been, on the average over the past 5 years, approximately equal to a rate of return of 12% per year. Accordingly, 12% per year is being established as a tentative minimum requirement for investment alternatives whose results are primarily measurable in quantitative dollar terms. The principle being applied is that such alternatives should be expected to maintain or to improve overall return-on-investment performance for the XYZ Com-pany. The minimum requirement is based on overall XYZ financial results (as opposed to results for various parts of the company) in order to avoid situations in which investment alternatives of a given level of attractiveness are unknowingly undertaken in one part of the company and rejected in another.

The 12% minimum attractive rate-of-return standard is intended as a guide, rather than as a hard-and-fast decision rule. Furthermore, it is intended to apply to alternatives having risks of the kind usually associated with investments which are primarily in plant and facilities. For alternatives involving expenditures with substantially lower risks, such as those solely for inventories or those in buy-or-lease alternatives, a lower minimum requirement may apply.

Other approaches used to establish the *MARR* include the following:

1. Add a fixed percentage to the firm's cost of capital.
2. Average rate of return over the past 5 years is used as this year's *MARR*.
3. Use different *MARR* for different planning horizons.
4. Use different *MARR* for different magnitudes of initial investment.
5. Use different *MARR* for new ventures rather than for cost-improvement projects.
6. Use as a management tool to stimulate or discourage capital investments, depending on the overall economic condition of the firm.
7. Use the average stockholder's return on investment for all companies in the same industry group.

There are a number of different approaches used by companies in establishing the discount or hurdle rate to be used in performing economic analyses. The issue is not a simple one.

Kaplan [34] observed that the use of discounted cash flow methods "most often goes wrong when companies set arbitrarily high hurdle rates for evaluating new investment projects." He noted that "the discounting function serves only to make cash flows received in the future equivalent to cash flows received now. For this narrow purpose—the only purpose, really, of discounting future cash flows—companies should use a discount rate based on the project's opportunity cost of capital (that is, the return available in the capital markets for investments of the same risk)" [34, pp. 87–88].

Although some advocate increasing the value of the *MARR* as the risks associated with the project increase, an alternative approach to dealing with risk seems preferable. Specifically, as noted in Section 5.7, risk analyses can be performed explicitly. In such cases, probabilities are assigned to the various possible outcomes of the economic investment, and, using a "normal" *MARR*, the expected present worth or the probability of the rate of return being greater than the *MARR* is determined. The *explicit* consideration of risk is preferred to the *implicit* approach of arbitrarily increasing the value of the *MARR*, based on a qualitative assessment of risk.

Regardless of how one chooses to deal with the risks associated with investing capital in anticipation of future returns, the proper determination of the discount rate is not a simple process. Even though it has been the subject of controversy for many years, in some ways we are only slightly closer to agreement today than we were 30 years ago. In many cases, a particular alternative will be preferred over a range of possible discount rates; in other cases, the alternative preferred will be quite sensitive to the discount rate used. Hence, depending upon the situation under study, it might not be necessary to specify a particular value for the discount rate—a range of possible values might suffice. We examine this situation in more depth in Section 5.7.

Subsequently, it will be convenient to refer to the interest rate, hurdle rate, or discount rate used as the *MARR* and to interpret its value using the opportunity cost concept. The argument will be made that money should not be invested in an alternative if it cannot earn a return at least as great as the *MARR*, since it is reasoned that other opportunities for investment exist that will yield returns equal to the *MARR*.

In the case of the public sector, a different interpretation is required in determining the discount rate to use. Since this chapter emphasizes economic analysis in the private sector, Chapter Seven discusses establishing the discount rate for the public sector.

5.6 COMPARING THE INVESTMENT ALTERNATIVES

We are now ready to begin comparing investment alternatives on the basis of economic considerations. This chapter has dealt with structuring investment alternatives, establishing a planning horizon, specifying the *MARR*, and defining the cash flow profiles. The time value of money and one or more methods of measuring the worth of investments, both developed in Chapter Four, may now be applied to select the preferred alternative. Recall that the measures of investment worth include (1) present worth, (2) annual worth, (3) future worth, (4) internal rate of return, (5) external rate of return, (6) savings/investment ratio, (7) payback period, and (8) capitalized worth.

Up to this point, there has been little mention of income taxes and depreciation. Certainly, these affect the cash flow patterns of our alternatives. In order to simplify the comparison of investment alternatives, however, we will assume that either the cash flows are after-tax cash flows or a before-tax study is desired. In Chapter Six we address the subject of income taxes and their effects on the preference among investment alternatives.

To illustrate the use of the various methods of comparing investment alternatives, consider the four alternatives having the cash flow profiles given in Table 5.7 for a planning horizon of 5 years. Note that to do nothing is presented here

Table 5.7 Cash Flow Profiles for Four Investment Alternatives

| End of year, t | Net Cash Flows for Alternatives | | | |
	A_{0t}	A_{1t}	A_{2t}	A_{3t}
0	$0	$ 0	−$50,000	−$75,000
1	0	4,500	20,000	20,000
2	0	4,500	20,000	25,000
3	0	4,500	20,000	30,000
4	0	4,500	20,000	35,000
5	0	4,500	20,000	40,000

as an explicit alternative. A minimum attractive rate of return of 12% will be used in subsequent analyses.

5.6.1 Present Worth Method

The *present worth method* can be used to compare investment alternatives by simply computing the present worth of each investment alternative and recommending the one having the greatest present worth. Alternatively, you can compute the present worth of each *increment of investment* and select or reject "increments" depending upon whether or not they have positive or negative present worths. A flowchart of the incremental approach is given in Figure 5.5.

EXAMPLE 5.9

Ranking Approach

For the data given in Table 5.7, the present worth of Alternative 0, the do-nothing alternative, is zero. For Alternative 1, the present worth is

$$PW_1(12\%) = \$4500(P|A\ 12,5)$$
$$= \$4500(3.6048)$$
$$= \$16,222$$

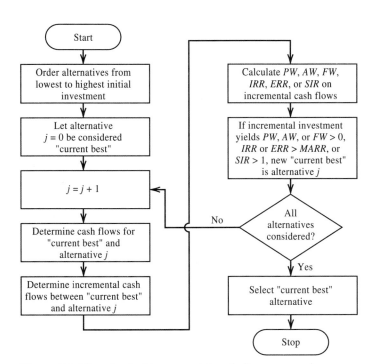

Figure 5.5 Flowchart of incremental cash flow approach.

For Alternative 2, the present worth is

$$PW_2(12\%) = -\$50,000 + \$20,000(P|A\ 12,5)$$
$$= -\$50,000 + \$20,000(3.6048)$$
$$= \$22,096$$

For Alternative 3, the present worth is

$$PW_3(12\%) = -\$75,000 + \$20,000(P|A\ 12,5) + \$5000(P|G\ 12,5)$$
$$= -\$75,000 + \$20,000(3.6048) + \$5000(6.3970)$$
$$= \$29,081$$

Since Alternative 3 has the greatest present worth, it would be recommended.

Incremental Approach

Using an incremental approach, since no other alternative requires a smaller initial investment than Alternative 0, it will be used as the base for incremental comparison. Based on a rank ordering of alternatives by the size of the initial investment required, the first incremental comparison will be between Alternatives 0 and 1. If the present worth of the incremental difference is positive, then Alternative 1 will be preferred to Alternative 0, and it will become the new base for incremental comparison; if the present worth of the incremental difference is negative, then Alternative 0 will remain the base, and Alternative 1 will be discarded from further consideration.

$$PW_{1-0}(12\%) = \$4500(P|A\ 12,5)$$
$$= \$4500(3.6048)$$
$$= \$16,222$$

Since $PW_{1-0} > 0$, Alternative 1 is preferred to Alternative 0. Next, we look at the increment of investment required to go from Alternative 1 to Alternative 2. The present worth of the incremental investment will be

$$PW_{2-1}(12\%) = -\$50,000 + \$15,500(P|A\ 12,5)$$
$$= \$5,874$$

Since $PW_{2-1} > 0$, Alternative 2 is preferred to Alternative 1. Next, we look at the incremental investment needed to go to Alternative 3 from Alternative 2. The incremental present worth will be

$$PW_{3-2}(12\%) = -\$25,000 + \$5000(P|G\ 12,5)$$
$$= \$6,985$$

Since $PW_{3-2} > 0$, Alternative 3 is preferred to Alternative 2. Since no additional alternatives are available, we conclude that Alternative 3 is the most economic choice.

Notice that the incremental approach always compares the "next available alternative," based on rank ordering by initial investment, against the best alternative available to date—what we called the *base alternative*. If, in the example, PW_{2-1}

had been less than zero, then Alternative 3 would have been compared to Alternative 1; if the incremental present worth equals zero, then we are indifferent between the two alternatives.

5.6.2 Annual Worth Method

The *annual worth method* is used in the same way as the present worth method, except annual worths are computed rather than present worths. It would be used when it is preferable to think in terms of annual worths or dollars per year.

EXAMPLE 5.10

Ranking Approach

For the data given in Table 5.7, the annual worth of Alternative 0, the do-nothing alternative, is zero. For Alternative 1, the annual worth is $4500. For Alternative 2, the annual worth is

$$AW_2(12\%) = -\$50,000(A|P\,12,5) + \$20,000$$
$$= -\$50,000(0.2774) + \$20,000$$
$$= \$6,130/\text{year}$$

For Alternative 3, the annual worth is

$$AW_3(12\%) = -\$75,000(A|P\,12,5) + \$20,000 + \$5000(A|G\,12,5)$$
$$= -\$75,000(0.2774) + \$20,000 + \$5000(1.7746)$$
$$= \$8,068/\text{year}$$

Alternative 3 is recommended since it has the greatest annual worth.

Incremental Approach

$$AW_{1-0}(12\%) = \$4500 > 0 \quad \text{(Prefer Alternative 1 to 0)}$$
$$AW_{2-1}(12\%) = -\$50,000(A|P\,12,5) + \$15,500$$
$$= \$1,630/\text{year} > 0 \quad \text{(Prefer Alternative 2 to 1)}$$
$$AW_{3-2}(12\%) = -\$25,000(A|P\,12,5) + \$5000(A|G\,12,5)$$
$$= \$1,938/\text{year} > 0 \quad \text{(Prefer Alternative 3 to 2)}$$

Since no additional alternatives are available, we conclude that Alternative 3 is the most economic choice.

5.6.3 Future Worth Method

The *future worth method* is used in the same way as the present worth and annual worth methods, except future worths are computed instead of present worths and annual worths. Although it is easy to apply, it is not used as often as the present worth and annual worth methods for economic analyses of investment alternatives.

EXAMPLE 5.11

Ranking Approach

For the data given in Table 5.7, the future worth of Alternative 0, the do-nothing alternative, is zero. For Alternative 1, the future worth is $28,588.

$$FW_1(12\%) = \$4500(F|A\ 12,5)$$
$$= \$28,588$$

For Alternative 2, the future worth is

$$FW_2(12\%) = -\$50,000(F|P\ 12,5) + \$20,000(F|A\ 12,5)$$
$$= \$38,941$$

For Alternative 3, the future worth is

$$FW_3(12\%) = -\$75,000(F|P\ 12,5) + \$20,000(F|A\ 12,5)$$
$$+ \$5000(P|G\ 12,5)(F|P\ 12,5)$$
$$= \$51,251$$

We prefer Alternative 3 since it has the greatest future worth.

Incremental Approach

$$FW_{1-0}(12\%) = \$28,588 > 0 \quad \text{(Prefer Alternative 1 to 0)}$$
$$FW_{2-1}(12\%) = -\$50,000(F|P\ 12,5) + \$15,500(F|A\ 12,5)$$
$$= \$10,353 > 0 \quad \text{(Prefer Alternative 2 to 1)}$$
$$FW_{3-2}(12\%) = -\$25,000(F|P\ 12,5) + \$5000(P|G\ 12,5)(F|P\ 12,5)$$
$$= \$12,310 > 0 \quad \text{(Prefer Alternative 3 to 2)}$$

Alternative 3 is the last surviving alternative; therefore, it is preferred.

5.6.4 Internal Rate-of-Return Method

The *internal rate-of-return method* is one of the most popular, yet misused, methods of comparing alternatives. Although we recommend using an incremental cash flow approach in applying the *IRR* method, an alternative approach is possible; it is referred to as the *aggregate cash flow approach*.

The philosophy underlying the aggregate cash flow approach is to maximize the aggregate (or total) return obtained by investing in one of the investment alternatives *and the "reserve account."* The reserve account consists of the uninvested funds, which are assumed to be earning a return at a rate equal to the *MARR*.

EXAMPLE 5.12

To illustrate the difference in the aggregate cash flow approach and the incremental cash flow approach, suppose two investment alternatives, A and B, are being considered: A requires an investment of $10,000, and B requires an investment of $50,000. Suppose A will return $2000/year forever; hence, an *IRR* of 20% is obtained. Suppose B will return $8000/year forever; in this case, an *IRR* of 16% is obtained. Now, suppose the *MARR* is 12%.

The aggregate cash flow approach proceeds in the following way. For B to be a feasible investment alternative, $50,000 must be available for investment. Investing in A means that $10,000 will return $2000/year and the reserve account will return 12% of the difference in the amount invested and the amount available for investment (i.e., (0.12)($40,000) or $4800/year). By investing in A, the aggregate return will be $6800/year from an aggregate investment of $50,000. The aggregate rate of return will be 13.6%. By investing the full amount in B, the aggregate rate of return will be 16%. In this case, B is preferred because it has the greatest aggregate rate of return.

Using an incremental cash flow approach, we begin with the alternative requiring the smallest initial investment, A, and compute its *IRR*; if it is greater than the *MARR*, then A is an acceptable base for incremental comparison. In this case, $IRR_A = 20\% > MARR$; so, A is an acceptable base. An incremental investment of $40,000 is required to step up to B from A; the incremental returns resulting from the incremental investment will be $6000/year; thus, the incremental *IRR* will be 15%, which is greater than the *MARR* of 12%. Hence, B is preferred to A. (From the results obtained, it should be obvious that A would have been preferred if the *MARR* had been 15% or greater.)

Now, let us apply the incremental cash flow approach in selecting the preferred alternative from among those described in Table 5.7.

EXAMPLE 5.13

Between Alternatives 1 and 0, there is no incremental investment, but there are incremental returns of $4500 in each of years 1 to 5. The incremental *IRR* is therefore $i_{1-0} = \infty\%$ (Prefer 1 over 0).

Between Alternatives 2 and 1,

$$FW_{2-1}(i) = \$0 = -\$50,000(F|P\,i,5) + \$15,500(F|A\,i,5)$$
$$@i = 15\% \quad \$0 \neq -\$50,000(2.0114) + \$15,500(6.7424) = \$3937.20$$
$$@i = 20\% \quad \$0 \neq -\$50,000(2.4883) + \$15,500(7.4416) = -\$9070.20$$

Interpolating, $i_{2-1} = 16.51\% > MARR$ of 12% (Prefer 2 over 1).

Between Alternatives 3 and 2.

$$FW_{3-2}(i) = \$0 = -\$25,000(F|P\,i,5) + \$5,000(P|G\,i,5)(F|P\,i,5)$$
$$@i = 18\% \quad \$0 \neq -\$25,000(2.2878) + \$5,000(5.2312)(2.2878) = \$2644.70$$
$$@i = 20\% \quad \$0 \neq -\$25,000(2.4883) + \$5,000(4.9061)(2.4883) = -\$1168.26$$

Interpolating, $i_{3-2} = 19.39\% > 12\%$ *MARR* (Prefer 3 over 2).

We again prefer Alternative 3 because it prevailed in the pairwise comparison of alternatives in terms of IRR on incremental cash flows. Interestingly, the IRR for Alternative 2 is 28.65%, and for Alternative 3 it is 24.89%. Hence, although Alternative 3 did not have the highest incremental IRR, it is the preferred alternative. In the pairwise comparison, the IRR of 19.35% on the incremental investment required to "step up" from Alternative 2 to Alternative 3 is greater than the MARR of 12%, and that is all that counts!

As mentioned earlier, decision makers often use the *IRR* method incorrectly. They compute the *IRR* for individual alternatives and choose the one having the greatest rate. In so doing, they ignore the fact that any remaining funds will earn only the *MARR*. In the case of the above example, an infinite return is earned on Alternative 1 since no investment is required; the individual rate of return for Alternative 2 is approximately 28.73%; and the individual rate of return for Alternative 3 is approximately 24.90%. Thus, by ranking alternatives on the basis of their individual rates of return, Alternative 1 would be preferred to Alternative 2, which would be preferred to Alternative 3. On the other hand, by examining the incremental returns and investing so long as each increment is justified, we found (correctly) that Alternative 3 was the most profitable alternative.

The incremental cash flow approach is recommended when using the *IRR* method. It is not only correct, but much easier to use than the aggregate cash flow approach.

5.6.5 External Rate-of-Return Method

The *external rate-of-return method* is rarely used, but it has advantages over the *IRR* method.

EXAMPLE 5.14

Let us now compare the alternatives of Example 5.9, Table 5.7 using the external rate-of-return method and incremental cash flows approaches. The positive net cash flows will be assumed to be reinvested at the $MARR = 12\%$ and Equation 4.15 will be employed.

Between Alternatives 1 and 0, there is no incremental investment, but there are incremental returns of $4500 in each of years 1 to 5. The incremental *ERR* is therefore $i'_{1-0} = \infty\%$ (Prefer 1 over 0).

Between Alternatives 2 and 1,

$$\$15,500(F|A\ 12,5) = \$50,000(1 + i')^5$$
$$\$15,500(6.3528) = \$50,000(1 + i')^5$$

Solving, $i'_{2-1} = 14.52\% > MARR$ (Prefer 2 over 1).

Between Alternatives 3 and 2,

$$\$5000(P|G\ 12,5)(F|P\ 12,5) = \$25,000(1 + i')^5$$
$$\$5000(6.3970)(1.7623) = \$25,000(1 + i')^5$$

Solving, $i' = 17.66\% > MARR$ (Prefer 3 over 2).

Again, Alternative 3 is preferred on the basis of its winning the series of pairwise comparisons using the ERR method and the incremental cash flow approach. As with the IRR method, Alternative 3 does not have a larger ERR than Alternative 2; computing the individual ERR values yields 20.51% for Alternative 2 and 19.59% for Alternative 3. Again, what counts is the rate of return on the incremental investment required to step up from Alternative 2 to Alternative 3; since the ERR of 17.66% on the $25,000 incremental investment is greater than the 12% MARR, we are far better off investing the $25,000 and obtaining Alternative 3 than leaving it in the "opportunity cost pool" earning only 12%.

5.6.6   Savings/Investment Ratio Method

The *savings/investment ratio method* is most often used in public sector projects; it is also known as the benefit/cost ratio.

EXAMPLE 5.15

We will now analyze the four alternatives of Example 5.9, Table 5.7, using the *SIR* formulation given in Equation 4.11. Again, the *MARR* = 12%.

Between Alternatives 1 and 0, there is no incremental investment, but there are incremental returns of $4500 in each of the years 1 to 5. The incremental *SIR* is therefore $SIR_{1-0}(12\%) = \infty$ (Prefer 1 to 0).

Between Alternatives 2 and 1,

$$SIR_{2-1}(12\%) = \frac{\$15,500(P|A\ 12,5)}{\$50,000}$$

$$= \frac{\$15,500(3.6048)}{\$50,000}$$

$$= 1.12 > 1.00 \quad \text{(Prefer 2 to 1)}$$

Between Alternatives 3 and 2,

$$SIR_{3-2}(12\%) = \frac{\$5,000(P|G\ 12,5)}{\$25,000}$$

$$= \frac{\$5,000(6.3970)}{\$25,000}$$

$$= 1.28 > 1.00 \quad \text{(Prefer 3 to 1)}$$

Alternative 3 is preferred since it won the final pairwise comparison by having an *SIR* greater than 1.00 on the incremental investment.

5.6.7   Payback Period Method

The *payback period method* is not consistent with the previous six methods presented, and as such may yield entirely different selections or rankings of alternatives. Neither the total nor the incremental cash flow approach as presented is applicable. The PBP method is recommended for use only as an auxilliary or secondary criterion in alternative selection.

EXAMPLE 5.16

Let us now use the payback period method to analyze the alternatives presented in Example 5.9, Table 5.7.

For Alternative 0, it is not meaningful to talk about payback. Likewise, since Alternative 1 requires no investment, its payback is instantaneous. For Alternatives 2 and 3, as shown in Table 5.8, the payback period is 3 for each. Likewise, the discounted payback period is 4 for each. If you assume the cash flows are distributed uniformly over each year, then fractional values can be obtained by using interpolation.

5.6.8   Capitalized Worth Method

The *capitalized worth method* is applicable only if there is reason to believe that a series of cash flows will repeat indefinitely into the future. As such, it is not a method comparable to those used previously in this section. Neither the total nor the incremental cash flow approach as presented is applicable.

EXAMPLE 5.17

Let us assume that the cash flows are presented in Example 5.9, Table 5.7 actually do repeat indefinitely into the future. We will continue using the *MARR* of 15%. Also, we will make use of the annual worths calculated in Example 5.10.

For Alternative 0,

$$CW_0(12\%) = \$0$$

For Alternative 1,

$$CW_1(12\%) = \frac{AW_1(12\%)}{.12}$$

$$= \frac{\$4500}{.12}$$

$$= \$37,500$$

For Alternative 2

$$CW_2(12\%) = \frac{AW_2(12\%)}{.12}$$

$$= \frac{\$6,130}{.12}$$

$$= \$51,083$$

Table 5.8 Computing Payback Period and Discounted Payback Period for Example 5.16

Time (t)	A_{2t}	Cumulative Cash Flow	Cumulative Present Worth	A_{3t}	Cumulative Cash Flow	Cumulative Present Worth
0	($50,000)	($50,000)	($50,000)	($75,000)	($75,000)	($75,000)
1	$20,000	($30,000)	($32,609)	$20,000	($55,000)	($57,609)
2	$20,000	($10,000)	($17,486)	$25,000	($30,000)	($38,705)
3	$20,000	$10,000	($4,335)	$30,000	$0	($18,980)
4	$20,000	$30,000	$7,100	$35,000	$35,000	$1,032
5	$20,000	$50,000	$17,043	$40,000	$75,000	$20,919
PBP =		3 yrs			3 yrs	
DPBP =			4 yrs			4 yrs

For Alternative 3,

$$CW_3(12\%) = \frac{AW_3(12\%)}{.12}$$

$$= \frac{\$8,068}{.12}$$

$$= \$67,233$$

We prefer Alternative 3 on the basis that it has the highest capitalized worth. Had we been calculating the capitalized cost where costs are positive, we would have preferred the smallest value.

5.7 PERFORMING SUPPLEMENTARY ANALYSES

The sixth step in performing economic comparisons of engineering investment alternatives is to perform supplementary analyses. In the previous sections we assumed all of the values of the parameters of the economic models were known with certainty. In particular, correct estimates of the values for the length of the planning horizon, the minimum attractive rate of return, and each of the individual cash flows were assumed to be available. In this section we consider the consequences of estimating the parameters incorrectly.

The discussion will concentrate on answering a number of "what if . . ." questions concerning the effects of different parameter values on the measure of economic effectiveness of interest. For example, when we are completely *uncertain* of the possible values a parameter can take on, we will be interested in determining the set of values for which an investment alternative is justified economically and the set of values for which an alternative is not justified; this process is called *break-even analysis*. Alternatively, when we are reasonably sure of the possible values a parameter can take on, but *uncertain* of their chances of occurrence, we will be interested in the sensitivity of the measure of merit to various parameter values; this process is referred to as *sensitivity analysis*. Finally, when probabilities can be assigned to the occurrence of the various values of the parameters, we can make probability statements concerning the values of the measure of merit for the various alternatives; this process is referred to as *risk analysis*.

The conditions that lead to break-even and sensitivity analyses are described in the economic analysis literature as conditions under *uncertainty*, since one is completely uncertain of the chances of a parameter taking on a given value. When the conditions are such that probabilities can be assigned to the various parameter values, the decision environment that results is said to be a decision under *risk*; hence, the term risk analysis is used.

One of the most perplexing (and frustrating) experiences for the engineering economic analyst is to encounter an individual who refuses to analyze systematically the economic performances of investment alternatives because "it's not possible to estimate future cash flows exactly." Two responses come to mind. First, failure to analyze the alternatives systematically implies a qualitative judgment will be made, based on even more imperfect information than would be available using a systematic approach. Second, the selection decision might not require precise estimates of the cash flows. In the latter case, the selection decision might be relatively insensitive to a broad range of possible values for the cash flows.

The purpose of break-even and sensitivity analyses is to *reduce* the amount of information needed to make a good decision. Instead of needing a "point" estimate for, say, the salvage value of a piece of equipment, an "interval" estimate might be sufficient. Break-even analysis can reduce the forecasting requirement to determining if the magnitude of a cash flow will exceed a specific value.

It is important to recognize that break-even, sensitiviy, and risk analyses are intended to make more realistic the economic comparison of investment alternatives. Without using such approaches, one must make decisions under the assumption of perfect information regarding the future.

5.7.1   Break-Even Analysis

Although we did not label the process as such, in Chapter Two we performed a number of break-even analyses; there we referred to the process as *equivalence.* In particular, in presenting the concept of equivalence, a situation was posed, and you were asked to determine the value of a particular parameter in order for two cash flow profiles to be equivalent. Another way of stating the problem could have been "Determine the value of X that will yield a break-even situation between the two alternatives." In this case, X denotes the parameter whose value is to be determined. Additionally, when the cash flow profiles for two alternatives are equivalent, a break-even situation can be said to exist between the two alternatives.

Another application of break-even analysis occurs in the use of the rate-of-return method. Specifically, the internal rate of return can be interpreted as the break-even value of the reinvestment rate, since such a reinvestment rate will yield a zero future worth for either an individual alternative or the differences in two alternatives.

Break-even analysis is certainly not an unfamiliar concept. Furthermore, the information obtained from a break-even analysis can be of considerable aid to anyone faced with an investment alternative involving a degree of uncertainty concerning the value of some parameter. The term *break-even* is derived from the desire to determine the value of a given parameter that will result in neither a profit nor a loss.

EXAMPLE 5.18

Suppose a firm is considering manufacturing a new product and the following data have been provided:

Sales price	$12.50/unit
Equipment cost	$200,000
Overhead cost	$50,000/year
Operating and maintenance cost	$25/operating hour
Production time/1000 units	100 hours
Planning horizon	5 years
Minimum attractive rate of return	15%

Assuming a zero salvage value for all equipment at the end of 5 years, and letting X denote the annual sales for the product, the annual worth for the investment alternative can be determined as follows:

$$AW(15\%) = -\$200,000(A\,|\,P\,15,5) - \$50,000 - 0.100(\$25)X + \$12.50X$$
$$= -\$109,660 + \$10.00X$$

Solving for X yields a *break-even* sales value of 10,966/year. If it is felt that annual sales of at least 10,966 units can be achieved each year, then the alternative appears to be worthwhile economically. Even though one does not know with certainty how many units of the new product will be sold annually, it is felt that the information provided by the break-even analysis will assist management in deciding whether or not to undertake the new venture.

A graphical representation of the example is given in Figure 5.6. The chart is referred to as a *break-even* chart, since one can determine graphically the *break-even point* by observing the value of X when annual revenue equals annual cost.

EXAMPLE 5.19

Consider a contractor who experiences a seasonal pattern of activity for compressors. The manager currently owns eight compressors and suspects that this number will not be adequate to meet the demand. The contractor realizes that there will arise situations when more than eight compressors will be required, and is considering purchasing an additional compressor for use during heavy demand periods.

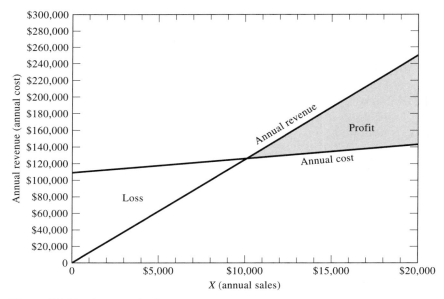

Figure 5.6 Break-even chart.

A local equipment rental firm will rent compressors at a cost of $50/day. Compressors can be purchased for $6000. The difference in operating and maintenance costs between owned and rented compressors is estimated to be $3000/year.

Letting X denote the number of days a year that more than eight compressors are required, the following break-even analysis is performed. A planning horizon of 5 years, zero salvage values, and 20% minimum attractive rate of return are assumed.

Annual worth (purchasing compressor)

$$AW_1(20\%) = -\$6000(A \,|\, P\, 20,5) - \$3000$$
$$= -\$5006.40$$

Annual worth (renting compressor)

$$AW_2(20\%) = -\$50X$$

Setting the annual worths equal for the two alternatives yields a *break-even value* of $X = 100.128$ days/year.

Hence, if the contractor anticipates that a demand will exist for an additional compressor more than 100 days/year over the next 5 years, then an additional compressor should be purchased.

5.7.2 Sensitivity Analysis

Break-even analysis is normally used when an accurate estimate of a parameter cannot be provided, but intelligent judgments can be made as to whether or not the parameter's value is less than or greater than some break-even value. Sensitivity analysis is used to analyze the effects of making errors in estimating parameter values.

Although the analysis techniques employed in break-even analysis and sensitivity analysis are quite similar, there are some subtle differences in the objectives of each. Because of the similarities in the two, it is not uncommon to see the terms used interchangeably.

EXAMPLE 5.20

To illustrate what we mean by a sensitivity analysis, consider the investment alternative depicted in Figure 5.7. The alternative is to be compared against the do-nothing alternative, which has a zero present worth. If errors are made in estimating the size of the required investment ($10,000), the magnitude of the annual receipts ($3000), the during of the project (5 years), the minimum attractive rate of return (12%), and/or the form of the series of receipts (uniform versus nonuniform), then the economic desirability of the alternative might be affected. It is anticipated that the future states (possible values) for each parameter will be contained within an interval having a range from -40 to $+40\%$ of

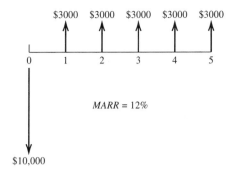

Figure 5.7 Cash-flow diagram.

the initial estimate. Hence, in the uncertain environment we have defined, an infinite number of future states are to be considered in the continuum from -40 to $+40\%$ of the initial estimate.

If it is assumed that all estimates are correct except the estimate of annual receipts, the annual worth for the alternative can be given as

$$AW(12\%) = -\$10,000(A\,|\,P\ 12,5) + \$3000(1 + X)$$

where X denotes the percent error (decimal equivalent) in estimating the value for annual receipts. Plotting annual worth as a function of the percent error in estimating the value of annual receipts yields the straight line having positive slope in Figure 5.8. Performing similar analyses for the initial investment required, the duration of the investment (planning horizon) and the minimum attractive rate of return yields the results given in Figure 5.8.

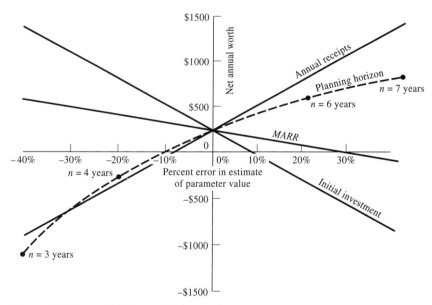

Figure 5.8 Deterministic sensitivity analysis.

As shown in Figure 5.8, the net annual worth for the investment is affected differently by errors in estimating the values of the various parameters. The net annual worth is relatively insensitive to changes in the minimum attractive rate of return; in fact, as long as the *MARR* is less than approximately 15.25%, the investment will be recommended. A *break-even* situation exists if either the annual receipts decrease by approximately 7.47% to $2774/year or the required investment increases by approximately 8.14%, or $10,814. If the project life is 4 years or less, then the investment will not be profitable.

The analysis depicted in Figure 5.8 examines the sensitivity of individual parameters one at a time. In practice, estimation errors can occur for more than one parameter. In such a situation, instead of a sensitivity curve, a sensitivity surface is needed.

EXAMPLE 5.21

Consider an investment alternative involving the modernization of a warehousing operation in which automated storage and retrieval equipment is to be installed in a new warehouse facility. The building has a projected life of 30 years, and the equipment has a projected life of 15 years. A minimum attractive rate of return of 15% is to be used in the economic analysis. It is anticipated that the new warehouse system will require 70 fewer employees than the present system. Each employee costs approximately $18,000/year. The building is estimated to cost $2,500,000, and the equipment is estimated to cost $3,500,000. Annual operating and maintenance costs for the building and equipment are estimated to be $150,000/year more than the current operation. Existing equipment and buildings not included in the new warehouse have terminal salvage values totaling $600,000. Investing in the new warehouse will negate the need to replace existing equipment in the future; the present-worth savings in replacement cost is estimated to be $200,000. The estimated cash flow profile for the investment alternative is given in Table 5.9.

A 30-year planning horizon is to be employed in the analysis. Since equipment life is estimated to be 15 years, it is assumed that identical replacement equipment will be purchased after 15 years. Futhermore, since constant worth dollar estimates are being used in contending with inflationary effects, it is assumed that the replacement equipment will have cash flows that are identical to those that occur during the first 15 years.

The architectural and engineering estimate of $2,500,000 for the building is believed to be quite accurate, as are the estimates of $150,000 for annual operating and maintenance costs, terminal salvage values totaling $600,000, and savings of $200,000 in replacement costs. However, it is felt that the estimate of $3,500,000 for equipment and the labor savings estimate of 70 employees are subject to error. A sensitivity analysis for these two parameters is to be performed on a before-tax basis.

Letting x denote the percent error in the estimate of equipment cost and y denote the percent error in estimating the annual labor savings, it can be seen

Table 5.9 Estimated Cash Flows for Warehouse System, Example 5.21

EOY	Building	Equipment	Labor Savings	Operating and Maintenance	Salvage[a]	Total
0	−$2,500,000	−$3,500,000			$800,000	−$5,200,000
1–15			$1,260,000	−$150,000		1,110,000
15		− 3,500,000				− 3,500,000
16–30			1,260,000	− 150,000		1,110,000
30	0	0				0

[a] Includes present worth of savings in equipment replacement.

that the warehouse modernization will be justified economically, if

$$PW = -\$2,500,000 - \$3,500,000(1 + x)[1 + (P|F\,15,15)]$$
$$-\$150,000(P|A\,15,30) + \$18,000(70)(1 + y)(P|A\,15,30)$$
$$+\$800,000 \geq 0$$

or if

$$PW = \$1,658,110 - \$3,930,150x + \$8,273,160y \geq 0$$

Solving for y gives

$$y \geq -0.2004 + 0.47505x$$

Plotting the equation, as shown in Figure 5.9, indicates the favorable region ($PW > 0$) lies above the break-even line, and the unfavorable region ($PW < 0$) lies below the break-even line.

If no errors are made in estimating the equipment cost, that is, $x = 0$, then up to a 20.04% reduction in annual labor savings can be tolerated. Likewise, if no errors are made in estimating the annual labor savings, that is, $y = 0$, then up to a 42.185% increase in equipment cost could occur, and the warehouse modernization would continue to be justified economically.

Since very little of the ±20% estimation error zone results in a negative present worth, it appears that the recommendation to modernize the warehouse is insensitive to errors in estimating either equipment cost, labor savings, or both. However, the decision is more sensitive to errors in estimating labor savings than in estimating equipment costs.

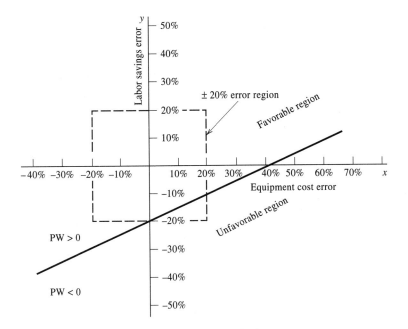

Figure 5.9 Multiparameter sensitivity analysis.

(This particular example illustrates the difficulties in distinguishing between break-even and sensitivity analyses. However, no matter what you call it, the analysis can be quite beneficial in gaining added understanding of the possible outcomes associated with a new venture involving a large capital investment.)

5.7.3 Risk Analysis

Risk analysis will be defined as the process of incorporating explicitly random variation in the estimates of measures of merit for an investment proposal. Using either analytical or simulation approaches, either exact values or estimates will be developed for the expected value and standard deviation of the measure of merit; also, the probabilities of present worth, annual worth, or future worth being greater than zero and the internal rate of return or external rate of return being greater than the MARR are of interest in risk analyses.

The magnitudes of cash flows, the duration of the planning horizon, and the value of the minimum attractive rate of return are candidates for probabilistic estimates. The cash flow occurring in a given year is often a function of a number of other factors such as selling prices, size and share of the market, market growth rate, investment required, inflation rate, tax rates, operating costs, fixed costs, and salvage values of all assets. The values of a number of these random variables can be correlated with each other, as well as autocorrelated.[1] Consequently, an analytical development of the probability distribution for the measure of merit is not easily achieved in most real-world situations. Thus simulation is widely used in performing risk analyses.

5.7.3.1 *Distributions*

The risk analysis procedure was developed to take into consideration the imprecision in estimating the values of the inputs required in making economic evaluations. The imprecision is represented in the form of a probability distribution.

Probability distributions for the random variables are usually developed on the basis of subjective probabilities. Typically, the further an event is into the future, the less precise is our estimate of the value of the outcome of the event. Hence, by letting the variance reflect our degree of precision, we would expect the variance of the probability distributions to increase with time.

Among the theoretical probability distributions commonly used in risk analysis are the normal distribution and the beta distribution. Examples of these distributions are depicted in Figures 5.10 and 5.11. For a discussion of a number of distributions and their process generators in the context of simulation, see [3, 42, 53, and 54], among others.

In some situations the subjective probability distribution cannot be represented accurately using a well-known theoretical distribution. Instead, one must estimate directly the probability distribution for the random variable.

One approach that can be used to estimate the subjective probability distribution is to provide optimistic, pessimistic, and most likely estimates for the random variable. The optimistic and pessimistic values should be ones that you do not anticipate will be exceeded with a significant probability (e.g., 1% chance). Given

---

[1] *Autocorrelation* means correlated with itself over time.

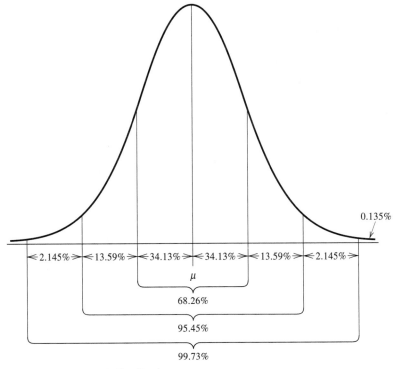

0.135%

2.145%   13.59%   34.13%   34.13%   13.59%   2.145%

μ

68.26%

95.45%

99.73%

Figure 5.10 Normal distribution.

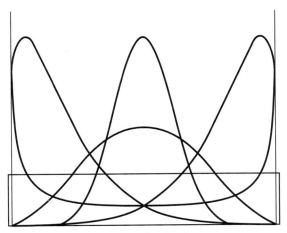

Figure 5.11 Sample beta distributions.

the practical limits on the range of values anticipated for the random variable, an estimate is provided of the chance that the most likely estimate will not be exceeded. Next, a smooth curve is passed through the three points obtained; the resulting cumulative distribution function for the random variable is divided into an appropriate number of intervals and the individual probabilities estimated.

EXAMPLE 5.22

To illustrate the process, suppose we wish to develop the probability distribution for, say, the anticipated salvage value of a machine in 5 years. We estimate the salvage value will range from $0 to $3000, with the most likely value being $1250. We estimate that there is a 40% chance of salvage value being less than $1250. Since we believe the extreme values of $0 and $3000 are not likely to occur, we use an S-shaped curve to represent the cumulative distribution function, as depicted in Figure 5.12a. Using intervals of $500, the cumulative distribution function is transformed into the probability distribution function shown in Figure 5.12b. Letting the midpoints of the intervals represent the probabilities associated with the intervals yields the probability mass function given in Figure 5.12c. Depending on the use to be made of the probabilities obtained, either of the three representations of the probability distribution could be used.

The process described above is quite subjective, since a number of different curves could be used to describe the cumulative distribution function. However, the probability distribution itself is subjectively based. What is sought is a probability distribution that best describes one's beliefs about the outcomes of the random variable. If the probability distribution obtained above does not reflect your beliefs, the process should be repeated until a satisfactory probability distribution is obtained.

The critics of risk analysis cite the degree of subjectivity involved in developing probability distributions. However, it is argued by those who favor the technique that the only alternative to using subjectively based probabilities is to use the traditional single-estimate approach, which implies that conditions of certainty exist. Interestingly, the final distribution obtained for the measure of merit is often quite insensitive to deviations in the shapes of the distributions for the parameters.

5.7.3.2 *Risk Aggregation*

Given the essential factors and their associated probability distributions, we are in a position to aggregate the distributions and obtain the probability distribution for the measure of merit. Three measures of merit have been mentioned: present worth, annual worth, and rate of return. In practice, a combination of the rate of return and either the present-worth or annual-worth measures of merit are often used. Each method of comparing investment alternatives has its limitations. For example, if the rate of return is being determined, the explicit reinvestment

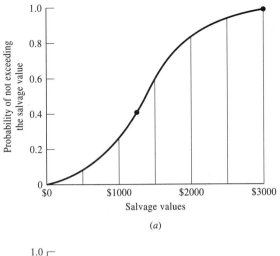

(a)

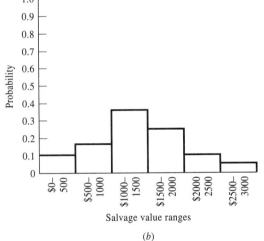

Salvage value ranges

(b)

Salvage value	Probability estimate
$ 250	0.10
$ 750	0.15
$1250	0.35
$1750	0.25
$2250	0.10
$2750	0.05

(c)

Figure 5.12 Developing the subjective probability distribution.

rate of return method is recommended, since multiple rates of return can arise when using the internal rate of return method.

Risk aggregation is achieved in basically two ways: by using simulation and analytically. Analytic approaches can be used in a number of simple cases. For more complex situations involving a large number of variables, simulation is used.

Simulation Approaches: Simulation, in the general sense, may be thought of as performing experiments on a model. Basically, simulation is an "if . . . , then . . ." device (i.e., *if* a certain input is specified, *then* the output can be determined). Some of the major reasons for using simulation in risk analysis follow:

1. Analytic solutions are impossible to obtain without great difficulty.
2. Simulation is useful in selling a system modification to management.
3. Simulation can be used as a verification of analytical solutions.

4. Simulation is very versatile.
5. Less background in mathematical analysis and probability theory is generally required.

Some of the major disadvantages of simulation are:

1. Simulations can be quite time consuming.
2. Simulations introduce a source of randomness not present in analytic solutions (sampling error).
3. Simulations do not reproduce the input distributions exactly (especially the tails of the distribution).
4. Validation is easily overlooked in using simulation.
5. Simulation is so easily applied it is often used when analytic solutions can be easily obtained at considerably less cost.

EXAMPLE 5.23

To illustrate the simulation approach in performing a risk analysis, consider an investment of $10,000 over a 4-year period that returns R_t at the end of year t, with R_t being a statistically independent random variable. Using a *MARR* of 20% it is desired to use simulation to estimate the probability of the investment being unprofitable (or risky!). The following probability distribution is assumed for R_t.

R_t	Probability	Random Number
$2,000	0.10	0
3,000	0.20	1, 2
4,000	0.30	3, 4, 5,
5,000	0.40	6, 7, 8, 9

(One-digit random numbers are assigned to the values of the random variables in proportion to their probability of occurrence.)

Consulting the random numbers given in Table 5.10 and beginning with column 1, row 1, the following sequence of four random digits is obtained: 9, 7, 5, and 9. Letting the first random digit (9) represent the return in the first year ($5000), the second random digit (7) represent the return in the second year ($5000), and so on yields simulated returns of $5000, $5000, $4000, and $5000 over the 4-year period for a present worth of $2364.80. Obviously, one simulated investment is not sufficient to estimate the probability of the investment being profitable. Continued simulation yields the results given in Table 5.11. For the 20 simulated investments, 6 had negative present worths; thus, the probability of the investment being unprofitable might be estimated to be 0.30. However, one should be cautioned against drawing conclusions from only 20 simulated investments. In practice, it would be desirable to develop a computer program to perform the simulation and obtain thousands of simulated investments before estimating the probability of a negative present worth. (Note: An estimate of 0.3806 resulted

Table 5.10 Two-Digit Random Numbers, Example 5.23

90	43	78	83	82	99	54	02
78	31	58	98	68	09	87	80
51	81	42	35	21	42	03	62
93	97	15	95	07	56	60	39
27	37	12	63	31	35	66	93
79	39	44	22	83	96	51	00
89	61	73	29	43	84	91	34
29	38	30	84	90	18	00	10
97	64	33	29	17	48	26	04
07	64	15	02	44	32	92	99
82	13	50	83	35	39	50	51
59	83	21	30	86	90	16	09
04	46	19	63	60	53	33	97
96	54	91	43	44	40	09	02
31	27	71	78	03	65	53	62
03	45	70	42	22	16	67	13
08	35	45	92	79	97	46	02
37	60	80	55	05	35	75	57
90	43	63	17	56	21	69	09
22	07	69	85	38	74	02	58
05	33	79	00	69	29	67	08
48	97	91	14	53	00	03	42
94	68	64	58	97	32	27	80
15	39	85	87	82	38	52	16
09	37	81	73	37	01	66	84

from 5000 simulated investments. The exact value of 0.3763 can be obtained by enumerating the 256 possible combinations of cash flows.)

EXAMPLE 5.24

As a second illustration of risk analysis, consider the following situation. An individual is planning on purchasing a used microcomputer for $1000 and performing certain billing and accounting functions for several retail businesses in the neighborhood. It is anticipated that the business will last only 4 years because of the growth of competition. If income in the third year exceeds expenses by more than $400, operations will continue the fourth year. However, if income is less than or equal to $400 more than expenses in the third year, the computer and software will probably be sold at the end of the third year. For simplicity, a probability of 0.70 is assigned to the possibility of selling out at the end of the

Table 5.11 Simulation Results for Example 5.23

Investment Trial	Year 1		Year 2		Year 3		Year 4		Present Worth, PW(20%)
	RN	Value	RN	Value	RN	Value	RN	Value	
1	9	$5,000	7	$5,000	5	$4,000	9	$5,000	$2,364.80
2	2	3,000	7	5,000	8	5,000	2	3,000	312.30
3	9	5,000	0	2,000	8	5,000	5	4,000	378.00
4	0	2,000	9	5,000	3	4,000	0	2,000	− 1,582.00
5	0	2,000	3	4,000	9	5,000	2	3,000	− 1,215.40
6	0	2,000	4	4,000	9	5,000	1	3,000	− 1,215.40
7	0	2,000	0	2,000	8	5,000	1	3,000	− 2,604.20
8	3	4,000	7	5,000	9	5,000	1	3,000	2,110.20
9	9	5,000	7	5,000	7	5,000	9	5,000	1,978.90
10	9	5,000	4	4,000	6	5,000	2	3,000	1,284.50
11	3	4,000	8	5,000	7	5,000	1	3,000	663.30
12	2	3,000	5	4,000	7	5,000	0	2,000	100.20
13	5	4,000	9	5,000	8	5,000	4	4,000	1,049.20
14	8	5,000	9	5,000	4	4,000	3	4,000	1,882.50
15	6	5,000	3	4,000	3	4,000	3	4,000	2,249.10
16	1	3,000	8	5,000	6	5,000	6	5,000	215.90
17	2	3,000	4	4,000	4	4,000	5	4,000	3.80
18	4	4,000	0	2,000	3	4,000	6	5,000	− 551.70
19	6	5,000	3	4,000	3	4,000	9	5,000	1,188.10
20	1	3,000	1	3,000	7	5,000	7	5,000	− 111.90

third year, given income is not greater than $400 above expenses in the third year. Let

$$E_j = \text{expense for year } j$$
$$I_j = \text{income for year } j$$
$$n = \text{life of the investment}$$
$$S_n = \text{salvage value based on an } n\text{-year life}$$

Assuming a zero discount rate for simplicity of calculations, the present worth of the investment is given as

$$PW = -\$1000 + \sum_{j=1}^{n} (I_j - E_j) + S_n$$

The probability distributions assumed to hold for this example are provided in Table 5.12. As in the previous example, a computer should be used to perform the simulation. However, to illustrate the technique, we will manually perform 10 simulations of the investment. The table of two-digit random numbers given in Table 5.10 will be used. Using the worksheet given in Table 5.13, 10 simulations of the investment yielded an average present worth of $805 for the investment. Of course, 10 simulations is not an adequate number of trials to draw strong conclusions concerning the investment. However, the example does demonstrate the simulation approach.

To illustrate the approach taken, the first random number selected will provide the simulated value for expenses in the first year. A random number of 90 was obtained from row 1, column 1 in Table 5.10. Consulting Table 5.17, it is seen that a random number of 90 represents an expense of $400; hence $400 is entered appropriately on the worksheet given in Table 5.13. The second random number is selected from row 2, column 1 of Table 5.10 to generate the income for year 1. A random number of 78 is obtained and, from Table 5.12, a simulated income of $600 is obtained for the first year. Continuing through the third year, it is found that income exceeds expenses by $500; hence the business will continue through the fourth year. A random number of 97 is drawn to generate the salvage value of $400 for the investment.

Note that in the second trial it was decided that the business should be discontinued after 3 years. Furthermore, income was never less than expenses in any year; this illustrates insufficient observations (trials) have been obtained since, in year 2, expense can exceed income with probability

$$Pr(E_2 = 600 \text{ and } I_2 = 500) = 0.10(0.20) = 0.02$$

and in year 4 expense can exceed income with probability

$$Pr(E_4 = 700 \text{ and } I_4 = 600) + Pr(E_4 = 800 \text{ and } I_4 = 600) = 0.125$$

As an alternative to using the discrete probability distributions given in Table 5.12, we might employ continuous distributions such as the normal, gamma, or beta distributions to represent income and expenses. If such distributions are to be used, appropriate techniques for generating simulated values of the random

Table 5.12 Data for the Risk Analysis, Example 5.24

E_1	$p(E_4)$	RN	I_1	$p(I_1)$	RN
$200	0.25	00-24	$ 400	0.50	00-49
300	0.50	25-74	600	0.50	50-99
400	0.25	75-99			

E_2	$p(E_2)$	RN	I_2	$p(I_2)$	RN
$300	0.10	00-09	$ 500	0.20	00-19
400	0.40	10-49	750	0.40	20-59
500	0.40	50-89	1,000	0.40	60-99
600	0.10	90-99			

E_3	$p(E_3)$	RN	I_3	$p(I_3)$	RN
$400	0.20	00-19	$ 800	0.30	00-29
500	0.30	20-49	1,000	0.50	30-79
600	0.30	50-79	1,200	0.20	80-99
700	0.20	80-99			

E_4	$p(E_4)$	RN	I_4	$p(I_4)$	RN
$500	0.25	00-24	$ 600	0.25	00-24
600	0.25	25-49	800	0.25	25-49
700	0.25	50-74	1,000	0.25	50-74
800	0.25	75-99	1,200	0.25	75-99

N	$p(N|I_3 - E_3 \leq \$400)$	RN
3	0.70	00-69
4	0.30	70-99

N	$p(N|I_3 - E_3 > \$400)$
4	1.00

S_3	$p(S_3)$	RN	S_4	$p(S_4)$	RN
$400	0.50	00-49	$ 300	0.60	00-59
500	0.50	50-99	400	0.40	60-99

variables are required. Since a detailed treatment of simulation is beyond the scope of this text, we refer you to texts devoted to simulation for additional discussion [3, 42, 53, 54].

Analytic Approaches: As an illustration of the use of analytic approaches in developing the probability distribution for present worth, consider the following present-worth relation.

$$PW = \sum_{j=0}^{n} C_j(1 + i)^{-j}$$

(5.6)

Suppose the cash flows, C_j, are *random variables* with expected values $E(C_j)$ and variances $V(C_j)$. Since the *expected value* of a sum of random variables is given by the sum of the expected values of the random variables, the expected present worth is given by

$$E(PW) = \sum_{j=0}^{n} E[C_j(1 + i)^{-j}] \qquad (5.7)$$

Furthermore, since the expected value of the product of a constant and a random variable is given by the product of the constant and the expected value of the random variable,

$$E[C_j(1 + i)^{-j}] = E(C_j)(1 + i)^{-j} \qquad (5.8)$$

Substituting Equation 5.8 into Equation 5.7 yields

$$E(PW) = \sum_{j=0}^{n} E(C_j)(1 + i)^{-j} \qquad (5.9)$$

Hence, we see that the expected present worth of a series of cash flows is found by summing the present worths of the expected values of the individual cash flows.

EXAMPLE 5.25

Recall Example 5.23 involving an investment of $10,000 for a 4-year period. The expected value of the return in a given year is given by

$$E[\text{Return}] = \$2000(0.10) + \$3000(0.20)$$
$$+ \$4000(0.30) + \$5000(0.40)$$
$$= \$4000$$

Thus, the expected present worth for the investment situation is

$$E(PW) = \$4000(P|A\ 20,4) - \$10,000$$
$$= \$354.80$$

(From Table 5.11, a computation of the average simulated present worth yields $425.01. A value of $323.94 resulted from 5000 simulated investments. This serves to illustrate the need for many simulated trials.)

To determine the variance of the present worth, we first recall that if X_1, X_2, X_3, and Y are random variables related as follows,

$$Y = X_1 + X_2 + X_3 \qquad (5.10)$$

Table 5.13 Simulation of the Sample Investment Problem in Example 5.24

Trial	Year	RN(E)	E	RN(I)	I	I − E	RN(N)	RN(S)	S	PW
1	1	90	$400	78	$ 600	$200				
	2	51	500	93	1,000	500				
	3	27	500	79	1,000	500				
	4	89	800	29	800	0	—	97	$400	$ 600
2	1	07	200	82	600	400				
	2	59	500	04	500	0				
	3	96	700	31	1,000	300	03	08	400	100
	4	—	—	—	—	—				
3	1	37	300	90	600	300				
	2	22	400	05	500	100				
	3	48	500	94	1,200	700				
	4	15	500	09	600	100	—	43	300	500
4	1	31	300	81	600	300				
	2	97	600	37	750	150				
	3	39	500	61	1,000	500				
	4	38	600	64	1,000	400		64	400	750
5	1	13	200	83	600	400				
	2	46	400	54	750	350				
	3	27	500	45	1,000	500				
	4	35	600	60	1,000	400	—	43	300	950

6	1	07	200	33	400	200				
	2	97	600	68	1,000	400				
	3	39	500	37	1,000	500				
	4	78	800	58	1,000	200	—	42	300	600
7	1	15	200	12	400	200				
	2	44	400	73	1,000	600				
	3	30	500	33	1,000	500				
	4	15	500	50	1,000	500	—	21	300	1,100
8	1	19	200	91	600	400				
	2	71	500	70	1,000	500				
	3	45	500	80	1,200	700				
	4	63	700	69	1,000	300	—	79	400	1,300
9	1	91	400	64	600	200				
	2	85	500	81	1,000	500				
	3	83	700	98	1,200	500				
	4	35	600	95	1,200	600	—	63	400	1,200
10	1	22	200	29	400	200				
	2	84	500	29	750	250				
	3	02	400	83	1,200	800				
	4	30	600	63	1,000	400	—	43	300	950
										8,050

then the variance of Y is given by

$$\boxed{\text{Var}(Y) = E(Y^2) - E(Y)^2} \tag{5.11}$$

or

$$\boxed{\text{Var}(Y) = E[(X_1 + X_2 + X_3)^2] - E(X_1 + X_2 + X_3)^2}$$

Expanding and collecting terms yields

$$
\begin{aligned}
\text{Var}(Y) = {}& E(X_1^2 + X_2^2 + X_3^2 + 2X_1X_2 + 2X_1X_3 + 2X_2X_3) \\
& - [E(X_1) + E(X_2) + E(X_3)]^2 \\
= {}& E(X_1^2) + E(X_2^2) + E(X_3^2) + 2E(X_1X_2) + 2E(X_1X_3) + 2E(X_2X_2) \\
& - E(X_1)^2 - E(X_2)^2 - E(X_3)^2 - 2E(X_1)E(X_2) - 2E(X_1)E(X_3) \\
& - 2E(X_2)E(X_3) \\
= {}& E(X_1^2) - E(X_1)^2 + E(X_2^2) - E(X_2)^2 + E(X_3^2) - E(X_3)^2 \\
& + 2[E(X_1X_2) - E(X_1)E(X_2)] + 2[E(X_1X_3) - E(X_1)E(X_3)] \\
& + 2[E(X_2X_3) - E(X_2)E(X_3)]
\end{aligned}
\tag{5.12}
$$

Since $E(X_k^2) - E(X_k)^2$ defines the variance of the random variable X_k and $E(X_pX_k) - E(X_p)E(X_k)$ defines the covariance of the random variables X_p and X_k, we see that

$$\boxed{\begin{aligned}\text{Var}(Y) = {}& \text{Var}(X_1) + \text{Var}(X_2) + \text{Var}(X_3) + 2\,\text{Cov}(X_1X_2) \\ & + 2\,\text{Cov}(X_1X_3) + 2\,\text{Cov}(X_2X_3)\end{aligned}} \tag{5.13}$$

where $\text{Cov}(X_pX_k)$ denotes the covariance of X_p and X_k. If the random variables X_p and X_k are statistically independent, then $\text{Cov}(X_pX_k) = 0$.

Generalizing to the sum of $(n + 1)$ random variables, if

$$\boxed{Y = \sum_{j=0}^{n} X_j} \tag{5.14}$$

then the expected value and variance of Y are given by

$$\boxed{E(Y) = \sum_{j=0}^{n} E(X_j)}$$

$$\boxed{\text{Var}(Y) = \sum_{j=0}^{n} \text{Var}(X_j) + 2\sum_{j=0}^{n-1}\sum_{k=j+1}^{n} \text{Cov}(X_jX_k)} \tag{5.15}$$

Additionally, if a_1 and a_2 are constants, and X_1, X_2, and Y are random variables related by

$$\boxed{Y = a_1X_1 + a_2X_2} \tag{5.16}$$

recall that

$$\boxed{\mathrm{Var}(Y) = a_1^2\,\mathrm{Var}(X_1) + a_2^2\,\mathrm{Var}(X_2) + 2a_1 a_2\,\mathrm{Cov}(X_1 X_2)} \qquad \textbf{(5.17)}$$

Combining the relationships given in Equations 5.15 and 5.17, the variance of present worth is found to be

$$\boxed{\begin{aligned} \mathrm{Var}(PW) &= \sum_{j=0}^{n}\mathrm{Var}(C_j)(1+i)^{-2j} \\ &\quad + 2\sum_{j=0}^{n-1}\sum_{k=j+1}^{n}\mathrm{Cov}(C_j C_k)(1+i)^{-(j+k)} \end{aligned}} \qquad \textbf{(5.18)}$$

EXAMPLE 5.26

Continuing Example 5.25 to compute the variance of present worth, recall from Example 5.23 that the random variables are statistically independent. Thus, $\mathrm{Cov}(C_j C_k) = 0$ for all j and k. The variance of an annual return is determined as follows:

$$\mathrm{Var}[\text{Return}] = E[\text{Return}^2] - E[\text{Return}]^2$$

and

$$\begin{aligned} E[\text{Return}^2] &= (\$2000)^2(0.10) + (\$3000)^2(0.20) \\ &\quad + (\$4000)^2(0.30) + (\$5000)^2(0.40) \\ &= 17 \times 10^6 \end{aligned}$$

Thus

$$\begin{aligned} \mathrm{Var}[\text{Return}] &= (17 \times 10^6) - (4000)^2 \\ &= 1 \times 10^6 \end{aligned}$$

The standard deviation is the square root of the variance and equals

$$SD[\text{Return}] = \$1000$$

To compute the variance of present worth, from Equation 5.18

$$\begin{aligned} \mathrm{Var}(PW) &= \mathrm{Var}[\text{Investment}] + \sum_{t=1}^{4}\mathrm{Var}(R_t)(1.20)^{-2t} \\ &= 0 + \sum_{t=1}^{4}(1 \times 10^6)(1.20)^{-2t} \\ &= (1 \times 10^6)\sum_{t=1}^{4}(P\,|\,F\,20,2t) \\ &= (1 \times 10^6)(0.6944 + 0.4823 + 0.3349 + 0.2326) \\ &= 1.7442 \times 10^6 \end{aligned}$$

Hillier [6] argues that it is probably unrealistic to expect investment analysts to develop accurate estimates for covariances in Equation 5.18. Consequently, he suggests that the net cash flow in any year be divided into those components of cash flow that are reasonably independent from year to year and those that are correlated over time. Specifically, it is assumed that

$$C_j = X_j + Y_{j1} + Y_{j2} + \cdots + Y_{jm} \tag{5.19}$$

where the X_j values are mutually independent over j but, for a given value of h, $Y_{0h}, Y_{1h}, \ldots, Y_{nh}$ are *perfectly* correlated.

When two random variables X and Y are perfectly correlated, one can be expressed as a linear function of the other. Hence, if

$$Y = a + bX \tag{5.20}$$

where a and b are constants, then the covariance of X and Y is given by

$$\text{Cov}(XY) = E(XY) - E(X)E(Y) \tag{5.21}$$

However, from Equation 5.20, we note that

$$E(Y) = a + bE(X) \tag{5.22}$$

and

$$\text{Var}(Y) = b^2 \, \text{Var}(X) \tag{5.23}$$

Substituting Equations 5.20 and 5.22 into Equation 5.21 gives

$$\begin{aligned}
\text{Cov}(XY) &= E(aX + bX^2) - E(X)[a + bE(X)] \\
&= aE(X) + bE(X^2) - aE(X) - bE(X)^2 \\
&= b \, \text{Var}(X)
\end{aligned} \tag{5.24}$$

Recalling that the *standard deviation* of a random variable is defined as the square root of the variance of the random variable, we see that Equation 5.24 can be expressed as

$$\text{Cov}(XY) = bSD(X)SD(X) \tag{5.25}$$

where $SD(X)$ denotes the standard deviation of X. However, from Equation 5.23, we see that $SD(Y)$ equals $bSD(X)$. Hence,

$$\text{Cov}(XY) = SD(X)SD(Y) \tag{5.26}$$

when X and Y are perfectly correlated.

The model suggested by Hillier yields the following expressions for the expected present worth and variance of present worth:

$$E(PW) = \sum_{j=0}^{n} E(X_j)(1 + i)^{-j} + \sum_{j=0}^{n} \sum_{h=1}^{m} E(Y_{jh})(1 + i)^{-j} \tag{5.27}$$

and

$$\text{Var}(PW) = \sum_{j=0}^{n} \text{Var}(X_j)(1 + i)^{-j} + \sum_{h=0}^{m} \left[\sum_{j=0}^{n} SD(Y_{jh})(1 + i)^{-j} \right]^2 \quad \text{(5.28)}$$

where $SD(Y_{jh})$ denotes the standard deviation of the random variable Y_{jh}.

A very important theorem from probability theory, the *central limit theorem*, is usually invoked at this point. The central limit theorem establishes under very general conditions that the sum of independently distributed, random variables tends to be distributed normally as the number of terms in the summation approaches infinity. Hence, it is argued that present worth, as defined by Equation 5.6, is normally distributed with mean and variance, as given by Equations 5.27 and 5.28. Of course, this assumes that the C_j are statistically independent. However, the view is usually taken that the normal distribution is a reasonable approximation to the distribution of present worth.

Given that present worth is assumed to be normally distributed, one can compute for each investment alternative the probability of achieving a given aspiration level. For example, one is usually interested in knowing the probability that present worth is less than zero. Some analysts interpret this to be a measure of the risk associated with an investment alternative.

In performing an analysis of the risk associated with an investment alternative, we have treated the simplest situation; only the cash flows were considered to be random variables, and the measure of merit employed was present worth, not rate of return. When either the discount rate or the planning horizon are treated as random variables and when the probability distribution for rate of return is desired, simulation approaches are usually employed.

EXAMPLE 5.27

As an illustration of the analytic approach in developing the probability distribution for *PW*, consider the following example problem based on one given by Giffin [27].

A flight school operator is considering the alternatives of purchasing a utility category training aircraft versus purchasing an acrobatic version of the same aircraft. Having been in the flight training business for a number of years, the operator has reason to believe that income and expense from a utility category aircraft will be nearly independent from year to year. The pertinent data are summarized in Table 5.14.

Investment in the acrobatic aircraft is a more risky but promising investment. The acrobatic aircraft is also expected to have a life of 5 years. The operator feels that maintenance costs will be nearly independent year to year with this aircraft. However, since the flight school has never offered an acrobatic course before, there is some uncertainty regarding the demand for time in such an aircraft. It is felt that the net cash flow from the sale of flight time for each of the 5 years will be perfectly correlated. The pertinent data are summarized in Table 5.15. An interest rate of 10% is to be used in the analysis.

Table 5.14 Estimated Net Cash Flows for a Utility Model Aircraft,[a] Example 5.27

Year	Source	Symbol	Expected Value	Range	Standard Deviation
0	Purchase	C_0	−$11,000	$9,500–12,500	$500
1	Income-expense	C_1	2,200	2,050– 2,350	50
2	Income-expense	C_2	2,200	1,900– 2,500	100
3	Income-expense	C_3	2,200	1,900– 2,500	100
4	Income-expense	C_4	2,000	1,700– 2,300	100
5	Income-expense	C_{51}	1,000	700– 1,300	100
5	Salvage	C_{52}	6,000	4,800– 7,200	400

[a] The data have been modified slightly from those given by Giffin [27].

From Equations 5.27 and 5.28, we obtain the following:

$$E(PW_1) = \$190 \qquad E(PW_2) = \$355$$
$$\mathrm{Var}(PW_1) = 334{,}741 \qquad \mathrm{Var}(PW_2) = 11{,}631{,}930$$

Assuming normally distributed PW, the probability of an equivalent present worth less than zero is found to be

$$Pr(PW_1 < 0 \mid i = 10\%) = 0.37$$

for the utility model aircraft and

$$Pr(PW_2 < 0 \mid i = 10\%) = 0.46$$

for the acrobatic model aircraft. Here we are faced with a choice between the alternative having the greatest expected value and the alternative having the

Table 5.15 Estimated Net Cash Flows for an Acrobatic Model Aircraft,[a] Example 5.27

Year	Source	Symbol	Expected Value	Range	Standard Deviation
0	Purchase	X_0	−$14,000	$13,100–14,900	$ 300
1	Expense	X_1	− 10,000	9,100–10,900	300
2	Expense	X_2	− 10,000	9,100–10,900	300
3	Expense	X_3	− 11,000	10,100–11,900	300
4	Expense	X_4	− 12,000	10,800–13,200	400
5	Expense	X_{51}	− 12,000	10,800–13,200	400
1	Income	Y_{11}	12,500	9,500–15,500	1,000
2	Income	Y_{21}	13,500	10,500–16,500	1,000
3	Income	Y_{31}	13,500	10,800–16,200	900
4	Income	Y_{41}	14,500	12,100–16,900	800
5	Income	Y_{51}	13,500	11,400–15,600	700
5	Salvage	X_{52}	7,500	6,000– 9,000	500

[a] The data have been modified slightly from those given by Giffin [27].

smallest variance. The choice will depend on the owner's attitudes toward risk. At this point we leave the flight school operator "up in the air" relative to a choice between the investment alternatives. If the decision were yours to make, which model aircraft would you choose? Why?

Risk analysis offers a number of important advantages over traditional deterministic approaches. Klausner [38] summarized some of the most significant advantages as follows:

1. **Uncertainty Made Explicit.** The uncertainty that an estimator feels about the estimate of an element value is brought out into the open and incorporated into the investment analysis. The analysis technique permits maximum information utilization by providing a vehicle for the inclusion of "less likely" estimates in the analysis.

2. More **Comprehensive Analysis.** This technique permits a determination of the effect of simultaneous variation of all the element values on the outcome of an investment. This approximates the "real world" conditions under which an actual investment's outcome will be determined. The Probabilistic Cash Flow Simulation generates an overall indication of potential variation in outcome and project risk. This indicator, in the form of a probability distribution, accounts statistically for element interaction.

3. **Variability of Outcome Measured.** One of the most significant advantages of this analysis technique is that it gives a measure of the dispersion around the investment outcome based on the expected cash flow. This dispersion, or variability, is an important consideration in the comparison of alternative investments. Other things being equal, lower variability for the same return is usually desirable. The probability distribution associated with each investment's outcome gives a clear picture of this important evaluation consideration.

4. **Promotes More Reasoned Estimating Procedures.** By requiring that element values be given as probability distributions rather than as single values, more reasoned consideration is given to the estimating procedure. Judgment is applied to the individual element values rather than to the investment's outcome, which is jointly determined by all of the elements. Thinking through the uncertainties in a project and recognizing what is known and unknown will go far toward ensuring the best investment decision. Understanding and dealing effectively with uncertainty and risk is the key to rational decision making.

Risk analysis is a technique that has been used by a number of firms to improve the decision-making process in a risk environment. When applied properly, risk analysis can enhance significantly the manager's understanding of the risks associated with an investment alternative. The major premise underlying risk analysis is the belief that a manager can make better decisions when provided a fuller understanding of the implications of the decision.

Too often, management will not accept estimates of cost savings or items not easily quantified and measured. As a result, savings and benefits due to inventory reductions, energy savings, quality improvements, space reductions, safety improvements, cycle-time reductions, throughput improvement, lead-time reductions, and increased flexibility are often ignored. Every effort should be made to capture the monetary impact of these benefits. Even though they cannot be measured precisely, their value is certainly greater than zero. As Kaplan observed, "Although intangible benefits are difficult to quantify, there is no reason to value them at zero in a capital expenditure analysis. Zero is, after all, no less arbitrary than any other number. Conservative accountants who assign zero values to many intangible benefits prefer being precisely wrong to being vaguely right. Managers need not follow their example" [34, p. 92]. In the end a rough estimate of intangible benefits is better than no estimate at all! Supplementary analyses provide one mechanism for assessing the impact of the selection decision on the values of difficult-to-measure benefits.

As noted, many firms have well-defined procedures for performing economic justifications. Hence, a supplementary analysis might not be required. Even so, we strongly recommend that they be performed—if for no one else's benefit than yours. Supplementary analyses are performed in order to answer a number of "what if?" questions that will either be asked by management or be of concern to management; in a sense, they provide an "insurance policy"; they are performed to provide a higher level of confidence in the results of the economic justifications.

In summary, we have found it to be particularly useful to do more than *just what was required by the client or the company* in performing economic justifications. The benefits of performing supplementary analyses included increasing our own confidence level regarding our recommendations, as well as increasing management's confidence in our analysis.

5.8 SELECTING THE PREFERRED ALTERNATIVE

The seventh, and final, step in performing a comparison of engineering investment alternatives involves the selection of the preferred alternative. The final selection decision might be quite different from the one recommended as "the economic choice." Multiple criteria typically exist, rather than the single criterion of maximizing the economic worth. (However, it should be noted that any long-term decision process that consistently selects alternatives that do not earn a rate of return at least equal to the cost of capital will result in financial ruin for the firm. Hence, even when multiple criteria exist, the alternative ultimately selected should not be uneconomic.)

The presence of multiple criteria, coupled with the risks and uncertainties associated with estimating future outcomes, result in the selection process being quite complicated. To make the process easier, the engineer is encouraged to address as many as possible of management's concerns in comparing the investment alternatives. To the extent that management's concerns have been addressed, the selection decision will agree with the engineer's recommendation.

Our discussion in this chapter has concentrated solely on the economic factor; we have been concerned with determining the most economical alternative. The final decision may be based on a host of criteria instead of on the single criterion of economics. Despite attempts to measure all benefits in economic terms, it is likely that some intangible factors or attributes will not be reduced to dollars. Consider such factors as improved safety, reduced cycle times, improved quality, improved safety, increased flexibility, increased customer service, improved employee morale, being the first in the industry to utilize a technology, and increased market visibility. Clearly, some of these factors are more readily measured in economic terms than others. A body of literature has evolved treating the subject of multiattribute decision making [5, 6, 10, 15, 19, 22, 41, 56].

For the purposes of this text, we will consider the use of weighted factor comparisons in coping with multiple attributes. To perform a weighted factor comparison, numerical values or weights are assigned proportionally to each factor reflecting the degrees of importance for the factors. A numerical score is then assigned to each alternative based on its performance against a particular factor. The scores are multiplied by the weights, and the products are summed over all factors to obtain a total weighted score. The alternative with the highest score is deemed the preferred choice. Although there are scaling difficulties associated with the technique, it is quite popular.

In an attempt to ensure consistency and to minimize the chances of a halo effect, a paired comparison is recommended. The halo effect is a phenomenon that occurs when a high or low ranking on one factor carriers over and influences the ranking on other factors.

To illustrate the paired comparison approach, suppose there are five investment alternatives (A, B, C, D, and E). Further, suppose the following preferences are obtained by comparing the alternatives two-at-a-time for a factor that is difficult to quantify, such as improved customer service: $A < B$, $A < C$, $A > D$, $A < E$, $B > C$, $B > D$, $B > E$, $C > D$, $C > E$, and $E > D$, where $A < B$ means that A ranks lower than B and $A > D$ means A ranks higher than D. Combining the paired comparisons yields the following ranking: $B > C > E > A > D$. Next, numerical values are assigned to each alternative in direct proportion to the performance of the alternative for the factor in question. Clearly, this is the most difficult and most subjective aspect of the process.

Boucher and MacStravic [10] developed software to facilitate the assignment of numerical values to each alternative for each factor. Their software is based on a methodology called nontraditional capital investment criteria (NCIC) and draws upon the work of Saaty [56] in the development of the analytic hierarchy process (AHP). Canada and Sullivan [15] address the use of multiattribute decision processes in capital budgeting and in advanced manufacturing systems. A case study of the use of NCIC in food processing illustrates the multiattribute NCIC process [11].

Figure 5.13 provides a weighted factor comparison form for evaluating up to five investment alternatives and twenty factors. In some applications of weighted factor comparison, pairwise comparisons are performed of the factors to facilitate the assignment of weights (Wt). In Figure 5.13, weights summing to 100 are assigned to the factors. For each factor, the alternatives are ranked and numerical values are assigned (Rt), ranging from zero to ten. The weight for the factor is multiplied by the numerical rating to obtain a score (Sc) for each alternative.

WEIGHTED FACTOR COMPARISON FORM

Company: Prepared by: Date:
Description of investment:

Factor	Wt	A		B		C		D		E	
		Rt	Sc	Rt	Sc	Rt	Sc	Rt	Sc	Rt	Sc
1.											
2.											
3.											
4.											
5.											
6.											
7.											
8.											
9.											
10.											
11.											
12.											
13.											
14.											
15.											
16.											
17.											
18.											
19.											
20.											
Totals	100										

Figure 5.13 Weighted factor comparison form.

EXAMPLE 5.28

Three investment alternatives (A, B, and C) are being considered by the Ajax Manufacturing Company. The present worths (PW) are $25,000, $20,000, and $18,000, respectively. The three alternatives perform quite differently in terms of product quality (Q), time required to fill a customer's order (T), and the reputation of the supplier of the technology (R).

A ranking of the factors (PW, Q, T, and R) yields the following weights being assigned: PW (30), Q (40), T (20), and R (10). In rating the three alternatives against the three factors and assigning rating values, the following resulted.

	A	B	C
Present worth (PW)	10	8	7.2
Product quality (Q)	8	10	5
Fill time (T)	3	10	7
Supplier reputation (R)	8	5	10

The weighted factor comparison for the three investment alternatives is summarized in Figure 5.14. From the results of the analysis, alternative B has the highest score. If the weights and ratings truly reflect management's feelings, then B would be recommended. From the analysis, we can conclude that the improved performance of B over A in product quality and time to fill a customer's order is worth at least $5000, based on the difference in present worths.

Just as supplementary analyses were recommended for economic measures of merit, so should supplementary analyses be performed of the weights and ratings assigned in a weighted factor comparison. Relative to estimating values of cash flows, planning horizon, and MARR, the assignment of weights and ratings values appears to be far more subjective.

As pointed out in Chapter One, the selection or rejection of the recommended solution is heavily dependent on the sales ability of the individual presenting the recommendation to management. Since the corporate decision makers are normally presented with many more investment alternatives than can be funded, it is important to communicate effectively in order to compete favorably for the company's limited capital. Toward this end, Klausner [39] provides four specific suggestions for the engineer or systems analyst.

1. Recognize that the decision-makers' perspective is broad and develop the proposal accordingly. The project's capital requirements should be related to previous and estimated future capital requirements on similar investments. The project's discounted cash flow return or net present value should be compared to other projects with which the decision makers are familiar. The proposal should be shown to fit in with long-range corporate plans and support short-range objectives. Comparison should be made with similar investments by competition and competitive advantage (if any) shown.

WEIGHTED FACTOR COMPARISON FORM

Company: Ajax Manufacturing Company   Prepared by: MHA   Date: April 1, 2001
Description of investment: Order picking equipment for Chicago distribution center

Factor	Wt.	A Rt	A Sc	B Rt	B Sc	C Rt	C Sc	D Rt	D Sc	E Rt	E Sc
1. Present worths	30	10	300	8	240	7.2	216				
2. Product quality	40	8	320	10	400	5	200				
3. Fill time, customer order	20	3	60	10	200	7	140				
4. Supplier reputation	10	8	80	5	50	10	100				
5.											
6.											
7.											
8.											
9.											
10.											
11.											
12.											
13.											
14.											
15.											
16.											
17.											
18.											
19.											
20.											
Totals	100		760		890		656				

Figure 5.14 Weighted factor comparison for Example 5.27.

Ancillary marketing, public relations, and/or political benefits that the company will derive from the investment should be pointed out. The effect that the investment will have on other functional activities of the company should be noted and overall benefits stressed. In other words, the investment proposal should be related to the well-being of the total enterprise.

2. Recognize that this investment proposal is only one of several that the decision makers are reviewing and that not all proposals will be accepted. The engineer should know how the decision makers classify investments and what competition there is for available capital resources, as well as the relative strengths and weaknesses of the investment proposal vis-à-vis competing investment proposals and deal with each in the presentation. If the proposal investment is relatively risk-free, that point should be made strongly to possibly offset less desirable aspects of the proposal (e.g., low return on investment).

3. Know the decision makers and tailor the investment proposal accordingly. The engineer should, for example, know and use the measure of merit they prefer, and support it with any other measures of merit that may be necessary or helpful. For example, if the decision makers favor discounted cash flow return on investment and the proposal has one that barely meets minimum return standards but does have a large net present value, the engineer must make sure that the potential contribution to profit is clearly pointed out. If it is known that the decision makers still calculate the payout period of each investment proposal, the investment's payout period should be stated—either to support the proposal or to point out its irrelevance in hopes of minimizing its impact on the ultimate investment decision. The engineer should always use technical and economical terms that the decision makers will understand and relate to.

   The engineer should become familiar with the values that the decision makers attach to different aspects of investment proposals—particularly economic uncertainty. Low risk should be emphasized in proposals to decision makers who tend to be risk-avoiders; potential economic gain should be highlighted for the risk-taking decision maker. This, of course, does not suggest that the engineer should be less than honest in the proposal, but only makes the point that the communication of the investment proposal should be developed with the decision maker in mind and emphasis varied accordingly.

4. Do not oversell the technical engineering aspects of the investment proposal. It should be remembered that decision makers are primarily interested in the economic aspects of the proposal. The engineer must resist the temptation to overstate the complicated and sophisticated technology that might underlie a proposal. The less decision makers understand about the technical/engineering aspects of a proposal, the more uneasy they become. This apprehension becomes part of the uncertainty which the decision makers subjectively assign to the proposal and the result could be its rejection in favor of a proposal with which they are more familiar, or at least, more comfortable. The engineer should keep in mind the background and interests of the decision makers and use technical and commercial terms with which they are familiar.

5.9   ANALYZING ALTERNATIVES WITH NO POSITIVE CASH FLOWS

The previous analysis of an investment decision involved four mutually exclusive alternatives, including the do-nothing alternative, which included positive-valued cash flows. However, there exist solutions in which no positive-valued cash flows are present.

EXAMPLE 5.29

Consider a case in which two new cost reduction alternatives have been proposed. The present method, which we refer to as the *do-nothing* alternative, is also a feasible alternative. Thus, three mutually exclusive alternatives are considered. Cash flow profiles for the alternatives are given in Table 5.16.

Alternative 0 is the do-nothing alternative in which the present expenditure of $12,000/year is continued. Alternative 1 involves an initial investment of $10,000 in order to reduce the annual expenditures by $3000 over the 5-year period. Alternative 2 requires an initial investment of $15,000 in order to obtain decreasing annual expenditures. The minimum attractive rate of return is specified to be 10%.

The *PW*, *AW*, or *FW* methods using the ranking approach, or any measure of investment worth using the incremental cash flow approach, may be applied to this problem. Let us analyze these alternatives in three different ways.

Annual Worth Method—Total Cash Flow Approach

$$AW_0(10\%) = -\$12,000/\text{year}$$
$$AW_1(10\%) = -\$10,000(A|P\ 10,5) - \$9000$$
$$= -\$10,000(.2638) - \$9000$$
$$= -\$11,638/\text{year}$$
$$AW_2(10\%) = -\$15,000(A|P\ 10,5) - \$9000 + \$1000(A|G\ 10,5)$$
$$= -\$15,000(.2638) - \$9000 + \$1000(1.8101)$$
$$= -\$11,146.90/\text{year}$$

Table 5.16 Cash Flow Profiles for Three Investment Alternatives with No Positive Cash Flows

t	A_{0t}	A_{1t}	A_{2t}
0	0	-$10,000	-$15,000
1	-$12,000	- 9,000	- 9,000
2	- 12,000	- 9,000	- 8,000
3	- 12,000	- 9,000	- 7,000
4	- 12,000	- 9,000	- 6,000
5	- 12,000	- 9,000	- 5,000

Alternative 2 is preferred since it has the highest annual worth (or lowest annual cost).

Annual Worth Method—Incremental Cash Flow Approach

$$AW_{1-0}(10\%) = -\$10,000(A\,|\,P\,10,5) + \$3000$$
$$= -\$10,000(.2638) + \$3000$$
$$= \$362/\text{year} \qquad (\text{Prefer 1 over 0})$$
$$AW_{2-1}(10\%) = -\$5,000(A\,|\,P\,10,5) + \$1000(A\,|\,G\,10,5)$$
$$= \$5,000(.2638) + \$1000(1.8101)$$
$$= \$491.10/\text{year} \qquad (\text{Prefer 2 over 1})$$

Again, Alternative 2 is preferred.

Internal Rate of Return Method—Incremental Cash Flow Approach

$$FW_{1-0}(i) = \$0 = -\$10,000(F\,|\,P\,i,5) + \$3,000(F\,|\,A\,i,5)$$
$$@i = 15\% \ \$0 \neq -\$10,000(2.0114) + \$3,000(6.7424) = \$113.20$$
$$@i = 18\% \ \$0 \neq -\$10,000(2.2878) + \$3,000(7.1542) = -\$1415.40$$

Interpolating, $i_{1-0} = 15.22\% > MARR$ (Prefer 1 over 0).

$$FW_{2-1}(i) = \$0 = -\$5000(F\,|\,P\,i,5) + \$1000(P\,|\,G\,i,5)(F\,|\,P\,i,5)$$
$$@i = 18\% \ \$0 \neq -\$5000(2.2878) + \$1000(5.2312)(2.2878) = \$528.94$$
$$@i = 20\% \ \$0 \neq -\$5000(2.4883) + \$1000(4.9061)(2.4883) = -\$233.65$$

Interpolating, $i_{2-1} = 19.39\% > MARR$ (Prefer 2 over 1).

Again, Alternative 2 is preferred.

The *IRR* method deserves additional explanation. Since the sum of the cash flows (ignoring the time value of money) is negative for each alternative, positive-valued rates of return among the three alternatives do not exist. However, even though Alternative 1 does not have a rate of return by itself, in comparison with Alternative 0, it is clear that the investment of the $10,000 yields annual savings of $3000 in annual expenditures. Thus, on an incremental basis, the $10,000 incremental investment produces positive-valued cash flows of $3000/year for each of the 5 years.

The return on the incremental investment, using the internal rate-of-return method, is found to be approximately 15.22%. Since we would do better to invest the $10,000 in Alternative 1 and earn 15.22% than to choose Alternative 0 and only earn the minimum attractive rate of return on the remaining money available for investment, Alternative 1 is preferred to Alternative 0.

Of course, we have $15,000 to invest (otherwise Alternative 2 would be infeasible). Thus, at this point we are willing to invest $10,000 in Alternative 1 and earn 15.22% and invest the remaining $5000 in some other opportunity and earn the minimum attractive rate of return of 10%. The question now considered is, "Should we use the additional $5000 and pool it with the $10,000 to invest in Alternative 2?" The rate of return on the incremental investment required to obtain Alternative 2 is calculated to be approximately 19.39%.

At this point, the question is "Should we invest $10,000 in Alternative 1 and invest $5000 elsewhere at the *MARR*, or should we invest $15,000 in Alternative 2?" Investing in Alternative 2 yields a return of 15.22% on the $10,000 increment and 19.39% on the $5000 increment. Consequently, Alternative 2 would be preferred to the investment of $10,000 in Alternative 1 and investing the remaining $5000 to earn only 10%.

5.10   MORE ON DEALING WITH UNEQUAL LIVES

Recall that in Section 5.3 we recommended that a planning horizon be specified, cash flows over the planning horizon be given explicitly, and the evaluation be performed over the common planning horizon time frame. While this method is becoming better known and accepted, many still compare alternatives having unequal lives on the basis of the individual life cycles. This approach either implicitly assumes a least common multiple of lives approach, or it implicitly establishes a salvage value which may or may not be reasonable. Enough background material has now been presented to permit us to clarify these points and demonstrate their drawbacks.

EXAMPLE 5.30

Consider two mutually exclusive alternatives having cash flow profiles as depicted in Figure 5.15. Using a minimum attractive rate of return of 15%, the following annual worths are obtained:

$$AW_1(15\%) = -\$5000(A|P\,15,5) - \$2000 + \$1000(A|F\,15,5)$$
$$= -\$5000(.2983) - \$2000 + \$1000(.1483)$$
$$= -\$3343.20$$
$$AW_2(15\%) = -\$7000(A|P\,15,6) - \$1500 + \$1500(A|F\,15,6)$$
$$= -\$7000(.2642) - \$1500 + \$1500(.1142)$$
$$= -\$3178.10$$

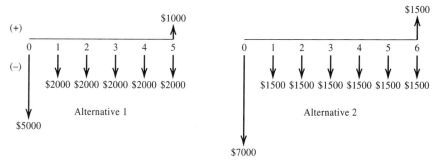

Figure 5.15 Cash flow diagrams for two investment alternatives.

Implicit in the comparison of these alternatives using annual worths based on individual life cycles is the assumption that a 30-year planning horizon is being used. As shown in Table 5.17, if the present worths are computed for a 30-year period, the following values will be obtained:

$$PW_1(15\%) = -\$21,952.10$$
$$PW_2(15\%) = -\$20,868.30$$

Converting the present worths to annual worths yields

$$AW_1(15\%) = -\$21,952.10(A|P\,15,30)$$
$$= -\$21,952.10(.1523)$$
$$= -\$3343.30$$
$$AW_2(15\%) = -\$20,868.30(A|P\,15,30)$$
$$= -\$20,868.30(.1523)$$
$$= -\$3178.24$$

which are the values obtained (except for round-off error) using individual life cycles.

An alternative assumption that could be made is that a 5-year planning horizon is being used and the salvage value for Alternative 2 is such that an annual worth of $-\$3178.10$ will still be obtained. Hence, S_5, the salvage value at the end of

Table 5.17 Cash Flows for Two Investment Alternatives, with a 30-year Planning Horizon

EOY	CF(1)	CF(2)
0	−$ 5,000	−$ 7,000
1-4	− 2,000	− 1,500
5	− 6,000	− 1,500
6	− 2,000	− 7,000
7-9	− 2,000	− 1,500
10	− 6,000	− 1,500
11	− 2,000	− 1,500
12	− 2,000	− 7,000
13-14	− 2,000	− 1,500
15	− 6,000	− 1,500
16-17	− 2,000	− 1,500
18	− 2,000	− 7,000
19	− 2,000	− 1,500
20	− 6,000	− 1,500
21-23	− 2,000	− 1,500
24	− 2,000	− 7,000
25	− 6,000	− 1,500
26-29	− 2,000	− 1,500
30	− 1,000	− 0
Present worth	−$21,952.10	−$20,868.30

the fifth year, for Alternative 2 must be

$$AW_2(15\%) = -\$7000(A|P\,15,5) - \$1500 + S_5(A|F\,15,5)$$
$$-\$3178.10 = -\$7000(.2983) - \$1500 + S_5(.1483)$$
$$S_5 = \$2764.67$$

This salvage value may or may not have any relationship to reality.

EXAMPLE 5.31

Consider the alternatives depicted in Figure 5.16. They will be used in a more glaring example to illustrate the shortcoming of an assumption that the salvage value for unused portions of an asset's life will be such that the annual worth will be unchanged. The annual worths for individual life cycles are found to be:

$$AW_1(15\%) = -\$5000(A|P\,15,5) - \$3000 + \$1000(A|F\,15,5)$$
$$= -\$5000(.2983) - \$3000 + \$1000(.1483)$$
$$= -\$4343.20$$
$$AW_2(15\%) = -\$6000(A|P\,15,6) - \$1000 - \$1000(A|G\,15,6)$$
$$+ \$1000(A|F\,15,6)$$
$$= -\$6000(.2642) - \$1000 - \$1000(2.0972) + \$1000(.1142)$$
$$= -\$4568.20$$

If a 5-year planning horizon is used, to obtain an annual worth of $-\$4568.20$ for Alternative 2 requires that the salvage value at the end of the fifth year, S_5, be such that

$$-\$4568.20 = -\$6000(A|P\,15.5) - \$1000 - \$1000(A|G\,15,5) + S_5(A|F\,15,5)$$
$$-\$4568.20 = -\$6000(.2983) - \$1000 - \$1000(1.7228) + S_5(.1483)$$
$$S_5 = -\$374.92$$

Thus, instead of having a $1000 salvage value at the end of the sixth year, a salvage value of $-\$374.92$ must exist at the end of the fifth year to yield the annual worth obtained by treating individual life cycles.

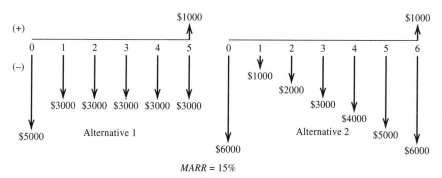

$MARR = 15\%$

Figure 5.16 Cash flow diagrams for two investment alternatives.

We do not recommend that alternatives having unequal lives be blindly compared on the basis of the annual worths for individual life cycles. Such an approach assumes *implicitly* that either a least common multiple of lives planning horizon is appropriate or the salvage values for unused portions of an asset's life are such that the annual worth is unchanged. We prefer to make *explicit* all assumptions concerning the planning horizon and the salvage values.

5.11   REPLACEMENT ANALYSIS

Replacement analysis is one of the most important and most common types of alternative comparisons encountered in practice. In a replacement analysis, one of the feasible alternatives involves maintaining the status quo; the remaining alternatives provide various replacement options that are available. In some organizations, replacement analyses are performed routinely in an effort to ensure that the best equipment and facilities are in use, compared to their possible successors.

The reasons for considering replacement are numerous. First, the current asset (defender) may have a number of deficiencies including high set-up cost, excessive maintenance, declining production efficiency, heavy energy consumption, and physical impairment. For example, when you are confronted with a car that is expensive to operate and maintain, or one soon to need a major overhaul, you begin to consider replacing the car.

Second, potential replacement assets (challengers) may take advantage of new technology and be easily set up, maintained at low cost, high in output, energy efficient, and possess increased capabilities, perhaps at a vastly reduced cost. For example, some new-generation, computer-controlled manufacturing equipment has rendered many old machines economically obsolete. Also, we can relate to the phenomenal accomplishments with which calculators and personal computers have resulted in increased capabilities, vastly lower prices, and economic obsolescence for equipment only a few years old.

Finally, the environment affects replacement decisions. Consumer preferences change, and present equipment may be unable to adapt to the new designs dictated by consumers. Demand levels may also cause equipment capacity to be relatively high or low, resulting in inefficiency or inability to perform. Rental firms specializing in equipment, automobiles, furniture, and so forth may affect ownership decisions by offering lease options. Also, outside contractors may be able to perform some tasks cheaper and better than in-house facilities. All of the above considerations and more may affect the decision to keep or replace an asset.

Replacement analyses are basically just another type of alternative comparison. That is, they follow the same systematic seven-step approach outlined and developed in this chapter. All of the six consistent measures of investment worth are also applicable. Unfortunately, replacement decisions are often confounded by our nearness to the existing asset. That is, we may have an "emotional" attachment to present equipment, particularly if we are the ones who recommended its installation previously. This attachment often results in attempts to recover *sunk costs,* which are past costs that are *unrecoverable*. Sunk costs should not be included in a replacement analysis (or any other alternative evaluation); to include sunk costs penalizes or burdens the potential replacement asset and unfairly favors the current asset.

Two approaches are commonly used in replacement analyses. One is the *cash flow approach* in which actual cash flows associated with keeping, purchasing, or leasing an asset are used directly. The other is the *outsider viewpoint approach* in which the cash flow profiles faced by an objective outsider are used. Which approach is used in a replacement analysis is strictly a matter of preference. Both are mathematically equivalent and yield consistent decisions.

5.11.1   Cash Flow Approach

The *cash flow approach* might properly be called the *insider viewpoint approach*. Often, there is no additional capital cost if the present asset is kept. If a replacement is purchased, there is often a trade-in allowance given for the present asset. In order to develop the cash flow profiles, the decision maker should ask "How much money will be spent and received if I adopt this alternative?" As noted above, past costs should be viewed from the proper perspective; unrecoverable past costs are *sunk costs* and are not to be included in economy studies that deal with the future, except as those sunk costs may affect income taxes if a present asset is disposed.

The planning horizon to be used is again at the discretion of the decision maker. As in any alternative evaluation, the current asset and its proposed replacements must be evaluated over a common planning horizon, with cash flow profiles extending throughout but not beyond that horizon for each alternative. Since the remaining life of the present asset is usually shorter than that of a new asset, the *shortest life among alternatives* is often chosen.

EXAMPLE 5.32

To illustrate a replacement analysis, consider a situation involving a chemical plant that owns a filter press that was purchased 3 years ago for $30,000. Actual operating and maintenance (O & M) expenses (excluding labor) for the press have been $4000, $5000, and $6000 each of the past 3 years, as depicted in Table 5.18. It is anticipated that the filter press can be used for 5 more years and salvaged for $2000 at that time. The undepreciated worth of the press as indicated on our accounting books is $12,600; however, technological developments of the past 3 years have resulted in excellent new competitive filter presses. As a result, the current market value for the used filter press is only $9000. If the old filter press is retained, annual operating and maintenance costs are anticipated to be as shown in Table 5.18.

A new filter press is available and can be purchased for $36,000. It has an anticipated life of 10 years and is expected to have annual operating and maintenance costs as also given in Table 5.18. Based upon historical data concerning salvage values of filter presses, estimated salvage values (S_t) for the new press are also given in the table. Alternative 1 is defined to be "keep the old press," while Alternative 2 is defined to be "replace the old press with the new press."

Table 5.18 Data for a Replacement Alternative

	Alterative 1		Alternative 2		
End of Year, t	Operating and Maintenance Costs	End of Year, t	Operating and Maintenance Costs	Salvage Value, S_t
−3	—	0	—	$36,000
−2	−$ 4,000	1	—	30,000
−1	− 5,000	2	−$1,000	24,600
0	− 6,000	3	− 2,000	19,800
1	− 7,000	4	− 3,000	15,600
2	− 8,000	5	− 4,000	12,000
3	− 9,000	6	− 5,000	9,000
4	− 10,000	7	− 6,000	6,600
5	− 11,000	8	− 7,000	4,800
		9	− 8,000	3,600
		10	− 9,000	3,000

Since the "old" press can only be used for 5 more years, a planning horizon of 5 years is specified; the salvage value for the "new" press is estimated to be $12,000 in 5 years. Cash flows for each alternative are given in Table 5.19. Note that the $9000 market value of the existing press is applied as a positive cash flow for the challenger since the old unit will be sold if the new unit is purchased. Computing the annual worths for each alternative using a minimum attractive rate of return of 15% yields the following results:

$$AW_1(15\%) = -\$7000 - \$1000(A\,|\,G\;15,5) + \$2000(A\,|\,F\,15,5)$$
$$= -\$7000 - \$1000(1.7228) + \$2000(.1483)$$
$$= -\$8426.20/\text{year}$$
$$AW_2(15\%) = -\$27,000(A\,|\,P\,15,5) - \$1000(A\,|\,G\;15,5) + \$12,000(A\,|\,F\,15,5)$$
$$= -\$27,000(.2983) - \$1000(1.7228) + \$12,000(.1483)$$
$$= -\$7997.30/\text{year}$$

On the basis of the annual worth difference of $428.90/year, it is recommended that the new filter press be purchased and the old press be sold.

In listing the cash flows for each alternative in Table 5.19, we ignored the operating and maintenance costs that will occur after the fifth year if the new filter press is purchased. The argument against including these operating and maintenance costs is based on the specification of a planning horizon of 5 years. The 5-year planning horizon was based on the maximum useful life for the old filter press. If the old press is retained, then it will have to be replaced in 5 years (if not before, because of the possible development of attractive replacement alternatives in the future). Consequently, in 5 years we might have available an alternative that will yield even greater operating and maintenance savings than the filter press currently being considered. Once the planning horizon has been

Table 5.19 Cash Flows for a Replacement Alternative—
Cash Flow Approach

End of Year, t	Alternative 1 Net Cash Flows, A_{1t}	Alternative 2 Net Cash Flows, A_{2t}	Difference in Cash Flows, $A_{2t} - A_{1t}$
0	$0	$-$36{,}000 + 9{,}000$	$-$27{,}000
1	$-$ 7{,}000$	0	7,000
2	$-$ 8{,}000$	$-1{,}000$	7,000
3	$-$ 9{,}000$	$-2{,}000$	7,000
4	$-$ 10{,}000$	$-3{,}000$	7,000
5	$-$11{,}000 + 2{,}000$	$-4{,}000 + 12{,}000$	17,000

established, it is not fair to include any cash flows that might occur later for one alternative without including similar estimates for the other alternative(s).

The above interpretation defines the planning horizon as a window through which only the cash flows that occur during the planning horizon can be seen. This window should *include a terminal (salvage) value for any alternative having a life longer than the planning horizon,* even though the alternative might not be physically replaced at that time. The end of the planning horizon defines a point in time at which another replacement study is planned. At that time, the future savings and costs will be compared against other available replacement candidates.

It should also be noted that neither the $30,000 first cost nor the $12,600 book value of the existing press appears as a cash flow. The $30,000 is a past cost which is now irrelevant (except as it affects taxes as seen in Chapter Six), and the $12,600 was never a cash flow. The difference between the value on our accounting books ($12,600) and the market value ($9000) is $3600. This $3600 is a sunk cost that analysts are often tempted to add to the first cost of the challenger in an attempt to recover it. To do so is absolutely incorrect, and biases the decision maker against the proposed replacement item. For example, if we burdened Alternative 2 by adding a sunk cost of $3600 to its first cost, its annual worth would be decreased by $-$3600(A\,|\,P\ 15{,}5) = -$3600(.2983) = -$1073.88/$year to a total of $-$7997.30/$year $-$ $1073.88/$year $= -$9071.18/$year. On this basis, we would mistakenly select Alternative 1 and continue to operate an economically inferior filter press.

EXAMPLE 5.33

If, in Example 5.32, a 10-year planning horizon is desired, we recommend that consideration be given to the replacement of the "old" filter press in 5 years. Based on a forecast of the growth of filter press technology, suppose we anticipate that at the end of 5 years a filter press will be available at a cost of $31,000. Net operating and maintenance costs, and a terminal salvage value are anticipated to be as depicted in Table 5.20.

Table 5.20 Cash Flows for a Replacement Alternative Using a 10-Year Planning Horizon-Cash Flow Approach

End of Year, t	Alternative 1 Net Cash Flows, A_{1t}	Alternative 2 Net Cash Flows, A_{2t}
0	$ 0	−$36,000 + $9,000
1	−$ 7,000	0
2	− 8,000	1,000
3	− 9,000	−2,000
4	− 10,000	−3,000
5	−11,000 + 2,000 − 31,000	−4,000
6	0	−5,000
7	− 1,000	−6,000
8	− 2,000	−7,000
9	− 3,000	−8,000
10	−4,000 + 15,000	−9,000 + 3,000

Calculating the annual worth of these alternatives results in:

$$
\begin{aligned}
AW_1(15\%) &= [-\$7000P|A\,15,5) + \$1000(P|G\,15,5) - \$29,000(P|F\,15,5) \\
&\quad -\$1000(P|G\,15,5)(P|F\,15,5) \\
&\quad + \$15,000(P|F\,15,10)](A|P\,15,10) \\
&= (-\$7000(3.3522) - \$1000(5.7751) - \$29,000(.4972) \\
&\quad -\$1000(5.7751)(.4972) + \$15,000(.2472)](.1993) \\
&= -\$8534.56/\text{year} \\
AW_2(15\%) &= -\$27,000(A|P\,15,10) - \$1000(A|G\,15,10) \\
&\quad + \$3,000(A|F\,15,10) \\
&= -\$27,000(.1993) - \$1000(3.3832) + \$3000(.0493) \\
&= -\$8616.40/\text{year}
\end{aligned}
$$

These computations establish that the difference in annual worths for the two alternatives is $81.84/year, with Alternative 1 being more economic. Thus, on the basis of technological forecasts of filter press alternatives in 5 years, it would appear advantageous to postpone the replacement. However, the degree of uncertainty in our forecast of the cash flows for a projected future replacement candidate would cause us to question the merits of postponing the replacement because of a difference of $81.84/year.

Often, there is a discrepancy between trade-in allowances if several potential replacement assets are being considered. Further, these trade-in values have little relationship to the true market value that could be realized if the existing asset were sold separately. In this case, when using the cash flow approach, the appropriate question to ask is still, "How much money will be spent and received if I adopt this alternative?"

EXAMPLE 5.34

In Example 5.32 Alternative 1 was to keep our existing filter press having a market value of $9000. Alternative 2 included selling the current press and buying a new one for $36,000. Suppose, however, that the dealer has offered to allow a $10,000 trade-in value for the old press. Also two additional alternatives have been identified. Alternative 3 is a new asset costing $40,000, having a salvage value after 5 years of $13,000, and having annual operating and maintenance costs as given in Table 5.21. A trade-in value of $12,000 is offered for the current press in Alternative 3. Alternative 4 is to lease a press for $7500/year payable at the beginning of each year during the 5-year horizon. If the lease is taken, the existing press will be sold on the open market. The complete cash flow profile for each alternative is given in Table 5.19.

Calculating the annual worths of these alternatives at a $MARR$ of 15% results in

$$AW_1(15\%) = -\$7000 - \$1000(A\,|\,G\,15,5) + \$2000(A\,|\,F\,15,5)$$
$$= -\$7000 - \$1000(1.7228) + \$2000(.1483)$$
$$= -\$8426.20/\text{year}$$

$$AW_2(15\%) = -\$26,000(A\,|\,P\,15,5) - \$1000(A\,|\,G\,15,5) + \$12,000(A\,|\,F\,15,5)$$
$$= -\$26,000(.2983) - \$1000(1.7228) + \$12,000(.1483)$$
$$= -\$7699.00/\text{year}$$

$$AW_3(15\%) = -\$28,000(A\,|\,P\,15,5) - \$500 - \$500(A\,|\,G\,15,5)$$
$$+ \$13,000(A\,|\,F\,15,5) = -\$28,000(.2983) - \$500$$
$$- \$500(1.7228) + \$13,000(.1483) = -\$7785.90/\text{year}$$

$$AW_4(15\%) = -\$7500(F\,|\,P\,15,1) + \$9000(A\,|\,P\,15,5) - \$800(A\,|\,G\,15,5)$$
$$= -\$7500(1.15) + \$9000(.2983) - \$800(1.7228)$$
$$= -\$7318.54/\text{year}$$

Based upon these calculations, the lease alternative (Alternative 4) appears to be economically most favorable.

5.11.2 Outsider Viewpoint Approach

The *outsider viewpoint approach* is preferred by many because it forces the decision maker to view both the existing asset and its challengers from an objec-

Table 5.21 Cash Flows for Several Replacement Alternatives—Cash Flow Approach

End of Year, t	Alternative 1 Net Cash Flows, A_{1t}	Alternative 2 Net Cash Flows, A_{2t}	Alternative 3 Net Cash Flows, A_{3t}	Alternative 4 Net Cash Flows, A_{4t}
0	0	−$36,000 + $10,000	−$40,000 + 12,000	$7,500 + 9,000
1	−$ 7,000	0	− 500	−7,500
2	− 8,000	−1,000	−1,000	− 7,500 − 800
3	− 9,000	−2,000	−1,500	− 7,500 − 1,600
4	− 10,000	−3,000	−2,000	− 7,500 − 2,400
5	−11,000 + 2,000	−4,000 + 12,000	−2,500 + 13,000	−3,200

tive point of view as would an "outsider." That is, the outsider is assumed to have no existing asset. The outsider is then free to choose either a used asset (the defender) available for the price of its market value, or any of the potential replacement assets (the challengers).

In essence, the outsider viewpoint approach considers the salvage value of the existing asset to be its investment cost if it is retained in service. Such an approach is consistent with the opportunity cost concept described in Chapter Eight. Since the retention of the defender is equivalent to a decision to forego the receipt of its salvage value, then an opportunity cost is assigned to the defender.

EXAMPLE 5.35

Let us again consider Example 5.32, this time using the outsider viewpoint. Recall that the present filter press has a market value of $9000, a life of 5 years, and a salvage value of $2000 at that time. The challenger has a cost of $36,000, a life of 10 years, and estimated salvage values at any point in time as given in Table 5.22. A 5-year planning horizon is to be used and the $MARR$ is 15%. Cash flows from the outsider's point of view are as given in Table 5.22.

We can see by comparing Tables 5.19 and 5.22 that the differences in cash flows between alternatives are the same; therefore, the cash flow approach and the outsider viewpoint approach are equivalent. As further verification, we can calculate the annual worth of each alternative.

$$AW_1(15\%) = -\$9000(A\,|\,P\,15,5) - \$7000 - \$1000(A\,|\,G\,15,5)$$
$$+\$2000(A\,|\,F\,15,5)$$
$$= -\$9000(.2983) - \$7000 - \$1000(1.7228) + \$2000(.1483)$$
$$= -\$11,110.90/\text{year}$$
$$AW_2(15\%) = -\$36,000(A\,|\,P\,15,5) - \$1000(A\,|\,G\,15,5) + \$12,000(A\,|\,F\,15,5)$$
$$= -\$36,000(.2983) - \$1000(1.7228) + \$12,000(.1483)$$
$$= -\$10,682.00/\text{year}$$

Note that the difference in annual worths is $428.90/year in favor of the new filter press. This is the identical conclusion reached in Example 5.32.

Table 5.22 Cash Flows for a Replacement Alternative—Outsider's Viewpoint Approach

End of Year, t	Alternative 1 Net Cash Flows, A_{1t}	Alternative 2 Net Cash Flows, A_{2t}	Difference in Cash Flows, $A_{2t} - A_{1t}$
0	−$ 9,000	−$36,000	−$27,000
1	− 7,000	0	7,000
2	− 8,000	− 1,000	7,000
3	− 9,000	− 2,000	7,000
4	− 10,000	− 3,000	7,000
5	−11,000 + 2,000	−4,000 + 12,000	17,000

In using the outsider viewpoint approach, we must be careful to use "rational" first costs for each alternative. Example 5.34 showed that the trade-in allowances may be different between alternatives, and that there may be little correspondence between the trade-in value and an asset's true market value. Usually, a high trade-in value indicates the seller is using an inflated selling price. In reality, this inflated selling price would most likely be decreased if a straight purchase with no trade-in were made. Therefore, for analysis purposes, when using the outsider's viewpoint approach, *we will decrease the inflated selling price of the asset by the difference between the trade-in value and a lower market value.* Presumably, no adjustment is necessary if the trade-in allowance is less than the market value since we would simply choose to dispose of the defender separately at the market value price.

EXAMPLE 5.36

Now, let us consider Example 5.34 from the outsider viewpoint. Recall that Alternatives 2 and 3 cost $36,000 and $40,000, respectively. Also, recall that the market value for the current asset is $9000, and the trade-in allowances under Alternatives 2 and 3 are $10,000 and $12,000, respectively. Since these trade-in allowances are inflated by $1000 and $3000, respectively, the corresponding first costs of Alternatives 2 and 3 will be reduced to $35,000 and $37,000. A planning horizon of 5 years and a *MARR* of 15% are still in effect. Table 5.23 presents the cash flow profiles from the outsider's point of view.

Calculating the annual worths of these alternatives results in

$$AW_1(15\%) = -\$9000(A\,|\,P\,15,5) - \$7000 - \$1000(A\,|\,G\,15,5)$$
$$+\$2000(A\,|\,F\,15,5)$$
$$= -\$9000(.2983) - \$7000 - \$1000(1.7228) + \$2000(.1483)$$
$$= -\$11,110.90/\text{year}$$
$$AW_2(15\%) = -\$35,000(A\,|\,P\,15,5) - \$1000(A\,|\,G\,15,5) + \$12,000(A\,|\,F\,15,5)$$
$$= -\$35,000(.2983) - \$1000(1.7228) + \$12,000(.1483)$$
$$= -\$10,383.70/\text{year}$$

Table 5.23 Cash Flows for Several Replacement Alternatives—
Outsider's Viewpoint Approach

End of Year, t	Alternative 1 Net Flows, Cash A_{1t}	Alternative 2 Net Flows, Cash A_{2t}	Alternative 3 Net Flows, Cash A_{3t}	Alternative 4 Net Flows, Cash A_{4t}
0	−$ 9,000	−$35,000	−$37,000	−$7,500
1	− 7,000	0	− 500	− 7,500
2	− 8,000	− 1,000	− 1,000	−7,500 − 800
3	− 9,000	− 2,000	− 1,500	−7,500 − 1,600
4	− 10,000	− 3,000	− 2,000	−7,500 − 2,400
5	−11,000 + 2,000	−4,000 + 12,000	−2,500 + 13,000	− 3,200

$$AW_3(15\%) = -\$37{,}000(A\,|\,P\,15{,}5) - \$500 - \$500(A\,|\,G\,15{,}5)$$
$$+\$13{,}000(A\,|\,F\,15{,}5)$$
$$= -\$37{,}000(.2983) - \$500 - \$500(1.7228) + \$13{,}000(.1483)$$
$$= -\$10{,}470.60/\text{year}$$

$$AW_4(15\%) = -\$7500(A\,|\,P\,15{,}1) - \$800(A\,|\,G\,15{,}5)$$
$$= -\$7500(1.15) - \$800(1.7228)$$
$$= -\$10{,}003.24/\text{year}$$

Again, the lease alternative is preferred based upon annual worth calculations. Note that the above annual worth differences between alternatives using the outsider viewpoint approach are *exactly* the same as the annual worth differences between alternatives in Example 5.34, in which the cash flow approach was used. This further substantiates the equivalence of the two approaches.

5.11.3   Optimum Replacement Interval

Although logical arguments can be given for treating a replacement decision as just another economic investment alternative, many firms fail to subject their existing equipment to careful scrutiny on a periodic basis to determine if replacement is required. Despite the fact that replacement studies can yield significant reductions in costs, it still remains that many firms postpone replacing assets beyond the "optimum" time for replacement, perhaps because a decision to replace an asset involves a change, and resistance to change is inherent in most individuals. For example, an engineer who 2 years ago successfully argued that compressor Z should be replaced by compressor Y may now find that compressor X is more economical than compressor Y. If X is now championed it may be viewed by management as an admission that the wrong compressor was selected as a replacement for Z.

Some reasons for delaying the replacement of assets beyond the economic replacement time are:

1. The firm is making a profit with its present equipment.
2. The present equipment is operational and is producing an acceptable quality product.
3. There is risk or uncertainty associated with predicting the expenses of a new machine, whereas one is relatively certain about the expenses of the current machine.
4. A decision to replace equipment is a stronger commitment, for a period of time into the future, than keeping the existing equipment.
5. Management tends to be conservative in decisions regarding the replacement of costly equipment.
6. There may be a limitation on funds available for purchasing new equipment, but less limitation on funds for maintaining existing equipment.
7. There may be considerable uncertainty concerning the future demands for the services of the equipment in question.
8. Sunk costs psychologically affect decisions to replace equipment.

9. An anticipation that technological improvements in the future might render obsolete equipment available currently; a wait-and-see attitude prevails.

10. Reluctance to be a pioneer in adopting new technology; instead of replacing now, wait for the competition to act.

As a unit of equipment ages, operating and maintenance costs increase. At the same time, the capital recovery cost decreases with prolonged use of the equipment. The combination of decreasing capital recovery costs and increasing annual operating and maintenance costs results in the equivalent uniform annual cost taking on a form similar to that depicted in Figure 5.17.

By forecasting the operating and maintenance costs for each year of service, as well as the anticipated salvage values for various replacement ages, one can determine the "optimum replacement interval" for equipment.

EXAMPLE 5.37

Suppose a small compressor can be purchased for $1000; the salvage value for the compressor is assumed to be negligible, regardless of the replacement interval. Annual operating and maintenance costs are expected to increase by $75/year, with the first year's cost anticipated to be $150. Using a minimum attractive rate of return of 20%, the following equivalent uniform annual costs are obtained:

$$EUAC(n = 1) = \$1000(A\,|\,P\,20,1) + \$150 + \$75(A\,|\,G\,20,1)$$
$$= \$1350.00/\text{year}$$
$$EUAC(n = 2) = \$1000(A\,|\,P\,20,2) + \$150 + \$75(A\,|\,G\,20,2)$$
$$= \$838.59/\text{year}$$
$$EUAC(n = 3) = \$1000(A\,|\,P\,20,3 + \$150 + \$75(A\,|\,G\,20,3)$$
$$= \$690.63/\text{year}$$
$$EUAC(n = 4) = \$1000(A\,|\,P\,20,4) + \$150 + \$75(A\,|\,G\,20,4)$$
$$= \$631.87/\text{year}$$

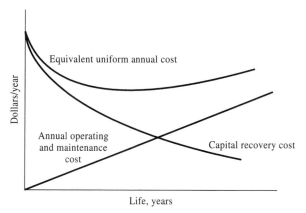

Figure 5.17 Portrayal of components of equivalent annual costs.

$$EUAC(n = 5) = \$1000(A\,|\,P\,20{,}5) + \$150 + \$75(A\,|\,G\,20{,}5)$$
$$= \$607.44/\text{year}$$

$$EUAC(n = 6) = \$1000(A\,|\,P\,20{,}6) + \$150 + \$75(A\,|\,G\,20{,}6)$$
$$= \$599.11/\text{year}$$

$$EUAC(n = 7) = \$1000(A\,|\,P\,20{,}7) + \$150 + \$75(A\,|\,G\,20{,}7)$$
$$= \$599.17/\text{year}$$

$$EUAC(n = 8) = \$1000(A\,|\,P\,20{,}8) + \$150 + \$75(A\,|\,G\,20{,}8)$$
$$= \$603.77/\text{year}$$

$$EUAC(n = 9) = \$1000(A\,|\,P\,20{,}9) + \$150 + \$75(A\,|\,G\,20{,}9)$$
$$= \$610.83/\text{year}$$

$$EUAC(n = 10) = \$1000(A\,|\,P\,20{,}10) + \$150 + \$75(A\,|\,G\,20{,}10)$$
$$= \$619.04/\text{year}$$

$$EUAC(n = 11) = \$1000(A\,|\,P\,20{,}11) + \$150 + \$75(A\,|\,G\,20{,}11)$$
$$= \$627.80/\text{year}$$

As n increases beyond 6 years, the equivalent uniform annual cost increases. Hence, for this example, a replacement interval of 6 years is indicated.

Certain assumptions are inherent in optimum replacement interval calculations. First, we implicitly assume that the planning horizon is an integer multiple of the replacement interval selected. (Recall our discussion of the use of the annual worth based on one life cycle and a planning horizon of the least common multiple of lives). Second, we have assumed that each time the compressor is replaced it will be replaced with a compressor having *equivalent uniform annual costs*. If neither assumption is valid, then the above approach is not valid.

EXAMPLE 5.38

Suppose in the previous case an 11-year planning horizon is appropriate. For simplicity, we will continue to assume that replacements will have identical cash flow profiles. If the original compressor is kept for 11 years, the present worth equivalent will be

$$PW(n = 11) = -\$627.80(P\,|\,A\,20{,}11)$$
$$= -\$2716.55$$

If the compressor is to be replaced at some intermediate point during the planning horizon, say after k years, and the replacement is to be kept until the end of the planning horizon, the following present worth calculations result for $k = 6$ or 7:

$$PW(k = 6) = -\$599.11(P\,|\,A\,20{,}6) - \$607.44(P\,|\,A\,20{,}5)(P\,|\,F\,20{,}6)$$
$$= -\$2600.72$$

$$PW(k = 7) = -\$599.17(P\,|\,A\,20{,}7) - \$631.87(P\,|\,A\,20{,}4)(P\,|\,F\,20{,}7)$$
$$= -\$2616.30$$

Thus, the compressor should be replaced after 6 years, and the replacement should be kept until the end of the planning horizon. Why is it unnecessary to

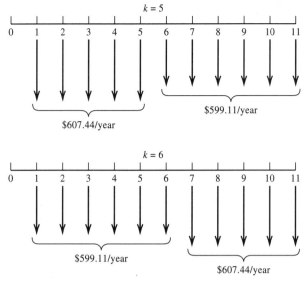

Figure 5.18 Comparison of replacement strategies involving replacement at the end of year 5 versus the end of year 6.

consider the remaining values of k (i.e., k = 5, 4, 3, 2, and 1)? In order to answer this, consider the case of k = 5 and compare the cash flow profiles over the 11-year planning horizon using k = 5 and k = 6. As shown in Figure 5.18, a consideration of the time value of money eliminates the possibility of k = 5 having a lower present worth than k = 6. Similar comparisons can be used to eliminate the cases of k = 1, 2, 3, and 4. Likewise, k = 8, 9, and 10 may be eliminated since costs have already begun to increase from k = 6 to k = 7.

For this particular example, failure to replace the equipment at precisely the optimum time will not result in a significantly lower annual *worth* or present *worth*. This is often the case with many replacement situations; the resulting measure of effectiveness is relatively insensitive to deviations from the optimum strategy. Hence, the firm might establish an operating policy of reviewing the *actual* operating and maintenance costs for the equipment at the anticipated "optimum" time of replacement and then perform a replacement study using the seven-step procedure for comparing investment alternatives.

The subject of replacement analysis is popular among engineering economists. In fact, sufficient literature is available on the subject to devote an entire book to it. For further treatment of replacement analysis, you may wish to consult the *Engineering Economist* publication.

5.12 SUMMARY

A systematic process of comparing investment alternatives was presented in this chapter. A cash flow approach was recommended, as was a planning horizon approach. Methods for comparing alternatives included the present worth, annual worth, future worth, internal rate of return, external rate of return, savings/investment ratio, payback period, and capitalized worth. The total cash flow and incremental cash flow approaches were considered in detail. Break-even, sensitivity, and risk analyses were described as supplementary analysis techniques. Replacement problems were treated as alternative comparison problems. The discussion of the comparison of investment alternatives did not consider either the effects of income taxes, the public sector requirements for measuring benefits in economic units, or the effects of risk and uncertainty; such considerations are reserved for the following chapters.

BIBLIOGRAPHY

1. American Telephone Company, *Engineering Economy,* 3rd Ed., McGraw Hill Book Co., New York, 1977.
2. Baldwin, R. H., "How to Assess Investment Proposals," *Harvard Business Review, 37* (3), 1959, pp. 98–99.
3. Banks, J., and Carson, II, J. S., *Discrete-Event System Simulation,* Prentice-Hall, Inc., Englewood Cliffs, NJ, 1984.
4. Banks, J., Burnette, B., Rose, J. D., and Kozloski, H., *SIMAN V and CINEMA V,* John Wiley & Sons, Inc., New York, 1995.
5. Belton, V., and Gear, T., "On a Shortcoming of Saaty's Method of Analytic Hierarchies," *Omega, 11* (3), 1983.
6. Belton, V., "A Comparison of the Analytic Hierarchy Process and a Simple Multi-Attribute Value Function," *European Journal of Operational Research, 26,* 1986, p. 7.
7. Beranek, W., "The Cost of Capital, Capital Budgeting, and the Maximization of Shareholder Wealth," *Journal of Financial and Quantitative Analysis, 10* (1), 1975, pp. 1–18.
8. Bernhard, R. H., "Discount Methods for Expenditure Evaluation—A Clarification of Their Assumptions," *Journal of Industrial Engineering, 18* (1), 1962, pp. 19–27.
9. Bernhard, R. H., "Unrecovered Investment, Uniqueness of the Internal Rate of Return and the Question of Project Acceptability," *Journal of Financial and Quantitative Analysis, 12* (1), 1977, pp. 33–38.
10. Boucher, T. O., and MacStravic, E. L., "Multiattribute Evaluation Within a Present Worth Framework and Its Relation to the Analytic Hierarchy Process," *The Engineering Economist, 37,* (1), Fall 1991, pp. 1–32.
11. Boucher, T. O., Luxhoj, J. T., Descovich, T., and Litman, N., "Multicriteria Evaluation of Automated Filling Systems: A Case Study," *Journal of Manufacturing Systems, 12* (5), 1993, pp. 357–378.
12. Buck, J. R., and Askin, R. G., "Partial Means in the Economic Risk Analysis of Projects," *The Engineering Economist, 31* (3) Spring 1986, pp. 189–212.
13. Bussey, L. E., *The Economic Analysis of Industrial Projects,* Prentice Hall, Inc., Englewood Cliffs, NJ, 1978.
14. Canada, J. R., "Annotated Bibliography on Justification of Flexible Manufacturing Systems," *The Engineering Economist, 31* (2), Winter 1986, p. 137.
15. Canada, J. R., and Sullivan, W. G., *Economic and Multiattribute Evaluation of Advanced Manufacturing Systems,* Prentice-Hall, Englewood Cliffs, NJ, 1989.
16. Canada, J. R., Sullivan, W. G., and White, J. A., *Capital Investment Decision Analysis for Management and Engineering,* 2nd Ed., Prentice-Hall, Englewood Cliffs, NJ, 1996.
17. Chandra, J., and Schall, S., "Economic Justification of Flexible Manufacturing Systems

Using the Leontief Input-Output Model," *The Engineering Economist, 34* (1), Fall 1988, p. 27.

18. DeGarmo, E. P., Sullivan, W. G., and Canada, J. R., *Engineering Economy,* 7th Ed., Macmillan, New York, 1984.

19. Dyer, J. S., "Remarks on the Analytic Hierarchy Process," *Management Service, 36* (3), 1990, p. 249.

20. *Economic Analysis Handbook,* Pub. P-442, Department of the Navy, Naval Facilities Engineering Command, Washington, DC, 1971.

21. Estes, C. B., and Jalali-Yazdi, A., "Now Replace Equipment on Actual Cost Basis," *Industrial Engineering, 10* (11), November 1978, pp. 34–37.

22. Falkner, C. H., and Benhajla, S., "Multiattribute Decision Models in the Justification of CIM Systems," *The Engineering Economist, 35* (2), Winter 1990, p. 91.

23. Fleischer, G. A., "Two Major Issues Associated with the Rate of Return Method for Capital Allocation: The 'Ranking Error' and 'Preliminary Selection,'" *The Journal of Industrial Engineering, 17* (4), 1996, pp. 202–208.

24. Fleischer, G. A., *Risk and Uncertainty: Non-Deterministic Decision Making in Engineering Economy,* AIIE Monograph Series, American Institute of Industrial Engineers, 1975.

25. Fleischer, G. A., *Engineering Economy, Capital Allocation Theory,* Brooks/Cole, Monterey, CA, 1984.

26. Grant, E. L., Ireson, W. G., and Leavenworth, R. S., *Principles of Engineering Economy,* 7th Ed. John Wiley and Sons, Inc., 1982.

27. Giffin, W. C., *Introduction to Operations Engineering,* R. D. Irwin, New York, 1971.

28. Hertz, D. B., "Risk Analysis in Capital Investments," *Harvard Business Review, 42* (1), January–February 1964.

29. Hillier, F. S., "The Derivation of Probabilistic Information for the Evaluation of Risky Investments," *Management Science, 9* (3), April 1963.

30. Hillier, F. S., "Supplement to 'The Derivation of Probabilistic Information Evaluation of Risky Investments," *Management Science, 11* (3), January 1965.

31. Hillier, F. S., *The Evaluation of Risky Interrelated Investments,* North-Holland Publishing, 1969.

32. Hillier, F. S., "A Basic Approach to the Evaluation of Risky Interrelated Investments," Chapter 1 in *Studies in Budgeting,* R. F. Byrne, W. W. Cooper, A. Charnes, O. A. Davis, and D. A. Gilford, eds., North-Holland Publishing, 1971.

33. Hillier, F. S., "A Basic Model for Capital Budgeting of Risky Interrelated Projects," *The Engineering Economist, 17* (1), 1971.

34. Kaplan, R. S., "Must CIM be Justified by Faith Alone?," *Harvard Business Review, 64* (2), March–April 1986, pp. 87–95.

35. Kaplan, S., "A Note on a Method for Precisely Determining the Uniqueness or Nonuniqueness of the Internal Rate of Return for a Proposed Investment," *Journal of Industrial Engineering, 26* (1), 1965, pp. 70–71.

36. Kaplan, S., "Computer Algorithms for Finding Exact Rates of Return," *Journal of Business, 40* (4), 1967, pp. 389–392.

37. Kaplan, S., and Barish, N., "Decision Making Allowing for Uncertainty of Future Investment Opportunities," *Management Science, 13* (10), June 1967.

38. Klausner, R. F., "The Evaluation of Risk in Marine Capital Investments," *The Engineering Economist, 14* (4), Summer 1969.

39. Klausner, R. F., "Communicating Investment Proposals to Corporate Decision-Makers," *The Engineering Economist, 17* (1), 1971, pp. 45–55.

40. Kulonda, D. J., "Replacement Analysis with Unequal Lives—The Study Period Method," *The Engineering Economist, 23* (3), 1978, pp. 171–80.

41. Lane, E. F., and Verdini, W. H., "A Consistency Test for AHP Decision Makers," *Decision Sciences, 20,* 1989, pp. 575–582.
42. Law, A. M., and Kelton, W. D., *Simulation Modeling and Analysis,* 2nd Ed., McGraw-Hill, Inc., New York, 1991.
43. Leautaud, J. L. L., "On the Fundamentals of Economic Evaluation," *The Engineering Economist, 19* (2), 1974, pp. 105–26.
44. Lifson, M., "Evaluation Modeling in the Context of the Systems Decision Process," *Decision and Risk Analysis: Powerful New Tools for Management,* Proceedings of the Sixth Triennial Symposium, published by *The Engineering Economist,* 1971.
45. Lohmann, J. R., "The IRR, NPV and the Fallacy of the Reinvestment Rate Assumptions," *The Engineering Economist, 33* (4), 1988, pp. 303–330.
46. Mao, J. C. T., "The Internal Rate of Return as a Ranking Criterion," *The Engineering Economist, 11* (4), 1966, pp. 1–13.
47. Mao, J. C. T., "An Analysis of Criteria for Investment and Financing Decisions Under Certainty: A Comment," *Management Science, 13* (3), 1966, pp. 289–91.
48. Mao, J. C. T., *Quantitative Analysis of Financial Decisions,* Macmillan, 1969.
49. May, K. O., "Intransitivity, Utility, and the Aggregation of Preference Patterns," *Econometrica, 22* (1), 1954, pp. 1–13.
50. Oakford, R. V., *Capital Budgeting: A Quantitative Evaluation of Investment.* Ronald Press, 1970.
51. Park, C. S. *Modern Engineering Economic Analysis,* Addison-Wesley, Reading, MA, 1990.
52. Park, C. S., and Sharp-Bette, G. P., *Advanced Engineering Economics,* John Wiley & Sons, Inc., New York, 1990.
53. Pegden, C. D., Shannon, R. E., and Sadowski, R. P., *Introduction to Simulation using SIMAN,* McGraw-Hill, Inc., New York, 1990.
54. Pritsker, A. A. B., *Introduction to Simulation and SLAM II,* 3rd Ed., John Wiley & Sons, Inc., New York, 1986.
55. Renshaw, E., "A Note on the Arithmetic of Capital Budgeting Decisions," *The Journal of Business, 30* (3), 1957, pp. 193–204.
56. Saaty, T. L., *The Analytic Hierarchy Process,* McGraw-Hill, New York, 1980.
57. Shore, B., "Replacement Decisions Under Capital Budgeting Constraints," *The Engineering Economist, 20* (4), 1975, pp. 243–56.
58. Solomon, E., "The Arithmetic of Capital Budgeting Decisions," *The Journal of Business, 29* (2), 1956, pp. 124–29.
59. Tarquin, A. J., and Blank, L. T., *Engineering Economy,* 2nd Ed., McGraw-Hill, 1983.
60. Teichroew, D., Robicheck, A. A., and Montalbano, M., "An Analysis of Criteria for Investment and Financing Decisions Under Certainty," *Management Science, 12* (3), 1965, pp. 151–79.
61. Thuesen, G. J., and Fabrycky, W. J., *Engineering Economy,* 7th Ed., Prentice-Hall, Englewood Cliffs, NJ, 1989.
62. Wabalickis, R. N., "Justification of FMS with the Analytic Hierarchy Process," *Journal of Manufacturing Systems, 7* (3), p. 175.
63. Wicks, E. L. M., and Boucher, T. O., *ISC's Users Manual for NCIC: A Decision Support Software Package for Investments in Integrated Automation and Advanced Systems,* Version 3.0, Integrated Systems and Controls Council, Material Handling Institute Division, Material Handling Industry, Charlotte, NC, 1993.
64. Wohl, M., "A New Ordering Procedure and Set of Decision Rules for the Internal Rate of Return Method," *The Engineering Economist, 30* (4), 1985, pp. 363–386.
65. Young, D., *Modern Engineering Economy,* John Wiley & Sons, Inc., New York, 1993.

PROBLEMS

1. Marlow Electronics has identified four investment proposals, A, B, C, and D. Proposals A and B are mutually exclusive and Proposals A and D are mutually exclusive. Proposal C is contingent on either Proposal A or B. To do nothing is not a viable alternative. Clearly specify all feasible alternatives to be considered. (5.2)
2. The ABC Company has identified three major proposals, A, B, and C, to enhance customer service. Proposals A and C are mutually exclusive; Proposal C is contingent on Proposal B. The initial investments required for Proposals A, B, and C are $70,000, $80,000, and $35,000, respectively. A total of $120,000 is available for investment. Clearly specify all feasible alternatives to be considered. (5.2)
3. Greasy Petroleum Company has available four proposals, A, B, C, and D. Proposal B is contingent on the acceptance of either Proposal A or Proposal D. Also, Proposal D is contingent on Proposal A, while Proposal A is contingent on either Proposal B or Proposal C. The initial investments for these proposals are $420,000, $60,000, $45,000, and $300,000, respectively. If Greasy has a budget limitation of $600,000, clearly identify all feasible alternatives to be considered. (5.2)
4. Consider the net cash flows (NCF) and salvage values (SV) for each of Alternatives 1 and 2 having lives of 3 and 5 years, respectively.

	Alternative 1		Alternative 2	
EOY	NCF_1	SV_1	NCF_2	SV_2
0	−$60,000	$60,000	−$100,000	$100,000
1	35,000	30,000	30,000	60,000
2	35,000	10,000	30,000	40,000
3	35,000	0	30,000	25,000
4			30,000	15,000
5			30,000	10,000

Assume each alternative can be renewed indefinitely with the same NCF and SV profiles.
   (a) If a least common multiple of lives approach is to be used, specify the planning horizon and the complete set of cash flows for each alternative.
   (b) Repeat part (a) using the shortest life among alternatives.
   (c) Repeat part (a) using the longest life among alternatives.
   (d) Repeat part (a) using a planning horizon of 2 years.
   (e) Repeat part (a) using a planning horizon of 4 years. (5.3)
5. Two alternatives have the following net cash flow (NCF) and salvage value (SV) profiles *for the first cycle of each.*

	Alternative 1		Alternative 2	
EOY	NCF_1	SV_1	NCF_2	SV_2
0	−$50,000	$50,000	−$80,000	$80,000
1	25,000	25,000	40,000	50,000
2	30,000	10,000	45,000	20,000
3	35,000	0	50,000	10,000

	Alternative 1		Alternative 2	
EOY	NCF_1	SV_1	NCF_2	SV_2
4			55,000	0
5			60,000	0

Alternatives 1 and 2 can both repeat; however, the cash flow profiles and salvage values will change. Renewal of Alternative 1 will cost 40% more for the initial investment, increasing to $70,000. Alternative 1 salvage values will decrease following the same pattern relative to the initial investment (e.g., $70,000, $35,000, $14,000, $0). Further renewals of Alternative 1 will cause initial investments and salvage values to increase 40% over the previous cycle's values, and salvage values will again decrease in the same proportions as illustrated in the table. Annual net revenues are projected to continue increasing at $5000/year (e.g., $40,000/year, $45,000/year, etc.) indefinitely. Renewal of Alternative 2 will cost 60% more for the initial investment with each renewal, and salvage values will decrease following the pattern shown. Also, annual net revenues will continue increasing at $5000/year.

a. Specify the complete set of cash flows for each alternative if a planning horizon of 2 years is to be used.

b. Repeat part (a) using a planning horizon of 3 years.

c. Repeat part (a) using a planning horizon of 4 years.

d. Repeat part (a) using a planning horizon of 5 years.

e. Repeat part (a) using a planning horizon of 10 years. (5.3)

6. Alternatives 1, 2, and 3 have lives of 3, 4, and 6 years, respectively. Their net cash flow (NCF) and salvage value (SV) profiles are as follows:

	Alternative 1		Alternative 2		Alternative 3	
EOY	NCF_1	SV_1	NCF_2	SV_2	NCF_3	SV_3
0	−$20,000	—	−$40,000	$40,000	−$70,000	$70,000
1	8,000	—	20,000	30,000	30,000	50,000
2	8,000	—	20,000	20,000	30,000	30,000
3	28,000	—	20,000	10,000	30,000	20,000
4			20,000	0	30,000	10,000
5					30,000	5,000
6					30,000	2,000

The NCF profile for Alternative 1 is due to a $20,000/year lease payable at the beginning of each year, plus an end-of-year net revenue of $28,000. This lease arrangement may be renewed indefinitely; however, premature cancellation of the lease costs $10,000. All other cash flows and salvage values are expected to repeat indefinitely as shown.

a. If a least common multiple of lives approach is to be used, specify the planning horizon and the complete set of cash flows for each alternative.

b. Repeat part (a) using the shortest life among alternatives.

c. Repeat part (a) using the longest life among alternatives.

d. Repeat part (a) using a planning horizon of 2 years.

e. Repeat part (a) using a planning horizon of 5 years.

     f. Repeat part (a) using a planning horizon of 9 years.

     g. Repeat part (a) using a (tentative) planning horizon of 18 years. (5.3)

  7. Four investment proposals, A, B, C, and D, are available. Proposals A and B are mutually exclusive, and Proposals A and C are mutually exclusive. Proposal D is contingent on either Proposal A or B. Funds available for investment are limited to $250,000.

EOY	NCF(A)	NCF(B)	NCF(C)	NCF(D)
0	−$80,000	−$125,000	−$100,000	−$80,000
1-5	30,000	40,000	35,000	32,000

     a. Specify the alternatives to be considered. (5.2)

     b. Clearly show the net cash flow profile associated with each alternative. (5.4)

  8. Three investment proposals have been selected for further evaluation. Proposals A and C are mutually exclusive; Proposal A is contingent on Proposal B; and Proposal B is contingent on Proposal C. A budget limitation of $200,000 exists. The cash flow profiles for the three investment proposals are given below for the 4-year planning horizon.

EOY	NCF(A)	NCF(B)	NCF(C)
0	−$60,000	−$40,000	−$80,000
1-4	20,000	10,000	15,000
4	—	15,000	20,000

     a. Specify the alternatives to be considered. (5.2)

     b. Clearly show the net cash flow profile associated with each alternative. (5.4)

  9. Metal Salvage, Inc. has available three investment proposals A, B, and C, having the cash flow profiles shown below. Proposals B and C are mutually exclusive and Proposal C is contingent on Proposal A being chosen.

	NCF(A)	NCF(B)	NCF(C)
Initial investment	$400,000	$600,000	$300,000
Life	8 years	12 years	6 years
Annual receipts	$320,000	$380,000	$400,000
Annual disbursements	$230,000	$240,000	$300,000
Salvage value	$100,000	$200,000	$100,000

     The firm is willing to use a planning horizon of 24 years and accepts the assumption of identical cash flow profiles for successive life cycles of a proposal.

     a. Determine the set of feasible alternatives. (5.2)

     b. Determine the net cash flow profile for each alternative. (5.4)

  10. A firm has available four proposals, A, B, C, and D. Proposal A is contingent on acceptance of either Proposal C or Proposal D. In addition, Proposal C is contingent on Proposal D, while Proposal D is contingent on either Proposal A or Proposal B. The firm has a budget limitation of $500,000. The cash flows are as follows:

EOY	NCF(A)	NCF(B)	NCF(C)	NCF(D)
0	−$250,000	−$350,000	−$50,000	−$38,000
1	85,000	119,000	10,000	3,000
2	85,000	122,000	10,000	8,000
3	85,000	125,000	10,000	13,000
4	85,000	128,000	10,000	18,000
5	150,000	170,000	10,000	23,000

a. Specify the alternatives to be considered. (5.2)

b. Clearly show the net cash flow profile associated with each alternative. (5.4)

11. ABC Limited is faced with three investment proposals, A, B, and C, having the cash flow profiles shown below over the planning horizon of 5 years. Proposals B and C are mutually exclusive, and Proposal C is contingent on Proposal A being selected. A budgetary limitation of $100,000 exists on the amount that can be invested initially.

EOY	NCF(A)	NCF(B)	NCF(C)
0	−$50,000	−$25,000	−$10,000
1	10,000	10,000	3,000
2	15,000	10,000	3,000
3	20,000	10,000	3,000
4	25,000	10,000	3,000
5	30,000	10,000	7,500

Specify the cash flow profiles for all feasible investment alternatives. (5.4)

12. Four investment proposals, W, X, Y, and Z, are being considered by the Ajax Corporation. Proposals X and Z are mutually exclusive. Proposal Y is contingent on either X or Z. Proposals W and Y are mutually exclusive. A budget limitation of $400,000 exists. Either Proposal X or Proposal Z must be included in the alternative selected.

EOY	NCF(W)	NCF(X)	NCF(Y)	NCF(Z)
0	−$200,000	−$250,000	−$180,000	−$200,000
1-8	60,000	100,000	50,000	40,000
8	80,000	20,000	240,000	200,000

a. Determine the set of feasible alternatives. (5.2)

b. Determine the net cash flow for each alternative. (5.4)

13. Rex Electric Company has borrowed money from First National Bank on a short-term loan at 18% compounded monthly. The company's tax rate is 36%. What is the after-tax cost of capital for this money? (5.5.1)

14. Ajax Machinery borrows money at 16% compounded semiannually. Assume a tax rate of 36%.

a. Determine the effective before-tax cost of capital in percent.

b. Determine the effective after-tax cost of capital in percent. (5.5.1)

15. Jericho Steel issues bonds promising semiannual interest payments at an annual nominal interest rate of 11%. The bonds are all sold for face value, and the costs of preparing and selling the bonds are relatively negligible. What is the after-tax cost of this debt funding to Jericho Steel if their tax rate is 36%? (5.5.1)

16. Jericho Steel calculates that they must pay interest of $30 each quarter to holders of bonds having $1000 face value. Using a 36% tax rate, clearly and explicitly state any necessary assumptions and
    a. Calculate the before tax interest rate of this debt funding.
    b. Calculate the after-tax interest rate of this debt funding. (5.5.1)

17. ABC Limited is a publicly held firm having a present trading price of $35 per share of common stock. The Board of Directors has just voted a dividend of $3 per share to be paid to all shareholders. What is the cost of this equity capital in percent? Is this a before-tax cost or an after-tax cost of capital? (5.5.1)

18. Dunbar Chemicals plans to invest its retained earnings in capital improvements and expansion ventures. The company's stock trades for $76 per share on the financial market, and the dividends recently paid were $6 per share. Is there a cost to Dunbar of using its own retained earnings? If so, what is the cost of capital in percent? (5.5.1)

19. The capital makeup of ABC Limited is as follows:

	Amount	Cost of Capital, Percent
Loans	$ 800,000	9.6
Bonds	$1,400,000	8.2
Stock	$2,100,000	11.9
Retained earnings	$ 700,000	11.9

What is the overall average cost of capital for ABC? (5.5.1)

20. Ajax Manufacturing Company has the following capital structure:

	Amount	Other Information
Loans	$2,600,000	14% before-tax rate, semiannual compounding
Bonds	$4,100,000	12% bond interest rate, quarterly payment
Stock	$5,000,000	$6.50 dividends on selling price of $65 per share
Retained earnings	$3,300,000	

Ajax's tax rate is 36%. Determine the firm's overall average after-tax cost of capital. (5.5.1)

21. Two investment alternatives are to be evaluated and the economically better alternative specified. The cash flow profiles for the alternatives are shown below.

EOY	NCF(1)	NCF(2)
0	−$10,000	−$15,000
1	1,000	1,500
2	1,500	2,000
3	—	500
4	2,500	3,000
5	3,500	4,000
6	—	500
7	4,500	5,000
8	6,000	11,500

a. Assuming $15,000 is available to invest, determine the total cash flow associated with each alternative over each year of the planning horizon. Any excess capital is invested at $MARR = 15\%$, earning annual returns over the planning horizon of 8 years. (5.6.1)

b. Determine the incremental cash flow profile between Alternatives 2 and 1 for each year of the planning horizon. (5.6.2)

22. Consider the following investment decision.

	Machine 1	Machine 2
First cost	$15,000	$20,000
Estimated life	5 years	10 years
Estimated annual revenues	$12,000	$14,000
Estimated annual operative costs	$ 6,000	$ 8,000
Estimated salvage value at end of 5 years	$ 1,500	$ 5,000

a. Assuming $20,000 is available to invest, determine the total cash flow associated with each machine over each year of the 5-year planning horizon. Any excess capital is invested at $MARR = 12\%$ earning annual returns over the planning horizon of 5 years. (5.6.1)

b. Determine the incremental cash flow profile between Machine 2 and Machine 1 for each year of the planning horizon of 5 years. (5.6.2)

23. Two mutually exclusive proposals, each with a life of 5 years, are under consideration. $MARR$ is 12%. Each proposal has the following cash flow profile:

EOY	NCF(A)	NCF(B)
0	−$30,000	−$42,000
1	9,300	12,625
2	9,300	12,625
3	9,300	12,625
4	9,300	12,625
5	9,300	12,625

a. Clearly specify the alternatives available and their net cash flow profiles. (5.4)
b. Determine which alternative the decision maker should select. Use a *ranking approach* and the present worth method. (5.6.3)
c. Determine which alternative the decision maker should select. Use an *incremental cash flow approach* and the present worth method (5.6.3)
d. Repeat part (b) using the annual worth method. (5.6.4)
e. Repeat part (c) using the annual worth method. (5.6.4)
f. Repeat part (b) using the future worth method. (5.6.5)
g. Repeat part (c) using the future worth method. (5.6.5)
h. Repeat part (c) using the internal rate of return method. (5.6.6)
i. Repeat part (c) using the external rate of return method. (5.6.7)
j. Repeat part (c) using the savings/investment ratio method. (5.6.8)
k. Repeat part (b) using the payback period method. (5.6.9)
l. Repeat part (b) using the discounted payback period method. (5.6.9).

24. The ABC Company is considering three investment proposals, A, B, and C. Proposals A and B are mutually exclusive, and Proposal C is contingent on Proposal B. The cash flow data for the investments over a 10-year planning horizon are given below. The ABC Company has a budget limit of $1 million for investments of the type being considered currently. *MARR* = 15%.

	NCF(A)	NCF(B)	NCF(C)
Initial investment	$600,000	$800,000	$470,000
Planning horizon	10 years	10 years	10 years
Salvage values	$ 70,000	$130,000	$ 65,000
Annual receipts	$400,000	$600,000	$260,000
Annual disbursements	$130,000	$270,000	$ 70,000

a. Clearly specify the alternatives available and their net cash flow profiles. (5.4)
b. Determine which alternative ABC's decision maker should select. Use a *ranking approach* and the present worth method. (5.6.3)
c. Determine which alternative ABC's decision maker should select. Use an *incremental cash flow approach* and the present worth method. (5.6.3)
d. Repeat part (b) using the annual worth method. (5.6.4)
e. Repeat part (c) using the annual worth method. (5.6.4)
f. Repeat part (b) using the future worth method. (5.6.5)
g. Repeat part (c) using the future worth method. (5.6.5)
h. Repeat part (c) using the internal rate of return method. (5.6.6)
i. Repeat part (c) using the external rate of return method. (5.6.7)
j. Repeat part (c) using the savings/investment ratio method. (5.6.8)
k. Repeat part (b) using the payback period method. (5.6.9)
l. Repeat part (b) using the discounted payback period method. (5.6.9).

25. A firm has available two mutually exclusive investment proposals, A and B. Their net cash flows are as shown in the table over a 10-year planning horizon. *MARR* is 12%.

EOY	NCF(A)	NCF(B)
0	−$40,000	−$30,000
1	8,000	9,000
2	8,000	8,500
3	8,000	8,000
4	8,000	7,500
5	8,000	7,000
6	8,000	6,500
7	8,000	6,000
8	8,000	5,500
9	8,000	5,000
10	8,000	4,500

a. Clearly specify the alternatives available and their cash flow profiles. (5.4)
b. Determine which alternative the decision maker should select. Use a *ranking approach* and the present worth method. (5.6.3)
c. Determine which alternative the decision maker should select. Use an *incremental cash flow approach* and the present worth method. (5.6.3)
d. Repeat part (b) using the annual worth method. (5.6.4)

e. Repeat part (c) using the annual worth method. (5.6.4)
f. Repeat part (b) using the future worth method. (5.6.5)
g. Repeat part (c) using the future worth method. (5.6.5)
h. Repeat part (c) using the internal rate of return method. (5.6.6)
i. Repeat part (c) using the external rate of return method. (5.6.7)
j. Repeat part (c) using the savings/investment ratio method. (5.6.8)
k. Repeat part (b) using the payback period method. (5.6.9)
l. Repeat part (b) using the discounted period method. (5.6.9)

26. A firm is faced with four investment proposals, A, B, C, and D, having the cash flow profiles shown below. Proposals A and C are mutually exclusive, and Proposal D is contingent on Proposal B being chosen. Currently, $750,000 is available for investment, and the firm has stipulated a *MARR* of 10%.

	NCF(A)	NCF(B)	NCF(C)	NCF(D)
Initial investment	$400,000	$400,000	$600,000	$300,000
Planning horizon	10 years	10 years	10 years	10 years
Annual receipts	$205,000	$215,000	$260,000	$230,000
Annual disbursements	$110,000	$125,000	$120,000	$150,000
Salvage value	$ 50,000	$ 50,000	$100,000	$ 50,000

a. Clearly specify the alternatives available and their net cash flow profiles. (5.4)
b. Determine which alternative the decision maker should select. Use a *ranking approach* and the present worth method. (5.6.3)
c. Determine which alternative the decision maker should select. Use an *incremental cash flow approach* and the present worth method. (5.6.3)
d. Repeat part (b) using the annual worth method. (5.6.4)
e. Repeat part (c) using the annual worth method. (5.6.4)
f. Repeat part (b) using the future worth method. (5.6.5)
g. Repeat part (c) using the future worth method. (5.6.5)
h. Repeat part (c) using the internal rate of return method. (5.6.6)
i. Repeat part (c) using the external rate of return method. (5.6.7)
j. Repeat part (c) using the savings/investment ratio method. (5.6.8)
k. Repeat part (b) using the payback period method. (5.6.9)
l. Repeat part (b) using the discounted period method. (5.6.9)

27. Rich N. Smug is considering two mutually exclusive investment alternatives, each requiring an investment of $5000. Alternative 1 returns $1000 after 1 year and $6000 after 2 years. This alternative is not renewable. Alternative 2 returns $1000/year for the first 3 years and $5000 after 4 years. Mr. Smug has established a *MARR* of 10%. He is committed to make one of the two investments.

a. Clearly specify the alternatives available and their net cash flow profiles. (5.4)
b. Determine which alternative Mr. Smug should select. Use a *ranking approach* and the present worth method. (5.6.3)
c. Determine which alternative Mr. Smug should select. Use an *incremental cash flow approach* and the present worth method. (5.6.3)
d. Repeat part (b) using the annual worth method. (5.6.4)
e. Repeat part (c) using the annual worth method. (5.6.4)
f. Repeat part (b) using the future worth method. (5.6.5)

   g. Repeat part (c) using the future worth method. (5.6.5)
   h. Repeat part (c) using the internal rate of return method. (5.6.6)
   i. Repeat part (c) using the external rate of return method. (5.6.7)
   j. Repeat part (c) using the savings/investment ratio method. (5.6.8)
   k. Repeat part (b) using the payback period method. (5.6.9)
   l. Repeat part (b) using the discounted period method. (5.6.9)

28. Insulation is being considered for installation on a steam pipe. Either 1-inch or 2-inch insulation is being figured. The annual heat loss from the pipe at present is estimated to be $2.50 per foot of pipe. The 1-inch insulation costs $1/foot and will reduce the heat loss by 65%; the 2-inch insulation costs $1.90/foot and will reduce the heat loss by 79%. The insulation will last 5 years with no salvage value. *MARR* is 25%.
   a. Clearly specify the alternatives available and their net cash flow profiles. (5.4)
   b. Determine which kind of insulation the decision maker should select. Use a *ranking approach* and the present worth method. (5.6.3)
   c. Determine which kind of insulation the decision maker should select. Use an *incremental cash flow approach* and the present worth method. (5.6.3)
   d. Repeat part (b) using the annual worth method. (5.6.4)
   e. Repeat part (c) using the annual worth method. (5.6.4)
   f. Repeat part (b) using the future worth method. (5.6.5)
   g. Repeat part (c) using the future worth method. (5.6.5)
   h. Repeat part (c) using the internal rate of return method. (5.6.6)
   i. Repeat part (c) using the external rate of return method. (5.6.7)
   j. Repeat part (c) using the savings/investment ratio method. (5.6.8)
   k. Repeat part (b) using the payback period method. (5.6.9)
   l. Repeat part (b) using the discounted period method. (5.6.9)

29. Three investment alternatives have the cash flow profiles shown below. A *MARR* of 20% is used in the selection of a preferred alternative.

EOY	NCF(1)	NCF(2)	NCF(3)
0	−$40,000	−$25,000	$0
1	16,500	5,000	0
2	16,500	10,000	0
3	16,500	15,000	0
4	16,500	20,000	0
5	16,500	25,000	0

   a. Determine which alternative the decision maker should select. Use a *ranking approach* and the present worth method. (5.6.3)
   b. Determine which alternative the decision maker should select. Use an *incremental cash flow approach* and the present worth method. (5.6.3)
   c. Repeat part (b) using the annual worth method. (5.6.4)
   d. Repeat part (c) using the annual worth method. (5.6.4)
   e. Repeat part (b) using the future worth method. (5.6.5)
   f. Repeat part (c) using the future worth method. (5.6.5)
   g. Repeat part (c) using the internal rate of return method. (5.6.6)
   h. Repeat part (c) using the external rate of return method. (5.6.7)
   i. Repeat part (c) using the savings/investment ratio method. (5.6.8)

j. Repeat part (c) using the savings/investment ratio method. (5.6.8)
k. Repeat part (b) using the payback period method. (5.6.9)
l. Repeat part (b) using the discounted period method. (5.6.9)
30. Consider the net cash flows of the following three alternatives. None is renewable. *MARR* is 12%.

EOY	NCF(1)	NCF(2)	NCF(3)
0	−$3,000	−$4,000	−$6,000
1	1,000	4,000	3,000
2	1,500	2,000	3,000
3	0	0	3,000
4	4,000	0	0
5	6,000	0	0

a. Compare these alternatives on the basis of the present worth method. (5.6.3)
b. Compare these alternatives on the basis of the payback period method. (5.6.7)
31. Two alternatives for a recreational facility are being considered. Their cash flow profiles are as follows for lives of 5 and 3 years, respectively.

EOY	NCF(1)	NCF(2)
0	−$11,000	−$5,000
1	5,000	2,000
2	4,000	3,000
3	3,000	4,000
4	2,000	
5	1,000	

Each alternative is expected to repeat forever with roughly the same net cash flow profile. Using the capitalized worth method and a *MARR* of 10%, select the preferred alternative. (5.6.8)
32. Solve problem 24 using capitalized worth, assuming the life cycles repeat indefinitely. (5.6.8)
33. Solve problem 26 using capitalized worth, assuming the life cycles repeat indefinitely. (5.6.8)
34. Three bridge designs are under consideration for a Payne County road. Wood, steel, and concrete underlie the three designs. The wood design costs $50,000 installed, with $4000/year maintenance and $30,000 rework at the end of every 10 years. Steel costs $70,000 installed, with $3000/year maintenance. Concrete costs $80,000 installed, but costs only $2000/year in maintenance. If *MARR* is 8%, which alternative should be selected on the basis of capitalized cost, assuming profiles remain constant indefinitely? (5.6.8)
35. Two 100-horsepower motors are under consideration by the Mighty Machinery Company. Motor Q costs $5000 and operates at 90% efficiency. Motor R costs $3500 and is 88% efficient. Annual operating and maintenance costs are estimated to be 15% of the initial purchase price. Power costs 3.2¢/kilowatt-hour. How many hours of full-load operation are necessary each year in order to justify the purchase of motor Q?

Use a 15-year planning horizon; assume that salvage values will equal 20% of the initial purchase price; and let the *MARR* be 15%. (*Note*: 0.746 kilowatts = 1 horsepower.) (5.7.1)

36. An aluminum extrusion plant manufactures a particular product at a variable cost of $0.04 per unit, including material cost. The fixed costs associated with manufacturing the product equal $30,000/year. Determine the break-even value for annual sales if the selling price per unit is (a) $0.40, (b) $0.30, and (c) $0.20. (5.7.1)

37. Owners of a nationwide motel chain are considering locating a new motel in Snyder, Arkansas. The complete cost of building a 150-unit motel (excluding furnishings) is $5 million; the firm estimates that the furnishings in the motel must be replaced at a cost of $1,875,000 every 5 years. Annual operating and maintenance cost for the facility is estimated to be $125,000. The average rate for a unit is anticipated to be $55/day. A 15-year planning horizon is used by the firm in evaluating new ventures of this type; a terminal salvage value of 20% of the original building cost is anticipated; furnishings are estimated to have no salvage value at the end of each 5-year replacement interval; land cost is not to be included. Determine the break-even value for the daily occupancy percentage based on a *MARR* of (a) 0%, (b) 10%, (c) 15%, and (d) 20%. (Assume that the motel will operate 365 days/year.) (5.7.1)

38. A consulting engineer is considering two pumps to meet a demand of 15,000 gallons/minute at 12 feet total dynamic head. The specific gravity of the liquid being pumped is 1.50. Pump A operates at 70% efficiency and costs $12,000; pump B operates at 75% efficiency and costs $18,000. Power costs $0.015/kilowatt-hour. Continuous pumping for 365 days/year is required (i.e., 24 hours/day). Using a *MARR* of 10% and assuming equal salvage values for both pumps, how many years of service are required for pump B to be justified economically? (*Note:* Dynamic head times gallon/minute times specific gravity divided by 3960 equals horsepower required. Horsepower times 0.746 equals kilowatts required.) (5.7.1)

39. A business firm is contemplating the installation of an improved material-handling system between the packaging department and the finished goods warehouse. Two designs are being considered. The first consists of an automated guided vehicle system (AGVS) involving three vehicles on the loop. The second design consists of a pallet conveyor installed between packaging and the warehouse. The AGVS will have an initial equipment cost of $280,000 and annual operating and maintenance costs of $50,000. The pallet conveyor has an initial cost of $360,000 and annual operating and maintenance costs of $35,000. The firm is not sure what planning horizon to use in the analysis; however, the salvage value estimates given in the following table have been developed for various planning horizons. Using a *MARR* of 10%, determine the break-even value for *N*, the planning horizon. (5.7.1)

Salvage Value Estimates		
N	AGVS	Pallet Conveyor
1	$230,000	$300,000
2	185,000	245,000
3	145,000	200,000
4	110,000	160,000
5	80,000	125,000
6	55,000	95,000

	Salvage Value Estimates	
N	AGVS	Pallet Conveyor
7	35,000	70,000
8	20,000	50,000
9	10,000	35,000
10	5,000	25,000
11	—	20,000
12	—	20,000

40. A manufacturing plant in Michigan has been contracting snow removal at a cost of $400/day. The past 3 years have produced heavy snowfalls, resulting in the cost of snow removal being of concern to the plant manager. The plant engineer has found that a snow-removal machine can be purchased for $25,000; it is estimated to have a useful life of 6 years, and a zero salvage value at that time. Annual costs for operating and maintaining the equipment are estimated to be $5,000. Determine the break-even value for the number of days per year that snow removal is required in order to justify the equipment, based on a *MARR* of (a) 0%, (b) 10%, and (c) 15%. (5.7.1)

41. The motor on a gas-fired furnace in a small foundry is to be replaced. Three different 15-horsepower electric motors are being considered. Motor X sells for $2500 and has an efficiency rating of 90%; motor Y sells for $1750 and has a rating of 85%; motor Z sells for $1000 and is rated to be 80% efficient. The cost of electricity is $0.065/kilowatt-hour. An 8-year planning horizon is used, and zero salvage values are assumed for all three motors. A *MARR* of 12% is to be used. Assume that the motor selected will be loaded to capacity. Determine the range of values for annual usage of the motor (in hours) that will lead to the preference of each motor. (*Note:* 0.746 kilowatts = 1 horsepower.) (5.7.1)

42. A machine can be purchased at $t = 0$ for $20,000. The estimated life is 5 years, with an estimated salvage value of zero at that time. The average annual operating and maintenance expenses are expected to be $5500. If *MARR* = 10%, what must the average annual revenues be in order to be indifferent between (a) purchasing the machine, or (b) doing nothing? (5.7.1)

43. Two condensers are being considered by the Ajax Company. A copper condenser can be purchased for $5000; annual operating and maintenance costs are estimated to be $500. Alternatively, a ferrous condenser can be purchased for $3500; since the Ajax company has not had previous experience with ferrous condensers, they are not sure what annual operating and maintenance cost estimate is appropriate. A 5-year planning horizon is to be used, salvage values are estimated to be 15% of the original purchase price, and a *MARR* of 20% is to be used. Determine the break-even value for the annual operating and maintenance cost for the ferrous condenser. (5.7.1)

44. In problem 39 suppose a 10-year planning horizon is specified. Perform a sensitivity analysis comparable to that given in Figure 5.8 to determine the effect of errors in estimating the *differences* in initial investment, the *differences* in annual operating and maintenance costs, and the *differences* in salvage values for the two investment alternatives (5.7.2)

45. In problem 37 suppose the following pessimistic, most likely, and optimistic estimates are given for building cost, furnishings cost, annual operating and maintenance costs and the average rate per occupied unit.

	Pessimistic	Most Likely	Optimistic
Building cost	$7,500,000	$5,000,000	$4,000,000
Furnishings cost	3,000,000	1,875,000	1,000,000
Annual operating and			
maintenance costs	200,000	125,000	75,000
Average rate	35/day	55/day	75/day

Determine the pessimistic and optimistic limits on the break-even value for the daily occupancy percentage based on a *MARR* of 12%. Assume the motel will operate 365 days/year. (5.7.2)

46. A warehouse modernization plan requires an investment of $3 million in equipment; at the end of the 10-year planning horizon, it is anticipated the equipment will have a salvage value of $600,000. Annual savings in operating and maintenance costs due to the modernization are anticipated to total $1,500,000/year. A *MARR* of 10% is used by the firm. Perform a sensitivity analysis to determine the effects on the economic feasibility of the plan due to errors in estimating the initial investment required, and the annual savings. (5.7.2)

47. An investment of $15,000 is to be made into a savings account. The interest rate to be paid each year is uncertain; however, it is estimated that it is twice as likely to be 6% as it is to be 4%, and it is equally likely to be either 6% or 8%. Determine the probability distribution for the amount in the fund after 3 years, assuming the interest rate is not autocorrelated. (5.7.3)

48. Assume an initial investment of $12,000, annual receipts of $4500, and an uncertain life for the investment. Use a 15% *MARR*. Let the probability distribution for the life of the investment be given as follows:

N	p(N)
1	0.10
2	0.15
3	0.20
4	0.25
5	0.15
6	0.10
7	0.05

   a. Perform 20 simulations and estimate the probability of the investment being profitable.
   b. Analytically determine the probability of the investment being profitable. (5.7.3)

49. In problem 48 suppose the *MARR* is not known with certainty and the following probability distribution is anticipated to hold:

i	p(N)
0.10	0.20
0.12	0.60
0.15	0.20

Use analytical methods to determine the probability of the investment being profitable. (5.7.3)

50. In problem 48 suppose the magnitude of the annual receipts (R) is subject to random variation. Assume that each annual receipt will be identical in value and the annual receipt has the following probability distribution:

R	$p(R)$
$5000	0.20
$4500	0.60
$4000	0.20

Use analytical methods to determine the probability of the investment's being profitable. (5.7.3)

51. In problem 48 suppose the minimum attractive rate of return is distributed as given in problem 49 and suppose the annual receipts are distributed as given in problem 50. Analytically determine the probability that the investment will be profitable. (5.7.3)

52. In problem 51 suppose the initial investment is equally likely to be either $9,000 or $11,000. Analytically determine the probability that the investment will be profitable. (5.7.3)

53. Suppose $n = 4$, $i = 0\%$, a $11,000 investment is made, and the receipt in year j, $j = 1, \ldots , 4$, is statistically independent and distributed as in problem 50. Determine the probability distribution for present worth. (5.7.3)

54. Consider an investment alternative having a 6-year planning horizon and expected values and variances for statistically independent cash flows as given below:

j	$E(C_j)$	$V(C_j)$
0	-$22,500	625×10^4
1	4,000	16×10^4
2	5,000	25×10^4
3	6,000	36×10^4
4	7,000	49×10^4
5	8,000	64×10^4
6	9,000	81×10^4

Using a discount rate of 10%, determine the expected values and variances for both present worth and annual worth. Based on the central limited theorem, compute the probability of a positive present worth; compute the probability of a positive annual worth. (5.7.3)

55. Solve problem 54 using a discount rate of (a) 0%, (b) 15%. (5.7.3)

56. Solve problem 54 when the C_j, $j = 1, \ldots , 6$, are perfectly correlated. (5.7.3)

57. Two investment alternatives are being considered. Alternative A requires an initial investment of $15,000 in equipment; annual operating and maintenance costs are anticipated to be normally distributed, with a mean of $5000 and a standard deviation of $500; the terminal salvage value at the end of the 8-year planning horizon is anticipated to be normally distributed, with a mean of $2000 and a standard deviation of $800. Alternative B requires end-of-year annual expenditures over the planning horizon. The

annual expenditure will be normally distributed, with a mean of $8000 and a standard deviation of $750. Using a *MARR* of 15%, what is the probability that Alternative A is the most economic alternative? (5.7.3)

58. In problem 57, suppose the *MARR* were 10% with probability 0.25, 12% with probability 0.50, and 15% with probability 0.25, what is the probability that Alternative A is the most economic alternative? (5.7.3)

59. Company W is considering investing $12,500 in a machine. The machine will last *N* years, at which time it will be sold for *L*. Maintenance costs for this machine are estimated to increase by 10%/year over its life. The maintenance cost for the first year is estimated to be $1500. The company has 10% *MARR*. Based on the probability distributions given below for *N* and *L*, what is the expected equivalent uniform annual cost for the machine? (5.7.3)

N	L	$p(N)$
6	$5,000	0.2
8	3,000	0.4
10	1,000	0.4

60. Two investment alternatives are under consideration. The data for the alternatives are given below. It is assumed the cash flows are not autocorrelated. (5.7.3)

EOY	$E[CF(A)]$	$SD[CF(A)]$	$E[CF(B)]$	$SD[CF(B)]$
0	−$10,000	$1,000	−$15,000	$1,500
1	4,000	400	8,000	800
2	5,000	500	8,000	800
3	6,000	600	8,000	800
4	7,000	700	8,000	800
5	8,000	800	8,000	800
6	9,000	900	8,000	800

a. For each alternative, determine the mean and standard deviation for present worth and annual worth using a *MARR* of 15%.

b. For each alternative, based on the central limit theorem, compute the probability of a positive present worth using a *MARR* of 15%.

c. Develop the mean and standard deviation for the *incremental* present worth using a *MARR* of 15%.

d. Based on the central limit theorem, compute the probability of a positive incremental present worth using a *MARR* of 15%.

61. Three investment alternatives (A, B, and C) are being considered. The discounted present worth (PW) for the alternatives are $35,000, $30,000, and $37,000, respectively. The three alternatives have different rankings in terms of employee morale (EM) and vendor reputation (VR). The following weights have been assigned to the three factors (PW, EM, and VR): 40, 35, and 25. On a scale from 1 to 10, the following ratings have been assigned to the three alternatives for the three factors:

	A	**B**	**C**
PW	9.2	8	10
EM	9	10	8.5
VR	9.5	10	9

Using the weighted factor comparison method, which investment alternative would be recommended? (5.8)

62. In problem 61, what change, if any, would occur in the recommendation if the ratings on
    a. PW had been 9.5, 8.5, and 10?
    b. EM had been 10, 9.5, and 9?
    c. VR had been 7.5, 10, and 8.5? (5.8)

63. What do you find to be the greatest weakness in using the weighted factor comparison method? (5.8)

64. Given the present worths in problem 61, what weights would you have assigned to the various alternatives for the present worth factor? What logical process did you use to arrive at your weights? (5.8)

65. Four investment alternatives (W, X, Y, and Z) are under conditions. The discounted present worths (PW) for the alternatives were $125,000, $200,000, $150,000, and $175,000. The payback periods (PP) for the alternatives were 3 years, 4 years, 2 years, and 5 years. The levels of risk (RL) associated with the investments are quite different, with W being the most risky, Z being the least risky, and X and Y being equally risky. The following weights have been assigned to the three factors (PW, PP, and RL): 35, 40, and 25. On a scale from 1 to 10, the following ratings have been assigned to the four alternatives for the three factors:

	W	**X**	**Y**	**Z**
PW	8.5	10	9	9.5
PP	9.7	8.9	10	8.5
RL	7.5	9	9	10

Using weighted factor comparison, which would be recommended? What is the imputed value, in discounted present worth, for the differences in payback periods and levels of risk for the alternatives? (5.8)

66. In problem 65, which investment would be recommended if the payback periods were all the same? (5.8)

67. In problem 65, which investment would be recommended if the risk levels were equal for all alternatives? (5.8)

68. The motor on a numerically controlled machine tool must be replaced. Two different 20-horsepower (output) electric motors are being considered. Motor U sells for $1360 and has an efficiency rating of 90%; Motor V sells for $300 and has a rating of 84%. The cost of energy is $0.055/kilowatt-hour. An 8-year planning horizon is used, and zero salvage values are assumed for both motors. Annual usage of the motor averages approximately 4800 hours. A *MARR* of 12% is to be used. Assume the motor selected will be loaded to capacity. (*Note:* 0.746 kilowatt = 1 horsepower.) (5.9)
    a. Clearly specify the net cash flow profiles for each mutually exclusive alternative.
    b. Determine which motor should be selected. Use a *ranking approach* and the present worth method.
    c. Repeat part (b) using the annual worth method.
    d. Repeat part (b) using the future worth method.

   e. Determine which motor should be selected. Use an *incremental cash flow approach* and the internal rate of return method.
   f. Repeat part (e) using the external rate of return method.
   g. Repeat part (e) using the savings/investment ratio method.

69. A consulting engineer is designing an irrigation system and is considering two pumps to meet a pumping demand for 15,000 gallons/minute at 15 feet total dynamic head. The specific gravity of the fertilizer mixture being pumped is 1.25. Pump A operates at 78% efficiency and costs $11,500; Pump B operates at 86% efficiency and costs $16,250. Energy costs $0.060/kilowatt-hour. Continuous pumping for 200 days/year is required (i.e., 24 hours/day). *MARR* is 10%, the planning horizon is 5 years, and equal salvage values are estimated for the two pumps. (*Note:* Dynamic head × gallons/minute × specific gravity × 0.746 ÷ 3960 = kilowatts.) (5.9)
   a. Clearly specify the net cash flow profiles for each mutually exclusive alternative.
   b. Determine which pump should be selected. Use a *ranking approach* and the present worth method.
   c. Repeat part (b) using the annual worth method.
   d. Repeat part (b) using the future worth method.
   e. Determine which pump should be selected. Use an *incremental cash flow approach* and the internal rate of return method.
   f. Repeat part (e) using the external rate of return method.
   g. Repeat part (e) using the savings/investment ratio method.

70. A chemical plant is considering the installation of a storage tank for water. The tank is estimated to have an initial cost of $213,000, annual costs for maintenance are estimated to be $3200/year. As an alternative, a holding pond can be provided some distance away at an initial cost of $90,000 for the pond, plus $45,000 for pumps and piping; annual operating and maintenance costs for the pumps and holding pond are estimated to be $8500. The planning horizon is 20 years with neither alternative having any salvage value. *MARR* is 15%. (5.9)
   a. Clearly specify the net cash flow profiles for each mutually exclusive alternative.
   b. Determine which alternative should be selected. Use a *ranking approach* and the present worth method.
   c. Repeat part (b) using the annual worth method.
   d. Repeat part (b) using the future worth method.
   e. Based on incremental cash flows, what is the discounted payback period?
   f. Determine which alternative should be selected. Use an *incremental cash flow approach* and the internal rate of return method.
   g. Repeat part (f) using the external rate of return method.
   h. Repeat part (f) using the savings/investment ratio method.

71. The state civil defense organization wishes to establish a communications network to cover the state. It is desired to maintain a specified minimum signal strength at all points in the state. Two alternatives have been selected for detailed consideration. Design I involves the installation of five transmitting stations of low power. The investment at each installation is estimated to be $40,000 in structure and $25,000 in equipment. Design II involves the installation of two transmitting stations of much higher power. Investment in structure will be approximately $130,000/installation; equipment at each installation will cost $180,000. Annual operating and maintenance costs per installation are anticipated to be $25,000 for Design I and $55,000 for Design II. Structures are anticipated to last for 20 years; equipment life is estimated to be 10

years, with replacements assumed to be identical in costs and performance. The planning horizon is 20 years, salvage values are nil at $t = 20$, and *MARR* Is 12%. (5.9).

a. Clearly specify the net cash flow profiles for each mutually exclusive alternative.
b. Determine which design should be selected. Use a *ranking approach* and the present worth method.
c. Repeat part (b) using the annual worth method.
d. Repeat part (b) using the future worth method.
e. Based on incremental cash flows, what is the discounted payback period?
f. Determine which design should be selected. Use an *incremental cash flow approach* and the internal rate of return method.
g. Repeat part (f) using the external rate of return method.
h. Repeat part (f) using the savings/investment ratio method.

72. Two compressors are being considered by the Ajax Company. Compressor A can be purchased for $9700; annual operating and maintenance costs are estimated to be $3500. Alternatively, Compressor B can be purchased for $7500; annual operating and maintenance costs are estimated to be $4200. An 8-year planning horizon is to be used; salvage values are estimated to be 15% of the original purchase price; a *MARR* of 12% is to be used. (5.9)

a. Clearly specify the net cash flow profiles for each mutually exclusive alternative.
b. Determine which compressor should be selected. Use a *ranking approach* and the present worth method.
c. Repeat part (b) using the annual worth method.
d. Repeat part (b) using the future worth method.
e. Based on incremental cash flows, what is the discounted payback period?
f. Determine which compressor should be selected. Use an *incremental cash flow approach* and the internal rate of return method.
g. Repeat part (f) using the external rate of return method.
h. Repeat part (f) using the savings/investment ratio method.

73. The Ajax Manufacturing Company wishes to choose one of the following machines:

	Machine 1	Machine 2
First cost	$10,000	$12,000
Planning horizon	6 years	6 years
Salvage value	$ 1,000	$ 1,000
Operating and maintenance cost for year k	$800 + 80k$	$200 + 100k$

MARR is 12%. (5.9)

a. Clearly specify the net cash flow profiles for each mutually exclusive alternative.
b. Determine which machine should be selected. Use a *ranking approach* and the present worth method.
c. Repeat part (b) using the annual worth method.
d. Repeat part (b) using the future worth method.
e. Based on incremental cash flows, what is the discounted payback period?
f. Determine which machine should be selected. Use an *incremental cash flow approach* and the internal rate of return method.
g. Repeat part (f) using the external rate of return method.
h. Repeat part (f) using the savings/investment ratio method.

74. The manager of the distribution center for the southeastern territory of a major pharmaceutical manufacturer is contemplating installing an improved material handling system linking receiving and storage, as well as storage and shipping. Two designs are being considered. The first consists of a conveyor system that is tied into an automated storage/retrieval system. Such a system is estimated to cost $1,125,000 initially, have annual operating and maintenance costs of $72,000, and a salvage value of $75,000 at the end of the 10-year planning horizon.

The second design consists of manually operated narrow-aisle, high-stacking lift trucks. To provide service comparable to that provided by the alternative design, an initial investment of $490,000 is required. Annual operating and maintenance costs of $160,000 are anticipated. An estimated salvage value of $30,000 is expected at the end of the planning horizon. *MARR* is 12%. (5.9)

a. Clearly specify the net cash flow profiles for each mutually exclusive alternative.
b. Determine which alternative should be selected. Use a *ranking approach* and the present worth method.
c. Repeat part (b) using the annual worth method.
d. Repeat part (b) using the future worth method.
e. Based on incremental cash flows, what is the discounted payback period?
f. Determine which alternative should be selected. Use an *incremental cash flow approach* and the internal rate of return method.
g. Repeat part (f) using the external rate of return method.
h. Repeat part (f) using the savings/investment ratio method.

75. A manufacturing plant in Minnesota has been contracting snow removal at a cost of $400/day. The past 4 years have yielded unusually heavy snowfalls, resulting in the cost of snow removal being of concern to the plant manager. The plant engineer has found that a snow-removal machine can be purchased for $53,000; it is estimated to have a useful life of 10 years, and a zero salvage value at that time. Annual costs for operating and maintaining the equipment are estimated to be $15,000, *MARR* is 10% and the estimated demand for the equipment is equal to 80 days/year. (5.9)

a. Clearly specify the net cash flow profiles for each mutually exclusive alternative.
b. Determine which alternative should be selected. Use a *ranking approach* and the present worth method.
c. Repeat part (b) using the annual worth method.
d. Repeat part (b) using the future worth method.
e. Based on incremental cash flows, what is the discounted payback period?
f. Determine which alternative should be selected. Use an *incremental cash flow approach* and the internal rate of return method.
g. Repeat part (f) using the external rate of return method.
h. Repeat part (f) using the savings/investment ratio method.

76. A firm is considering two compressors. One must be chosen; the relevant data follow:

	Compressor 1	Compressor 2
Initial investment	$15,000	$30,000
Life	6 years	8 years
Salvage value	$ 1,000	$ 6,000
Annual disbursements	$ 8,000	$ 4,500

MARR is 10% and the cash flow profiles are believed to be repeatable over a planning horizon of 24 years. (5.9)

a. Clearly specify the net cash flow profiles for each mutually exclusive alternative.

b. Determine which compressor should be selected. Use a *ranking approach* and the present worth method.

c. Repeat part (b) using the annual worth method.

d. Repeat part (b) using the future worth method.

e. Based on incremental cash flows, what is the discounted payback period?

f. Determine which compressor should be selected. Use an *incremental cash flow approach* and the internal rate of return method.

g. Repeat part (f) using the external rate of return method.

h. Repeat part (f) using the savings/investment ratio method.

77. Two submersible pumps are under consideration. Their cost data follow:

	Pump 1	Pump 2
First cost	$15,000	$21,000
Salvage value	$ 1,000	$ 2,000
Annual operating cost	$ 3,000	$ 2,000
Life	5 yrs.	10 yrs.

Either pump is needed over a 10-year planning horizon. The company uses a *MARR* of 15%. (5.9).

a. Clearly specify the net cash flow profiles for each mutually exclusive alternative.

b. Determine which pump should be selected. Use a *ranking approach* and the present worth method.

c. Repeat part (b) using the annual worth method.

d. Repeat part (b) using the future worth method.

e. Based on incremental cash flows, what is the discounted payback period?

f. Determine which pump should be selected. Use an *incremental cash flow approach* and the internal rate of return method.

g. Repeat part (f) using the external rate of return method.

h. Repeat part (f) using the savings/investment ratio method.

78. Two numerically controlled drill presses are being considered by the production department of a major corporation; one must be selected. Both machines meet the quality and safety standards of the firm. Comparative data are as follows. *MARR* is 15%. (5.9)

	Drill Press X	Drill Press Y
Initial investment	$28,000	$39,000
Estimated life	6 years	6 years
Salvage value	$ 4,000	$ 7,000
Annual operating cost	$12,000	$ 8,000
Annual maintenance cost	$ 3,000	$ 5,000

a. Clearly specify the net cash flow profiles for each mutually exclusive alternative.

b. Determine which alternative should be selected. Use a *ranking approach* and the present worth method.

c. Repeat part (b) using the annual worth method.
d. Repeat part (b) using the future worth method.
e. Based on incremental cash flows, what is the discounted payback period?
f. Determine which alternative should be selected. Use an *incremental cash flow approach* and the internal rate of return method.
g. Repeat part (e) using the external rate of return method.
h. Repeat part (f) using the savings/investment ratio method).

79. The Ajax Manufacturing Company wishes to choose one of the following machines.

	Machine 1	Machine 2	Machine 3
First cost	$10,000	$15,000	$18,000
Planning horizon	5 years	5 years	5 years
Salvage value	—	$ 2,000	$ 3,000
Operating and maintenance cost for year k, $k = 1, \ldots, n$	$600(1.10)^{k-1}$	$400(1.08)^{k-1}$	$200 + 100k$

MARR is 12% and the planning horizon is 5 years. Based upon the annual worth method, determine the preferred machine. (5.9)

80. Machine 1 initially costs $160,000. Its resale value at the end of the kth year of service equals $160,000 − $12,000k$ for $k = 1, 2, \ldots, 10$. Operating and maintenance expenses equal $35,000/year. Machine 2 initially costs $76,000. Its resale value at the end of the kth year, S_k, equals $76,000(0.80)^k$ for $k = 1, 2, \ldots, 5$. Operating and maintenance expenses equal $40,000/year. Using the annual worth method and a *MARR* of 10%, determine which machine is preferred assuming machine cash flows are repeatable. (5.9)
a. Use a planning horizon of 5 years.
b. Use a planning horizon of 20 years.

81. A firm will either lease an office copier at an end-of-year cost of $10,000 for a 6-year period, or they will purchase the copier at an initial cost of $66,117. If purchased, the copier will have a zero salvage value at the end of its 6-year life. No other costs are to be considered. *MARR* is 12%. Should the firm lease or buy? Justify your answer on the basis of the following. (5.9)
a. The present worth method.
b. The internal rate of return method.

82. A company can construct a new warehouse for $1 million, or it can lease an equivalent building for $100,000/year for 25 years with the option of purchasing the building for $1 million at the end of the 25-year period. Lease payments are due at the *beginning* of each year. The company can earn 15%/year before taxes on its invested capital. Indicate the preferred alternative based upon the following. (5.9)
a. The annual worth method.
b. The savings/investment ratio method.

83. In Problem 82, for what beginning-of-year lease payment will the company be indifferent between leasing and buying? (5.9)

84. E-M Systems Company must decide if they should purchase a small computer or lease the computer. The computer costs $64,000 initially and will last 5 years, having a $5000 salvage value at that time. If the computer is purchased, all maintenance costs must be paid by E-M. Maintenance costs are $2,500/year over the life of the equipment.

The E-M Company uses an interest rate of 12% in evaluating investment alternatives. For what beginning-of-year annual leasing charge is the firm indifferent between purchasing and leasing over the 5-year period? (5.9)
a. Use the annual worth method.
b. Use the external rate of return method.

85. A firm is considering either leasing or buying a microcomputer system. If purchased, the initial cost will be $250,000; annual operating and maintenance costs will be $80,000/year. Based on a 6-year planning horizon, it is anticipated the computer will have a salvage value of $30,000 at that time. If the computer is leased, annual operating and maintenance costs in excess of the annual lease payment will be $60,000/year. Based on an interest rate of 10%, what annual beginning-of-year lease payment will make the firm be indifferent between leasing and buying? (5.9)
a. Use the present worth method.
b. Use the internal rate of return method.

86. Thermo-D, Inc. is considering investing $14,000 in a heat exchanger. The heat exchanger will last 6 years, at which time it will be sold for $2000. Maintenance costs for the exchanger are estimated to increase by $300/year over its life. The maintenance cost for the first year is estimated to be $2000. As an alternative, the company may lease the equipment for $X/year, including maintenance. For what value of X should the company lease the heat exchanger? The company expects to earn 12% on its investments. Assume beginning-of-year lease payments. (5.9)
a. Use the future worth method.
b. Use the external rate of return method.

87. Apricot Computers is considering replacing its material handling system and either purchasing or leasing a new system. The old system has an annual operating and maintenance cost of $32,000, a remaining life of 8 years, and an estimated salvage value of $5000 at that time.

   A new system can be purchased for $250,000; it will be worth $25,000 in 8 years; and it will have annual operating and maintenance costs of $18,000/year. If the new system is purchased, the old system can be traded in for $20,000.

   Leasing a new system will cost $26,000/year, payable at the beginning of the year, plus operating costs of $9000/year, payable at the end of the year. If the new system is leased, the old system will be sold for $10,000.

   MARR is 15%. Compare the annual worths of keeping the old system, buying a new system, and leasing a new system based upon a planning horizon of 8 years.
   a. Use the cash flow approach. (5.11.1)
   b. Use the outsider viewpoint approach. (5.11.2)

88. Fluid Dynamics Company owns a pump that it is contemplating replacing. The old pump has annual operating and maintenance costs of $8000/year: it can be kept for 4 years more and will have a zero salvage value at that time.

   The old pump can be traded in on a new pump. The trade-in value is $4000. The new pump will cost $18,000 and have a value of $9000 in 4 years and will have annual operating and maintenance costs of $4500/year.

   Using a *MARR* of 10%, evaluate the investment alternative based upon the present worth method and a planning horizon of 4 years.
   a. Use the cash flow approach. (5.11.1)
   b. Use the outsider viewpoint approach. (5.11.2)

89. A company owns a 5-year old turret lathe that has a book value of $25,000. The

present market value for the lathe is $16,000. The expected decline in market value is $2000/year to a minimum market value of $4000. Maintenance plus operating costs for the lathe equal $4200/year.

A new turret lathe can be purchased for $45,000 and will have an expected life of 8 years. The market value for the turret lathe is expected to equal $45,000(0.70)^k at the end of year k. Annual maintenance and operating cost is expected to equal $1600. Based on a 12% before-tax *MARR*, should the old lathe be replaced now? Use an equivalent uniform annual cost comparison, a planning horizon of 7 years, and the outsider viewpoint approach. (5.11.2)

90. Esteez Construction Company has an overhead crane that has an estimated remaining life of 7 years. The crane can be sold for $14,000. If the crane is kept in service it must be overhauled immediately at a cost of $6000. Operating and maintenance costs will be $5000/year after the crane is overhauled. After overhauling it, the crane will have a zero salvage value at the end of the 7-year period. A new crane will cost $36,000, will last for 7 years, and will have a $8000 salvage value at that time. Operating and maintenance costs are $2500 for the new crane. Esteez uses an interest rate of 15% in evaluating investment alternatives. Should the company buy the new crane based upon an annual cost analysis?
a. Use the cash flow approach. (5.11.1)
b. Use the outsider viewpoint approach. (5.11.2)

91. A small foundry is considering the replacement of a No. 1 Whiting cupola furnace that is capable of melting gray iron only with a reverberatory-type furnace that can melt gray iron and nonferrous metals. Both furnaces have approximately the same melting rates for gray iron in pounds per hour. The foundry company plans to use the reverberatory furnace, if purchased, primarily for melting gray iron, and the total quantity melted is estimated to be about the same with either furnace. Annual raw material costs would therefore be about the same for each furnace. Available information and cost estimates for each furnace is given below.

Cupola furnace. Purchased used and installed 8 year ago for a cost of $20,000. The present market value is determined to be $8000. Estimated remaining life is somewhat uncertain but, with repairs, the furnace should remain functional for 7 years more. If kept 7 years more, the salvage value is estimated as $2000 and average annual expenses expected are:

Fuel	$35,000
Labor (including maintenance)	$40,000
Payroll taxes	10% of direct labor costs
Taxes and insurance on furnace	1% of purchase price
Other	$16,000

Reverberatory furnace. This furnace costs $32,000. Expenses to remove the cupola and install the reverberatory furnace are about $2400. The new furnace has an estimated salvage value of $3000 after 7 years of use and annual expenses are estimated as:

Fuel	$29,000
Labor (operating)	$30,000
Payroll taxes	10% of direct labor costs
Taxes and insurance on furnace	1% of purchase price
Other	$16,000

In addition, the furnace must be relined every 2 years at a cost of $4000/occurrence. If the foundry presently earns an average of 15% on invested capital before income taxes, should the cupola furnace be replaced by the reverberatory furnace?
a. Use the cash flow approach. (5.11.1)
b. Use the outsider viewpoint approach. (5.11.2)

92. Metallic Peripherals, Inc. has received a production contract for a new product. The contract lasts for 5 years. To do the necessary machining operations, the firm can use one of its own lathes, which was purchased 3 years ago at a cost of $16,000. Today the lathe can be sold for $8000. In 5 years the lathe will have a zero salvage value. Annual operating and maintenance costs for the lathe are $4000/year. If the firm uses its own lathe it must also purchase an additional lathe at a cost of $12,000, its value in 5 years will be $3000. The new lathe will have annual operating and maintenance costs of $3500/year.

As an alternative, the presently owned lathe can be traded in for $10,000 and a new lathe of larger capacity purchased for a cost of $24,000; its value in 5 years is estimated to be $8000, and its annual operating and maintenance costs will be $6000/year.

An additional alternative is to sell the presently owned lathe and subcontract the work to another firm. Company X has agreed to do the work for the 5-year period at an annual cost of $12,000/end-of-year.

Using a 15% interest rate, determine the least-cost alternative for performing the required production operations.
a. Use the cash flow approach. (5.11.1)
b. Use the outsider viewpoint approach (5.11.2)

93. A machine was purchased 5 years ago at $12,000. At that time, its estimated life was 10 years with an estimated end-of-life salvage value of $1200. The average annual operating and maintenance costs have been $14,000 and are expected to continue at this rate for the next 5 years. However, average annual revenues have been and are expected to be $20,000. Now, the firm can trade in the old machine for a new machine for $5000. The new machine has a list price of $15,000, an estimated life of 10 years, annual operating plus maintenance costs of $7500, annual revenues of $13,000, and salvage values at the end of the jth year according to

$$S_j = \$15,000 - \$1500j, \quad \text{for } j = 0, 1, 2, 3, 4, 5, 6, 7, 8, 9, 10$$

Determine whether to replace or not by the annual worth method using a *MARR* equal to 15% compounded annually. Use a 5-year planning horizon and the outsider viewpoint approach. (5.11.2)

94. A building supplies distributor purchased a gasoline-powered fork-lift truck 4 years ago for $8000. At that time, the estimated useful life was 8 years with a salvage value of $800 at the end of this time. The truck can now be sold for $2500. For this truck, average annual operating expenses for year j have been

$$C_j = \$2000 + \$400(j - 1)$$

Now the distributor is considering the purchase of a smaller battery-powered truck for $6500. The estimated life is 10 years, with the salvage value decreasing by $600 each year. Average annual operating expenses are expected to be $1200. If a *MARR* of 10% is assumed and a 4-year planning horizon is adopted, should the replacement be made now? (5.11.2)

95. Kwik-Kleen Car Wash has been experiencing difficulties in keeping its equipment operational. The owner is faced with the alternative of overhauling the present equipment or replacing it with new equipment. The cost of overhauling the present equipment is $8500. The present equipment has annual operating and maintenance costs of $7500. If it is overhauled, the present equipment will last for 5 years more and be scrapped at zero value. If it is not overhauled, it has a trade-in value of $3200 toward the new equipment.

New equipment can be purchased for $28,000. At the end of 5 years the new equipment will have a resale value of $12,000. Annual operating and maintenance costs for the new equipment will be $3000.

Using a *MARR* of 12%, what is your recommendation to the owner of the car wash? Base your recommendation on a present worth comparison and the outsider viewpoint approach. (5.11.2)

96. A firm is contemplating replacing a computer they purchased 3 years ago for $400,000. It will have a salvage value of $20,000 in 4 years. Operating and maintenance costs have been $75,000/year. Currently the computer has a trade-in value of $100,000 toward a new computer that costs $300,000 and has a life of 4 years, with a salvage value of $50,000 at that time. The new computer will have annual operating and maintenance costs of $80,000.

If the current computer is retained, another small computer will have to be purchased in order to provide the required computing capacity. The smaller computer will cost $150,000, has a salvage value of $20,000 in 4 years, and has annual operating and maintenance costs of $30,000.

Using an annual worth comparison before taxes, with a *MARR* of 15%, determine the preferred course of action. (5.11.2)

97. National Chemicals has an automatic chemical mixture that it has been using for the past 4 years. The mixer originally cost $18,000. Today the mixer can be sold for $10,000. The mixer can be used for 10 years more and will have a $2500 salvage value at that time. The annual operating and maintenance costs for the mixer equal $6000/year.

Because of an increase in business, a new mixer must be purchased. If the old mixer is retained, a new mixer will be purchased at a cost of $25,000 and have a $4000 salvage value in 10 years. This new mixer will have annual operating and maintenance costs equal to $5000/year.

The old mixer can be sold and a new mixer of larger capacity purchased for $32,000. This mixer will have a $6000 salvage value in 10 years and will have annual operating and maintenance costs equal to $8000/year.

Based on a *MARR* of 15%, what do you recommend? (5.11.2)

98. The Ajax Specialty Items Corporation has received a 5-year contract to produce a new product. To do the necessary machining operations, the company is considering two alternatives.

Alternative A involves continued use of the currently owned lathe. The lathe was purchased five years ago for $20,000. Today the lathe is worth $8000 on the used machinery market. If this lathe is to be used, special attachments must be purchased at a cost of $3500. At the end of the 5-year contract, the lathe (with attachments) can be sold for $2000. Operating and maintenance costs will be $7000/year if the old lathe is used.

Alternate B is to sell the currently owned lathe and buy a new lathe at a cost of $25,000. At the end of the 5-year contract, the new lathe will have a salvage value of $13,000. Operating and maintenance costs will be $4000/year for the new lathe.

Using an annual worth analysis, should the firm use the currently owned lathe or buy a new lathe? Base your analysis on a minimum attractive rate of return of 15%. (5.11.2)

99. The Telephone Company of America purchased a numerically controlled production machine 5 years ago for $300,000. The machine currently has a trade-in value of $70,000. If the machine is continued in use, another machine, X, must be purchased to supplement the old machine. Machine X costs $200,000, has annual operating and maintenance costs of $40,000, and will have a salvage value of $30,000 in 10 years. If the old machine is retained, it will have annual operating and maintenance costs of $55,000 and will have a salvage value of $15,000 in 10 years.

As an alternative to retaining the old machine, it can be replaced with Machine Y. Machine Y costs $400,000, has anticipated annual operating and maintenance costs of $70,000, and has a salvage value of $140,000 in 10 years.

Using a *MARR* of 15% and a present worth comparison, determine the preferred economic alternative. (5.11.2)

100. A highway construction firm purchased a particular earth-moving machine 3 years ago for $125,000. The salvage value at the end of 8 years was estimated to be 35% of first cost. The firm earns an average annual gross revenue of $105,000 with the machine and the average annual operating costs have been and are expected to be $65,000.

The firm now has the opportunity to sell the machine for $70,000 and subcontract the work normally done by the machine over the next 5 years. If the subcontracting is done, the average annual gross revenue will remain $105,000 but the subcontractor charges $85,000/end-of-year for these services.

If a 15% rate of return before taxes is desired, determine by the annual worth method whether or not the firm should subcontract. (5.11.2)

101. Bumps Unlimited, a highway contractor, must decide whether to overhaul a tractor and scraper or replace it. The old equipment was purchased 5 years ago for $130,000; it has a projected life of 12 years, with a $15,000 salvage value at that time. If traded in on a new tractor and scraper, it can be sold for $60,000. Overhauling the old equipment will cost $20,000. If overhauled, operating and maintenance costs will be $18,000/year, and the overhauled equipment will have a newly projected salvage value of $0 in 7 years.

Using an annual worth comparison with a *MARR* of 15%, should the equipment be overhauled or replaced? (5.11.2)

102. A new automatic insertion machine has just been introduced on the market. The Electronics Products Company (EPC) has learned that it will do the work of four existing machines. Currently EPC has 8 automatic insertion machines. Four operators and a supervisor are responsible for servicing the machines. The current annual labor cost for an operator is $24,000 and a supervisor is $32,000. Labor costs are expected to increase at an annual rate of 10%. The current machines require maintenance totaling $80,000/year, with the cost expected to increase at an annual rate of 12%.

If the eight current machines are replaced with two new machines, only one operator will be required; the supervisor will no longer be required, since the operator would be assigned to another supervisor. Also, replacement of the current machines will release 3000 square feet of floor space needed for surface mount machines. Floor space costs $2.50 per square foot per year.

Currently, the average investment in work in process inventory totals $1 million. With the new machines, the average inventory investment will be reduced to $600,000; however, due to inflation, it is anticipated that the investment in work in process inventory will increase at an annual rate of 6%. It is anticipated that the maintenance costs for the new machines will total $20,000/year during the first year and increase at an annual ràte of 10%.

The old machines can be sold for a total of $50,000. The two new machines can be installed for a total cost of $500,000. Using a 5-year planning horizon, the cumulative salvage value for the two machines will be approximately $50,000 in 5 years.

EPC believes the discount rate used with then-current estimates of cash flows should be 15%, including inflation. If constant worth cash flows are used, then a *MARR* of 8% should be used.

Determine the economic viability of purchasing the two new automatic insertion machines, first, by ignoring the inventory savings and, second, by including the savings realized by reducing work in process inventory. (5.11.2)

103. A firm is presently using a machine that has a market value of $11,000 to do a specialized production job. The requirement for this operation is expected to last only 6 years more, after which it will no longer be done. The predicted costs and salvage values for the present machine are:

Year	1	2	3	4	5	6
Operating cost	$1,500	$1,800	$2,100	$2,400	$2,700	$3,000
Salvage value	8,000	6,000	5,000	5,000	3,000	2,000

A new machine has been developed that can be purchased for $17,000 and has the following predicted cost performance.

Year	1	2	3	4	5	6
Operating cost	$ 1,000	$ 1,100	$ 1,200	$1,300	$1,400	$1,500
Salvage value	13,000	11,000	10,000	9,000	8,000	7,000

If interest is at 0%, when should the new machine be purchased? (5.11.3)

104. A particular unit of production equipment has been used by a firm for a period of time sufficient to establish very accurate estimates of its operating and maintenance costs. Replacements can be expected to have identical cash flow profiles in successive life cycles if constant worth dollar estimates are used. The appropriate discount rate is 15%. Operating and maintenance costs for a unit of equipment in its tth year of service, denoted by C_t, are as follows:

t	C_t	t	C_t
1	$6,000	6	$15,000
2	7,500	7	17,250
3	9,150	8	19,650
4	10,950	9	22,200
5	12,900	10	24,900

Each unit of equipment costs $45,000 initially. Because of its special design, the unit of equipment cannot be disposed of at a positive salvage value following its purchase; hence, a zero salvage value exists, regardless of the replacement interval used.

a. Determine the optimum replacement interval assuming an infinite planning horizon. (Maximum feasible interval = 10 years.)

b. Determine the optimum replacement interval assuming a finite planning horizon of 15 years, with $C_{t+1} = C_t + \$1500 + \$150(t - 1)$ for $t = 10, 11, \ldots$

c. Solve parts (a) and (b) using a discount rate of 0%.

d. Based on the results obtained, what can you conclude concerning the effect the discount rate has on the optimum replacement interval? (5.11.3)

105. Given an infinite planning horizon, identical cash flow profiles for successive life cycles, and the following functional relationships for C_t, the operating and maintenance cost for the tth year of service for the unit of equipment in current use, and F_n, the salvage value at the end of n years of service:

$$C_t = \$4000(1.10)^t \qquad t = 1, 2, \ldots, 12$$
$$F_n = \$44,000(0.50)^n \quad n = 0, 1, 2, \ldots, 12$$

Determine the optimum replacement interval assuming a MARR of (a) 0%, (b) 10%. (Maximum life = 12 years.) (5.11.3)

Chapter 6

Income Taxes
and Depreciation

6.1 INTRODUCTION

Taxes can make a world of difference in an engineering economic analysis. Tax dollars *are* cash flows, and therefore it is necessary to explicitly consider them, just as costs of wages, equipment, materials, and energy are included. One of the major factors affecting taxes is depreciation. Although depreciation allowances are not cash flows, their magnitudes and timing do affect taxes. Proper knowledge and application of the tax laws can make the economic difference between accepting and rejecting a project, as well as between profit and loss on the corporate bottom line.

While the Omnibus Reconciliation Act of 1993 rewrote the tax rates, the Tax Reform Act of 1986 virtually rewrote the book on the method of calculating depreciation and income taxes. This followed the major revisions of the Economic Recovery Tax Act of 1981, which itself vastly changed the pre-1981 depreciation and tax laws. Because of these changes, companies today may have assets that are simultaneously being depreciated under very different systems.

This chapter presents first the big picture on the calculation of taxes in an economic analysis. It then focuses on the details of depreciation analysis, beginning with the traditional (pre-1981) depreciation methods that underlie current depreciation tables. Then, the current Modified Accelerated Cost Recovery System (MACRS—1986 to present) is presented in some detail. This is followed by a brief treatment of the Accelerated Cost Recovery System (ACRS—1981 through 1986), and other still-allowable depreciation strategies are presented. The chapter then illustrates the effects of depreciation method, recovery period, interest on borrowed money, capital gains and losses, disposition, tax credit, and the expense deduction on after-tax cash flows. Inflation considerations and depletion complete the chapter.

It is pretty clear that this is a long chapter! It is designed to give breadth and detail on topics of interest to those who are likely to become involved in after-tax engineering economic analyses. You need not read every section of the chapter to gain useful insights about taxes, depreciation, and their effects. At the most basic level, you should study Sections 6.1 to 6.6, 6.8, 6.15, and 6.22. These sections alone will provide a working knowledge of many typical situations needing economic evaluation. For more detail on items of current interest, Sections 6.11 to 6.14, and 6.17 to 6.21 should be added. Items that are currently obsolete, but which provide perspective and background in the development of the depreciation and taxation structure, are presented in Sections 6.7, 6.9, 6.10, and 6.16. An asterisk (*) has been placed by each such section; occasional reference to material in these sections should not cause any readability problems. Homework problems are keyed to the various sections of the chapter, and an asterisk has been placed by those relating to the obsolete material.

No attempt is made in this chapter to discuss the minute details of tax law. Rather, the objective is to communicate the major aspects of corporate tax treatment so that you can perform economic analyses on an after-tax basis and recognize when you need help. You are definitely advised to seek assistance from corporate tax personnel when performing formal after-tax economic analyses.

6.2   TAX CONCEPTS

The taxes paid by a corporation represent a real cost of doing business and, consequently, affect the cash flow profile. For this reason, it is wise to perform economic analyses on an *after-tax* basis. After-tax analysis procedures are identical to the before-tax evaluation procedures studied already; however, the cash flows are adjusted for taxes paid or saved.

There are numerous kinds of taxes. Examples include *ad valorem* (property), *sales, excise* (a tax on the manufacture, sale, or consumption of a commodity), and *income taxes.* Income taxes are usually the only significant taxes to be considered in an economic analysis. Income taxes are assessed on gross income less certain allowable deductions, incurred both in the normal course of business as well as on gains resulting from the disposal of property.

Federal and state income tax regulations are not only detailed and intricate, but they are subject to change over time. There is a tendency for the tax laws to be changed in an effort to affect and improve the state of the economy. For this reason, only the general concepts and procedures for calculating after-tax cash flow profiles and performing after-tax analyses are emphasized here. Furthermore, only federal income tax will be considered because of the diversity of state laws. Practitioners involved in an actual analysis should seek the assistance of corporate tax counsel regarding any uncertainties about tax laws in effect at that time.

6.3   CORPORATE INCOME TAX—ORDINARY INCOME

A corporation, for income tax purposes, includes associations, business trusts, joint stock companies, insurance firms, and trusts and partnerships that actually operate as associations or corporations. Corporate income tax, however, is not limited to traditional business organizations. Engineers, doctors, lawyers, and other professional people may be treated as corporations if they have formally organized under state professional association acts.

Ordinary federal income tax imposed on corporations is presented in Table 6.1. The tax rates shown are those current as of 1997 for tax years beginning January 1, 1993 and beyond. Small businesses with taxable incomes less than $75,000 are subject to very low tax rates. Note the "strange" rates of 39% from $100,000 to $335,000, and 38% from $15,000,000 to $18,333,333. The 39% rate effectively eliminates the benefit of graduated rates for midsized corporations with taxable income of at least $335,000; they will, in effect, pay a flat tax of 34% on every dollar up to $10,000,000. The 38% rate has the same effect on large corporations, with taxable income of at least $18,333,333; they will, in effect, pay a flat tax of 35% on every dollar of taxable income.

In the remainder of this chapter, an applicable tax rate of 35% will be used unless otherwise specified. This implicitly assumes we are dealing with a company already having over $18,333,333 in taxable income. This will yield results not far from those achieved using a tax rate of 34%, assuming a midsized company with taxable income from $335,000 to $10,000,000.

EXAMPLE 6.1

Our small company is currently forecasting a taxable income of $50,000 for this year. We are considering another investment that will increase taxable income by $45,000. If we take on the project, what will be our *increased* income tax liability? What would it be if we were currently forecasting a taxable income of $400,000 for this year?

Answer 1: Without the project the tax is 0.15($50,000) = $7500. With the project the tax is 0.15($50,000) + 0.25($25,000) + 0.34($20,000) = $20,550. The tax increase is $20,550 − $7500 = $13,050 or 29% of the $45,000.

Answer 2: Without the project the tax is 0.15($50,000) + 0.25($25,000) + .34($25,000) + 0.39($235,000) + 0.34($65,000) = $136,000 or 34% of $400,000. Because every dollar of the additional $45,000 will be taxed at 34%, the increase will be 0.34($45,000) = $15,300 for a total tax of $151,300.

Table 6.1 Corporate Income Tax Rates for Tax Years Beginning January 1, 1993 and Beyond

Taxable Income (TI), in $	Tax Rate (t)	Income Tax (T)
$0 < \text{TI} \le 50{,}000$	0.15	0.15(TI)
$50{,}000 < \text{TI} \le 75{,}000$	0.25	$7500 + 0.25(\text{TI} - 50{,}000)$
$75{,}000 < \text{TI} \le 100{,}000$	0.34	$13{,}750 + 0.34(\text{TI} - 75{,}000)$
$100{,}000 < \text{TI} \le 335{,}000$	0.39(0.34 + 0.05)	$22{,}250 + 0.39(\text{TI} - 100{,}000)$
$335{,}000 < \text{TI} \le 10{,}000{,}000$	0.34	$113{,}900 + 0.34(\text{TI} - 335{,}000)$
$10{,}000{,}000 < \text{TI} \le 15{,}000{,}000$	0.35	$3{,}400{,}000 + 0.35(\text{TI} - 10{,}000{,}000)$
$15{,}000{,}000 < \text{TI} \le 18{,}333{,}333$	0.38(0.35 + 0.03)	$5{,}150{,}000 + 0.38(\text{TI} - 15{,}000{,}000)$
$18{,}333{,}333 < \text{TI}$	0.35	0.35(TI)

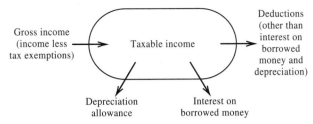

Figure 6.1 Pictorial representation of taxable income.

Taxable income must first be determined before any tax rate can be applied. Taxable income is gross income less allowable deductions. Gross income is income in a general sense less any monies specifically exempt from tax liability. Corporate deductions are subtracted from gross income and commonly include items such as salaries, wages, repairs, rent, bad debts, taxes (other than income), charitable contributions, casualty losses, interest, and depreciation. Interest and depreciation are of particular importance, because we can control them to some extent through financing arrangements and accounting procedures.

Taxable income is represented pictorially in Figure 6.1, which shows that taxable income for any year is what is left after deductions, including interest on borrowed money and depreciation allowance, are subtracted from gross income. These components are not all cash flows, since the depreciation allowance is simply treated as an expense in determining taxable income.

6.4 AFTER-TAX CASH FLOW

We have now looked at the basic elements needed to calculate after-tax cash flows. These elements are summarized in Figure 6.2. This shows that the after-tax cash flow is the amount remaining after income taxes and deductions, including interest but excluding depreciation allowance, are subtracted from gross income.

In many of the following tables, we simplify our terminology by speaking of *before-tax cash flows*. The term before-tax cash flow is used when no borrowed money is involved, and it equals gross income less deductions, not including depreciation. The term before-tax and loan cash flow is used when borrowed money is involved, and it equals gross income less deductions, not including either depreciation or principal or interest on the loan.

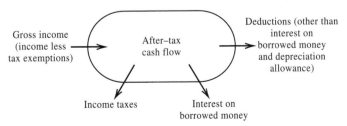

Figure 6.2 Pictorial representation of after-tax cash flow.

EXAMPLE 6.2

Let us consider a surface mount placement (SMP) machine used in the manufacture of electronic components. In the first year, the reduced operating expenses (gross income less deductions other than interest and depreciation) will be $25,000, no money is borrowed, and depreciation is $19,600 for year 1. The effective tax rate is 35%. What are the (1) taxable income and (2) after-tax cash flow values for year 1?

Taxable income is $25,000 − $19,600 = $5400. Tax is $5400 × 0.35 = $1890. So, after-tax cash flow is $25,000 − $1890 = $23,110 for the year.

EXAMPLE 6.3

Again consider Example 6.2. In year 2, the reduced operating expenses (gross income less deductions other than interest and depreciation) are again $25,000, no money is borrowed, but depreciation is $31,360 for year 2. The effective tax rate is 35%. What are the (1) taxable income and (2) after-tax cash flow values for year 2?

Taxable income is $25,000 − $31,360 = −$6360. Tax is −6360 × 0.35 = −$2226. So, after-tax cash flow is $25,000 − (−$2226) = $27,226 for the year.

In Example 6.3, we see that an alternative's taxable income and tax can be negative for a particular year. It is assumed that such values are used to offset positive tax liabilities from other corporate activities. When there are insufficient positive taxable incomes to be offset, the taxable income losses can be carried back 3 years or forward 15 years in order to reduce positive taxable income. That is, if an operating loss occurs in 1999, we can reopen books beginning in 1996, then 1997, and 1998. If the loss is not fully absorbed against positive taxable incomes, the remainder may be used in 2000, 2001, and so on for as many as 15 years. We will assume that negative taxable incomes are used to offset positive values for the year in question.

Now, before moving into a study of depreciation, let us see one last example that clearly shows the importance of explicitly considering taxes and performing an after-tax cash flow analysis.

EXAMPLE 6.4

Part 1: Again consider a surface mount placement (SMP) machine used in the manufacture of electronic components. The first cost including installation is $98,000, with an estimated salvage value of $7000 at the end of a useful life of

7 years. Reduced operating expenses are estimated to be $25,000 each year, no money is borrowed, and the desired rate of return is 15%. Determine the desirability of buying the machine based upon a present worth analysis. Table 6.2 shows the before-tax cash flow profile, the present worth of each year's cash flow, and the $8642.05 sum of present worths found by taking $PW(15) = -\$98,000 + \$25,000(P|F\ 15,1) + \cdot \cdot \cdot + \$25,000(P|F\ 15,7) + \$7000(P|F\ 15,7) = \$8642.05$. Clearly, the investment in the SMP machine appears attractive.

Part 2: Now, repeat Part 1 of this example using an after-tax cash flow analysis. Suppose depreciation allowances are $19,600, $31,360, $18,816, $11,289.60, $11,289.60, $5644.80, and $0, respectively, over the 7 years. These are based upon the Modified Accelerated Cost Recovery System (MACRS) depreciation method you will study later in this chapter. The effective tax rate is 35%. Determine the desirability of buying the machine based upon a present worth analysis of the after-tax cash flows. Table 6.3 shows the before-tax cash flow (BTCF) profile, the MACRS depreciation allowed, taxable income, tax, after-tax cash flow (ATCF), the present worth of each year's ATCF, and the $-\$5010.01$ sum of present worths found by taking $PW(15) = -\$98,000 + \$23,110(P|F\ 15,1) + \cdot \cdot \cdot + \$16,250(P|F\ 15,7) + \$4550(P|F\ 15,7) = -\$5010.01$. Clearly, the investment in the SMP machine does not appear attractive, a different decision than reached from the BTCF analysis!

In Tables 6.2 and 6.3, Year 7 is listed twice, once to record the normal cash flows and once to show the salvage value transaction. Note that the $7000 salvage value is considered a "depreciation recapture" and taxed as ordinary income.

Table 6.2 Before-Tax Cash Flow Profile, Example 6.4 Part 1

End of Year, A	Before-Tax Cash Flow, B	PW OF BTCF, C
0	$-98,000.00	$-98,000.00
1	25,000.00	21,739.13
2	25,000.00	18,903.59
3	25,000.00	16,437.91
4	25,000.00	14,293.83
5	25,000.00	12,429.42
6	25,000.00	10,808.19
7	25,000.00	9,398.43
7	7,000.00 (Salvage)	2,631.56
	$PW(15\%) =$	$8.642.05

Table 6.3 After-Tax Cash Flow Profile Using MACRS Depreciation, Example 6.4 Part 2

End of Year, A	Before-Tax Cash Flow, B	MACRS Deduction, C	Taxable Income, B-C D	Tax, D × 0.35 E	After-Tax Cash Flow, B-E F	PW of ATCF G
0	$-98,000.00				$-98,000.00	$-98,000.00
1	25,000.00	$19,600.00	$5,400.00	1,890.00	23,110.00	20,095.65
2	25,000.00	31,360.00	-6,360.00	-2,226.00	27,226.00	20,586.77
3	25,000.00	18,816.00	6,184.00	2,164.40	22,835.60	15,014.78
4	25,000.00	11,289.60	13,710.40	4,798.64	20,201.36	11,550.19
5	25,000.00	11,289.60	13,710.40	4,798.64	20,201.36	10,043.65
6	25,000.00	5,644.80	19,355.20	6,774.32	18,225.68	7,879.46
7	25,000.00	0.00	25,000.00	8,750.00	16,250.00	6,108.98
7	7,000.00 (Salvage)	0.00	7,000.00 (Depreciation Recapture)	2,450.00	4,550.00	1,710.51
					PW(15%) =	-$5,010.01

This is because the entire cost basis ($P = \$98,000$) had already been "recovered" through MACRS deductions. Depreciation recapture is covered later in this chapter.

6.5 THE MEANING OF DEPRECIATION

Most property decreases in value with use and time. That is, it *depreciates*. In determining taxable income, the law permits deduction of a reasonable allowance for wear and tear, natural decay or decline, exhaustion, or obsolescence of property used in a trade or business or of property held for the production of income [1]. The amount of the annual allowance using the current *Modified Accelerated Cost Recovery System* (*MACRS*) depreciation method depends on several factors, including (1) cost basis or investment in the property, (2) property class and recovery period, (3) date placed in service, (4) which convention (e.g., half-year convention) is used, and (5) method of depreciation [2]. Similar factors could be listed for the older ACRS depreciation method. Other depreciation methods (e.g., traditional methods) require the (6) useful life, and (7) estimated salvage value.

We saw in Figure 6.1 that federal taxable income results by deducting certain items such as expenses from gross income. For tax purposes, investment in a depreciable asset is treated as a prepaid expense, and the depreciation allocates that expense over time. It is important to note that depreciation is not a cash flow, but is merely *treatable* as an expense for income tax purposes. A larger deduction in a year decreases net taxable income and hence income taxes, making more money available for reinvestment or dividends. Because of the time value of money, it is generally desirable to take larger allowable deductions in the early years and smaller deductions in the later years of an asset's life.

6.6   SOME DEFINITIONS USED IN DEPRECIATION ACCOUNTING

Depreciable property must meet these requirements:

1. It must be used in business or held for the production of income.
2. It must have a life that can be determined, and that life must be longer than 1 year.
3. It must be something that wears out, decays, gets used up, becomes obsolete, or loses value from natural causes [3, 4, 5].

Depreciable property may be tangible or intangible. *Tangible property* can be seen or touched. *Intangible property*, such as a copyright or franchise is not tangible. Tangible property may be personal or real. *Personal property* is property, such as cars, trucks, machinery, furniture, equipment, and anything that is tangible except real property. *Real property* is land and generally anything that is erected on, growing on, or attached to land. Land itself, however, is never depreciable [3, 4, 5]. Most engineering economic analyses involve tangible personal property; some include tangible real property.

The allowable depreciation deduction is determined by the MACRS for most tangible property placed in service *after* 1986. *The cost basis* of the property is recovered over the *recovery period* of the asset. The cost basis is essentially the taxpayer's investment. In most cases, this is the cost of the property plus the cost of additions to that property, including installation cost. It is also known as the *unadjusted basis*. The cost basis less capital recovered, such as depreciation deductions, is known as the *unrecovered investment, adjusted basis,* or *book value.* The recovery period is simply the time over which the cost basis can be recovered; it is currently keyed to 3, 5, 7, 10, 15, 20, 27.5, and 39 years, depending on the type of property involved [3]. It is usually shorter than the inherent physical life of the asset.

The allowable depreciation deduction is determined by the *Accelerated Cost Recovery system* (*ACRS*) on items placed in service from 1981 through 1986. Depreciation on assets placed in service before 1981, and assets specifically excluded from the MACRS and ACRS, requires an explicit estimate of the salvage value and the useful life. *Salvage value* is an estimate of the market value at the end of the asset's useful life. *Useful life* is an estimate of time during which the asset may actually be used in trade or business. These estimates are used explicitly by the traditional methods of depreciation based upon years, such as straight-line, declining balance, and sum of the years' digits depreciation methods. The useful life is not used in either the MACRS or ACRS method. Technically speaking, salvage value is used in the MACRS calculations, but always assumes a value of zero.

Nearly all engineering economic analyses involving new depreciable property make use of the MACRS. Imbedded within the MACRS-*General Depreciation System* (*GDS*) is the use of the 200% and 150% declining balance and straight-line methods. The MACRS-*Alternative Depreciation System (ADS)*, a straight-line method, is required for some properties and is optional for others. Detailed discussions of the GDS and ADS methods follow in Section 6.8. Replacement analyses could involve ACRS and pre-1981 depreciation methods, depending on the date an asset was placed in service. Straight-line depreciation still applies in the case of some intangible property. Due to the strong reliance on traditional depreciation methods such as the straight-line and declining balance methods, we will start with those in order to build a strong foundation for the MACRS.

*6.7 TRADITIONAL (PRE-1981) DEPRECIATION METHODS BASED ON YEARS

Current MACRS depreciation rates are based on the traditional methods of depreciation. In order to understand the mathematical rationale underlying MACRS, the traditional methods of depreciation must be studied. Prior to 1981, the three most commonly used depreciation strategies included the (1) straight-line, (2) declining balance, and (3) sum of the years' digits methods. All three methods utilize a useful life (n) instead of a recovery period, and they require an estimate of the salvage value (F) at the end of the useful life. Under each of these methods, the total depreciation deductions taken over the useful life of the property may not exceed the cost basis less the salvage value ($P - F$) [2].

*6.7.1 Straight-Line Depreciation

The *straight-line method* provides for the uniform write-off of an asset. The depreciation deduction at the end of each year t is equal throughout the property's useful life and is given by

$$D_t = \frac{P - F}{n} \tag{6.1}$$

The unrecovered investment at the end of year t is given by

$$B_t = P - \left(\frac{P - F}{n}\right) t \tag{6.2}$$

The straight-line method is very simple and was therefore popular for much property placed in service prior to 1981. Any of those assets remaining continue to be depreciated as originally set up. Straight-line depreciation is still required for depreciating intangible property.

EXAMPLE 6.5

Reconsider the SMP machine of Example 6.4 in which the cost basis is $98,000 with an estimated salvage value of $7000 and a useful life of 7 years. Had it been placed in service prior to 1981, the straight-line deduction and unrecovered investment for each year would be as given in Table 6.4.

*6.7.2   Declining Balance Depreciation

The *declining balance method* is known for its accelerated write-off of assets. That is, it provides relatively high depreciation deduction allowances in the early years and lower allowances throughout the rest of the property's useful life. In this method, the depreciation deduction at the end of each year t is a constant fraction (p) of the unrecovered investment at the end of the previous year. That is,

$$D_t = pB_{t-1} \tag{6.3}$$

The unrecovered investment at the end of each year t is given by

$$B_t = P(1 - p)^t \tag{6.4}$$

Substituting Equation 6.4 into Equation 6.3 allows us to calculate the year t depreciation directly as

$$D_t = pP(1 - p)^{t-1} \tag{6.5}$$

Note that in the declining balance method of depreciation, the estimated salvage value need not come into play in figuring the deduction; however, the unrecovered investment must never fall below the salvage value.

Twice the straight-line rate, or $2/n$, is known as *200% declining balance* depreciation. It was used only if it was elected for new pre-1981 tangible property having a useful life of 3 years or more. For used tangible property, either *150%* or *125% declining balance* depreciation was the maximum permitted.

Declining balance depreciation is sometimes used alone; however, switching from either the 200% or 150% declining balance method to straight-line depreciation is permitted. The optimum switch takes place whenever straight-line depreci-

Table 6.4 Straight-Line Depreciation and Unrecovered Investment, Example 6.5

End of Year, t	SL Deduction, D_t	Unrecovered Investment, B_t
0		$98,000.00
1	$13,000.00	85,000.00
2	13,000.00	72,000.00
3	13,000.00	59,000.00
4	13,000.00	46,000.00
5	13,000.00	33,000.00
6	13,000.00	20,000.00
7	13,000.00	7,000.00

ation on the unrecovered portion of the asset exceeds the declining balance allowance. That is, a switch to straight-line occurs at the first year for which

$$\frac{B_{t-1} - F}{n - (t - 1)} > pB_{t-1} \tag{6.6}$$

The estimated salvage value is used in determining the straight-line depreciation component, even though it is neglected in the double declining balance method. Switching to straight-line depreciation is never desirable if the estimated salvage value F exceeds the declining balance unrecovered investment for the last year B_n, causing depreciation deductions to be truncated.

EXAMPLE 6.6

Let us again work with the SMP machine of Example 6.4 in which $P = \$98,000$, $F = \$7000$, and $n = 7$. Had it been placed in service prior to 1981, the 200% declining balance with a switch to straight-line depreciation and unrecovered investment for each year would be as given in Table 6.5. The value of p is $\frac{2}{7}$.

*6.7.3 Sum of the Years' Digits Depreciation

The *sum of the years' digits method*, like the declining balance method, is known for its accelerated write-off of assets. The name *sum of the years' digits* comes from the fact that the sum

$$1 + 2 + \cdots + (n - 1) + n = \frac{n(n + 1)}{2}$$

is used directly in the calculation of allowable depreciation. The depreciation deduction during any year t is expressed as

$$D_t = \frac{n - (t - 1)}{n(n + 1)/2}(P - F) \tag{6.7}$$

The unrecovered investment at the end of each year t is given by

$$B_t = P - \sum_{j=1}^{t} \frac{n - (j - 1)}{n(n + 1)/2}(P - F)$$

which reduces to

$$B_t = (P - F)\frac{(n - t)(n - t + 1)}{n(n + 1)} + F \tag{6.8}$$

Table 6.5 200% Declining Balance Switching to Straight-Line Depreciation and Unrecovered Investment, Example 6.6

End of Year, t	200% Declining Balance Deduction, D_t	Straight-Line Deduction on Remaining Life, D_t	Unrecovered Investment, B_t
0			$98,000.00
1	$28,000.00[a]	$13,000.00	70,000.00
2	20,000.00[a]	10,500.00	50,000.00
3	14,285.71[a]	8,600.00	35,714.29
4	10,204.08[a]	7,178.57	25,510.20
5	7,288.63[a]	6,170.07	18,221.57
6	5,206.16	5,610.79[a]	12,610.79
7	3,603.08	5,610.79[a]	7,000.00

[a] Indicates depreciation deduction actually used. Switching to straight line occurs in year 6.

Sum of the years' digits depreciation was allowed if it was elected for new pre–1981 tangible property having a useful life of 3 years or more.

EXAMPLE 6.7

Again let us consider the SMP machine of Example 6.4 having $P = \$98,000$, $F = \$7000$, and $n = 7$. Had it been placed in service prior to 1981, the sum of the years' digits depreciation and unrecovered investment for each year would be as given in Table 6.6.

Table 6.6 Sum of the Years' Digits Depreciation and Unrecovered Investment, Example 6.7

End of Year, t	Value of $[n - (t - 1)]/[n(n + 1)/2]$	SOYD Deduction, D_t	Unrecovered Investment, B_t
0			$98,000.00
1	7/28	$22,750.00	75,250.00
2	6/28	19,500.00	55,750.00
3	5/28	16,250.00	39,500.00
4	4/28	13,000.00	26,500.00
5	3/28	9,750.00	16,750.00
6	2/28	6,500.00	10,250.00
7	1/28	3,250.00	7,000.00

6.8   MODIFIED ACCELERATED COST RECOVERY SYSTEM (MACRS)

The MACRS applies to most depreciable property placed in service after 1986. MACRS consists of two systems that determine how property is depreciated. The main system is the General Depreciation System (GDS) while the second system is called the Alternative Depreciation System (ADS). The MACRS-GDS is ordinarily used, and for much property is based on declining balance switching to straight-line depreciation. The MACRS-ADS uses a longer recovery period and uses only the straight-line method.

Both the MACRS-GDS and MACRS-ADS have preestablished recovery periods for most property. Under MACRS-GDS, most property is assigned to eight property classes. These property classes establish the number of years over which the cost basis of an item is recovered.

6.8.1   MACRS-GDS Property Classes

Depreciable tangible property may be assigned to one of the following eight classes [3]:

1. *Three-Year Property.* Included, with some important exceptions, is qualifying property with a class life[1] of *4 years or less.* This involves tractor units for over-the-road use; special handling devices for the manufacture of food and beverages; special tools for the manufacture of rubber, finished plastic, glass, and fabricated metal products; and special tools for the manufacture of motor vehicles. Specifically *excluded* are automobiles and light, general-purpose trucks, which had been 3-year property under the ACRS, but are now 5-year property under the MACRS-GDS.

2. *Five-Year Property.* Included is property with a class life of *more than 4 years but less than 10 years,* plus automobiles; light, general-purpose trucks; and certain research and experimentation, alternative energy, and biomass property. Some properties included in the class life interval are computers and peripheral equipment; data-handling equipment (typewriters, calculators, copiers, and so on); heavy, general-purpose trucks; offshore oil and gas well-drilling assets; construction assets; timber cutting and sawing assets; computer-based telephone central office switching equipment, satellite space segment property; and many assets used for the manufacture of knitted goods, carpets, apparel, medical and dental supplies, chemicals, and electronic components.

3. *Seven-Year Property.* Included is property with a class life of *10 years or more but less than 16 years,* property without any class life and not specifically included in the 27.5- and 39-year categories to follow, and several specifically named items. Some assets in this interval include office furniture, fixtures

---

[1] "Class life" refers to the class life guideline under the old Class Life Asset Depreciation Range (CLADR) system. The CLADR Class Life, and both GDS and ADS Recovery Periods are reconciled in [3].

and equipment; assets used in the exploration for and production of petroleum and natural gas deposits; theme and amusement park assets; and most assets used for manufacturing such things as food products, spun yarn, wood products and furniture, pulp and paper, rubber products, finished plastic products, leather products, glass products, foundry products, fabricated metal products, motor vehicles, aerospace products, athletic goods, and jewelry.

4. *Ten-Year Property.* This class includes property with a class life of *16 years or more but less than 20 years.* Some assets included are vessels, tugs, and similar water transportation equipment; assets used in petroleum refining; and assets used in the manufacture of grain, sugar, vegetable oil products, and substitute natural gas-coal gasification.

5. *Fifteen-Year Property.* This class includes property with a class life of *20 years or more but less than 25 years.* Some assets included in the interval are land improvements such as sidewalks, roads, drainage facilities, sewers, bridges, fencing, and landscaping; assets used in the manufacture of cement; some water and pipeline transportation assets; telephone distribution plant assets; certain electric and gas utility property; some liquefied natural gas assets; and municipal wastewater treatment plants.

6. *Twenty-Year Property.* Property with a class life *of 25 years or more, other than real property with a class life of 27.5 years or more, plus municipal sewers,* is included. Such assets are farm buildings; some railroad structures and electric generating equipment; certain transmission lines, pole lines, buried cable, and repeaters; and much other utility property.

7. *Residential Rental Property.* Included is a rental home or structure for which 80% or more of the gross rental income for the tax year is rental income from dwelling units. A dwelling unit is a house or an apartment used to provide living accommodations in a building or structure, but *not* a unit in a hotel, motel, inn, or other establishment in which more than one-half of the units are used on a transient basis. The recovery period for this property is 27.5 years.

8. *Nonresidential Real Property.* Depreciable property that has a class life of *27.5 years or more* and is not residential rental property generally falls into this class. The recovery period is 39 years for property placed in service today, a marked increase from past years.

It is obvious from the preceding abbreviated property class descriptions that determining the class to which a given property belongs is a complex task. Professional tax guidance and the complete Table of Class Lives and Recovery Periods are recommended.

6.8.2   Calculating the MACRS-GDS Deductions for 3-, 5-, 7-, 10-, 15-, and 20-Year Property

For property in the 3-, 5-, 7-, and 10-year classes, the primary depreciation method used is based on *200% declining balance switching to straight-line depreciation with a half-year convention over the GDS recovery period (200% DBSLH-GDS).* This

Table 6.7 Depreciation Methods Chart

Property Class	Method-Recovery Period	Referred to in Text as
3-, 5-, 7-, 10-year	200%DBSLH-GDS	MACRS-GDS
	*150%DBSLH-ADS	not used
	*SLH-GDS	not used
	*SLH-ADS	Elective MACRS-ADS
15-, 20-year	150%DBSLH-GDS	MACRS-GDS
	*SLH-GDS	not used
	*SLH-ADS	Elective MACRS-ADS
Residential rental	SLM-GDS	MACRS-GDS
Nonresidential real	*SLH-ADS	Elective MACRS-ADS
Tax-exempt use	SL-ADS	not used
Tax-exempt bond—financed		
Imported		
Foreign use		

* Elective method

primary depreciation method for these four property classes is known in this text as MACRS-GDS. It is tabulated and widely used. Three other, less attractive, methods may be elected. These include (1) 150% DBSLH-ADS, (2) straight-line depreciation with a half-year convention over the GDS recovery period (SLH-GDS), and (3) straight-line depreciation with a half-year convention over the ADS recovery period (SLH-ADS). These are summarized in Table 6.7.

The primary depreciation method for the 15- and 20-year classes is similar, but uses a 150% rate (*150% DBSLH-GDS*). This primary depreciation method for these two property classes is known in this text as MACRS-GDS. It is tabulated and widely used. Two other less attractive methods may be elected, including (1) SLH-GDS, and (2) SLH-ADS. The ADS recovery periods are typically longer than those of the GDS. These are also summarized in Table 6.7.

To figure the 200% declining balance part of 200% DBSLH-GDS depreciation used in the MACRS-GDS tables, first determine the *rate* of depreciation, $1/n$, where n is the recovery period. For a \$100,000 property with a 7-year recovery period, n is 7 and the rate is $\frac{1}{7}$ or 14.29%. This rate is then multiplied by 2 for 200% DBSLH-GDS, resulting in $p = 0.2857$ or 28.57%, the *allowable percentage* for the 7-year property class. The basis of the property at time zero, say $P = \$100,000$, is then multiplied by half this figure, or 14.29%, due to the half-year convention. The deduction for year 1, D_1, is then $0.5pP = 0.5(0.2857)(\$100,000) = \$14,285$. The adjusted cost basis or book value at the end of year 1 is then $B_1 = P - D_1 = \$100,000 - \$14,285 = \$85,715$. Allowable depreciation for year 2 is $D_2 = p(P - D_1) = pB_1 = 0.2857(\$85,715) = \$24,489$. Likewise, $D_3 = p(P - D_1 - D_2) = pB_2$, and so on. Since the half-year convention treats year 1 as only one-half year, it is necessary to carry depreciation over into year 8 in order to realize the full recovery period. In general, $n + 1$ years are involved in full depreciation of a property in any of the first six classes. The above discussion applies fully to 150% DBSLH-GDS, except the rate is multiplied by 1.5 instead of 2.

Because each year's depreciation involves multiplying B_t times p, the book value will never reach zero under the 200% declining balance method alone. However, the MACRS requires switching to straight-line depreciation at the optimum year t', which will ensure depreciation to a book value of zero. The optimum year is when the straight-line deduction first exceeds the 200% declining balance allowance. Straight-line depreciation for year $t = 2$ through n is simply the remaining book value divided by the number of years left in the recovery period. That is, $D_t = B_{t-1}/(n - t + 1.5)$. Note that $n - t + 1.5$ is the recovery period remaining, considering the half-year convention, counting year t. Once straight-line depreciation is begun, D_t remains the same each year through year n, followed by $0.5D_n$ in year $n + 1$.

EXAMPLE 6.8

Consider the $100,000 7-year property discussed in the text above. Using the approach described, calculate the 200% DBSLH-GDS depreciation for each year of the recovery period. It is already established that $p = 0.2857$. The 200% declining balance and straight-line components are shown separately in Table 6.8. The year 6 200% DB depreciation is calculated as $0.2857(\$22,312.13) = \$6374.89$. Straight-line depreciation for year 6 is calculated as $\$22,312.13/(7 - 6 + 1.5) = \$8924.85$.

The MACRS expressions for 200% declining balance depreciation may be summarized as follows:

$$D_1 = 0.5pP = d_1P$$
$$D_2 = pB_1 = p(P - D_1) = p(P - d_1P) = p(1 - d_1)P = d_2P$$
$$D_3 = pB_2 = p(P - D_1 - D_2) = p(P - d_1P - d_2P)$$
$$= p(1 - d_1 - d_2)P = d_3P$$

and, in general,

$$D_t = p\left(1 - \sum_{j=1}^{t-1} d_j\right)P = d_tP; \qquad t = 1, 2, 3, \ldots \qquad (6.9)$$

Note that D_t is the actual amount of depreciation, and d_t is the depreciation rate or fraction. For straight-line depreciation at the optimal time t',

$$D_{t'} = B_{t'-1}/(n - t' + 1.5) = \left(1 - \sum_{j=1}^{t'-1} d_j\right)P/(n - t' + 1.5) = d_{t'}P$$

Since D_t will be the same from t' through n,

$$D_{t=} d_{t'}P = d_tP; \qquad t = t', t' + 1, t' + 2, \ldots, n \qquad (6.10)$$

Table 6.8 200% Declining Balance Switching to Straight-Line Depreciation with a Half-Year Convention, Example 6.8

End of Year, t	200% Declining Balance Depreciation, D_t	Straight-Line Depreciation on Remainder of Recovery Period, D_t	Book Value, B_t
0			$100,000.00
1	$14,285.71[a,b]	$7,142.86[b]	85,714.29
2	24,489.80[a]	13,186.81	61,224.49
3	17,492.71[a]	11,131.73	43,731.78
4	12,494.79[a]	9,718.17	31,236.98
5	8,924.85[a]	8,924.85	22,312.13
6	6,374.89	8,924.85[a]	13,387.28
7	3,824.94	8,924.85[a]	4,462.43
8		4,462.43[a,b]	0.00

[a] Indicates depreciation allowance actually used. Switching to straight-line occurs in year 6.
[b] Half-year convention used in year 1 and year 8.

and

$$D_t = 0.5d_{t'}P = d_{n+1}P; \qquad t = n + 1 \tag{6.11}$$

Note that each year's depreciation can be expressed as a factor d_t times the original cost basis. This observation holds true for both 200% and 150% DBSLH depreciation.

The MACRS-GDS deduction may be easily calculated by applying the appropriate percentage ($d_t \times 100\%$), for a specific property class and year, to the cost basis P. These percentages are found in Table 6.9. The allowable deduction in year t, D_t, is given by

$$D_t = d_t P \tag{6.12}$$

Note that the percentages in Table 6.9 add up to 100% for each property class. This means the entire cost basis of a property can be depreciated over its recovery period, even if its useful life is much longer. The unrecovered investment B_t at the end of year t is given by

$$B_t = P - \sum_{j=1}^{t} D_j = P\left(1 - \sum_{j=1}^{t} d_j\right) \tag{6.13}$$

This represents the amount of the cost basis yet to be recovered.

Table 6.9 MACRS-GDS Percentages ($d_t \times 100\%$)
for 3-, 5-, 7-, and 10-year Property are 200% DBSLH,
and 15- and 20-year Property are 150% DBSLH

EOY	3-Year Property	5-Year Property	7-Year Property	10-Year Property	15-Year Property	20-Year Property
0						
1	33.33%	20.00%	14.29%	10.00%	5.00%	3.750%
2	44.45	32.00	24.49	18.00	9.50	7.219
3	14.81	19.20	17.49	14.40	8.55	6.677
4	7.41	11.52	12.49	11.52	7.70	6.177
5		11.52	8.93	9.22	6.93	5.713
6		5.76	8.92	7.37	6.23	5.285
7			8.93	6.55	5.90	4.888
8			4.46	6.55	5.90	4.522
9				6.56	5.91	4.462
10				6.55	5.90	4.461
11				3.28	5.91	4.462
12					5.90	4.461
13					5.91	4.462
14					5.90	4.461
15					5.91	4.462
16					2.95	4.461
17						4.462
18						4.461
19						4.462
20						4.461
21						2.231

EXAMPLE 6.9

A company is purchasing a surface mount placement machine. First cost including installation will be $98,000 with an estimated salvage value of $7000 at the end of a projected useful life of 7 years. This is 5-year property, and the percentages in Table 6.9 have been used to figure depreciation. The allowable MACRS deduction and resulting unrecovered investment for each year are given in Table 6.10.

Note in Example 6.9 that neither the estimated salvage value nor the projected useful life are involved at this point. We will have to account for the actual salvage value received, but this will be illustrated later. Further, the entire cost basis is recovered over 6 years, rather than over the 7-year useful life.

In Example 6.9 it is assumed that the asset is kept throughout the recovery period. If it is sold or otherwise disposed of, the depreciation allowance in the

Table 6.10 MACRS Deduction and Unrecovered Investment on SMP Machine, Example 6.9

End of Year, t	MACRS-GDS Deduction, D_t	Unrecovered Investment, B_t
0		$98,000.00
1	$19,600.00	78,400.00
2	31,360.00	47,040.00
3	18,816.00	28,224.00
4	11,289.60	16,934.40
5	11,289.60	5,644.80
6	5,644.80	0.00

year of disposition may be affected. For 3-, 5-, 7-, 10-, 15-, and 20-year property, a half-year convention applies in the year of disposition.

EXAMPLE 6.10

In Example 6.9, the MACRS deduction for our $98,000 SMP machine in year 4 is $11,289.60. What would be the year 4 deduction if our computer were sold sometime later in year 4? Immediately after year 4?
Answers: $5644.80 and $11,289.60, respectively.

6.8.3 Calculating MACRS-GDS Deductions for Residential Rental (27.5-Year) and Nonresidential Real (39-Year) Property

For residential rental and nonresidential real property, the primary depreciation method used is based on straight-line depreciation with a mid-month convention over the GDS recovery period (SLM-GDS). This primary depreciation method for these two property classes is known in this text as MACRS-GDS. It is tabulated and widely used. One other, less attractive, method may be elected. It is SLM-ADS. These are summarized in Table 6.7.

To figure the straight-line depreciation with a mid-month convention, for use with 27.5-year residential rental or 39-year nonresidential real property, first determine the *rate* of depreciation, $1/n$, where n is the recovery period. For a $1 million nonresidential real property, n is 39, and the rate is $\frac{1}{39} = 0.025641$ or 2.5641%. The rate 2.564 will be the MACRS-GDS percentage ($d_t \times 100\%$) applied to the cost basis to determine allowable depreciation in all years except the first and the last of the recovery period. If the property is placed in service during the fourth month of the tax year, the required mid-month convention allows a $8.5/12 = 0.7083$ share of a whole year's deduction during the first year. Therefore, the allowable percentage in the first year is 0.7083(2.564%) = 1.816%. Equations (6.12) and (6.13) hold for SLM depreciation. The allowable percentages for 27.5- and 39-year properties are given in Table 6.11.

EXAMPLE 6.11

A firm having a tax year of January 1 to December 31 (a calendar year taxpayer) will place a $220,000 27.5-year residential rental property in service on May 1. The allowable MACRS deductions for each year are given in Table 6.12.

In Example 6.11, recovery takes place over a period of 28 years. Note that the entire cost basis is recovered. This would have been true even if an estimated salvage value had been declared.

In Example 6.11, it is assumed that the asset is kept throughout the recovery period. If it is sold or otherwise disposed of, the depreciation allowance in the year of disposition may be affected. If so, for 27.5- and 39-year property, a mid-month convention applies.

EXAMPLE 6.12

In Example 6.11, the MACRS deduction in year 8 is $7999.20. What would be the year 8 deduction if the rental property were sold on October 1?
    Answer: (9.5/12) × $7999.20 = $6332.70.

6.8.4   MACRS-Alternative Depreciation System (ADS) Election

A taxpayer may elect to claim MACRS-ADS deductions instead of the regular allowances presented earlier. The MACRS-ADS method is simply straight-line depreciation with either a half-year (SLH) or a mid-month (SLM) convention as appropriate. For most property, the straight-line method is applied over the MACRS-ADS recovery period[2] of the property with a half-year convention. For residential rental and nonresidential real property, a 40-year period is used with a mid-month convention. Some other properties do not follow the guidelines and are assigned an explicit period.

The MACRS-ADS method is *required* for use on some property, including property (1) used predominately outside the U.S., (2) having any tax-exempt use, (3) financed by tax-exempt bonds, or (4) imported and covered by executive order of the President. Only in rare circumstances would a taxpayer elect to use MACRS-ADS depreciation if not otherwise required to do so.

---

[2] The MACRS-ADS recovery period was discussed in the presentation of MACRS property classes. It is generally longer than the MACRS-GDS recovery period.

Table 6.11 a. MACRS-GDS Percentages ($d_t \times 100\%$) for 27.5-year Residential Rental Property Using Mid-Month Convention

Year		Month in Tax Year Property Placed in Service											
		1	2	3	4	5	6	7	8	9	10	11	12
	1	3.485%	3.182%	2.879%	2.576%	2.273%	1.970%	1.667%	1.364%	1.061%	0.758%	0.455%	0.152%
	2–9	3.636	3.636	3.636	3.636	3.636	3.636	3.636	3.636	3.636	3.636	3.636	3.636
Even	10–26	3.637	3.637	3.637	3.637	3.637	3.637	3.636	3.636	3.636	3.636	3.636	3.636
Odd	11–27	3.636	3.636	3.636	3.636	3.636	3.636	3.637	3.637	3.637	3.637	3.637	3.637
	28	1.970	2.273	2.576	2.879	3.182	3.485	3.636	3.636	3.636	3.636	3.636	3.636
	29							0.152	0.455	0.758	1.061	1.364	1.667

b. MACRS-GDS Percentages ($d_t \times 100\%$) for 39-year Nonresidential Real Property Using Mid-Month Convention

Year	Month in Tax Year Property Placed in Service											
	1	2	3	4	5	6	7	8	9	10	11	12
1	2.461%	2.247%	2.033%	1.819%	1.605%	1.391%	1.177%	0.963%	0.749%	0.535%	0.321%	0.107%
2–39	2.564	2.564	2.564	2.564	2.564	2.564	2.564	2.564	2.564	2.564	2.564	2.564
40	0.107	0.321	0.535	0.749	0.963	1.177	1.391	1.605	1.819	2.033	2.247	2.461

Table 6.12 MACRS Deduction and
Unrecovered Investment on 27.5-year
Residential Rental Property (Example 6.11)

End of Year, t		MACRS Deduction, D_t	Unrecovered Investment, B_t
	0		220,000.00
	1	5,000.60	214,999.40
	2–9	7,999.20	.
Even	10–26	8,001.40	.
Odd	11–27	7,999.20	.
	.	.	
	.	.	
	.	.	
	28	7,000.40	0.00

In general, SLH depreciation over an n-year period may be expressed as having rate

$$d_t = 1/n \tag{6.14}$$

and depreciation allowances of

$$D_t = 0.5 d_t P; \qquad t = 1, n + 1 \tag{6.15}$$

and

$$D_t = d_t P; \qquad t = 2, 3, \ldots, n \tag{6.16}$$

If the property is disposed of during the period, standard MACRS rules apply regarding the half-year and mid-month conventions.

EXAMPLE 6.13

A press forming machine is purchased for the manufacture of glass windows for $250,000. Its MACRS-ADS recovery period is 14 years, even though it is 7-year property. The firm uses Elective MACRS-ADS in order to defer depreciation deductions into more profitable years. The firm's allowable recovery during each year is given in Table 6.13. Note that $d_t = \frac{1}{14} = 0.071429$.

EXAMPLE 6.14

Reconsider Example 6.11 with the 27.5-year residential rental property purchased for $220,000 on May 1 by a calendar year taxpayer. If we use Elective MACRS-

Table 6.13 Electric MACRS-ADS
Deductions on Press Forming Machine,
Example 6.13 ($n = 14$)

End of Year, t	MACRS-ADS Deduction, D_t
0	—
1	$ 8,928.57
2–14	17,857.14
15	8,928.57

ADS and then sell the property on October 1 of year 8, the schedule of allowable deductions is as presented in Table 6.14. Recall that a 40-year period must be used and, therefore, $d_t = \frac{1}{40} = .025$.

*6.9 ACCELERATED COST RECOVERY SYSTEM (ACRS)

Capital costs for most depreciable property placed in service after 1980 and before 1987 was recovered using the *Accelerated Cost Recovery System (ACRS)*. When first put in place, this was a major departure from prior practice which used the traditional depreciation methods based upon years (Sec. 6.7) consisting of straight line, declining balance, and sum of the years' digits depreciation.

The ACRS approach preceded the MACRS and used table of depreciation percentages applied to the cost basis to determine depreciation allowance in a given year. Another major change was in the period over which depreciation could take place. The ACRS defined "Recovery Property Classes" based upon the Class Life Asset Depreciation Range (CLADR) tables. These tables dramatically reduced the "recovery period," or the time over which depreciation could

Table 6.14 Elective MACRS-ADS Deductions on Residential Rental Property, Example 6.14 ($n = 40$)

End of Year, t	Elective MACRS-ADS (SLM) Deduction, D_t
0	—
1	$(7.5/12)(1/40)($220,000) = $3,437.50$
2–7	5,500.00
8	$(9.5/12)(1/40)($220,000) = $4,354.17$

take place. This was a benefit for companies, in that depreciation could be taken earlier, thereby reducing taxable income in earlier years and deferring taxes until later years. The time value of money implications were enormous. In the MACRS approach which followed ACRS, many recovery periods were lengthened.

The ACRS made use of five recovery property classes, including, 3-, 5-, and 10-year property, 15-year public utility property, and 15-, 18-, or 19-year real property. All assets under the ACRS 3-, 5-, and 10-year property labels are now fully depreciated. The lingering public utility property or real property that is still being depreciated, if germane to an economic analysis, will require one to locate and use the ACRS tables and Recovery Property Class guidelines [2].

The method for calculating ACRS deductions was the same as for MACRS deductions. An ACRS percentage was multiplied by a cost basis to determine the deduction in a given year. For 3-, 5-, 10-, and 15-year public utility property, no deduction was allowed in the year of disposition, whereas MACRS allows a half-year's deduction. Clearly, disposition timing was more important when using the ACRS, as compared to the MACRS.

As might be expected, there was also an Alternative ACRS method that a taxpayer might elect to claim ACRS straight-line deductions instead of the regular allowances. This alternative approach used optional straight-line recovery periods that gave the user three choices. For example, normal 5-Year Recovery Property could use 5, 12, or 25 years for straight-line depreciation, with a half-year convention, under the Alternative ACRS method.

For a substantive treatment of the ACRS, with examples, the reader is referred to the third edition of this text.

*6.10   COMPARISON OF DEPRECIATION METHODS

The performance of MACRS can be compared to that of other depreciation methods by looking at the profile of the unrecovered investment (B_t) of each over time. We desire to have the unrecovered investment drop quickly in the early years. This is accomplished by taking larger depreciation deductions in the early years to create higher after-tax cash flows in those years. This results in a higher present worth of after-tax cash flows. Remember, however, that this only defers taxes until later years. The before-tax equivalent uniform annual cost remains the same, regardless of the method of depreciation.

*6.10.1   Value-Time Curves

Figure 6.3 shows the unrecovered investment (B_t) using the surface mount placement (SMP) machine addressed in Examples 6.5, 6.6, 6.7, and 6.9. These examples cover straight line (SL), 200% declining balance switching to straight line (200% DB/SL), sum of the years' digits, and MACRS-GDS, respectively. The depreciation under ACRS, the predecessor to MACRS, is also graphed for reference. Note that the method of depreciation determines the path of unrecovered investments which can vary widely.

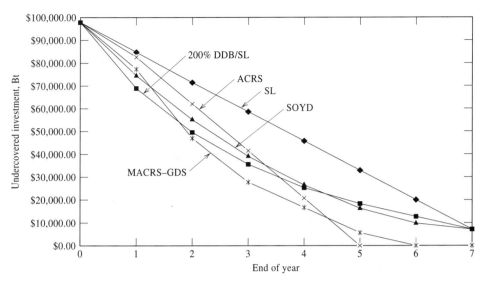

Figure 6.3 Value-time curves for five different depreciation methods.

*6.10.2 Depreciation and the *EUAC*

Calculation of the equivalent uniform annual cost (*EUAC*) is presented in previous chapters. The *EUAC* represents the equivalent uniform annual cost of capital recovered plus a return on the unrecovered investment. Using the surface mount placement (SMP) machine having $P = \$98,000$, $F = \$7000$, $n = 7$, and interest rate $i = 15\%$, the annual cost is[3]

$$EUAC = P(A|P\,15,7) - F(A|F\,15,7)$$
$$= \$98,000(.2404) - \$7000(.0904)$$
$$= \$22,922.79$$

Note that *EUAC* is calculated here without respect to any depreciation method.

Table 6.15 presents the straight-line capital recovered plus return on the unrecovered investment from Example 6.5, covering the 7-year useful life of the SMP machine.

Calculating the uniform annual cost of capital recovered plus return at $i = 15\%$ yields

$$A = [27,700](P|F\,15,1) + 25,750(P|F\,15,2) + \cdot\cdot\cdot + 16,000(P|F\,15,7)]$$
$$[A|P\,15,7]$$
$$= [27,700(.8696) + 25,750(.7561) + \cdot\cdot\cdot + 16,000(.3759)][.2404]$$
$$= \$22,922.79/\text{year}$$

---

[3] In Chapter Six, some calculations are performed using interest factors determined and retained within a computer for reasons of precision. Use of the interest tables in the Appendix will provide answers which differ insignificantly.

Table 6.15 Straight-Line Capital Recovery Plus Return

End of Year, t	SL Depreciation (Capital Recovered), D_t	Book Value (Unrecovered Investment), B_t	Return on Unrecovered Investment, iB_{t-1}	Capital Recovered Plus Return, $D_t + iB_{t-1}$
0		$98,000.00		
1	$13,000.00	85,000.00	$14,700.00	$27,700.00
2	13,000.00	72,000.00	12,750.00	25,750.00
3	13,000.00	59,000.00	10,800.00	23,800.00
4	13,000.00	46,000.00	8,850.00	21,850.00
5	13,000.00	33,000.00	6,900.00	19,900.00
6	13,000.00	20,000.00	4,950.00	17,950.00
7	13,000.00	7,000.00	3,000.00	16,000.00

This is the same as the *EUAC*! Had the *SOYD* or 200% *DB/SL* methods of Examples 6.6 and 6.7 been used, the result would have been the same, $A = \$22,922.79$/year.

Even using the MACRS-GDS method of Examples 6.9 will yield the same result if we are careful. Remember, the MACRS-GDS allows us to recover the entire cost basis $P = \$98,000$, rather than just $P - F = \$91,000$. This $7000 over-recovery will eventually be lost in the form of "depreciation recapture" (discussed in Sec. 6.15) at the time of sale. If this $7000 is subtracted from the MACRS-GDS deduction (D_t) and capital recovered plus return $(D_t + iB_{t-1})$ in year 7, the uniform annual cost of capital recovered plus return will be $A = \$22,922.79$/year.

In summary, the equivalent uniform annual cost of an asset before taxes remains the same, regardless of the depreciation method used. It is most easily calculated using the formula for *EUAC*.

6.11   OTHER DEPRECIATION METHODS NOT BASED ON YEARS

Most depreciation methods are tied to the passage of time. Occasionally, however, recovery of cost over time would have little relation to an asset's use or its production of income. That is why the IRS may allow taxpayers to utilize other consistent methods of depreciation. Some recognized methods are briefly presented below. Assets that we elect to depreciate in any of these ways may require IRS approval and are specifically excluded from the definition of property to which the MACRS applies.

6.11.1   Units of Production Method

This procedure allows equal depreciation per each unit of output, regardless of the lapse of time involved. The allowance for year t is equal to the total depreciable amount $(P - F)$ times the ratio of units produced during the year (U_t) to the total units that are expected to be produced during the useful life of the asset (U). That is,

$$D_t = (P - F)\frac{U_t}{U} \qquad (6.17)$$

This method has proven suitable for depreciating equipment used in exploiting natural resoures such as mines, wells, etc.

6.11.2   Operating Day (Hour) Method

This is similar to the previous method, in that year t depreciation is based on the ratio of days (hours) used during the year (Q_t) to total days (hours) expected in a useful life (Q). Depreciation is expressed as

$$D_t = (P - F)\frac{Q_t}{Q} \qquad (6.18)$$

6.11.3   Income Forecast Method

This method is applicable to depreciate the cost of rented property such as video tapes, sound recordings, and motion picture films. The ratio of year t rental income (R_t) to the total useful life income (R) is multiplied by the total lifetime depreciation, or

$$D_t = (P - F)\frac{R_t}{R} \qquad (6.19)$$

6.12   EFFECT OF DEPRECIATION METHOD AND RECOVERY PERIOD

When we look at the taxable income and after-tax cash flow representations in Figures 6.1 and 6.2, it is natural to wonder about the tax effects of different depreciation methods, recovery periods, and interest on borrowed money. We will show that these factors have a substantial effect on taxes and thus on cash flow profiles.

We have seen how some depreciation methods provide for a higher depreciation deduction during the early years of an asset's life and a correspondingly lower allowance in later years. This places a lower tax burden on the asset during the early years followed by a higher yearly burden. *In most cases the total undiscounted ordinary income tax dollars paid will be the same*, regardless of the depreciation method used. This is always true when the tax rate remains constant and the total depreciation allowance for each method is the same or an adjustment for under- or over-recovery is made at ordinary income tax rates as was done in Example 6.4.

In order to illustrate the effects of depreciation method and recovery period, examples using the same basic numbers, but applying MACRS-GDS, Elective MACRS-ADS, Elective MACRS-ADS under a longer recovery period, and pre-1981 sum of the years' digits depreciation methods are presented.

EXAMPLE 6.15

Consider a surface mount placement (SMP) machine used in the manufacture of electronic components. The first cost including installation is $98,000, with an estimated salvage value of $7000 at the end of a useful life of 7 years. Reduced operating expenses are now estimated to be $28,000 each year, no money is borrowed, and the desired rate of return is 15%. Determine the desirability of buying the machine based upon a present worth analysis. Table 6.16 shows the before-tax cash flow (BTCF) profile, the MACRS-GDS deduction allowed, taxable income, tax, after-tax cash flow (ATCF), and the present worth of each year's ATCF.

Note that the assumed salvage value recovered in year 7 is taxed. This is because MACRS-GDS has already recovered all $98,000 of the investment in the machine. The $7000 salvage value represents over-recovery and is considered to be "depreciation recapture" which is taxed. Finally, note that the sum of present worths of after-tax cash flows is $3102.81. Clearly, the investment in the SMP machine appears attractive.

EXAMPLE 6.16

Now, consider the same SMP machine from Example 6.15 in which the first cost is $98,000, salvage value is $7000, useful life is 7 years, reduced operating expenses are $28,000 each year, no money is borrowed, and the desired rate of return is

Table 6.16 After-Tax Cash Flow Profile Using MACRS-GDS, Example 6.15

End of Year, A	Before-Tax Cash Flow, B	MACRS-GDS Deduction, C	Taxable Income, B-C D	Tax, D × 0.35 E	After-Tax Cash Flow, B-E F
0	−$98,000.00				−$98,000.00
1	28,000.00	$19,600.00	$8,400.00	$2,940.00	25,060.00
2	28,000.00	31,360.00	−3,360.00	−1,176.00	29,176.00
3	28,000.00	18,816.00	9,184.00	3,214.40	24,785.60
4	28,000.00	11,289.60	16,710.40	5,848.64	22,151.36
5	28,000.00	11,289.60	16,710.40	5,848.64	22,151.36
6	28,000.00	5,644.80	22,355.20	7,824.32	20,175.68
7	28,000.00	0.00	28,000.00	9,800.00	18,200.00
7	7,000.00 (Salvage)	0.00	7,000.00 (Depreciation recapture)	2,450.00 PW(15%) =	4,550.00 $3,102.81

Table 6.17 After-Tax Cash Flow Profile Using MACRS-ADS ($n = 6$), Example 6.16

End of Year, A	Before-Tax Cash Flow, B	MACRS-ADS Deduction, ($n = 6$) C	Taxable Income, B-C D	Tax, D × 0.35 E	After-Tax Cash Flow, B-E F
0	−$98,000.00				−$98,000.00
1	28,000.00	$8,166.67	$19,833.33	$6,941.67	21,058.33
2	28,000.00	16,333.33	11,666.67	4,083.33	23,916.67
3	28,000.00	16,333.33	11,666.67	4,083.33	23,916.67
4	28,000.00	16,333.33	11.666.67	4,083.33	23,916.67
5	28,000.00	16,333.33	11,666.67	4,083.33	23,916.67
6	28,000.00	16,333.33	11,666.67	4,083.33	23,916.67
7	28,000.00	8,166.67	19,833.33	6,941.67	21,058.33
7	7,000.00 (Salvage)	0.00	7,000.00 (Depreciation recapture)	2,450.00 PW(15%) =	4,550.00 −$346.18

15%. This time, however, let it be depreciated using Elective MACRS-ADS ($n = 6$). The alternative depreciation system calls for a recovery period of 6 years, and straight-line depreciation using a half-year convention must be used. Table 6.17 illustrates the calculations.

Note that the Elective MACRS-ADS depreciation extends over 7 years due to the half-year convention in years 1 and 7. Also, the deduction is clearly straight line, with a rate of 1/6 or 16.67%. All other calculations are routine, but the present worth of after-tax cash flows is not! It is −$346.18, compared to the $3102.81 from Example 6.15. Not only is the present worth less by the amount of $3448.99, but it is negative. This means that if the Elective MACRS-ADS depreciation method were used, the machine would be rejected on the basis of not meeting the 15% desired rate of return, a different decision from that reached when MACRS-GDS was used in Example 6.15!

EXAMPLE 6.17

Suppose instead of the SMP machine we have been discussing, we are interested in a different asset used in the manufacture of medical and dental supplies. Let it, like our SMP machine, have a cost basis of $98,000, and an estimated salvage value of $7000 at the end of a useful life of 7 years. Also, it will reduce operating expenses (and therefore increase before-tax cash flows) by $28,000/year. It, too, is a MACRS-GDS 5-year property, but its Elective MACRS-ADS ($n = 9$) recovery period is 9 years. The cash flow calculations are given in Table 6.18.

Table 6.18 After-Tax Cash Flow Profile Using MACRS-ADS ($n = 9$), Example 6.17

End of Year, A	Before-Tax Cash Flow, B	MACRS-ADS Deduction, ($n = 9$) C	Taxable Income, B-C D	Tax, D × 0.35 E	After-Tax Cash Flow, B-E F
0	−$98,000.00				−$98,000.00
1	28,000.00	$5,444.44	$22,555.56	$7,894.44	20,105.56
2	28,000.00	10,888.89	17,111.11	5,988.89	22,011.11
3	28,000.00	10,888.89	17,111.11	5,988.89	22,011.11
4	28,000.00	10,888.89	17,111.11	5,988.89	22,011.11
5	28,000.00	10,888.89	17,111.11	5,988.89	22,011.11
6	28,000.00	10,888.89	17,111.11	5,988.89	22,011.11
7	28,000.00	5,444.44	22,555.56	7,894.44	20,105.56
7	7,000.00 (Salvage)	32,666.67	−25,666.67 (Depreciation recapture)	−8,983.33 $PW(15\%) =$	15,983.33 −$2,789.19

Calculating the present worth of the after-tax cash flows at $i = 15\%$, we have −$2789.19.

Table 6.18 is different from Tables 6.16 and 6.17 in several respects. The manufacturing asset is assumed to be sold toward the end of year 7, before the recovery period is completed. According to MACRS rules, the half-year convention is applied in the year of sale. Also, only $5444.44(2) + $10,888.89(5) = $65,333.33 is recovered through Elective MACRS-ADS ($n = 9$) deductions, leaving $32,666.67 unrecovered. Another $7000 is recovered through salvage. The remaining $98,000 − $65,333.33 − $7000 = $25,666.67 is considered a Section 1231 loss and is assumed to be applied against other Section 1231 gains elsewhere in the company, resulting in a $25,666.67(0.35) = $8983.33 tax savings or negative tax. Section 1231 gains and losses are covered later in this chapter (see Sec. 6.15).

A quick look at Tables 6.17 and 6.18 clearly shows the effect of the recovery period on the present worth of after-tax cash flows. The only difference between them is the use of a 6-versus 9-year recovery period. Yet, the difference from −$346.18 and −$2789.19, or $2443.01, is substantial. The decision resulting from the analyses in both Tables 6.17 and 6.18 is to reject the purchase of the asset based upon a negative present worth. But, in each case, the asset could have been depreciated as in Table 6.16, which resulted in a present worth of $3102.81—a quite acceptable number for investment! In Tables 6.16, 6.17, and 6.18, the total tax paid is exactly the same. Note that the MACRS-GDS method (Table 6.16) provided higher depreciation in the early years, resulting in lower taxable income, lower taxes, and higher after-tax cash flows in those early years. Only the depreciation methods and recovery periods are different, resulting in a time value of money effect which is seen in the present worth of the after-tax cash flows.

Table 6.19 After-Tax Cash Flow Profile Using Sum of the Years' Digits Depreciation, Example 6.18

End of Year, A	Before-Tax Cash Flow, B	SOYD Deduction, C	Taxable Income, B-C D	Tax, D × 0.35 E	After-Tax Cash Flow, B-E F
0	−$98,000.00				−$98,000.00
1	28,000.00	$22,750.00	$5,250.00	$1,837.50	26,162.50
2	28,000.00	19,500.00	8,500.00	2,975.00	25,025.00
3	28,000.00	16,250.00	11,750.00	4,112.50	23,887.50
4	28,000.00	13,000.00	15,000.00	5,250.00	22,750.00
5	28,000.00	9,750.00	18,250.00	6,387.50	21,612.50
6	28,000.00	6,500.00	21,500.00	7,525.00	20,475.00
7	28,000.00	3,250.00	24,750.00	8,662.50	19,337.50
7	7,000.00 (Salvage)	7,000.00	0.00 (Depreciation recapture)	0.00	7,000.00
				$PW(15\%) =$	$1,884.68

EXAMPLE 6.18

Let us compare our post-1986 depreciation methods of Example 6.15 to 6.17 with a popular pre-1981 method, sum of the years' digits depreciation. Again, P = $98,000, F = $7000, n = 7, and the before-tax cash flow is $28,000/year. The cash flow calculations are given in Table 6.19. Calculating the present worth of the after-tax cash flows at i = 15% results in $1884.68.

In this example, the $7000 salvage value is not taxed. The total SOYD depreciation is only P − F = $91,000, and the last $7000 is recovered through the salvage value. Since there is no over- or under-recovery, no extra tax treatment is necessary.

It is interesting to note that the present worth of after-tax cash flows in Example 6.15 through 6.18 ranges from a high of $3102.81 under MACRS-GDS with its 5-year recovery period to a low of −$2789.19 under Elective MACRS-ADS (n = 9) with its 9-year recovery period. Similar calculations using regular ACRS and pre-1981 straight-line depreciation result in present worths of after-tax cash flows of $2045.52 and −$718.89, respectively. Again, nothing about the SMP machine is different; only the depreciation strategy has changed. By now, it should be clear why depreciation methods are of such concern to corporate taxpayers.

Examples 6.15 to 6.18 illustrate the important points that the method of depreciation and the recovery period affect cash flows and, therefore, the economic desirability of a project. It is interesting to note that the total capital recovery ($98,000), total taxes ($36,750), and total undiscounted after-tax cash flows ($68,250) are the same in each of Table 6.16 to 6.19. Can we make any generalized statements *from these examples* about the preferability of depreciation methods and recovery periods with respect to effects on taxes and after-tax cash flows? No, because these examples are worked only for a particular set of conditions. It can, however, be shown mathematically that certain depreciation strategies are superior to others in that they provide a higher present worth of tax savings, assuming that the effective tax rate remains the same from year to year. For example, the MACRS-GDS precentages as given in Tables 6.9 and 6.11 are always preferable to those for MACRS-ADS. Further, a shorter recovery period is always superior to a longer recovery period. For pre-1981 property, both double declining balance switching to straight-line and sum of the years' digits are preferable to straight-line depreciation; however, double declining balance switching to straight-line is not always better than the sum of the years' digits method.

6.13 EFFECT OF INTEREST ON BORROWED MONEY

Investment alternatives may be financed using equity (owner's) funds or debt (borrowed) funds. Until now we have implicitly assumed all financing to be through equity, although many companies use a mix of debt and equity for financing plant and equipment. Borrowed funds, including both principal and interest, must be repaid. The interest repaid each year affects taxable income and, consequently, taxes. Both the principal and interest payments affect after-tax cash flows.

There are four common (and many less common) ways in which money can be repaid. First is the periodic payment of interest over the stipulated repayment period with the entire principal being repaid at the end of that time. The second requires a periodic payment that uniformly repays the principal and also covers the periodic interest. In this method, the payments decrease as the interest on the unrepaid principal decreases. Third is the method requiring a uniform periodic payment for the sum of principal plus interest. In each payment the proportion of principal gradually increases as the proportion of interest decreases. The fourth method repays nothing, neither interest nor principal, until the end of a specified period.

Loan interest rates have varied widely in recent years. For example, during 1 calendar year, the prime interest rate charged by banks to their best corporate customers varied in the range from about 11% to 22%. More recently, the prime rate has had single digits. Since neither the banks nor the customers want to get "stuck" in an unfavorable long-term agreement, loans with adjustable interest rates have become commonplace. For example, a typical small business loan might span 1 to 2 years, with interest adjusted to the current market rate and paid monthly, and with the entire principal due at the end of the loan period.

EXAMPLE 6.19

Let us illustrate the four basic plans for repaying principal and interest on borrowed money. Assume that a business borrows $50,000 to be used in financing

Table 6.20 Illustration of Four Common Methods of Principal and Interest Repayment, Example 6.19

End of Year	Interest Accrued During Year	Total Money Owed Before Yearly Payment	Interest Payment	Principal Payment	Total Payment	Total Money Owed After Yearly Payment
Method 1						
0						$50,000.00
1	$9,000.00	$59,000.00	$9,000.00	$0.00	$9,000.00	50,000.00
2	9,000.00	59,000.00	9,000.00	0.00	9,000.00	50,000.00
3	9,000.00	59,000.00	9,000.00	0.00	9,000.00	50,000.00
4	9,000.00	59,000.00	9,000.00	0.00	9,000.00	50,000.00
5	9,000.00	59,000.00	9,000.00	50,000.00	59,000.00	0.00
Method 2						
0						$50,000.00
1	$9,000.00	$59,000.00	$9,000.00	$10,000.00	$19,000.00	40,000.00
2	7,200.00	47,200.00	7,200.00	10,000.00	17,200.00	30,000.00
3	5,400.00	35,400.00	5,400.00	10,000.00	15,400.00	20,000.00
4	3,600.00	23,600.00	3,600.00	10,000.00	13,600.00	10,000.00
5	1,800.00	11,800.00	1,800.00	10,000.00	11,800.00	0.00
Method 3						
0						$50,000.00
1	$9,000.00	$59,000.00	$9,000.00	$6,988.89	$15,988.89	43,011.11
2	7,742.00	50,753.11	7,742.00	8,246.89	15,988.89	34,764.22
3	6,257.56	41,021.77	6,257.56	9,731.33	15,988.89	25,032.88
4	4,505.92	29,538.80	4,505.92	11,482.97	15,988.89	13,549.91
5	2,438.98	15,988.89	2,438.98	13,549.91	15,988.89	0.00
Method 4						
0						$50,000.00
1	$9,000.00	$59,000.00	$0.00	$0.00	$0.00	59,000.00
2	10,620.00	69,620.00	0.00	0.00	0.00	69,620.00
3	12,531.60	82,151.60	0.00	0.00	0.00	82,151.60
4	14,787.29	96,938.89	0.00	0.00	0.00	96,938.89
5	17,449.00	114,387.89	64,387.89	50,000.00	114,387.89	0.00

an alternative, and the interest rate on this loan is 18% compounded annually. The stipulated repayment period is 5 years.

A summary of all relevant components of our example is presented in Table 6.20. In method 1, the interest equals $50,000(.18) = $9000/year. Only the interest is paid, and only the principal of $50,000 is owed after each year's payment. Method 2 repays the principal in equal amounts of $50,000/5 = $10,000 as well as the interest for year t, which is given by $[\$50,000 - \$10,000(t - 1)]0.18$. Clearly, the interest payment, total payment, and total money owed after yearly payments are a decreasing gradient series. Method 3 requires equal annual total payments. This annual payment is equal to $\$50,000(A|P\ 18,5) = \$15,988.89$. The principal component of this annual payment for year t can be found quickly as $\$15,988.89(P|F\ 18,5 - t + 1)$, using the method given in Chapter Three. In method 4, the interest accrued during each year is added to the principal such that the total amount owed after t years is $\$50,000(1.18)^t$. When payment is made at the end of year 5, everything over $\$50,000$ is considered interest.

We have seen that the interest on borrowed money is deductible for tax purposes, whereas the principal repayment does not enter into taxable income. In addition, both the interest and principal portions of a payment are real and must be taken into account when calculating cash flows.

EXAMPLE 6.20

Let us illustrate the effect of borrowed money by recalling the SMP machine of Example 6.15 having $P = \$98,000$, $F = \$7000$, a useful life of 7 years, and before-tax and loan cash flows of $\$28,000$. Assume that $\$50,000$ of the $\$98,000$ paid for the machine is through debt funding. The loan is to be repaid in equal annual installments (method 3) at 18% over 5 years. The remaining $\$48,000$ will be equity money. Applicable depreciation and loan activity (method 3) have previously been calculated in Examples 6.15 and 6.19, respectively. The resulting after-tax cash flow profile is detailed in Table 6.21.

Calculating the present worth of the after-tax cash flows at $i = 15\%$, we have

$$PW(15) = -\$48,000 + \$12,221.11(P|F\ 15,1) + \$15,896.81(P|F\ 15,2)$$
$$+ \cdot \cdot \cdot + (\$18,200 + \$4550)(P|F\ 15,7) = \$7059.78$$

The value of our effectiveness measure jumped to $\$7059.78$ for this particular example. Note that the present worth calculation on cash flows was made using a discount rate equal to our 15% *MARR*. The 18% loan rate was used *only* in sizing the loan repayments. If we actually need to borrow the $\$50,000$ to implement the project, our estimates indicate that a handsome monetary return will be received on the $\$48,000$ equity investment. If, on the other hand, we really have at least the other $\$50,000$ available, borrowing allows us to invest that money as equity capital in another alternative that will earn a return at least equal to the *MARR*.

A relatively high rate of interest, 18%, was used for the loan in Example 6.20 for two reasons. First, it appears illogical for company management to borrow at 18% when they require only a 15% *MARR*. Remember though, the 15% *MARR* is an *after-tax* rate of return. Since loan interest is a deductible item, it is being paid with *before-tax* dollars. The after-tax interest rate or "cost of capital"

Table 6.21 After-Tax Cash Flow Profile Using the MACRS-GDS, Example 6.20

End of Year, A	Before-Tax and Loan Cash Flow, B	Loan Principal Payment, C	Loan Interest Payment, D	MACRS-GDS Deduction, E	Taxable Income B-D-E, F	Tax F × 0.35, G	After-Tax Cash Flow B-C-D-G, H
0	−$98,000.00	−$50,000.00					−$48,000.00
1	28,000.00	6,988.89	$9,000.00	$19,600.00	−$600.00	−$210.00	12,221.11
2	28,000.00	8,246.89	7,742.00	31,360.00	−11,102.00	−3,885.70	15,896.81
3	28,000.00	9,731.33	6,257.56	18,816.00	2,926.44	1,024.25	10,986.85
4	28,000.00	11,482.97	4,505.92	11,289.60	12,204.48	4,271.57	7,739.54
5	28,000.00	13,549.91	2,438.98	11,289.60	14,271.42	4,995.00	7,016.11
6	28,000.00	0.00	0.00	5,644.80	22,355.20	7,824.32	20,175.68
7	28,000.00	0.00	0.00	0.00	28,000.00	9,800.00	18,200.00
7	7,000.00 (Salvage)				7,000.00 (Depreciation recapture)	2,450.00	4,550.00
						$PW(15\%) =$	7,059.78

for the loan is 18% × (1 − tax rate) = 18(1 − 0.35) = 11.70%. The second reason is to illustrate that borrowed money, even at relatively high rates, can often "leverage" equity capital such that a nice return is earned, as shown in Example 6.20.

We cannot conclude from the example that borrowing money is always favorable. The desirability of borrowed funds depends on the terms of the loan, including method of repayment, interest, and repayment period. Furthermore, collateral, which may be lost (after legal action) if the principal and interest cannot be paid on schedule, is frequently required. In summary, each alternative investment and financing strategy should be compared on its own merits.

We have used the same basic example to illustrate the implications of depreciation and financing strategies thus far throughout the chapter. The present worth of after-tax cash flows was seen to vary from −$2789.19 using equity financing under elective MACRS-ADS over 9 years, to $7059.78 using a more sophisticated mix of debt and equity financing and the MACRS-GDS with a 5-year recovery period. Nothing about the basic incomes or costs relating to the asset were changed. By now the reader should have a good idea how to calculate and assess after-tax cash flows under differing depreciation schedules and financing arrangements.

6.14   CAPITAL GAINS, LOSSES, AND TAX TREATMENT

Capital gains have been accorded favored tax treatment from time to time in the past. Capital gains currently receive preferential tax rates, being less severely taxed than ordinary income. Taxation of capital gains is always a controversial item, as is the rate at which to tax them. Since the capital gains provision has a history of being "on" and "off," it is covered here because it might be relevant to an after-tax economic analysis.

6.14.1   Capital Gains and Losses

Examples of *capital assets* are stocks, stock rights, bonds, and stock received as a dividend. Real property and property used for the production of income, not used in a trade or business, are capital assets. Patents held for investment, inventions in the hands of the inventor, life estates, inherited jewelry, bank accounts, and cotton acreage allotments, for example, are also considered capital assets. Gold, silver, stamps, coins, gems, and so on, expand the list.

Specifically *excluded* are the things of most interest in a corporate engineeering economic analysis—real property and depreciable property used in a trade or business! [3] That is, machinery, plant, and equipment are not capital assets. Instead, they are classified as Section 1231 property (defined in the next section), which receives favored capital asset treatment.

Before applying capital treatment, the extent of the capital gain or loss must be determined. This requires a balancing of long- and short-term gains and losses. The excess of selling price over cost basis of a capital asset is a *capital gain*. If the selling price is less than the cost basis, then a *capital loss* has occurred. If the period during which the taxpayer holds the asset is 1.5 year or less, the gain or loss is referred to as a *short-term gain or loss*. If the asset is held longer than 1.5 year, a *long-term gain or loss* has occurred. A *net short-term gain (loss)* is the sum of the short-term gains minus the sum of short-term losses. Similarly, a *net long-term gain* (loss) is the sum of long-term gains minus long-term losses.

EXAMPLE 6.21

A corporation has the following capital gains and losses during a tax year. The net short- and long-term gains and losses are summarized as follows:

Short-term capital gains	$19,300
Short-term capital losses	($27,600)
Net short-term capital loss	($ 8,300)
Long-term capital gains	$77,500
Long-term capital losses	($18,750)
Net long-term capital gain	$58,750

6.14.2   Tax Treatment of Capital Assets

Short- and long-term capital gains may be consolidated for tax purposes. The result may be a net capital gain, or a net capital loss. The consolidated result

Table 6.22 Results and Corporate Tax Treatment of Consolidated Long- and Short-Term Capital Gains and Losses

Net Short-Term Capital Gains and Losses	Net Long-Term Capital Gains and Losses		
	None	Long-Term Gain	Long-Term Loss
None	No capital treatment	Net capital gain taxed as capital gain	Net capital loss may be carried back or over
Short-term gain	Net capital gain taxed as ordinary income	Net capital gain. Long-term gain taxed as capital gain *and* short-term gain taxed as ordinary income	If a net capital loss, may be carried back or over. If a net capital gain, taxed as ordinary income
Short-term loss	Net capital loss may be carried back or over	If a net capital gain, taxed as capital gain. If a net capital loss, may be carried back or over	Net capital loss may be carried back or over

Note: Capital losses carried back and over are treated as short-term capital losses.

dictates whether the tax treatment will be capital gains, ordinary income, or a loss to be carried back to a previous tax year or over to a future tax year. The consolidation rules and tax treatments are summarized in Table 6.22.

EXAMPLE 6.22

In Example 6.21 we determined that a corporation has a net short-term capital loss of $8300 and a net long-term capital gain of $58,750. Consolidating these in accordance with Table 6.22, we have a net capital gain of $58,750 − $8300 = $50,450 that is taxed as a capital gain.

As this chapter is written, the excess of net long-term capital gains over net short-term capital losses is excluded and taxed at a rate of only 28%, a lot lower than the ordinary income tax rates of up to 35%. Actually, the taxpayer has the choice of treating capital gains as either ordinary income (the *regular method*) or segregating the net capital gain to be taxed at 28% (the *alternate method*). This choice is useful when taxable income is sufficiently low such that the regular tax rate (e.g., 15%) is less than the capital gains rate of 28%.

EXAMPLE 6.23

Suppose a corporation has a $400,000 taxable ordinary income, plus $10,000 that is a net capital gain taxed as a capital gain. What would be the tax on the net capital gain under *current* law?

Answer: Alternative 1 − $10,000(0.34) = $3400; Alternative 2 − $10,000(0.28) = $2800. Obviously, Alternative 2 would be selected, saving $600 in taxes.

EXAMPLE 6.24

Suppose a corporation has a $40,000 taxable ordinary income, plus $10,000 that is a net capital gain taxed as a capital gain. What would be the tax on the net capital gain under *current* law?

Answer: $10,000(0.15) = $1500.

When capital losses are subtracted from capital gains and the result is negative, Table 6.22 shows that the net loss may be carried back or over. The carry-back period is 3 years and the carry-forward period is 5 years, during which time the capital loss is treated as a short-term capital loss and can be used to offset capital gains. Capital losses cannot be used to offset ordinary income.

6.15   DISPOSITION OF DEPRECIABLE PROPERTY

A disposition is the permanent withdrawal of property from use in trade or business or in the production of income. A withdrawal may be made by sale, exchange, retirement, abandonment, or destruction. There is generally a gain or loss to be recognized on the disposition of an asset by sale.

6.15.1   Section 1231 Gains and Losses and Depreciation Recapture

Much of the law regarding disposition refers to Section 1231 property. *Section 1231 property* includes property used in a trade or business or held for the production of rents or royalties and held for more than 1 year. Property used in trade or business includes real property and depreciable personal property, but not property held for sale to customers. As such, nearly *any property we would be concerned with in an engineering economic analysis is Section 1231 property*.

When a property is disposed of, a gain occurs if the selling price exceeds the unrecovered investment, adjusted basis, or book value of the asset. If the reverse is true, the result is a loss. All gain on 3-, 5-, 7-, 10-, 15-, and 20-year property is treated as ordinary income so long as it does not exceed the total depreciation taken on the asset. This gain is called *depreciation recapture*. Any gain in excess of depreciation taken is called a *Section 1231 gain*. A loss on such property is called a *Section 1231 loss*.

Gain on a disposition of residential rental or nonresidential real property is not recaptured as ordinary income. Rather, any gain or loss from disposition of such property is a Section 1231 gain or loss, respectively.

To help understand Section 1231 gains and losses and depreciation recapture, Figures 6.4 and 6.5 and Table 6.23 are presented in which the following notation is used:

P = cost basis or orginal investment.

B_t = unrecovered investment at time t; this is the cost basis less allowable MACRS deductions.

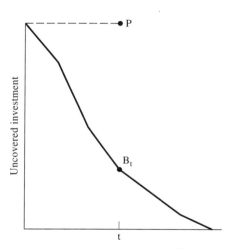
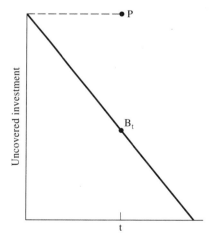

Figure 6.4 Unrecovered investment versus time for tangible property, other than residential rental or non-residential real.

Figure 6.5 Unrecovered investment versus time for residental rental and nonresidential real property.

F_t = salvage value actually received at time t.

t = any time, during or beyond the recovery period.

These figures and table summarize gains and losses of Section 1231 property.

6.15.2   Tax Treatment of Depreciable Properties

Gains and losses upon disposition of property can have a significant effect upon taxes paid. A special procedure, grouping all Section 1231 gains and losses, is used to determine how dispositions will be treated for tax purposes in a given year. But, since it is impossible to know the performance of all Section 1231 assets in advance, a reasonable rule is necessary in order to treat dispositions of property in engineering economic analyses. Therefore, Section 1231 losses are treated herein as ordinary losses for tax purposes, while Section 1231 gains are treated as capital gains, taxable at the current rate of 28%. Of course, depreciation recapture is treated as ordinary income.

EXAMPLE 6.25

In Examples 6.9 and 6.15, the SMP machine's cost basis P = $98,000 was fully recovered under the MACRS by the end of year 6. How should the $7000 salvage value received at year 7 be treated for tax purposes?

The machine is tangible, depreciable, 5-year property. Referring to Tables 6.16 and 6.23 and Figure 6.4, B_7 = $0 and F_7 = $7000. Therefore, $F_7 - B_7$ = $7000 is depreciation recapture and should be taxed as ordinary income in the amount of $7000(0.35) = $2450.

Table 6.23 Rules for Gains and Losses on Depreciable Properties

Section 1231 depreciable property, other than residential rental or nonresidential real.
Rules (see Figure 6.4)

If	Then
$F_t > P$	$F_t - P$ is Section 1231 gain, and
	$P - B_t$ is depreciation recapture treated as ordinary income.
$B_t < F_t < P$	$F_t - B_t$ is depreciation recapture treated as ordinary income.
$F_t < B_t$	$B_t - F_t$ is Section 1231 loss.

Section 1231 residential rental or nonresidential real property.
Rules (see Figure 6.5)

If	Then
$B_t < F_t$	$F_t - B_t$ is Section 1231 gain.
$F_t < B_t$	$B_t - F_t$ is Section 1231 loss.

EXAMPLE 6.26

Let us reconsider Example 6.17 in which we used MACRS-ADS depreciation over a 9-year recovery period. Since the useful life of the machine was 7 years, only \$65,333.33 of the cost basis $P = \$98,000$ had been recovered through depreciation. How should the disposal of the asset in year 7 be handled?

The machine is used in the manufacture of medical and dental supplies and is 5-year property. Referring to Tables 6.18 and 6.22 and Figure 6.4, $B_7 = \$98,000 - \$65,333.33 = \$32,667.67$ and $F_7 = \$7000$. Therefore, $B_7 - F_7 = \$32,666.67 - \$7000 = \$25,666.67$ is a Section 1231 loss. Since we have no other information regarding other Section 1231 gains and losses, this loss will be assumed to apply against ordinary income. As such, there should be a tax of $-\$25,666.67(0.35) = -\$8,983.33$, or a negative tax to apply against other positive taxes in year 7. This agrees with the treatment in Example 6.17.

EXAMPLE 6.27

Although unrealistic, suppose our SMP machine of Example 6.15 having $P = \$98,000$, a useful life of 7 years, and MACRS-GDS recovery over 6 years is sold at $t = 7$ for \$120,000. How should this salvage value be treated for tax purposes?

The machine has a 5-year recovery period. Referring to Table 6.23 and Figure 6.4, $B_7 = \$0$ and $F_7 = \$120,000$. Therefore, $F_7 - P = \$120,000 - \$98,000 = \$22,000$ which is a Section 1231 gain. This gain will be assumed to be taxed as a capital gain. The amount $P - B_7 = \$98,000$, represents depreciation recapture and is taxed as ordinary income. Thus, the total tax should be $\$22,000(0.28) + \$98,000(0.35) = \$40,460$. Note that the \$22,000 received favorable treatment, being taxed at the capital gains rate of only 28%.

EXAMPLE 6.28

Consider the residential rental property of Examples 6.11 and 6.12, placed in service on May 1, in which $P = \$220,000$. After 7 years and 5 months (this will be in the 8th tax year), on October 1, the property is sold for \$150,000. How should this sale be taxed?

This is 27.5-year residential rental property. Referring to Table 6.12 and the calculations in Example 6.12, total depreciation to date is $\$5000.60 + \$7999.20(6) + \$6332.70 = \$59,328.50$. The unrecovered investment is $B_8 = \$220,000 - \$59,328.50 = \$160,671.50$. So, $F_8 - B_8 = \$150,000 - \$160,671.50 = -\$10,671.50$ which is a Section 1231 loss. We will tax the $-\$10,671.50$ as an ordinary income loss, resulting in $-\$10,671.50(0.35) = -\$3735.03$ tax.

*6.16   INVESTMENT TAX CREDIT

The *investment tax credit* (*ITC*) is meant to stimulate investment by providing reduced taxation in the year in which an asset is placed in service. The ITC has had a turbulent history, dating back to 1962. It has been on-again, off-again during that time, with varying rules, percentages, and so forth. The ITC was repealed for property placed in service after 1985. It was argued that with a lowering of the maximum tax rate from 46% to 34% at that time, a further tax break such as the ITC should be discontinued. It is possible that the ITC will reappear if legislators believe it is in the best economic interest of the United States, and/or their re-election efforts. For purposes of introducing readers to the concept of the ITC, a brief explanation, plus one "adapted" example will be presented. This introduction will be kept simple. But, beware! The rules governing the ITC were somewhat complex. For a much more thorough treatment of the ITC, see earlier editions of this text and/or pre-1986 tax law.

The investment tax credit permitted taxpayers to claim a credit, not just a deduction, against income taxes for qualified investment in certain new or used depreciable property. The regular credit allowable was 10% of the eligible investment. If the full ITC was taken, law required that half the amount of the ITC be used to reduce the cost basis of the asset, thereby lessening the allowable cost recovery through depreciation. For example, if the 10% ITC was claimed, the cost basis of the asset was to be reduced immediately by 5%.

EXAMPLE 6.29

Let us suppose that the ITC is immediately reinstated exactly as last in effect, but is using today's tax rates and depreciation methods. We desire to take advantage of the ITC with our SMP machine of Example 6.15. Recall that P = $98,000, F = $7000, useful life is 7 years, before-tax cash flow is $28,000, and $i = 15\%$. What will be the present worth of after-tax cash flows?

The ITC will be $98,000(0.10) = $9800. The cost basis must be reduced by $4900, to $93,100. Table 6.24 summarizes the cash flow calculations. The present worth of the after-tax cash flows is $10,440.92. Recall that the present worth in Example 6.15 without the ITC was only $3102.81.

It is obvious that the ITC could be a major economic factor for companies of any size. As such, there may be pressure on lawmakers to reinstate the ITC in the future.

6.17   SECTION 179 EXPENSE DEDUCTION

The Section 179 expense deduction is designed to stimulate investment, primarily by smaller firms, by providing reduced taxation in the first year of an asset's life. Taxpayers may elect to treat the cost of certain property as an expense rather than as a capital expenditure. This property, called Section 179 property, is depreciable property. It includes most tangible property, other than buildings, that would be of interest in an engineering economic analysis.

Table 6.24 After-Tax Cash Flow Profile Using the Pre-1986 Investment Tax Credit Concept, Example 6.29

End of Year, A	Before-Tax Cash Flow, B	MACRS-GDS Deduction, C	Taxable Income B-C, D	Tax D × 0.35, E	Investment Tax Credit, F	After-Tax Cash Flow, B-E+F, G
0	−$98,000.00					−$98,000.00
1	28,000.00	$18,620.00	$9,380.00	$3,283.00	$9,800.00	34,517.00
2	28,000.00	29,792.00	−1,792.00	−627.20	0.00	28,627.20
3	28,000.00	17,875.20	10,124.80	3,543.68	0.00	24,456.32
4	28,000.00	10,725.12	17,274.88	6,046.21	0.00	21,953.79
5	28,000.00	10,725.12	17,274.88	6,046.21	0.00	21,953.79
6	28,000.00	5,362.56	22,637.44	7,923.10	0.00	20,076.90
7	28,000.00	0.00	28,000.00	9,800.00	0.00	18,200.00
7	7,000.00		7,000.00	2,450.00		4,550.00
	(Salvage)		(Depreciation recapture)		PW(15%) =	$10,440.92

The election to expense must take place in the year the property is placed in service. The expense deduction is limited to $17,500 for qualifying property. There are some limits placed on the expense deduction. The $17,500 limit applies to each taxpayer, not to each business. Also, the amount eligible to be expensed for any tax year is reduced, dollar for dollar, by the amount that the aggregate cost of qualifying property exceeds $200,000. That is, if the amount of eligible property placed in service during a tax year is $200,000, $210,000, or $217,500 the maximum amount that may be expensed is $17,500, $7,500, or $0, respectively. Further, the amount eligible to be expensed cannot exceed the taxable income derived from the trade or business in which the property is used. Finally, conversion of the property to nonbusiness use before the end of the property's recovery period results in recapture as ordinary income of the excess of the amount expensed over the MACRS deduction that would have been allowed.

EXAMPLE 6.30

Let us replay Example 6.29, except we will use the Section 179 expense deduction instead of the ITC. Recall that $P = \$98,000$, $F = \$7000$, useful life is 7 years, before-tax cash flow is $28,000, and $i = 15\%$. Assume that this is the company's only asset purchased and placed in service during the tax year. Also, assume that the company's tax rate is 35%. (Note that it is very unlikely that a company paying taxes at 35% would purchase only one asset during the year.) What will be the present worth of after-tax cash flows?

The Section 179 expense deduction will be $17,500. The cost basis is therefore reduced to $80,500. Table 6.25 summarizes the cash flow calculations. The present worth of after-tax cash flows is $4201.64.

6.18   CARRY-BACK AND CARRY-FORWARD RULES

In several examples we saw that an alternative's taxable income and tax can be negative for a particular year. It was implicitly assumed that those values were used to offset positive tax liabilities from other corporate activities. When there are insufficient incomes or taxes to be offset, we can carry back or carry forward ordinary income losses and capital losses.

Ordinary income losses can be carried back 3 years, beginning with the earliest. If not entirely used to offset income in that year, it is carried to the second year back and so on. After applying losses to all 3 back years, remaining losses can be carried forward to as many as 15 years following the loss year. Capital losses may be carried back 3 years and forward 5 years. They are treated as short-term capital losses and used to offset capital gains. They cannot be used to offset ordinary income.

6.19   AFTER-TAX ANALYSES UNDER INFLATION

We were introduced to inflation in Section 3.2. We learned that there are two procedures which may be used to perform economic analyses under inflation:

Table 6.25 After-Tax Cash Flow Profile Using the Section 179 Expense Deduction, Example 6.30

End of Year, A	Before-Tax Cash Flow, B	Section 179 Expense Deduction, C	MACRS-GDS Deduction, D	Taxable Income B-C-D, E	Tax E × 0.35, F	After-Tax Cash Flow, B-F, G
0	-$98,000.00					-$98,000.00
1	28,000.00	$17,500.00	$16,100.00	-$5,600.00	-1,960.00	29,960.00
2	28,000.00	0.00	25,760.00	2,240.00	784.00	27,216.00
3	28,000.00	0.00	15,456.00	12,544.00	4,390.40	23,609.60
4	28,000.00	0.00	9,273.60	18,726.40	6,554.24	21,445.76
5	28,000.00	0.00	9,273.60	18,726.40	6,554.24	21,445.76
6	28,000.00	0.00	4,636.80	23,363.20	8,177.12	19,822.88
7	28,000.00	0.00	0.00	28,000.00	9,800.00	18,200.00
7	7,000.00 (Salvage)		0.00	7,000.00 (Depreciation recapture)	2,450.00	4,550.00
					PW(15%) =	$4,201.64

1. Express all cash flows in terms of their "then-current" amounts and use an interest rate (i) that combines the real discount rate (d) and the inflation rate (j): $i = d + j + dj$.
2. Express all cash flows in terms of their "constant worth" amounts and use the real discount rate (d) alone without an inflation rate component.

While these two procedures seem simple enough, it is easy to err in an aftertax economic analysis under inflation. A series of examples will illustrate the problems.

EXAMPLE 6.31

We will again consider our SMP machine example, but with some modification. The first cost of the machine installation is still $98,000 with an estimated salvage value of $7000 after a useful life of 7 years. Recall that the machine has a 5-year recovery period, and we will use MACRS-GDS depreciation. Again, MARR = 15%.

In past examples, the reduced operating expenses realized by purchasing the machine have been $28,000/year. Therefore, the before-tax cash flow column has shown this figure for each year of the 7-year useful life of the computer. Now, suppose we have reconsidered and changed the estimate of reduced operating expenses. We now believe they will be smaller than originally projected, having a value "today" of $18,500, but increasing at a 10% ($g = 0.10$) rate over the 7-year horizon. That is, the then-current before-tax cash flow will be $BTCF(t) = \$18,500(1.10)^t$, equivalent to Equation 3.60.

Up to now, we have given MARR = 15% little thought. Since we are considering then-current cash flows, it is apparent that MARR must be considered a combined rate that includes both the real discount rate and the inflation rate. Suppose the company's real discount rate is 8.49% and the projected inflation rate is 6.00%. Then, $i = d + j + dj = 0.0849 + 0.0600 + 0.0849(0.0600) = .1500$ or MARR = 15%. It is this 15% that will be used to discount all after-tax cash flows since they will be in then-current dollars. The calculations are shown in Table 6.26.

The present worth of the after-tax cash flows, properly discounted at 15%, is −$1874.82. This means the machine is not a good investment, owing to the revised BTCF estimates.

Note that we did nothing new or different in Table 6.26 to reflect an analysis under inflation. *This is because we have implicitly assumed all analyses in this chapter to be based on then-current dollars with a MARR that reflects both the real discount rate and the inflation rate.*

To provide confidence that the then-current analysis is correct, consider a constant-worth analysis. It should provide the same decision and, in fact, the same present worth of the after-tax cash flows.

Table 6.26 After-Tax Cash Flows Using Then-Current Procedure Under Inflation, Example 6.31

End of Year, A	Before-Tax Cash Flow, B	MACRS-GDS Deduction, C	Taxable Income B-C, D	Tax D × 0.35, E	After-Tax Cash Flow B-E, F
0	-$98,000.00				-$98,000.00
1	20,350.00	$19,600.00	$ 750.00	$ 262.50	20,087.50
2	22,385.00	31,360.00	-8,975.00	-3,141.25	25,526.25
3	24,623.50	18,816.00	5,807.50	2,032.63	22,590.88
4	27,085.85	11,289.60	15,796.25	5,528.69	21,557.16
5	29,794.44	11,289.60	18,504.84	6,476.69	23,317.74
6	32,773.88	5,644.80	27,129.08	9,495.18	23,278.70
7	36,051.27	0.00	36,051.27	12,617.94	23,433.32
7	7,000.00 (Salvage)	0.00	7,000.00 (Depreciation recapture)	2,450.00	4,550.00
				PW(15%) =	-$1,874.82

EXAMPLE 6.32

This example will show an INCORRECT constant-worth analysis to illustrate how easy it is to make a mistake and even a wrong decision using this analysis procedure. Suppose everything is exactly as described in Example 6.31—P = $98,000, F = $7000, useful life = 7 years, 5-year property, depreciation using MACRS-GDS, MARR = 15%, d = .0849, j = .0600, and then-current BTCFs of $18,500(1.10)^t$. To use a constant-worth analysis requires that the BTCF be expressed in constant-worth dollars as in Example 6.31. That is, $BTCF(t)$ = $18,500(1.10)^t(1.06)^{-t}$. Table 6.27 summarizes the cash flow calculations.

The present worth of the after-tax cash flow, properly discounted at the rate d = .0849, is $1950.94. But, this is different than the present worth determined in Example 6.31! Even worse, based on these results, we would decide that the machine is a good investment—we would be making an incorrect decision! What happened? What went wrong?

In a constant-worth analysis, all cash flows *and charges* must be expressed in constant-worth dollars. In Table 6.27, the depreciation charges were left unchanged. They were expressed in terms of the exact amounts to be subtracted from BTCFs in future years. *That is, depreciation charges were left in then-current dollars.*

EXAMPLE 6.33

This example will correct the errors made in Example 6.32. All details remain the same—P = $98,000, F = $7000, useful life = 7 years, 5-year recovery period,

Table 6.27 INCORRECT After-Tax Cash Flows While *attempting* to Use Constant-Worth Procedure Under Inflation, Example 6.32

End of Year, A	Before-Tax Cash Flow, B	MACRS-GDS Deduction, C	Taxable Income B-C, D	Tax D × 0.35, E	After-Tax Cash Flow B-E, F
0	−$98,000.00				−$98,000.00
1	19,198.11	$19,600.00	$ −401.89	$ −140.66	19,338.77
2	19,922.57	31,360.00	−11,437.43	−4,003.10	23,925.67
3	20,674.37	18,816.00	1,858.37	650.43	20,023.94
4	21,454.53	11,289.60	10,164.93	3,557.73	17,896.80
5	22,264.14	11,289.60	10,974.54	3,841.09	18,423.05
6	23,104.29	5,644.80	17,459.49	6,110.82	16,993.47
7	23,976.15	0.00	23,976.15	8,391.65	15,584.50
7	4,655.40 (Salvage)	0.00	4,655.40 (Depreciation recapture)	1,629.39 PW(8.49%) =	3,026.01 $ 1,950.94

depreciation using MACRS-GDS, MARR = 15%, $d = 0.0849$, $j = 0.0600$, and constant-worth BTCFs of $18,500(1.10)^t(1.06)^{-t}$. This time, however, depreciation charges are converted to a constant-worth amount $D_t(1.06)^{-t}$. The cash flow calculations are given in Table 6.28.

The present worth of the after-tax cash flows, properly discounted using d = .0849, is −$1874.82.

Table 6.28 Correct After-Tax Cash Flows Using Constant-Worth Procedure Under Inflation, Example 6.33

End of Year, A	Before-Tax Cash Flow, B	MACRS-GDS Deduction, C	Taxable Income B-C, D	Tax D × 0.35, E	After-Tax Cash Flow B-E, F
0	−$98,000.00				−$98,000.00
1	19,198.11	$18,490.57	$ 707.55	$ 247.64	18,950.47
2	19,922.57	$27,910.29	−7,987.72	−2,795.70	22,718.27
3	20,674.37	15,798.28	4,876.09	1,706.63	18,967.73
4	21,454.53	$ 8,942.42	12,512.11	4,379.24	17,075.29
5	22,264.14	8,436.25	13,827.89	4,839.76	17,424.37
6	23,104.29	3,979.36	19,124.93	6,693.73	16,410.57
7	23,976.15	0.00	23,976.15	8,391.65	15,584.50
7	4,655.40 (Salvage)	0.00	4,655.40 (Depreciation recapture)	1,629.39 PW(8.49%) =	3,026.01 −$1,874.82

The present worth of −$1874.82 is precisely the same amount realized with the then-current analysis in Example 6.31. Of course, this is as it should be, since the two methods of analysis are equivalent. Note that every entry in Table 6.28 is simply the corresponding entry in Table 6.26 divided by 1.06^t.

6.20   DEPLETION

Depletion is a gradual reduction of minerals, gas and oil, timber, and natural deposits. In a sense, depletion is closely akin to depreciation. The difference is that while a depleting asset is losing value by actually being removed and sold, a depreciable asset is losing value through wear, tear, and obsolescence in the manufacture of goods to be sold. Money recovered through the depletion allowance is likely to be used in the exploration and development of depletable assets, just as depreciation reserves are reinvested for new equipment. As in most tax-related matters, laws relating to lessors, lessees, royalties, and sales are complex and will probably require expert assistance. However, the importance of depletion and the alternative methods that can be used for figuring the allowance will be illustrated here.

Cost depletion is the basic method of computing the depletion deduction. First, for any given year, the number of units (tons, barrels, board-feet, etc.) remaining in the property must be estimated. Then, the adjusted cost basis (first cost less depletion deductions taken previously) is divided by the number of units remaining. This quotient, multiplied by the number of units sold during the year, determines the depletion allowance. It is expressed as

$$\text{Cost depletion deduction} = \frac{\text{Adjusted cost basis} \times \text{Number of units sold}}{\text{Number of units remaining in property}}$$

This calculated deduction may or may not be used, depending on whether or not the percentage depletion allowance provides a larger deduction.

Percentage depletion provides an allowance equal to a percentage of the gross income from the mineral property. The applicable percentage depends on the type of property depleted. Table 6.29 gives some recent depletion percentages. The percentage depletion deduction may not exceed 50% of the taxable income *before* allowance for depletion. The total of allowances under percentage depletion is in no way limited by the adjusted basis of the property. Thus, even though the basis of depleted property may be reduced to $0.00, percentage depletion is still allowed. Cost depletion, however, stops when the adjusted basis reaches $0.00. Whichever of the two depletion methods results in the higher allowance in a given year must be used.

EXAMPLE 6.34

We own a small independent oil company and have purchased the rights, drilled, and developed a 48,000-barrel oil well on NW/4 35/19N/1W, Payne County, Oklahoma, for $140,000. Operating expenses, based on past experience, are equal to $5 + $3(t − 1)$/barrel, where t is the year in which the oil is removed and

Table 6.29 Depletion Percentage for Some Minerals and Similar Resources[a]

Type	Depletion Percentage
Oil and gas (percentage depletion is not allowed, except for certain gas and oil production)	
a. Oil and gas—independent producers and royalty owners, with some limitations	15
b. Natural gas—regulated natural gas and gas sold under a fixed contract, with some limitations	22
Sulphur and uranium; and, if from U.S. deposits, asbestos, bauxite, chromite, graphite, mica, quartz crytals, cadmium, cobalt, lead, mercury, nickel, tin, zinc, and certain other ores and minerals	22
Gold, silver, oil shale, copper, and iron ore if from U.S. deposits	15
Numerous other minerals such as diatomaceous earth, granite, limestone, marble, and so forth	14
Coal, lignite, perlite, and sodium chloride	10
Clay and shale used for sewer pipe and brick	$7\frac{1}{2}$
Clay (for roofing tile, flower pots, etc.), gravel, peat, pumice, sand, and stone	5

[a] As of 1997.

sold. This increase in cost per barrel over time indicates the increased difficulty of recovering the oil as the field is depleted. Our geologist expects the field to last 6 years, yielding 15,000, 13,000, 10,000, 6000, 3000, and 1000 barrels of oil over that time. Each barrel will be worth $20 to our company. The cash flow calculations are given in Table 6.30.

Using the cost method, ($140,000/48,000)(15,000) = $43,750 will be our depletion allowance during year 1. From Table 6.29, our depletion percentage is seen to be 15%. This results in a $300,000(0.15) = $45,000 deducation in year 1, so this figure will be used since it is larger than $43,750.

Note that in year 5, cost depletion of $6000 is chosen. This is because the $9000 percentage depletion is limited in $4500, or 50% of taxable income before depletion.

6.21   INTERNATIONAL CONSIDERATIONS

Engineering economic analyses are certainly applicable on a worldwide basis, even though the materials covered in this chapter are focused only on United States tax law. There are three major international considerations of interest in the taxation of corporations (taxation of individuals is not a focus of this text). These include: (1) an organization taxed by its "home" country for income derived from trade or business in that country; (2) an organization taxed by a "non-home" country for income due to trade or business in the "non-home"

Table 6.30 Depletion Allowance, Tax, and After-Tax Cash Flows for Oil Well, Example 6.34

End of Year, A	Barrels Sold B	Gross Income C	Operating Cost D	Taxable Income Before Depletion C-D	50% of Taxable Income Before Depletion $E \times 0.5$ F	Percentage Depletion at 15% $C \times 0.15$ G	Cost Depletion H	Adjusted Cost Basis I	Barrels of Oil Remaining J	Taxable Income $E - \max\{[\min (F,G)], H\}$ K	Income Tax $K \times 0.34$ L	After-Tax Cash Flow $E - L$ M
0		-$140,000		-$140,000				$140,000	48,000			-$140,000
1	15,000	300,000	$ 75,000	225,000	$112,500	$45,000[a]	$43,750	95,000	33,000	$180,000	$61,200	163,800
2	13,000	260,000	104,000	156,000	78,000	39,000[a]	37,424	56,000	20,000	117,000	39,780	116,220
3	10,000	200,000	110,000	90,000	45,000	$30,000[a]	28,000	26,000	10,000	60,000	20,400	69,600
4	6,000	120,000	84,000	36,000	18,000	18,000[a]	15,600	8,000	4,000	18,000	6,120	29,880
5	3,000	60,000	51,000	9,000	4,500	9,000	6,000[a]	2,000	1,000	3,000	1,020	7,980
6	1,000	20,000	20,000	0	0	3,000	2,000[a]	0	0	-2,000	-680	680

[a] Indicates depletion allowance used.

$5 + 3(t-1)$

306

country; and (3) an organization taxed in its "home" country for trade or business in a "non-home" country [1].

The first case, in which an organization is taxed by its "home" country for income derived from trade or business in that country, should be of primary interest to those who plan to work for an organization doing business at "home." For example, this would include a small U.S. (or Japanese or Indian) organization doing business entirely in the "home" country or even a local region. In this case, the laws may be different from those studied in this chapter, but the concepts regarding deductions, depreciation schedules, treatment of interest, and "special" incentives such as Capital Gains Treatment, the Section 179 Expense Deduction, the Investment Tax Credit, and others, will likely be similar, depending on country. Those who have studied this chapter will be well armed in the fundamentals, but will need to obtain and study the subject country's specific tax-related rules.

The second case, in which an organization is taxed by a non-home country for income due to trade or business in the non-home country, should be of primary interest to those who plan to work for an organization that has income from trade or business outside its "home" boundaries. For example, this would include a U.S. (or Taiwanese or Peruvian) organization earning income from trade or business in the Middle East. The taxation imposed by the Middle Eastern country will depend on their laws, the nature of the business, the country's reciprocity agreements with the home country, etc. In any event, (1) the fundamentals of taxation will likely be similar to those studied, and (2) help will be needed by tax experts if this taxation is a major consideration in an engineering economy analysis. In many non-home countries, the tax scheme imposed upon an organization from some other home country will likely be close to that used for their own domestic organizations.

The third case, in which an organization is taxed in its "home" country for trade or business in a non-home country, is commonplace. For example, a U.S. (or Australian or Malaysian) company with income due to trade or business in France can expect to be taxed by its "home" country on the income earned from France. But, this quickly becomes very complex and is affected by tax rates of the respective countries, credits issued for foreign taxes paid, and so on.

The safest way to proceed in any after-tax economic analysis involving taxation from a non-home country is to seek competent tax assistance from those who regularly deal with such matters. Otherwise, incorrect decisions are possible, just as was shown in this chapter for the relatively simple case of a U.S. company being taxed for income due to trade or business in its "home" country, the United States. Bottom line: get help on international taxation if it can make a difference!

6.22 SUMMARY

This chapter has presented the most important elements of depreciation and tax law as they pertain to economic analyses. It is clear that depreciation (or depletion) method, recovery period, financing, various deductions, and inflation can have significant effects on the desirability of making an investment. Many of these factors are law-related, and changes are being made daily. Therefore, in cases of uncertainty regarding depreciation and tax treatment, it is wise to seek competent advice.

BIBLIOGRAPHY

1. Bittker, Boris, I., and Lokken, Lawrence, *Fundamentals of International Taxation*, Warren, Gorham & Lamont, Boston, MA, 1991.
2. Department of the Treasury, Internal Revenue Service, *Depreciating Property Placed in Service Before 1987*, Publication 534, Cat. No. 150640, November 1995.
3. Department of the Treasury, Internal Revenue Service, *How to Depreciate Property*, Publication 946, Cat. No. 13081F, 1996.
4. Department of the Treasury, Internal Revenue Service, *Tax Guide for Small Business*, Publication 334, Cat. No. 11063P, 1996.
5. Department of the Treasury, Internal Revenue Service, *Corrections to Publication 334, Tax Guide for Small Business*, Notice 1154, 1997.

PROBLEMS

1. What is the federal income tax for each of the following corporate taxable incomes? (6.3)
   (a) $25,000
   (b) $70,000
   (c) $95,000
   (d) $200,000
   (e) $1,000,000
   (f) $12,000,000
   (g) $17,000,000
   (h) $25,000,000
2. A consulting engineer is seeking a new project that will increase taxable income significantly. Determine the tax increase and the percentage of the new taxable income that goes to pay the tax increase. The present and additional new taxable incomes are, respectively (6.3).
   (a) $45,000 and $40,000
   (b) $90,000 and $40,000
   (c) $900,000 and $40,000
3. An air purifier for use in manufacturing semiconductors is placed in service with a first cost of $50,000. It will be used for 8 years, have an annual gross income less operating expenses of $14,000 and will have no salvage value. The effective tax rate is 35%.
   (a) Determine the after-tax cash flow for years 0–8 if depreciation allowances are $10,000, $16,000, $9600, $5760, $5760, $2880, $0, and $0 during the 8 years (treated as MACRS-GDS 5 year recovery property). (6.4)
   (b) Determine the after-tax cash flow for years 0–8 if depreciation allowances are $5000, $10,000, $10,000, $10,000, $10,000, $5000, $0, and $0 during the 8 years (treated as Elective MACRS-ADS 5 year recovery property). (6.4)
   (c) Determine the after-tax cash flow for years 0–8 if depreciation allowances are $10,000 for years 1–5 and $0 in years 6–8 (treated as traditional straight-line depreciation using Class Life 5 years). (6.4)

4. A special handling device for the manufacture of food is placed in service. It costs $30,000 and has a salvage value of $2000 after a useful life of 5 years. The device generates a savings of $14,000/year. The effective tax rate is 35%. Find the after-tax cash flow for each year if:

   (a) Depreciation allowances are $9999, $13,335, $4443, $2223, and $0 for years 1–5 (treated as MACRS-GDS 3 years recovery property). (6.4)

   (b) Depreciation allowances are $3750, $7500, $7500, $7500, and $3750 for years 1–5 (treated as Elective MACRS-ADS 4-year recovery property). (6.4)

   (c) Depreciation allowances are $7500 for each of years 1–4 and $0 in year 5 (treated as traditional straight-line depreciation using Class Life 4 years). (6.4)

5. A test instrument used in the manufacture of electronic components is purchased for $40,000. It is estimated to have a life of 7 years. Operating and maintenance costs increase by $500/year with the cost for the first year being $1500. A $5000 salvage value is anticipated. Assuming a 35% tax rate and a 15% after-tax MARR, determine the annual equivalent after-tax cost of the test instrument. (Hint: First, calculate the after-tax cash flows, letting gross income equal $0.)

   (a) Use annual depreciation allowances of $8000, $12,800, $7680, $4608, $4608, and $2304 (treated as MACRS-GDS 5-year recovery property). (6.4)

   (b) Use annual depreciation allowances of $3332, $6668, $6668, $6668, $6664, $6668, and $3332 (treated as Elective MACRS-ADS 6-year recovery property). (6.4)

6. A special fixture for the manufacture of fabricated metal products has the following net cash flows before taxes:

EOY	0	1	2	3	4
BTCF	−60,000	36,000	38,000	24,000	20,000 + 15,000 salvage

The effective tax rate is 35%, and the after-tax MARR is 18%. The company is profitable in its other activities. Find the after-tax cash flows and the present worth of those cash flows if:

   (a) Depreciation allowances are $19,998, $26,670, $8886, and $4446 for years 1–4 (treated as MACRS-GDS 3-year recovery property). (6.4)

   (b) Depreciation allowances are $10,002, $19,998, $19,998, and $10,002 for years 1–4 (treated as Elective MACRS-ADS 3-year recovery property). (6.4)

7. For each of the following assets, state whether the asset is tangible/intangible property, personal/real property, and depreciable/nondepreciable property. (6.6)

   (a) A Mooney viscometer used in a polymers lab.
   (b) A computer.
   (c) A trademark.
   (d) A metal high-bay storage building.
   (e) A plot of land.
   (f) A file cabinet.
   (g) A commercial van.
   (h) A fork truck for material handling.
   (i) A flatbed document scanner.
   (j) A melt indexer used in research.

*8. A mold for manufacturing medical supplies is purchased at the beginning of the fiscal year for $30,000. The estimated salvage value after 10 years is $3000. Calculate the depreciation deduction and the resulting unrecovered investment during each year of the asset's life.

(a) Use traditional pre-1981 straight-line depreciation. (6.7.1)

(b) Use traditional pre-1981 200% declining balance switching to straight-line depreciation. (6.7.2)

(c) Use traditional pre-1981 sum of the years' digits depreciation. (6.7.3)

*9. A panel truck is purchased for $17,000. The truck is expected to be of use to the company for 6 years, after which it will be sold for $2500. Calculate the depreciation deducation and the resulting unrecovered investment during each year of the asset's life.

(a) Use traditional pre-1981 straight-line depreciation. (6.7.1)

(b) Use traditional pre-1981 200% declining balance switching to straight-line depreciation. (6.7.2)

(c) Use traditional pre-1981 sum of the years' digits deprecation. (6.7.3)

*10. A digitally controlled plane for manufacturing furniture is purchased on April 1 by a calendar-year taxpayer for $66,000. It is expected to last 12 years and have a salvage value of $5000. Calculate the depreciation deduction during years 1, 4, and 8.

(a) Use traditional pre-1981 straight-line depreciation. (6.7.1)

(b) Use traditional pre-1981 200% declining balance switching to straight-line depreciation. (6.7.2)

(c) Use traditional pre-1981 sum of the years' digits depreciation. (6.7.3)

*11. A file server and peripherals are purchased in December by a calendar-year taxpayer for $8000. The server will be used for 6 years and be worth $200 at that time. Calculate the depreciation deduction during years 1, 3, and 6.

(a) Use traditional pre-1981 straight-line depreciation. (6.7.1)

(b) Use traditional pre-1981 200% declining balance switching to straight-line depreciation. (6.7.2)

(c) Use traditional pre-1981 sum of the year's digits depreciation. (6.7.3)

12. For each of the assets named in problem 7, state both the MACRS-GDS property class, if applicable, and the specific depreciation method to be used. (6.8.1)

13. A mold for manufacturing medical supplies (MACRS-GDS 5-year recovery period; Elective MACRS-ADS 9-year recovery period) is purchased at the beginning of the fiscal year for $30,000. The estimated salvage value after 10 years is $3000. Calculate the depreciation deduction and the resulting unrecovered investment during each year of the asset's life.

(a) Use MACRS-GDS allowances. (6.8.2)

(b) Use Elective MACRS-ADS allowances. (6.8.4)

14. A panel truck (MACRS-GDS 5-year recovery period; Elective MACRS-ADS 5-year recovery period) is purchased for $17,000. The truck is expected to be of use to the company for 6 years, after which it will be sold for $2500. Calculate the depreciation deduction and the resulting unrecovered investment during each year of the asset's life.

(a) Use MACRS-GDS allowances. (6.8.2)

(b) Use Elective MACRS-ADS allowances. (6.8.4)

15. A digitally controlled plane for manufacturing furniture (MACRS-GDS 7-year recovery period; Elective MACRS-ADS 10-year recovery period) is purchased on April 1 by a calendar-year taxpayer for $66,000. It is expected to last 12 years and have a salvage value of $5000. Calculate the depreciation deduction during years 1, 4, and 8.

(a) Use MACRS-GDS alowances. (6.8.2)

(b) Use Elective MACRS-ADS allowances. (6.8.4)

16. A file server and peripherals (MACRS-GDS 5-year recovery period; Elective MACRS-ADS 5-year recovery period) are purchased in December by a calendar-year taxpayer for $8000. The server will be used for 6 years and be worth $200 at that time. Calculate the depreciation deduction during years 1, 3, and 6.
    (a) Use MACRS-GDS allowances. (6.8.2)
    (b) Use Elective MACRS-ADS allowances. (6.8.4)

17. Material-handling equipment used in the manufacture of grain products (MACRS-GDS 10-year recovery period; Elective MACRS-ADS 17-year recovery period) is purchased and installed for $180,000. It is placed in service in the middle of the tax year. If it is removed just *before* the end of the tax year approximately 4.5 years from the date placed in service, determine the deprecision deduction during each of the tax years involved.
    (a) Use MACRS-GDS allowances. (6.8.2)
    (b) Use Elective MACRS-ADS allowances. (6.8.4)

18. Repeat problem 17 if the material-handling equipment is removed just after the end of the tax year.
    (a) Use MACRS-GDS allowances. (6.8.2)
    (b) Use Elective MACRS-ADS allowances. (6.8.4)

19. Electric utility transmission and distribution equipment (MACRS-GDS 20-year recovery period; Elective MACRS-ADS 30-year recovery period) is placed in service at a cost of $300,000. It is expected to last 30 years with a salvage value of $15,000. Determine the depreciation deduction and the resulting unrecovered investment during each of the first 4 tax years.
    (a) Use MACRS-GDS allowances. (6.8.2)
    (b) Use Elective MACRS-ADS allowances. (6.8.4)

20. A business building (MACRS-GDS 39-year recovery period; Elective MACRS-ADS 40-year recovery period) is placed in service by a calendar-year taxpayer on January 4 for $300,000. Calculate the depreciation deduction for years 1 and 10, assuming the building is kept longer than 10 years.
    (a) Use MACRS-GDS allowances. (6.8.3)
    (b) Use Elective MACRS-ADS allowances. (6.8.4)

21. A rental apartment complex (MACRS-DGS 27.5-year recovery period; Elective MACRS-ADS 40-year recovery period) is placed in service by a calendar-year taxpayer on January 4 for $220,000. If the apartments are kept for 5 years and 2 months (sold on March 6), determine the depreciation deduction during each of the 6 tax years involved.
    (a) Use MACRS-GDS allowances. (6.8.3)
    (b) Use Elective MACRS-ADS allowances. (6.8.4)

22. A building used for the overhaul of dewatering systems (MACRS-GDS 39-year recovery period; Elective MACRS-ADS 40-year recovery period) is placed in service on October 10 by a calendar-year taxpayer for $140,000. It is sold almost 4 years later on August 15. Determine the depreciation deduction during years 1, 3, and 5, and the unrecovered investment at the end of years 1, 3, and 5.
    (a) Use MACRS-GDS allowances. (6.8.3)
    (b) Use Elective MACRS-ADS allowances. (6.8.4)

23. A rental house (MACRS-GDS 27.5-year recovery period; Elective MACRS-ADS 40-year recovery period) is placed in service by a calendar-year taxpayer during July for

$70,000. It is sold in July, 3 years later. Determine the depreciation deduction and resulting unrecovered investment for each applicable year.

(a) Use MACRS-GDS allowances. (6.8.3)

(b) Use Elective MACRS-ADS allowances. (6.8.4)

24. A specially coated mold for manufacturing tires (MACRS-GDS 3-year recovery period; Elective MACRS-ADS 4-year recovery period) costs $35,000 and has a salvage value of $1750 after a useful life of 5 years. The mold generates a net savings of $14,000/year. The effective tax rate is 35%. Find the after-tax cash flow for each year.

(a) Use MACRS-GDS allowances. (6.8.2)

(b) Use Elective MACRS-ADS allowances. (6.8.4)

*(c) Use traditional pre-1981 straight-line depreciation. (6.7.1)

*(d) Use traditional pre-1981 sum of the years' digits depreciation. (6.7.3)

25. A firm may either invest $30,000 in a numerically controlled lathe for use in furniture manufacturing (MACRS-GDS 7-year recovery period; Elective MACRS-ADS 10-year recovery period) that will last for 11 years and have a zero salvage value at that time, or invest $X in a methods improvement study (this is not depreciable). Both investment alternatives yield an increase in income of $15,000/year for 11 years. With regular MACRS allowances for the investment, and with a 35% tax rate, for what value of X will the firm be indifferent between the two investment alternatives? Assume an after-tax MARR OF 15%. (6.8.2)

26. Two mutually exclusive alternatives, A and B (both MACRS-GDS 5-year recovery property), are available. Alternative A requires an original investment of $100,000, has a useful life of 6 years, annual operating costs of $2500, and a salvage value in year k given by $100,000(0.70)^k$. Alternative B requires an original investment of $150,000, has a life of 8 years, zero annual operating costs, and a salvage value in year k given by $150,000(0.8)^k$. The after-tax MARR is 15%; a 35% tax rate is applicable. MACRS-GDS allowances are used. Perform an annual worth comparison and recommend the least-cost alternative. (6.8.2)

(a) Use a planning horizon of 6 years.

(b) Use a planning horizon of 8 years.

(c) Use a planning horizon of 24 years.

27. A special tool used in the manufacture of cars (MACRS-GDS 3-year recovery period; Elective MACRS-ADS 3-year recovery period) is placed in service for $10,000 and has a useful life of 4 years, after which there is $0 salvage value. Interest is 15%. Determine for each year the capital recovered, the return on capital unrecovered, and the capital recovered plus return. Also, find the annual equivalent capital recovered plus return. (6.10.2)

(a) Use MACRS-GDS allowances.

(b) Use Elective MACRS-ADS allowances.

*(c) Use traditional pre-1981 straight-line depreciation.

*(d) Use traditional pre-1981 sum of the years' digits depreciation.

*28. An asset used in the printing business (MACRS-GDS 7-year recovery period; Elective MACRS-ADS 11-year recovery period) is placed in service for $50,000 and has a useful life of 12 years, after which there is a $5000 salvage value. Interest is 15%. Determine for each year the capital recovered, the return on capital unrecovered, and the capital recovered plus return. Also, find the annual equivalent capital recovered plus return. (6.10.2)

(a) Use MACRS-GDS allowances.
(b) Use Elective MACRS-ADS allowances.
(c) Use traditional pre-1981 straight-line depreciation.
(d) Use traditional pre-1981 sum of the years' digits depreciation.

29. A high-precision robotic welder is purchased for $90,000. Installation cost is $18,000 due to the extreme care and provisions required. It is estimated that the asset can be sold for $12,000 after a useful life of 12,000 hours of operation; however, before selling, it must be removed at a cost of $3000. Inspection and maintenance cost $30/operating hour. It requires 10 minutes to weld one unit with the robot, and 24,000 units are welded per year. The after-tax MARR is 15%, and the effective tax rate is 35%. Depreciation is based on operating hours. (6.11.1)
(a) What is the equivalent annual after-tax cost of the welder?
(b) What is the cost of welding per part?

30. An automatic control mechanism is estimated to provide 10,500 hours of service during its life. The mechanism costs $21,600 and has a salvage value of $600 after 10,500 hours of use. Its use is projected over a 4-year period as follows:

Year	Hours of Use
1	3000
2	2100
3	4500
4	900

Calculate the depreciation charge for each year using a method similar to the "operating day" method, but based on operating hours. (6.11.2)

31. A utility trailer costs $4000 and is rented out by the hour, day, or week. It is expected to depreciate to $0 by the time it has been rented out for a total of $20,000 in gross income. If its forecasted annual revenues are $6500, $6500, $5000, and $2000, what will be the appropriate annual depreciation charges? (6.11.3)

32. Equipment for the manufacture of dental supplies (MACRS-GDS 5-year recovery period; Elective MACRS-ADS 9-year recovery period) is purchased for $54,000 with an estimated salvage value of $9000 at the end of 10 years. Annual revenues and operating costs, excluding depreciation, are $58,500 and $34,500, respectively. Assume that MARR = 18% and the firm's tax rate is 35%. (6.12)
(a) Determine the total undiscounted amount of depreciation deductions for the 10-year period if (1) MACRS-GDS allowances are used, and (2) Elective MACRS-ADS allowances are used.
(b) Determine the present worth of the depreciation deductions described in (a).
(c) Determine the total undiscounted amount of taxes paid for the 10-year period if (1) MACRS-GDS allowances are used, and (2) Elective MACRS-ADS allowances are used.
(d) Determine the present worth of the taxes described in (c).
(e) What do your answers to (a) and (c) suggest about the relationship between depreciation method, the total amount of depreciation allowed, and the total amount of taxes paid?
(f) What do your answers to (b) and (d) suggest about the effect of the depreciation method used and the present worths of depreciation and taxes?

33. The Kamath company is considering the purchase of an asset for manufacturing apparel

(MACRS-GDS 5-year recovery period; Elective MACRS-ADS 9-year recovery period) that will cost $60,000, have a life of 10 years, and no salvage value. It will generate a gross income of $65,000/year. The annual cost of operating and maintaining the asset will be $45,000/year. The tax rate is 35% and MARR is 15% after taxes. (6.12)

(a) Determine the total undiscounted amount of depreciation deductions for the 10-year period if (1) MACRS-GDS allowances are used, and (2) if Elective MACRS-ADS allowances are used.

(b) Determine the present worth of the depreciation deductions described in (a).

(c) Determine the total undiscounted amount of taxes paid for the 10-year period if (1) MACRS-GDS allowances are used, and (2) Elective MACRS-ADS allowances are used.

(d) Determine the present worth of the taxes described in (c).

(e) What do your answers to (a) and (c) suggest about the relationship between depreciation method, the total amount of depreciation allowed, and the total amount of taxes paid?

(f) What do your answers to (b) and (d) suggest about the effect of the depreciation method used and the present worths of depreciation and taxes?

34. Bran Community Enterprises plans to invest in a water purification system for making cookies (MACRS-GDS 7-year recovery period; Elective MACRS-ADS 12-year recovery period) requiring $100,000 capital. The firm has $60,000 available and must borrow the additional $40,000 at an interest rate of 11% per year. The system will last 8 years with a salvage value of $0. The before-tax and loan cash flow for each of years 1 to 8 is $30,000. Only the loan interest is paid each year, and the entire principal is repaid at the end of year 3. MACRS-GDS depreciation is used; the applicable tax rate is 35%. Construct a table showing each of the following for each of the 8 years. (6.13)

(a) Before-tax and loan cash flow.

(b) Loan principal payment.

(c) Loan interest payment.

(d) MACRS-GDS deduction.

(e) Taxable income.

(f) Taxes.

(g) After-tax cash flow.

35. Rework problem 34 assuming the loan is paid back over 3 years as follows. (6.13)

(a) In equal annual amounts.

(b) The principal in equal annual amounts, plus yearly interest.

36. Turner Imprecision, Inc. purchased a new digitally controlled machine (MACRS-GDS 5-year recovery period; Elective MACRS-ADS 5-year recovery period) used to more precisely deposit materials on electronic chips. The machine cost $600,000 and has an expected life of 10 years, but Turner plans to keep it only 6 years with a $30,000 salvage value at that time. It will replace several manually operated machines, resulting in net productivity and quality increases of $180,000/year. It is financed at 12% for $240,000; the remaining $360,000 is equity money. Repayment occurs over 3 years with equal principal amounts of $80,000, plus interest on the unrecovered balance. Clearly and completely set up a table and determine the after-tax cash flow for each appropriate year. Be sure to include everything that helps determine the ATCF. The tax rate is 35%, and MACRS-GDS allowances are used. (6.13)

37. Complete the following table. (6.13)

End of Year	Before-Tax and Loan Cash Flow	Loan Principal Payment	Loan Interest Payment	MACRS-GDS Deduction	Taxable Income	Tax 0.35	After-Tax Cash Flow
0	(140,000.00)	(60,000.00)					(80,000.00)
1	28,000.00	0.00	6,600.00	20,006.00			
2	28,000.00	0.00	6,600.00	34,286.00			
3		0.00		24,486.00		3,819.90	
4		0.00				1,369.90	
5	56,000.00	0.00	6,600.00			12,914.30	
6	28,000.00	0.00	6,600.00				18,280.80
7		0.00	6,600.00		25,898.00		
8		0.00	6,600.00		43,156.00		
9	63,000.00	0.00	6,600.00				
10	42,000.00	60,000.00	6,600.00				

38. DeYong Corporation purchases an extruder for use in the manufacture of film in the chemical industry (MACRS-GDS 5-year recovery period; Elective MACRS-ADS 9.5-year recovery period) by borrowing the entire $25,000 purchase price. The loan is to be repaid with three equal annual payments at an annual compound rate of 10%. It is anticipated that the exchanger will be used for 6 years and then be sold for $2000. Annual operating and maintenance expenses are estimated to be $11,000/year. Assume MACRS-GDS depreciation, a 35% tax rate, and an after-tax MARR of 15%. Compute the after-tax equivalent uniform annual cost for the heat exchanger. (6.13)

39. SchuerMann Food Consumption Specialties, Inc., borrows $30,000, paying back $7500/year on the principal plus 9% on the unpaid balance at the end of each year. The $30,000 is used to purchase a special handling device (MACRS-GDS 3-year recovery period; Elective MACRS-ADS 4-year recovery period) that has a life of 5 years and a $3000 salvage value at that time. SchuerMann uses MACRS-GDS depreciation, requires a 15% after-tax MARR, and has a 35% tax rate. Determine the after-tax equivalent uniform annual cost of the device. (6.13)

40. Greene Research has the following capital gains and losses and ordinary taxable incomes:

Short-term capital gains	$50,000
Short-term capital losses	28,000
Long-term capital gains	173,000
Long-term capital losses	41,000
Ordinary taxable income	275,000

The capital gains rate is 28%. Determine the federal taxes to be paid by the company. (6.14.2)

(a) Use the regular method.

(b) Use the alternate method.

41. Melkote Materials Distribution has short-term capital gains of $23,000, short-term capital losses of $17,000, long-term capital gains of $65,000, and long-term capital losses of $17,000. Ordinary taxable income is $90,000. The capital gains rate is 28%. How much tax must be paid? (6.14.2)

    (a) Use the regular method.
    (b) Use the alternate method.
42. Melkote Materials Distribution in problem 41 had short-term capital losses of $40,000 (the other information stays the same). How much tax must be paid now? (6.14.2)
43. Determine the amount of tax (or other appropriate handling) for each of the following situations. The regular tax rate is 35%, and the capital gains rate is 28%. (6.14.2)

    (a) Net short-term capital losses $13,000
        Net long-term capital losses $7,000
    (b) Net short-term capital losses $75,000
        Net long-term capital gains $130,000
    (c) Net short-term capital losses $63,000
        Net long-term capital gains $30,000
    (d) Net short-term capital gains $45,000
        Net long-term capital gains $60,000
    (e) Net short-term capital gains $35,000
        Net long-term capital losses $21,000
    (f) Net short-term capital gains $36,000
        Net long-term capital losses $54,000

44. Shambles Video is considering a camera for CATV-program origination (MACRS-GDS 5-year recovery period; Elective MACRS-ADS 9-year recovery period). The purchase price is $100,000, and the camera will be sold during the sixth year for $15,000. How should the salvage value received be treated for tax purposes if the following methods are used and the tax rate is 35%? (6.15.2)
    (a) Use MACRS-GDS allowances.
    (b) Use Elective MACRS-ADS allowances.
45. Reconsider Shambles' camera in problem 44. Assume it is sold for $15,000 during the fourth year.
    (a) Use MACRS-GDS allowances. (6.8.3)
    (b) Use Elective MACRS-ADS allowances. (6.8.4)
46. J. P. Ratt left the accounting business to begin manufacturing athletic goods for the martial arts. An asset for this purpose (MACRS-GDS 7-year recovery period; Elective MACRS-ADS 12-year recovery period) is purchased for $120,000 and depreciated using MACRS-GDS allowances. Assuming we do not know the status of other Section 1231 assets in the company, determine the tax treatment of the asset's salvage value in the following cases. (6.15.2)
    (a) It is sold in year 3 for $140,000.
    (b) It is sold in year 3 for $96,000.
    (c) It is sold in year 3 for $6,000.
47. Rework problem 46 assuming J. P. Ratt plans to use Elective MACRS-ADS to recover the asset. (6.15.2)
48. Norma Jean Properties places a residential property into service in the first month of the tax year for $200,000 and recovers it using MACRS-GDS allowances. Assuming we do not know the status of other Section 1231 assets in the company, determine the tax treatment of the asset's salvage value in the following cases. (6.15.2)
    (a) It is sold after 3 years, 4 months for $220,000.
    (b) It is sold after 3 years, 4 months for $130,000.
49. Rework problem 48 assuming Norma Jean Properties plans to use Elective MACRS-ADS to recover the asset. (6.15.2)

50. Coleman Hair Specialties places in service a nonresidential real property (MACRS-GDS 39-year recovery period; Elective MACRS-ADS 40-year recovery period) for $250,000 and recovers it using MACRS-GDS. It is purchased in the fifth month of the tax year. Assuming we do not know the status of other Section 1231 assets in the company, determine the tax treatment of the asset's salvage value in the following cases. (6.15.2)

 (a) It is sold after 5 years, 4 months for $280,000.

 (b) It is sold after 5 years, 4 months for $240,000.

 (c) It is sold after 5 years, 4 months for $90,000.

51. Rework problem 50 assuming Coleman Hair Specialties plans to use Elective MACRS-ADS to recover the property. (6.15.2)

*52. DEMANDO Construction purchases a crane (MACRS-GDS 5-year recovery period; Elective MACRS-ADS 6-year recovery period) for $250,000. The crane will be sold during the sixth year for an estimated $20,000. Income less operating and maintenance costs for the crane is $80,000/year. Using a 35% income tax rate and MACRS-GDS deductions, compute the after-tax cash flows for the 6-year period. *Assume the investment tax credit has been reinstated, and the full 10% will be taken at the end of year 1*. (6.16)

*53. The C. B. Esteez Company proposes to invest $100,000 in an automatic control panel related to research and experimentation (MACRS-GDS 5-year recovery period). It is anticipated that this investment will cause a reduction in annual disbursements of $36,000 over 7 years with a $4000 salvage value at that time. Regular MACRS depreciation will be used, and the tax rate is 35%. MARR = 14%. Determine the equivalent present worth of after-tax cash flows for the panel. *Assume the investment tax credit has been reinstated, and the full 10% will be taken at the end of year 1*. (6.16).

54. Terrell Golf Products, Inc., is considering a $90,000 heat recovery incinerator (MACRS-GDS 7-year recovery period; Elective MACRS-ADS 10-year recovery period) which is expected to cause a reduction in net out-of-pocket costs of $37,500/year for 6 years. The incinerator will be depreciated using MACRS-GDS allowances, and no salvage value is expected. The tax rate is 35%, and the Section 179 expense deduction will be taken in year 1. If MARR is 18%, determine the present worth of the after-tax cash flows. (6.17)

55. K&D Enteprises plans to invest $65,000 in a marketing information system (MACRS-GDS 5-year recovery period; Elective MACRS-ADS 5-year recovery period) that will reduce annual costs by $26,000/year over a 6-year planning horizon, after which the salvage value will be $1000. The tax rate is 35%, and MARR = 15% after taxes. MACRS-GDS depreciation will be used and the full Section 179 expense deduction will be taken in year 1. What is the equivalent after-tax annual cost of the investment? (6.17)

56. Patsy Graphics is considering a high-end automatic color copier (MACRS-GDS 5-year recovery period; Elective MACRS-ADS 6-year recovery period) that will cost $50,000, have a life of 6 years, and no prospective salvage value at the end of that time. Because of this venture, there will be an increasing yearly gross income estimated to be $35,000 in the first year, increasing by $4000 each year thereafter. The operating cost during the first year will be $11,000, increasing by $1000 each year thereafter. The company contemplating the purchase of this asset has an effective tax rate of 35% and uses MACRS-GDS allowances. Determine the after-tax cash flow for each year and the present worth of this investment if MARR is a combined rate of 14%, including an inflation component of 5%. (6.19).

(a) Use a then-current analysis.

(b) Use a constant-worth analysis.

57. Bob's Construction invested $60,000 in a compressor (MACRS-GDS 5-year recovery period; Elective MACRS-ADS 6-year recovery period equipment) that will have a useful life of 7 years with a $4000 salvage value at that time. The firm's income will increase according to the estimate $25,000(1.10)^t$ in year t over the 7 years. The tax rate is 35%, the combined MARR is 16%, inflation is running at 5%, and MACRS-GDS allowances are used. Based on the present worth of the after-tax cash flows, should the company have invested? (6.19)

(a) Use a then-current analysis.

(b) Use a constant-worth analysis.

58. MLW Entertainment, Inc., is considering purchasing a computerized scheduling system (MACRS-GDS 5-year recovery period; Elective MACRS-ADS 10-year recovery period), installed and complete. The cost is $130,000 and the system is expected to last 7 years with a salvage value of $5000 at that time. Additional gross income due to the computer will be $35,000 the first year, increasing by $4000 each year thereafter. Operating and maintenance costs will be $15,000 in year 1, increasing by $2000 each year thereafter. MACRS-GDS allowances are used, the tax rate is 35%, the real discount rate is 10%, and the general inflation rate is 5%. Based on the present worth of the after-tax cash flows, should the company invest? (6.19)

(a) Use a then-current analysis.

(b) Use a constant-worth analysis.

59. K&D (see problem 55) are very smart managers and wish to leverage their money by financing $50,000 of the $65,000 investment at 11% over 4 years. The principal will be paid back in equal annual amounts each year, plus interest on the unrecovered balance. K&D believe cost reductions will be flat at $26,000/year over the 6 years. Inflation is projected to average 6%, and the MARR of 15% is a combined rate. Now, what is the equivalent after-tax annual worth of the investment? (6.19)

(a) Use a then-current analysis.

(b) Use a constant-worth analysis.

60. A gold mine that is expected to produce 20,000 ounces of gold is purchased for $1,600,000. The gold can be sold for $370/ounce; however, it costs $200/ounce for mining and processing costs. If 2500 ounces are produced this year, what will be the depletion allowance for (a) unit depletion, and (b) percentage depletion? (6.20)

61. A West Virginia coal mine having 9 million tons of coal has a first cost of $36 million. The gross income for this coal is $55/ton. Operating costs are $25/ton. If, during the first 2 years of operation, the mine yields 225,000 tons and 300,000 tons, respectively, determine the after-tax cash flow for each year, assuming that this mine is the owner's only venture. Use the better method of depletion each year. (6.20)

Chapter 7

Economic Analysis of Projects in the Public and Regulated Sectors

7.1 INTRODUCTION

Economic analyses are, or should be, performed when undertaken by: (1) governments—local, state, or federal; and (2) private firms such as public utilities which are subject to regulation. While economic evaluation in each of these sectors follows the same principles learned already in this book, each sector has its own peculiarities, methods, and language. Also, while the overall objective is still the sound investment of scarce resources, different but equivalent criteria are used from those seen earlier.

Government entities fund projects using money taken, usually in the form of taxes, from the public. They are then to provide goods or services to the public that would be infeasible for individuals to provide on their own. While they are not in business to make a profit, it is incumbent upon those government entities to make wise investment decisions, providing benefits that exceed costs. Evaluation of projects in the public sector is discussed extensively in Sections 7.2 through 7.7.

Private firms that are regulated by governments, boards or commissions are different, too. While they are in business to make a return for their shareholders, they are often not subjected to competition in the usual sense. As such, their performance and related costs are "regulated" by a "public service commission" or a "corporation commission." The economic evaluation of project investments then focuses not on maximization of present worth of after-tax cash flows, but upon minimization of revenues required from the public to provide the goods or services that meet minimum performance requirements. The use of revenue requirements analysis to evaluate projects in the regulated sector is presented in Section 7.8.

7.2 THE NATURE OF PUBLIC PROJECTS

Evaluating and selecting projects to be approved, paid for, and operated by government entities for the public welfare should represent a sound investment of tax money. Projects should provide benefits, for the greater good of the public, that exceed the costs of providing those benefits. It is natural that the most frequently used method in evaluating government (local, state, or federal) projects is benefit-cost analysis. Cost-effectiveness analysis is also used, but to a lesser extent. More emphasis will be devoted to benefit-cost methods, which require that benefits and costs be evaluated on a monetary basis. Cost effectiveness requires a numerical measure of effectiveness; however, that measure need not be in terms of money.

There are many types of government projects and many agencies involved. Four classes reasonably cover the spectrum of projects entered into by government. They include cultural development, protection, economic services, and natural resources. *Cultural development* is enhanced through education, recreation, and historic and similar institutions or preservations. *Protection* is achieved through military services, policy and fire protection, and the judicial system. *Economic services* include transportation, power generation, and housing loan programs. *Natural resource* projects might entail wildland management, pollution control, and flood control. Although these are obviously incomplete project lists in each class, it is not so obvious that some projects belong in more than one area. For example, flood control is certainly a form of assistance for some, provides transportation and power generation for others, and likely relates to natural resource benefits.

Government projects have a number of interesting characteristics that set them apart from projects in the private sector. Many government projects are huge, having first costs of tens of millions of dollars. They tend to have long lives, such as 50 years for a bridge or a dam. The multiple-use concept is common, as in wildland management projects, where economic (timber), wildlife preservation (deer, squirrel), and recreation projects (camping, hiking) are each considered uses of importance for the land. The benefits or enjoyment of government projects are often completely out of proportion to the financial support of individuals or groups. Also, there are often multiple government agencies that have an interest in a project. In fact, some governmental units support other governmental units in their investments, an example being the federal support of state road projects. Further, public sector projects are not easily evaluated, since it may be many years before their benefits are realized. Finally, there is usually not a clear-cut measure of success or failure in the public sector such as the rate of return or net present worth criteria.

7.3 OBJECTIVES IN PUBLIC PROJECT EVALUATION

If large, complex, lengthy, multiple-use projects of interest to several groups are to be evaluated for their desirability, the criteria for the evaluation must first be agreed on. Arrow, et al. [1], reporting on a blue-ribbon panel's findings, said in 1996: "Decisionmakers should not be precluded from considering the economic costs and benefits of different policies in the development of regulations. Laws that prohibit costs or other factors from being considered in administrative decisionmaking are inimical to good public policy." Prior to that, Campen [3] notes that on February 17, 1981, President Ronald Reagan signed Executive Order 12291, thereby formally making benefit-cost analysis a central element in his administration's regulatory policy. Henceforth, proponents of a regulation would have to demonstrate that the benefits from its adoption would outweigh the costs. Liberal critics were quick to denounce benefit-cost analysis for its pro-business bias, viewing it as one more tool for reducing the size and scope of big government. Sixteen years earlier, however, on August 25, 1965, President Lyndon Johnson announced that a new set of budgetary techniques was to be used throughout the civilian sector of the federal government. This time, it was liberals, including President Johnson and his budget director, who were principal advocates of the new techniques. Of course, this time it was conservatives who were concerned. The conservatives thought that it might be used as a tool for *expanding* government spending. Even this was not the first time benefit-cost analysis had been proposed.

Very applicable to today is the criterion specified in the Flood Control Act on June 22, 1936, which stated ". . . that the Federal Government should improve or participate . . . if the benefits to whomsoever they may accrue are in excess of the estimated costs . . . [18]." The setting for modern evaluation of government projects dates clear back to the River and Harbor Act of 1902, which "required a board of engineers to report on the desirability of Army Corps of Engineers' river and harbor projects, taking into account the amount of commerce benefited and the cost." But, the intellectual father of benefit-cost analysis is said to be the French economist and engineer Jules Dupuit, who in 1844 wrote the frequently cited study "On the Measure of Utility of Public Works [13]." Clearly, the idea of benefit-cost analysis has been around for a long time.

Prest and Turvey [14] give a short, reasonable definition of benefit-cost analysis:

> ". . . a practical way of assessing the desirability of projects where it is important to take a long view (in the sense of looking at repercussions in the further, as well as the nearer, future) and a wide view (in the sense of allowing for side-effects of many kinds on many persons, industries, regions etc.); that is, it implies the enumeration and evaluation of all the relevant costs and benefits."

The "long view" is nothing more than considering the entire planning horizon, which, granted, is generally far longer for projects in the public sector than for those in the private sector. The "wide view" and the notion of evaluating all "relevant costs and benefits" probably spell out the greatest single difference between government and private economic evaluation. That is, government projects often affect many individuals, groups, and things, either directly or indirectly, for better or for worse. In evaluating these projects, the analyst tries to capture these effects on the public, quantify them, and, where possible, make the measures in monetary terms. Positive effects are referred to as *benefits*, while negative effects are *disbenefits*. In contrast, the private sector "effects" of primary importance are those that relate to *income* being returned to the organization. Costs of construction, financing, operation, and maintenance are estimated in much the same way in both the public and private sectors.

7.4 GUIDELINES IN PUBLIC SECTOR EVALUATION

Believe it or not, an evaluation of benefits does not always take place when regulations, policies, or projects are undertaken by government entities using tax dollars. Arrow, et al. [2] in 1996 set out a number of guidelines for decision makers on using economic analysis to evaluate proposed policies. While that group focused on policies and regulations in environment, health, and safety, the guidelines transcend these areas and should be kept in mind on any government investment. The guidelines are summarized as:

1. *A benefit-cost analysis is a useful way of organizing a comparison of the favorable and unfavorable effects of proposed policies.* Benefit-cost analysis can help the decision maker better understand the implications of a decision and it should play an important role in informing the decision-making process.

2. *Economic analysis can be useful in designing regulatory strategies that achieve a desired goal at the lowest possible cost.* Economic analysis can highlight the extent to which cost savings can be achieved by using alternative, more flexible approaches that reward performance.

3. *Congress should not preclude decision makers from considering the economic costs and benefits of different policies in the development of regulations. At the very least, agencies should be encouraged to use economic analysis to help set regulatory priorities.* Current planning in most regulatory agencies places insufficient emphasis on the likely benefits and costs of regulations and excessive emphasis on politics and deadlines.

4. *Benefit-cost analysis should be required for all major regulatory decisions.* An important benefit of mandatory benefit-cost analysis is that it facilitates external monitoring of an agency's performance.

5. *Agencies should not be bound by a strict benefit-cost test, but should be required to consider available benefit-cost analyses. For regulations whose expected costs far exceed expected benefits, agency heads should be required to present a clear explanation justifying the reasons for their decision.* There may be factors other than economic benefits and costs that agencies will want to weigh when making decisions, such as equity within and across generations. In addition, a decision maker may want to place greater weight on particular characteristics of a decision, such as potential irreversible consequences.

6. *For legislative proposals involving major health, safety, and environmental regulations, the Congressional Budget Office should do a preliminary benefit-cost analysis that can inform legislative decisionmaking.* Because laws give rise to regulations, some kind of benefit-cost analysis is likely to be useful in informing the policy process.

7.5 BENEFIT-COST ANALYSIS

The notion of benefit-cost analysis is simple in principle. It follows the same systematic approach used in selecting between economic investment alternatives, including the following seven steps adapted from Chapter Five.

1. *Define the set of feasible, mutually exclusive, public sector alternatives to be compared.* Each viable alternative should be described thoroughly, specifying all of the good and bad effects that the alternative will have on the public. This includes effects directly on people, safety, environment, land values, recreation, etc. In addition, all aspects of project development, operation, maintenance, and eventual salvage should be stated.

2. *Define the planning horizon to be used in the benefit-cost study.* Defining the planning horizon is essential in order to define the period over which the best project(s) is to be selected. As in other economic analyses, a common planning horizon should be used to compare alternatives.

3. *Develop the cost-savings and benefit-disbenefit profiles in monetary terms for each alternative.* Quantifying all *costs and savings* refers to governmental expenditures and incomes received relating to a project over the planning horizon. Disbursements include all first and continuing costs of a project; income may result from tolls, fees, or other charges to the user public. Costs, unlike benefits, are quantified for public projects in a way similar to private projects. Each *benefit* or *disbenefit* during the planning horizon should be quantified insofar as possible. Benefits refer to desirable consequences on the public. Disbenefits are negative effects on the public. Unfortunately, placing dollar values on benefits received by a diverse public is often not an easy task.

4. *Specify the MARR to be used.* Deciding on the interest rate or MARR to be used is a classic problem in the public sector. Often the rate selected is far too low. An appropriate rate is suggested in [1] to be the average rate of return on private sector investments.

5. *Compare the alternatives using a specified measure of worth, such as the benefit-cost ratio or the present worth of benefits-costs.* A base measure such as annual equivalent benefits and costs or present worths of benefits and costs is then established, and the *measure of worth* is chosen. Benefit-cost analyses frequently use the benefit-cost ratio (B/C) or, to a lesser extent, a measure of benefits less costs (B − C). If

B_{jt} = public benefits associated with project j during year $t, t = 1, 2, . . ., n$

C_{jt} = governmental costs associated with project j during year $t, t = 0, 1, 2, . . ., n$

and

i = appropriate interest rate

then the *B/C* criterion may be expressed mathematically, using a present worth base measure, as

$$B/C_j(i) = \frac{\sum_{t-1}^{n} B_{jt}(1 + i)^{-t}}{\sum_{t-0}^{n} C_{jt}(1 + i)^{-t}} \qquad (7.1)$$

Note that the B/C ratio is an alternative name for the savings/investment ratio (SIR) treated in Chapter Four. In this case, however, the benefits replace net positive cash flows, and costs replace net negative cash flows. Otherwise, the B/C and SIR formulas are identical.

The $B - C$ criterion is expressed as

$$(B - C)_j(i) = \sum_{t-0}^{n} (B_{jt} - C_{jt})(1 + i)^{-t} \qquad \textbf{(7.2)}$$

which is similar to the present worth method described in Chapter Four.

When two or more project *alternatives are being compared* using a *B/C* ratio, the analysis should be done on an *incremental basis*. That is, first the alternatives should be ordered from lowest to highest cost. Then, the incremental benefits of the second alternative over the first, $\Delta B_{2-1}(i)$, are divided by the incremental costs of the second over the first, $\Delta C_{2-1}(i)$. That is,

$$\Delta B/C_{2-1}(i) = \frac{\Delta B_{2-1}(i)}{\Delta C_{2-1}(i)} = \frac{\sum\limits_{t=1}^{n} (B_{2t} - B_{1t})(1 + i)^{-t}}{\sum\limits_{t=0}^{n} (C_{2t} - C_{1t})(1 + i)^{-t}} \qquad \textbf{(7.3)}$$

Note that if the first alternative is to do nothing, the incremental *B/C* ratio is also the straight *B/C* ratio for the second alternative. As long as $\Delta B/C_{2-1}(i)$ exceeds 1.0, Alternative 2 is preferable to Alternative 1. Otherwise, Alternative 1 is preferred to Alternative 2. The winner of these is then compared on an incremental basis with the next most costly alternative. These pairwise comparisons continue until all alternatives have been exhausted, and only one "best" project remains. The procedure used is very similar to that specified for the rate-of-return method in Chapter Four.

With the $B - C$ criterion, an incremental basis may be used following the same rules as for the *B/C* ratio, but preferring Alternative 2 to Alternative 1 as long as the following condition holds:

$$\Delta(B - C)_{2-1}(i) = \Delta B_{2-1}(i) - \Delta C_{2-1}(i)$$

$$= \sum_{t=1}^{n} [B_{2t}(i) - B_{1t}(i)](1 + i)^{-t} \qquad \textbf{(7.4)}$$

$$- \sum_{t=0}^{n} [C_{2t}(i) - C_{1t}(i)](1 + i)^{-t} \geq 0$$

Where benefits and costs are known directly, the value of $(B - C)_j$ for each alternative j may be calculated and the maximum value selected.

6. *Perform supplementary analyses.* Supplementary analyses in the form of risk analysis, sensitivity analysis, and break-even analysis may be performed. These are useful in benefit-cost analyses due to the relative uncertainties in quantifying many benefits or disbenefits. If a project decision or selection

is very sensitive to, say, public benefits that are also somewhat unknown or intangible, it may be wise to do further research to ensure a good decision.

7. *Select the preferred alternative.* Not only is the preferred alternative selected, but all quantitative and qualitative supporting considerations should be recorded in detail. This is particularly true in the case of public sector projects, due to the wide base of interested parties.

These seven major steps in conducting a benefit-cost analysis, as well as the B/C and $B - C$ relationship to the present worth, annual equivalent, and rate of return criteria, are illustrated in the remainder of this section.

EXAMPLE 7.1

We are given the task of deciding between three highway alternatives to replace a winding, old, dangerous road. The length of the current route is 26 miles. Alternative A is to overhaul and resurface the old road at a cost of $3 million. Resurfacing will then be required at a cost of $2.5 million at the end of each 10-year period. Annual maintenance for Alternative A will cost $10,000/mile. Alternative B is to cut a new road following the terrain; it will be only 22 miles long. Its first cost will be $10 million, and surface renovation will be required every 10 years at a total cost of $2,250,000. Annual maintenance will be $10,000/mile. Alternative C also involves a new highway which, for practical considerations, will be built along a 20.5-mile straight line. Its first cost, however, will be $18 million, because of the extensive additional excavating necessary along this route. It, too, will require resurfacing every 10 years at a cost of $2,250,000. Annual maintenance will be $18,000/mile. This increase over route B is due to the additional roadside bank retention efforts that will be required.

Our task is to select one of these alternatives, considering a planning horizon of 30 years with negligible residual value for each of the highways at that time. One of these alternatives is required, since the old road has deteriorated below acceptable standards. We can calculate the annual equivalent first cost and resurfacing cost of each alternative using an interest rate of 8%:

Route A: $[\$3,000,000 + \$2,500,000(P|F\,8,10)$
$+ \$2,500,000(P|F\,8,20)](A|P\,8,30)$
$= [\$3,000,000 + \$2,500,000(0.4632) + \$2,500,000(0.2145)]$
$\times (0.0888)$
$= \$416,849/\text{year}$

Route B: $[\$10,000,000 + \$2,250,000(P|F\,8,10)$
$+ \$2,250,000(P|F\,8,20)](A|P\,8,30)$
$= [\$10,000,000 + \$2,250,000(0.4632) + \$2,250,000(0.2145)]$
$\times (0.0888)$
$= \$1,023,404/\text{year}$

Route C: $[\$18,000,000 + \$2,250,000(P|F\,8,10)$
$+ \$2,250,000(P|F\,8,20)](A|P\,8,30)$
$= [\$18,000,000 + \$2,250,000(0.4632) + \$2,250,000(0.2145)]$
$\times (0.0888)$
$= \$1,733,804/\text{year}$

Maintenance cost:

Route A: $\left(10,000\,\dfrac{\$}{\text{mile–year}}\right)(26\text{ miles}) = \$260,000/\text{year}$

Route B: $(\$10,000)(22) = \$220,000/\text{year}$

Route C: $(\$18,000)(20.5) = \$369,000/\text{year}$

Clearly, route A costs less than route B, which itself costs less than route C. Can we now conclude that route A should be selected based on cost to the government? Absolutely not, according to the criterion that seeks to maximize benefits to the public as a whole, minus costs. In fact, we have analyzed only one side of the problem. Now, we must attempt to quantify the benefits along each of the routes.

Traffic density along each of the three routes will fluctuate widely from day to day, but will average 4000 vehicles/day throughout the year. This volume is composed of 350 light commercial trucks, 250 heavy trucks, 80 motorcycles, and the remainder are automobiles. The average cost per mile of operation for these vehicles is $0.50, $0.85, $0.15, and $0.30, respectively.

There will be a time savings because of the different distances along each of the routes, as well as the different speeds which each of the routes will sustain. Route A will allow heavy trucks to average 35 miles/hour, while other traffic can maintain an average speed of 45 miles/hour. Routes B and C will allow heavy trucks to average 40 miles/hour, and the rest of the vehicles can average 50 miles/hour. The cost of time for all commercial traffic is valued at $22/vehicle/hour, and for noncommercial traffic, $8/vehicle/hour. Twenty-five percent of the automobiles and all of the trucks are considered commercial.

Finally, there is a significant safety factor that should be included. Along the old winding road, there has been an excessive number of accidents per year. Route A will reduce the number of vehicles involved in accidents to 105, and routes B and C are expected to involve only 75 and 70 vehicles in accidents, respectively, per year. The average cost per vehicle in an accident is estimated to be $9000, considering actual physical property damages, lost wages because of injury, medical expenses, and other relevant costs.

We now set about to analyze the various benefits in monetary terms. We have considered savings in vehical operation, time, and accident prevention. The costs incurred by the public for these items are calculated in the following steps.

Operational costs:

Route A: $\left(350 \dfrac{\text{light trucks}}{\text{day}}\right)\left(26 \dfrac{\text{miles}}{\text{light truck}}\right)\left(0.50 \dfrac{\$}{\text{mile}}\right)\left(365 \dfrac{\text{days}}{\text{year}}\right)$

$+ \left(250 \dfrac{\text{heavy trucks}}{\text{day}}\right)\left(26 \dfrac{\text{miles}}{\text{heavy truck}}\right)\left(0.85 \dfrac{\$}{\text{mile}}\right)\left(365 \dfrac{\text{days}}{\text{year}}\right)$

$+ \left(80 \dfrac{\text{motorcycles}}{\text{day}}\right)\left(26 \dfrac{\text{miles}}{\text{motorcycle}}\right)\left(0.15 \dfrac{\$}{\text{mile}}\right)\left(365 \dfrac{\text{days}}{\text{year}}\right)$

$+ \left(3320 \dfrac{\text{automobiles}}{\text{day}}\right)\left(26 \dfrac{\text{miles}}{\text{automobile}}\right)\left(0.30 \dfrac{\$}{\text{mile}}\right)\left(365 \dfrac{\text{days}}{\text{year}}\right)$

$= \$13,243,295$

Route B: $[350(\$0.50) + 250(\$0.85) + 80(\$0.15) + 3320(\$0.30)]$

$\times (22)(365)$

$= 11,205,865/\text{year}$

Route C: $[350(\$0.50) + 250(\$0.85) + 80(\$0.15) + 3320(\$0.30)]$

$\times (20.5)(365)$

$= \$10,441,829$

Time costs:

Route A: $\left(350 \dfrac{\text{light trucks}}{\text{day}}\right)\left(26 \dfrac{\text{miles}}{\text{light truck}}\right)\left(\dfrac{1\text{ hour}}{45\text{ miles}}\right)\left(365 \dfrac{\text{days}}{\text{year}}\right)$

$\times \left(22 \dfrac{\$}{\text{hour}}\right) + \left(250 \dfrac{\text{heavy trucks}}{\text{day}}\right)\left(26 \dfrac{\text{miles}}{\text{heavy truck}}\right)\left(\dfrac{1\text{ hour}}{35\text{ miles}}\right)$

$\times \left(365 \dfrac{\text{days}}{\text{year}}\right)\left(22 \dfrac{\$}{\text{hour}}\right) + \left(80 \dfrac{\text{motorcycles}}{\text{day}}\right)\left(26 \dfrac{\text{miles}}{\text{motorcycle}}\right)$

$\times \left(\dfrac{1\text{ hour}}{45\text{ miles}}\right)\left(365 \dfrac{\text{days}}{\text{year}}\right)\left(8 \dfrac{\$}{\text{hour}}\right) + \left(3320 \dfrac{\text{automobiles}}{\text{day}}\right)$

$\times \left(26 \dfrac{\text{miles}}{\text{automobile}}\right)\left(\dfrac{1\text{ hour}}{45\text{ miles}}\right)\left(365 \dfrac{\text{days}}{\text{year}}\right)$

$\times \left(0.25 \times 22 \dfrac{\$}{\text{hour}} + 0.75 \times 8 \dfrac{\$}{\text{hour}}\right) = \$11,301,837/\text{year}$

Route B: $\left[\dfrac{350}{50}(\$22) + \dfrac{250}{40}(\$22) + \dfrac{80}{50}(\$8)\right.$

$\left. + \dfrac{3320}{50}(0.25 \times \$22 + 0.75 \times \$8)\right]$

$\times (22)(365) = \$8,575,237/\text{year}$

Route C: $\left[\dfrac{350}{50}(\$22) + \dfrac{250}{40}(\$22) + \dfrac{80}{50}(\$8) \right.$

$\left. + \dfrac{3320}{50}(0.25 \times \$22 + 0.75 \times \$8) \right](20.5)(365)$

$= \$7,990,562/\text{year}$

Safety costs:

Route A: $\left(105 \dfrac{\text{vehicles}}{\text{year}} \right)\left(9000 \dfrac{\$}{\text{vehicle}} \right) = \$945,000/\text{year}$

Route B: $(75)(\$9000) = \$675,000/\text{year}$

Route C: $(70)(9000) = \$630,000/\text{year}$

All relevant government and public costs are summarized in Table 7.1.

We desire to compare these three alternative routes using benefit-cost criteria. Let our first criterion be the popular benefit-cost ratio. Since one of these alternatives must be selected, we will assume the lowest government cost alternative, route A, will be selected unless the extra expenditures for routes B or C prove more worthy. Since we have not defined "benefits" per se, user benefits will be taken as the incremental reduction in user costs from the less expensive to the more expensive alternatives. Since we are looking at incremental benefits, it makes sense to compare these against the respective incremental costs needed to achieve these additional benefits.

The incremental benefits and costs for route B as compared to route A for $i = 8\%$ are given as follows:

$$\Delta B_{B-A}(8) = \text{public costs}_A(8) - \text{public costs}_B(8)$$
$$\$25,490,132 - \$20,456,102 = \$5,034,030/\text{year}$$
$$\Delta C_{B-A}(8) = \text{government costs}_B(8) - \text{government costs}_A(8)$$
$$\$1,243,404 - \$676,849 = \$566,555/\text{year}$$

Table 7.1 Summary of Annual Equivalent Government and Public Costs, Example 7.1

	Route A	Route B	Route C
Government first cost of highway	$ 416,849/year	$ 1,023,404/year	$ 1,733,804/year
Government cost of highway maintenance	260,000/year	220,000/year	369,000/year
Public operational costs	$13,243,295/year	$11,205,865/year	$10,441,829/year
Public time costs	$11,301,837/year	$ 8,575,237/year	$ 7,990,562/year
Public safety cost	$ 945,000/year	$ 675,000/year	$ 630,000/year
Total government costs	$ 676,849/year	$ 1,243,404/year	$ 2,102,804/year
Total public costs	$25,490,132/year	$20,456,102/year	$19,062,391/year

That is, for an incremental expenditure of $566,555/year, the government can provide added benefits of $5,034,030/year for the public. The appropriate benefit cost ratio is then

$$\Delta B / C_{B-A}(8) = \frac{\Delta B_{B-A}(8)}{\Delta C_{B-A}(8)} = \frac{\$5,034,030}{\$566,555} = 8.89$$

This clearly indicates that the additional funds for route B are worthwhile, and we desire route B over route A.

Using a similar analysis, we now calculate the benefits, costs, and $\Delta B | C$ ratio to determine whether or not route C is preferable to route B.

$$\Delta B_{C-B}(8) = \$20,456,102 - \$19,062,391 = \$1,393,711/\text{year}$$
$$\Delta C_{C-B}(8) = \$2,102,804 - \$1,243,404 = \$859,400/\text{year}$$
$$\Delta B / C_{C-B}(8) = \frac{\$1,393,761}{\$859,400} = 1.62$$

This benefit-cost ratio, being greater than 1.00, indicates that the additional expenditure of $859,400/year to build and maintain route C would provide commensurate benefits to the public. In fact, the user savings over route B would be $1,393,711/year. Of the three alternative routes, route C is preferred.

The next benefit-cost criterion takes advantage of the fact that if $\Delta B / C > 1$, then $\Delta(B - C) > 0$. That is, the difference in incremental benefits and costs may be used in place of the incremental benefit-cost ratio.

EXAMPLE 7.2

Applying this measure to the routes in the previous example results in the following calculations for comparing routes A and B. We know that

$$\Delta B_{B-A}(8) = \$5,034,030/\text{year}$$
$$\Delta C_{B-A}(8) = \$\ 566,555/\text{year}$$

This results in

$$\Delta(B - C)_{B-A}(8) = \Delta B_{B-A}(8) - \Delta C_{B-A}(8) = \$5,034,030 - \$566,555$$
$$= \$4,467,475$$

Thus, we again conclude that route B is preferred to route A. Similarly,

$$\Delta(B - C)_{C-B}(8) = \Delta B_{C-B}(8) - \Delta C_{C-B}(8)$$
$$= \$1,393,711 - \$859,400 = \$534,311/\text{year}$$

which indicates that route C is worthy of the additional expenditure required; hence route C should be constructed.

The benefit-cost criteria are consistent with the methods of alternative evaluation presented in Chapter Five. For example, suppose it is desired to base a decision on the annual worth criterion, minimizing the sum of government construction and maintenance costs per year as well as public user costs per year.

EXAMPLE 7.3

Calculating the equivalent uniform annual cost using

$$AW_{\text{total}}(i) = AW_{\text{total government costs}}(i) + AW_{\text{total public costs}}(i)$$

we have

Route A: $AW_A(8) = \$676{,}849 + \$25{,}490{,}132 = \$26{,}166{,}981/\text{year}$
Route B: $AW_B(8) = \$1{,}243{,}404 + \$20{,}456{,}102 = \$21{,}699{,}506/\text{year}$
Route C: $AW_C(8) = \$2{,}102{,}804 + \$19{,}062{,}391 = \$21{,}165{,}195/\text{year}$

Route C is preferred.

It is interesting and logical to note that the differences in equivalent uniform annual costs for routes A,B and B,C are the same as the difference in incremental benefits minus incremental costs for routes A,B and B,C. That is,

$$AW_A(8) - AW_B(8) = \$26{,}166{,}981 - \$21{,}699{,}506 = \$4{,}467{,}475/\text{year}$$
$$\Delta B_{B-A}(8) - \Delta C_{B-A}(8) = \$5{,}034{,}030 - \$566{,}555 = \$4{,}467{,}475/\text{year}$$

and

$$AW_B(8) - AW_C(8) = \$21{,}699{,}506 - \$21{,}165{,}195 = \$534{,}311/\text{year}$$
$$\Delta B_{C-B}(8) - \Delta C_{C-B}(8) = \$1{,}393{,}711 - \$859{,}400 = \$534{,}311/\text{year}$$

Since the benefit-cost criteria can be related to the annual worth method, it is certainly related to, and consistent with, the present worth method. Typically, where government projects are involved, one of the benefit-cost criteria (B/C or $B - C$) is used. More often than not, the criterion is the benefit-cost ratio. This is unfortunate because, just as in rate of return analyses in the private sector, the benefit-cost ratio is easy to misuse and misinterpret, and it is very sensitive to the classification of problem elements as "benefits" or "costs." These problems will be discussed subsequently.

7.6 IMPORTANT CONSIDERATIONS IN EVALUATING PUBLIC PROJECTS

In Section 7.5 on benefit-cost analysis, it was stated that there are a number of pitfalls that can affect the analysis of government projects. Actually, opportunities for error pervade benefit-cost analyses, from the very initial philosophy, through to the interpretation of a B/C ratio. It is important to talk about the more significant of these, both to help prevent analysts from erring, and to help those who may be reviewing a biased evaluation.

The major topics to be considered include the following:

1. Point of view (national, state, local, individual).
2. Selection of the interest rate.
3. Assessing benefit-cost factors.
4. Overcounting.
5. Unequal lives.
6. Tolls and fees.
7. Multiple-use projects.
8. Problems with the B/C ratio.

7.6.1   Point of View

The stance taken by the engineer in analyzing a public venture can have an extensive effect on the economic "facts." The analyst may take any of several viewpoints, including those of:

1. An individual who will benefit or lose.
2. A particular governmental organization.
3. A local area such as a city or county.
4. A regional area such as a state.
5. The entire nation.

The first of these viewpoints is not particularly interesting from the standpoint of economic analysis. Nonetheless, all too frequently, an isolated road is paved, a remote stretch of water or sewer line is extended under exceptional circumstances, or a seemingly ideal location for a public works facility is suddenly eliminated from consideration. In these cases, the benefit-cost analysis, its review, and the implementation decision are usually made by a small, select group.

The other four viewpoints are, however, of considerable interest to those involved in public works evaluation. Analyzing projects or project components from viewpoint 2, that of a particular government agency, is analogous to economic comparisons in private enterprise. That is, only the gains and losses to the organization involved are considered. This viewpoint, which seems contrary to benefit-cost optimization to the public as a whole, may be appropriate under certain circumstances.

EXAMPLE 7.4

Consider a Corps of Engineers construction project in which the water table must be lowered in the immediate area so work can proceed. Any of several water cutoff or dewatering systems may be employed. Water cutoff techniques include driving a sheet pile diaphragm or using a bentonite slurry trench to cut off the flow of water to the construction area. Dewatering methods include deep-well turbines, an eductor system, or wellpoints for lowering the water level. It is sometimes appropriate for the Corps to evaluate these different techniques from an "organization" point of view, since each of the feasible methods provides the same service or outcome—a dry construction site. Therefore, the most economical decision from the Corps' point of view is also correct from the view of the public as a whole, since the benefits or contributions to the project are the same regardless of the method chosen.

The third point of view, that of a locality such as a city or county, is popular among local government employees and elected officials. Unfortunately, seemingly localized projects often impact a much wider range of the citizenry than is apparent.

EXAMPLE 7.5

County officials are to decide whether or not future refuse service should be county owned and operated, or whether a private contractor should be employed. The job requires front-end loader compaction trucks as well as roll on-off container capability. Primarily rural roads are traveled, and from 1 container (a roadside picnic area) to 50 containers (a large rurally located industrial plant) must be collected at each shop. Front-end loader containers range from 2 to 8 cubic yards, while roll on-off containers are sized from 15 to 45 cubic yards. Several trucks and drivers will be required, including a base for operations and maintenance.

The cost in dollars per ton of refuse collected, removed, and disposed of, is given below:

Personnel services	$ 9.12/ton
Materials, supplies, utilities	7.70/ton
Maintenance and repair	8.43/ton
Overhead	5.12/ton
Depreciation	5.42/ton
Five percent interest on half financed by bonds	2.70/ton
TOTAL COUNTY COST	$38.49/ton
Federal taxes foregone	$ 2.81/ton
State taxes foregone	0.30/ton
Property taxes foregone	2.31/ton
Eight percent return on half financed by tax money	4.32/ton
Not necessarily paid by county	$ 9.74/ton

County cost to provided refuse service will be $38.49/ton. The county is not, however, required to pay the additional $9.74/ton for federal and state taxes, *ad valorem* taxes, or a return on appropriated money, as would a private firm.

It is obvious from the example that a "local" county decision can affect a much wider public. Suppose, based on $38.49/ton, the county decided to own and operate the needed refuse service. Federal taxes of $2.81/ton that would have been paid by a private contractor will not be paid. Since the federal government will still have the same revenue requirements, the difference will be made up by passing on an infinitesimally small burden to the national public as a whole. Although this is easily rationalized at the local level—to spread a portion of the cost of county refuse service over the entire country—consider the result if every town, city, and county took this attitude.

An analogous argument follows for state income taxes foregone; however, now the burden is being spread over the people of the state. Even though this is much smaller than the national population, the burden per person to make up the lost state tax income is still very small. Again, providing refuse service at less cost to the local populace, at the expense of the state, is tempting from a parochial point of view.

If the county were to plan on not having to pay the *ad valorem* tax, the slack would be taken up by increasing the property tax rates in the county. Although this approach increases the burden on county property owners, that burden may be entirely disproportionate when compared to the refuse service each requires.

The last two points of view include a regional (e.g., a state) or national perspective. Ideally, a national outlook is preferable for local, regional, and national public works projects. Experience indicates, however, that the primary concern of public works officials and politicians is their particular constituency.

Perhaps the best advice for evaluators and decision makers in the public realm is to examine multiple viewpoints. That is, project evaluators are often not in a position to decide on a single specific point of view. They should instead present a thorough analysis clearly indicating any benefits or costs that depend on the perspective taken. Similarly, decision makers should require multiple points of view so they can be aware of the kind and degree of repercussions resulting from their actions.

7.6.2   Selection of the Interest Rate

The interest rate, discount rate, or minimum attractive rate of return is another premier factor to be decided upon when evaluating public works projects. In the route selection examples 7.1–7.3, the interest rate was taken at 8% with no question of the appropriateness of such a figure. Many do question the rate selected; unfortunately, arguments range from 0% up to perhaps 15% (and even higher), comparable to interest rates used in the private sector. Clearly, the interest rate has a significant effect on the net present worth or annual worth of cash flows in the private sector. Similarly, there is a significant effect on the net present worth of benefits minus costs or the benefit-cost ratio in public sector analyses.

EXAMPLE 7.6

Three projects each have investments of $50,000 required. The annual benefits are $15,000, $9000, and $5000 for 5, 10, and 20 years, respectively. Note that no project renewal will be performed, and benefits will cease after the time noted. The time horizon to be considered is 20 years. The projects have the following economic profile:

	Project		
	A	B	C
Initial investment	$50,000	$50,000	$50,000
Annual benefits	15,000	9,000	5,000
Life in years	5	10	20
Present value of benefits:			
At 0%	75,000	90,000	100,000
At 5%	64,943	69,495	62,311
At 10%	56,862	55,301	42,568

Example 7.6 shows that different decisions can be made, depending upon the interest rate used in the analysis to discount benefits and costs. Project C appears to be the best when evaluated at 0%, primarily based upon its long-lived but marginal benefits. Such projects appear more attractive than they really are, when evaluated at 0% interest. Prior to 1940, the Soviet Union did not use discount rates and consequently invested huge sums in long-lived, capital intensive projects [1]. Even in the United States, some years ago, the Water Resources Council evaluated 245 authorized Corps of Engineers projects. In about one-third of them, costs actually exceeded benefits when the discount rate was raised from 5.375% to 7%.

A 10% discount rate was specified in March 1972 by the U.S. Office of Management and Budget (OMB) Circular A-94, for use in federal government investments. For Example 7.6 above, Project A is the best of the three at this rate, due to the immediacy of its much higher benefits. Before considering several explicit schools of thought on the appropriate interest rate, it is helpful to know how public activities are financed.

Financing of Government Projects: There are several different ways that units of government finance public sector projects. The most obvious way is, of course, through taxation such as income tax, property tax, sales tax, and road user tax. A second approach is through the issuance of bonds or notes. A third type of fund raising includes income-generating activities such as a municipally owned power plant, a toll road, or other activity where a charge is made to cover (or partially offset) the cost of the service performed. Although these are the primary sources of government funds, there are a number of ways in which this money may be passed from one government authority to another by way of direct payments, loans, subsidies, and grants.

Because federal funds are raised through tax money and federal borrowing, federal projects may be financed through direct payment. In this case, no monetary return is expected by the government; however, the "return" is expressed through the benefits incurred by the public. Direct payment financing may be total as in the case of many Corps of Engineers projects, or partial, for example 90%, as in cost sharing with states for interstate highways.

Financing for projects of national interest and impact may also be available through no-interest or low-interest rate loans. Both are available for long periods of time, say up to 40 years, with terms more favorable than could normally be expected from conventional sources of money. Such loans are available for financing large projects such as university dormitories. There are also occasions when principal payment deferment is permitted during the early years of a project.

Other forms of federal financing include subsidies and federal loan insurance. Subsidies are used to encourge projects or services believed to be in the public's best interest, such as in the area of transportation. Loan insurance is used to eliminate the private lending institution's risk, allowing lower interest loans over longer periods than conventionally available. Insured loans began with the Federal Housing Administration.

State and local public projects are financed from taxes or bonds. There are, however, constraints on bond financing. First, bond issues must be approved by the voters. Often, there must be a 60% or even two-thirds vote in favor of the bond. Second, in order to prevent excessive borrowing, states have limited the amount of bond debt that may be undertaken. This is often a fraction of the property valuation assessment in the local area. Finally, there are also restrictions that govern the payback requirements and lives of the bonds. Considering these restrictions, the temperament of the public regarding bond issues, and the future needs that are likely to require bond financing, care is obviously needed in selecting public works projects to be implemented or put before the people.

Considerations in the Selection of the Interest Rate: Many arguments over the correct philosophy to use in selecting the interest rate have surfaced over time. For practical purposes, most of these philosophies are somewhat aligned with one of the following positions:

1. *A zero interest rate is approximate when tax monies are used for financing.* Advocates of a zero interest rate when tax money is used argue that current taxes require no principal or interest payment at all. Hence, current tax monies (e.g., highway user taxes) should be considered "free" money, and a no-interest or discount rate applied. Counterarguments to this stance point out that a zero (or even low) interest rate will allow very marginal projects or marginal "add-on" project enhancements to achieve a B/C ratio greater than 1. This, in turn, takes money away from other projects that are truly deserving.

2. *The interest value need only reflect society's rate of time preference (SRTP).* The "societal rate of time preference (SRTP)" reflects society's preference for consumption this year rather than next year [3]. This rate is, say, 7–13% for a house, 8–24% for a new or used car, and so on. It is noted that this rate is usually less than rates of return desired by society on additional investment. Henderson [8] says i is the rate that "reflects the government's judgment about the relative value which the community as a whole is believed to assign, or which the government feels it ought to assign, to present as opposed to future consumption at the margin." Howe [9] makes clear that the societal time preference rate "need bear no relation to the rates of return in the private sector, interest rates, or any other measurable market phenomena."

3. *The interest rate should match that paid by government for borrowed money.* Many people back the use of an interest rate that matches that paid by government for borrowed money. This seems reasonable in that government bonds are in direct competition with other investment opportunities avail-

able in the private sector. Of course, many government bond coupon rates tend to be lower because of their tax exempt status (the bond interest is not taxable) than their private industry counterparts. One way of obtaining an interest figure is to determine the rate on "safe" long-term federal bonds [9]. There are also good arguments that the cost of government money may be too low an interest rate. For example, the opportunities foregone by other government agencies or investors in the private sector may have provided a far higher "return" than investments approved using the cost of government money. In addition, using a rate equal to the cost of government borrowings includes no provision for risk, nor does it include the subsidizing effect of the tax exemption.

4. *The appropriate interest rate is dictated by the opportunity cost of those investments forgone by private investors who pay taxes or purchase bonds.* This philosophy calls for an opportunity cost approach, taking into account many of the factors not considered in the pure cost of government borrowed money. For example, Howe [9] says this philosophy is that "no public project should be undertaken that would generate a rate of return less than the rate of return that would have been experienced on the private uses of funds that would be precluded by the financing of the public project (say, through taxes or bonds)." How are these private use rates determined? Consider a situation in which private investments would yield an average annual return of $i_1\%$. Also consider the rate of return on consumption, which is measured by the rate that consumers are willing to pay to consume now instead of later. This rate is, say 7–13% for a house, 8–24% for a new or used car, and so on. Let this rate of return on consumption average out to $i_2\%$. Now, if a fraction α of government financing precludes private investments, while $1 - \alpha$ precludes consumption, the rate of return foregone to finance public projects is given as:

$$i = \alpha i_1 + (1 - \alpha)i_2 \qquad 0 \le \alpha \le 1 \qquad (7.5)$$

An empirical study conducted by Haveman [7] in 1966 showed the appropriate weighted average to be 7.4% *at that time.* With the many economic changes since that time, a rate of perhaps 10-15% is more appropriate for the early 2000s.

5. *The appropriate interest rate is dictated by the opportunity cost of those investments foregone by government agencies due to budget constraints.* This last philosophy also requires an opportunity cost approach in which an artificial interest rate reflects the rates of return foregone on government projects by virtue of having insufficient funds. That interest rate is found by continuing to increase the value of i until only the projects remain having $B/C > 1$ or $B - C > 0$ which can be afforded with money available.

What conclusions can be reached from these philosophical arguments? What guidelines are available for evaluators of public works projects? Clearly, no one answer is universally applicable. But, as a general rule, we recommend that the rate should approximate the average rate of return on private sector investments, similar to number (4) above. This was the consensus philosophy that led to the OMB's Circular A-94 which specified the rate of 10%, applicable for 1972.

7.6.3   Assessing Benefit-Cost Factors

The benefit-cost analyst knows, before starting a study, that placing a monetary figure on certain "social benefits" may be difficult. Actually, there is even a more fundamental problem—what factors to assess. Some insight is available by considering the following four types of factors discussed by Cohn [4]. Another breakout of factors is presented by Camper [3].

1. *"Internal"* effects are those which accrue directly or indirectly to the individual(s) or organization(s) with which the analyst is primarily concerned. Rides on Atlanta's MARTA transportation are direct benefits, while materials used in MARTA's construction are direct costs. These effects are always included in a benefit-cost analysis.
2. *"External" technological (or real)* effects are those which cause changes in the physical opportunities for consumption or production. For example, effects on navigation and water sport recreation due to a new hydroelectric plant fall into this category. These effects should be included in an analysis.
3. *"External" pecuniary* effects relate to changes in the distribution of incomes through changes in the prices of goods, services, and production factors. For example, increases in rents near a subway station are pecuniary in nature since the benefits to landlords in the form of higher rentals are exactly equal to the costs of tenants. Therefore, there is no net change. Many authors agree that these effects can safely be ignored.
4. *Secondary effects* involve changes in the demand for and supply of goods, services, resources, and production factors which arise from a particular project. As an example, phosphate mining in Idaho on government lands will bring instant population increases to nearby small towns. Secondary effects include increasing the incomes of various store owners; however, their increased sales are almost certainly offset by reduced sales of stores elsewhere. McKean [12] contends that only incremental income arising from such effects should be included.

EXAMPLE 7.7

A dam and reservoir are contemplated in an effort to reduce flood damage to homes and crops in a low area of northeastern Oklahoma. Annual damage to property averages approximately $230,000/year. The dam and reservoir contemplated should virtually eliminate damage to the area in question. No other benefits (e.g., irrigation, power generation, recreation) will be provided.

The engineer notes that the *primary* benefit to the public will be the $230,000/year damage prevented. However, the engineer argues that this will also cut back on money paid to contractors and servicemen for home and car repair, to health care units, for insurance premiums, and so forth. In other words, the building of a dam will provide disbenefits to those who would normally receive part of their livelihood from helping flood-damaged families. Should the engineer include in the evaluation only the direct benefits to the flood damage victims, or should the other effects be included as well? That depends on how the analyst chooses to handle *secondary* effects, as indicated in the following discussion.

The disbenefits to those who would lose income if the dam and reservoir were built are considered *secondary effects*. That is, there would be a decrease in the demand for and supply of post-flood restoration goods and services. There would however, be an increase in the demand for goods and services in constructing and maintaining the dam. It is argued that only the *incremental* incomes or incremental profits (losses or lost profits in this case) should be considered when secondary effects are involved.

Another argument calls upon the "ripple" effect of the economy. That is, every secondary effect disbursement by one person or organization is a receipt to another person or organization. Each receipt then contributes to another disbursement, and so on. If the ripple philosophy were tracked, the sum of receipts less disbursements would equal zero, and there would be no economic evaluation.

With either philosophy, the secondary effects of the example's flood control dam and reservoir would be small, if not negligible. This is intuitively reasonable, because the dam's main benefits represent the measure of the direct usefulness of the dam serving its intended purpose, whereas the diseconomies described are, in fact, secondary and diffuse.

EXAMPLE 7.8

Now, reconsider the previous flood control dam and reservoir of Example 7.7. Suppose the reservoir would cause a loss of agricultural land for grazing and crops. Should this loss be considered in the benefit-cost analysis? Yes, because it is an external real effect causing changes in the physical opportunities for consumption or production.

External technological effects are often well defined and, in practice, are usually included in benefit-cost analyses. External pecuniary effects, however, may not be vividly apparent, as illustrated in the following example.

EXAMPLE 7.9

A large irrigation project is being considered in the heart of cotton country. The irrigation will provide a significant effect on the quantity and quality of the cotton grown. This additional supply of cotton will, however, depress the price of cotton, lowering the profitability of other cotton growers. Also, the same effect will be felt throughout, say, the clothing industry and manufacturers of cotton substitutable products (products that may be used in place of cotton items) will likely have to reduce prices. At the same time, producers of cotton complementary goods (items that go well with or are used in conjunction with cotton products) will note increased demand, and may increase prices because of an insufficient supply.

Which of these effects would we include in an evaluation of the irrigation project? Of the factors mentioned above, none would be included in the analysis. Each of the effects described relates to changes in the distribution of incomes through changes in the prices of goods, services, and production factors. As such, they are considered external pecuniary effects, which are not "real" benefits or disbenefits and, hence, are not included.

In the above examples, internal and external technological effects were considered to be factors in an analysis, while external pecuniary and secondary effects were not included. There is, of course, no complete agreement on either of these classifications or whether or not they should be evaluated. As a guide, all identifiable effects of a project should be delineated. Some effects will clearly provide direct benefits or disbenefits to the public that should be counted. There may also be some factors that obviously should not be included. The third group—the controversial factors, if any—should be studied in depth and the reason for including or not considering these controversial factors should be stated in writing and become a part of the evaluation for the record.

7.6.4   Overcounting

A common dysfunction of trying to consider a large variety of effects in a benefit-cost analysis is to overcount, or unknowingly count some factors twice.

EXAMPLE 7.10

Reconsider Example 7.9 involving a cotton irrigation system. The increased quantity of cotton will require that additional gin and seed mill hands be employed, removing a significant number of persons from the welfare rolls. The amount of their new wage, equal to the sum of their old welfare payments plus some increase, represents an increase in real output and constitutes a legitimate national benefit of the project. To then add the reduction in welfare payments from the taxpayers to the unemployed would be to double-count welfare payments, once from the standpoint of the recipient, and once from the taxpayer's viewpoint.

7.6.5   Unequal Lives

When comparing one-shot public works projects having unequal lives, the planning horizon will commonly coincide with the longest lived alternative. When some projects are expected to have a long life while others have a shorter life, changes in the discount rate could change the attractiveness of some projects with respect to the alternatives.

EXAMPLE 7.11

Two projects each have first costs of $200,000, with annual operating costs of $30,000. The life of Project A is 15 years, while that of Project B is 30 years, and benefits accrued to the public are estimated at \$60,000/year and \$54,000/year, respectively. Each is a one-shot project; hence, Project A will have no benefits or costs after year 15. A 30-year planning horizon will be used. If the $MARR$ is set at 8%, which project is more attractive?

$$(B - C)_A(8) = \$60,000(P|A\ 8,15) - \$30,000(P|A\ 8,15) - \$200,000$$
$$= \$60,000(8.5595) - \$30,000(8.5595) - \$200,000$$
$$= \$56,785$$
$$(B - C)_B(8) = \$54,000(P|A\ 8,30) - \$30,000(P|A\ 8,30) - \$200,000$$
$$= \$54,000(11.2578) - \$30,000(11.2578) - \$200,000$$
$$= \$70,187$$

Over a 30-year planning horizon, and using an interest rate of 8%, Project B is clearly better. Now suppose that insufficient money is available to implement all of the many projects, B included, that a govenment agency would like to do. A decision is made to consider all projects using an interest rate that reflects the opportunity cost of investments foregone by government agencies. That is, the interest rate is increased until only those projects that can be afforded remain. Assuming the interest rate is up to 12%, now which project is more desirable?

$$(B - C)_A(12) = \$60,000(P|A\ 12,15) - \$30,000(P|A\ 12,15) - \$200,000$$
$$= \$60,000(6.8109) - \$30,000(6.8109) - \$200,000$$
$$= \$4327$$
$$(B - C)_B(12) = \$54,000(P|A\ 12,30) - \$30,000(P|A\ 12,30) - \$200,000$$
$$= \$54,000(8.0552) - \$30,000(8.0552) - \$200,000$$
$$= -\$6675$$

At the higher interest rate, Project A is more favorable because of the increased emphasis on early year net benefits as opposed to the heavily discounted net benefits during the latter years of the planning horizon.

When a planning horizon shorter than some project durations is selected, a residual value must be estimated for those projects. The residual value is handled in the same way as a salvage value.

EXAMPLE 7.12

Reconsidering the previous example and selecting a 15-year planning horizon, let us now determine the more favorable alternative, assuming project B has a residual value of 40% of first cost. Let $i = 12\%$.

$$(B - C)_A(12) = \$60,000(P|A\ 12,15) - \$30,000(P|A\ 12,15) - \$200,000$$
$$= \$60,000(6.8109) - \$30,000(6.8109) - \$200,000$$
$$= \$4327$$

$$(B - C)_B(12) = \$54{,}000(P|A\ 12{,}15) + 0.4(\$200{,}000(P|F\ 12{,}15)$$
$$- \$30{,}000(P|A\ 12{,}15) - \$200{,}000$$
$$= \$54{,}000(6.8109) + \$80{,}000(0.1827) - \$30{,}000(6.8109)$$
$$- \$200{,}000 = -\$21{,}922$$

7.6.6 Tolls, Fees, and User Charges

Tolls, fees, and user charges have an interesting effect on the fiscal aspects of public projects. If a toll, fee, or user charge is regarded as a payment or partial payment for benefits derived, it can be argued that net benefits received are reduced by the amount of the payment. Similarly, the amount of the payment decreases the cost of the project to the government. Thus, the B/C ratio will change, but the $B - C$ measure of merit will remain constant so long as total user benefits remain constant.

EXAMPLE 7.13

Suppose 10,000 people/year attend a public facility that has an equivalent uniform annual cost of $40,000. The people, on the average, receive recreational benefits in the amount of $6 each. The B/C ratio would be

$$B/C = \frac{\$6(10{,}000)}{\$40{,}000} = 1.5$$

and the $B - C$ measure of merit is

$$B - C = 6(\$10{,}000) - \$40{,}000 = \$20{,}000/\text{year}$$

Based on either criterion, the public facility appears worthwhile. Now suppose that a fee of $3.50/season is charged. The net benefits are now $6.00 - $3.50, or $2.50/person, and the government cost is reduced by $35,000/year. Thus, the B/C and $B - C$ measures are as follows.

$$B/C = \frac{\$60{,}000 - \$35{,}000}{\$40{,}000 - \$35{,}000} = 5$$
$$B - C = \$60{,}000 - \$35{,}000 - (\$40{,}000 - \$35{,}000) = \$20{,}000/\text{year}$$

Note that in the example B/C changed while $B - C$ did not. This phenomenon will be discussed in a subsequent section on "Problems With the B/C Ratio."

It might be concluded that tolls, fees, and user charges are irrelevant, at least with respect to the $B - C$ measure of merit. This, however, *is not* true if the number of users or degree of use is linked to the fee charged, as it almost always will be.

EXAMPLE 7.14

As an extreme, suppose that the 10,000 users of the public facility in the previous example receive different levels of benefits, but they average out to $6/person. The actual breakout is that 8000 persons perceive $3 worth of enjoyment, 1000 persons perceive $6 worth, and 1000 persons expect to derive $30 in recreational benefits. With a user fee of $3.50/person, it is assumed only 2000 will patronize the facility. Thus, the B/C and $B - C$ measures would be:

$$B/C = \frac{1000(\$6) + 1000(\$30) - 2000(\$3.50)}{\$40,000 - 2000(\$3.50)} = \frac{\$29,000}{\$33,000} = 0.88$$

$$B - C = [1000(\$6) + 1000(\$30) - 2000(\$3.50)]$$
$$- [\$40,000 - 2000(\$3.50)]$$
$$= -\$4000$$

In this case, the reaction of demand to a fee would cause the costs to exceed realized benefits. Thus, when tolls, fees, and user charges are expected, their effect on user demand, and hence total user benefits, must be determined and accounted for.

7.6.7 Multiple-Use Projects

Multiple-use projects receive a great deal of attention, both pro and con. Multiple uses, and hence multiple benefits, are often available at slight incremental costs over single-use projects. Of course, the incremental capital and net operating costs required for an additional use must provide at least a like worth of benefits.

EXAMPLE 7.15

An irrigation dam and reservoir will provide present-worth benefits of $25 million over the next 50 years. The present-worth cost of construction, operation, and maintenance of the irrigation facility will be $14,500,000. A single-purpose flood control dam providing present-worth benefits of $6 million would cost a present worth of $9 million. Suitable design modifications can be made to the irrigation dam and reservoir to provide the flood control benefits, too, at a total package present worth cost of $18,500,000. Funds permitting, what should be done?

First, it must be understood that the benefits and costs discounted to the present were discounted at the *MARR* deemed suitable by the decision maker. Assuming this to be true, it is clear that the irrigation project is worthwhile, providing a benefit-cost ratio of

$$B/C_{\text{irrigation}} = \frac{\$25,000,000}{\$14,500,000} = 1.72[1]$$

---

[1] This is also the incremental B/C ratio of irrigation over doing nothing.

As a single-purpose facility, a flood control dam would not provide benefits commensurate with its costs, yielding a B/C ratio of

$$B/C_{\text{flood control}} = \frac{\$6,000,000}{\$9,000,000} = 0.67^1$$

As a multiple-use facility, however, the flood control benefits may be provided at a sufficiently low incremental cost to be justifiable. The incremental B/C ratio is

$$\Delta B/\Delta C_{\substack{\text{irrigation plus} \\ \text{flood control}}} = \frac{\$6,000,000}{\$18,500,000 - \$14,500,000} = 1.5$$

Thus, a multiple-use facility should be built.

The example illustrates how multiple uses can draw on each other, providing benefits economically that could never have been provided using a single-purpose facility (e.g., the B/C flood control ratio of 0.67). Multiple purpose projects also have their problems. For example, it is frequently desirable to "allocate" the costs of a project to its various uses.

EXAMPLE 7.16

A city's refuse is used to fire a power-generation facility owned by a municipality. Not only is electrical power supplied to a segment of the city, but burning of the refuse after processing has substantially reduced the need for an expensive landfill operation. Since this project is self-supporting, construction, operation, and maintenance costs must be allocated between the (1) power, and (2) disposal benefits provided in order to determine the user charge for electrical energy and refuse disposal. Arguments for cost allocation range from (1) no costs should be allocated to refuse disposal because the refuse is being used in place of fuel oil or coal and, in fact, a credit should be issued, to (2) refuse disposal should receive sufficient cost allocation to raise rates above those for conventional disposal to include the aesthetic benefits of no unsightly public landfill.[2]

7.6.8   Problems with the B/C Ratio

There are two frequent problems with the B/C ratio that require an explanation and warning. Either can give misleading results that may cause an otherwise perfect analysis to point toward the wrong project.

---

[2] These extremes in arguments have actually been used by public officials of one major city.

First, it is sometimes difficult to decide whether an item is a benefit to the public or a cost savings to the government. Similarly, there is often uncertainty between disbenefits and costs.

EXAMPLE 7.17

A project provides annual equivalent benefits of $100,000, disbenefits of $60,000/year, and annual costs of $5000. What is the benefit-cost ratio?

Let us first calculate the B/C ratio for the problem as stated.

$$B/C = \frac{\$100,000 - \$60,000}{\$5000} = 8$$

Such a high ratio leads one to believe that the project is outstanding.

Now another analyst notes that the government will likely reimburse those incurring damages from the project in an amount equivalent to $60,000/year. Thus, the analyst concludes that the public disbenefits have been compensated for by the government and calculates a B/C ratio of

$$B/C = \frac{\$100,000}{\$5000 + \$60,000} = 1.54$$

which is considerably lower.

Example 7.17 shows that a wide range of B/C ratios may reasonably be obtained on a single project simply by interpreting certain elements of the problem differently. The resolution of this problem is not difficult. The analyst should simply calculate the net benefits less the net costs.

EXAMPLE 7.18

Let us reconsider Example 7.17 and calculate the annual net benefits less net costs. The first analyst would have calculated

$$B - C = (\$100,000 - \$60,000) - (\$5000) = \$35,000/\text{year}$$

and the second analyst would have calculated

$$B - C = (\$100,000) - (\$5000 + \$60,000) = \$35,000/\text{year}$$

Calculating $B - C$ eliminates the inherent bias in the B/C ratio and does not require an incremental approach between alternatives where benefits and costs are known directly for each alternative. That is, if mutually exclusive alternatives are involved, over the same time horizon, the one having the highest $B - C$ value should be selected. Unfortunately, the B/C ratio is by far the more popular criterion of the two.

The potential for the other B/C problem was illustrated in the dam and reservoir irrigation example. That is, when the B/C ratio is used, it should be based on *incremental benefits* and *incremental costs*.[3] Simply to calculate the B/C ratio of each alternative and take the one with the largest ratio is incorrect and will frequently lead to errors in project selection.

EXAMPLE 7.19

In the dam and reservoir of Example 7.15 we calculated the B/C irrigation ratio to be 1.72. To compare this value against a total project $B/C_{\text{irrigation+flood control}}$ ratio of

$$\frac{\$25,000,000 + \$6,000,000}{\$18,500,000} = 1.68$$

would cause us to select irrigation only, in error.

A related error is to require the incremental B/C ratio to be above that for the previous incremental B/C ratio. In the dam and reservoir example, had the incremental B/C ratio, $\Delta B/\Delta C_{\text{irrigation plus flood control}} = 1.5$ been compared against the $B/C_{\text{irrigation}} = 1.72$ ratio, again an incorrect conclusion would have resulted. As long as the incremental B/C ratio exceeds 1, the incremental benefits justify the incremental costs. In this regard, the B/C ratio criterion is closely akin to the rate-of-return criterion discussed in Chapter Four.

7.7 COST-EFFECTIVENESS ANALYSIS

To this point it has been assumed that public project effects were measurable, either directly or indirectly, in monetary terms. There are circumstances, however, when project outputs are not measurable monetarily and must be expressed in physical units appropriate to the project. In these cases, *cost-effectiveness analysis* has proven to be a useful technique for deciding between projects or systems for the accomplishment of certain goals. Although cost-effectiveness is most often associated with the economic evaluation of complex defense and space systems, it has also proven useful in the social and economic sectors. In fact, it has roots dating back to Arthur M. Wellington's *The Economic Theory of Railway Location* in 1887 [17].

Cost-effectiveness analyses require that three conditions be met, according to Kazanowski [10]. They are:

1. Common goals or purposes must be identifiable and attainable.
2. There must be alternative means of meeting the goals.
3. There must be perceptible constraints for bounding the problem.

---

[3] Actually, the B/C irrigation ratio in the referenced example is a ratio of incremental benefits to incremental costs, where the pairwise comparison is between irrigation and doing nothing.

Common goals are required in order to have a basis for comparison. For example, it would not make sense to compare (1) a submarine, with (2) a sophisticated single sideband communication network. Obviously, alternative methods of accomplishing the goals must be available in order to have a comparison. Finally, reasonable bounds for constraining the problem by time, cost, and/or effectiveness are necessary to limit and better define the alternatives to be considered.

7.7.1   The Standardized Approach

Kazanowski [10] presents 10 standardized steps that constitute a correct approach to cost-effectiveness analyses. They are, in their usual order, presented below.

1. *Define the goals,* purpose, missions, and so forth, that are to be met. Cost-effectiveness analysis will identify the best alternative way of meeting these goals.
2. *State the requirements* necessary for attainment of the goals. That is, state any requirements that are essential if the goals are to be attained.
3. *Develop alternatives* for achieving the goals. There must be at least two alternative ways of meeting or exceeding the goals.
4. *Establish* evaluation *measures* which relate capabilities of alternatives to requirements. An excellent list of evaluation criteria is contained in Kazanowski [11]. Typical measures are performance, availability, reliability, maintainability, and so forth.
5. *Select* the fixed-effectiveness or fixed-cost *approach.* The fixed-effectiveness criterion is the minimum alternative cost to achieve the specified goals or effectiveness levels. Alternatives failing to achieve these levels may either be eliminated or assessed penalty costs. The fixed-cost criterion is the amount of effectiveness achieved at a given cost. "Cost" is usually taken to mean a present worth or annual equivalent of "life-cycle cost," which includes research and development, engineering, construction, operation, maintenance, salvage, and other costs incurred throughout the life cycle of the alternative.
6. *Determine capabilities* of the alternatives in terms of the evaluation measures.
7. *Express* the alternatives and their *capabilities* in a suitable manner.
8. *Analyze* the various *alternatives* based upon the effectiveness criteria and cost considerations. Often, some alternatives are clearly dominated by others and should be removed from consideration.
9. *Conduct a sensitivity analysis* to see if minor changes in assumptions or conditions cause significant changes in alternative preferences.
10. *Document all* considerations, analyses, and decisions from the above nine steps.

Clearly, one cannot become a cost-effectiveness expert by reading these 10 steps. They do, however, give an indication of what is involved in a cost-effectiveness study. Excellent detailed presentations on cost effectiveness are available in English [5] and Fabrycky and Blanchard [6].

Table 7.2 Evaluation of Candidate
Propulsion Systems for Example 7.20

Propulsion System	Life Cycle Cost in Millions	Reliability
1	24	0.99
2	24	0.98
3	20	0.98
4	20	0.97

EXAMPLE 7.20

Four different propulsion systems are under consideration. The life-cycle cost is not to exceed $24 million. A single effectiveness measure is decided on, that being reliability. Reliability is defined as "the probability that the propulsion system will perform without failure under given conditions for a given period of time." Four contractors submit candidate systems, which are evaluated as shown in Table 7.2. Since there is only one effectiveness measure, the results given in Table 7.2 may be expressed graphically as shown in Figure 7.1.

If we look at these systems in pairs, comparing just 1 and 2 is equivalent to a fixed-cost comparison in which we prefer system 1 because of its higher reliability. Similar reasoning leads us to prefer system 3 over system 4. If we were to compare only systems 2 and 3, this would be a fixed-effectiveness comparison, in which case we prefer system 3 due to its lower cost. Clearly, system 1 dominates 2, 3 dominates 2, and 3 dominates 4. We are left with making the decision between systems 1 and 3. This choice will depend on whether or not an additional percent reliability justifies the expenditure of an additional $4 million.

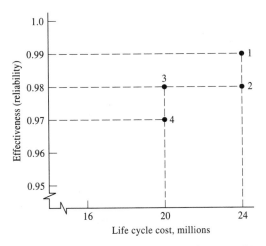

Figure 7.1 Effectiveness (reliability) versus life-cycle cost.

In this example, it is tempting to go the next step and say that based on the ratios of reliability to cost, 0.99/$2.4 million and 0.98/$2.0 million, system 3 should be selected. Unfortunately, apparent as such a decision may seem, the correct decision may depend on many other considerations, not the least of which are the payload (perhaps human life) and the consequences of system failure (perhaps total destruction with no payroll recovery).

7.8   REVENUE REQUIREMENTS METHOD

Engineering economic analyses take on a different and somewhat "inverted" approach for regulated firms such as public utility companies. Public utilities include suppliers of electricity, gas, telephone, water, cable, and so on. These companies are usually given a certain territory to cover, and little or no competition is allowed, giving the company a virtual monopoly (note that this is changing in several areas). When this occurs, the customer is no longer protected by competition and freedom of choice. Rather, the federal, state, or local regulatory commission governs such things as performance standards and rate charged to users.

Even though public utilities could use conventional economic analyses for the evaluation of alternative investments, it is rarely done. This is because traditional economic analyses focus on the selection of alternatives, using limited financial resources, that will improve the wealth of the owners of the firm. In a utility, the emphasis is more focused on providing desired services and minimizing the requirements of revenues that must be obtained by customers. As a result, the method commonly used by privately owned public utility companies is called *the minimum revenue requirements* method [15]. Of course, we cannot forget the owners of the public utility. In using the revenue requirements method, a fair return is explicitly included for equity holders. That is, the utility should be allowed to recover, through rates for products and services, all costs plus a fair return on their investment.

7.8.1   Definition of Terms

To develop the equation for determining the revenue requirements, several terms must be defined:

P Initial investment
BV_t Unrecovered investment at end of year t
Db_t Book depreciation in year t used to determine unrecovered investment
Dt_t Tax depreciation in year t used to determine taxable income
r_e Return on equity capital
r_d Cost of debt capital
r_o Overall after-tax cost of capital
c Debt ratio (fraction of capital borrowed)
C_t Annual expenses in year t associated with project
t' Tax rate
T_t Income taxes in year t

FC_t Fixed cost of investment in year t
RR_t Revenue requirement in year t

Given this notation, we can develop a mathematical representation of revenue requirements.

Figure 7.2 illustrates the components of revenue which make up the total revenue requirement in a given year, t. These include:

1. Annual expenses associated with the project, C_t
2. Book depreciation deduction, Db_t
3. Interest paid on debt, I_t
4. Income taxes paid, T_t
5. Return to owners, R_t

The first three components of revenue are straightforward. The income tax paid, however, ultimately depends upon the taxable income. But, from Figure 7.2, taxable income depends on the amount of income taxes paid! So, a roundabout way of writing the relevant equations is required.

7.8.2 Determining the Minimum Revenue Requirement

Use of a fixed-charge rate concept is common in the utility industry. The fixed-charge rate represents the cost of the investment, expressed as a fraction of the unrecovered investment. The fixed charges include items 2–5 of the list of components making up the revenue requirement. We can write:

$$FC_t = Db_t + r_d \times c \times BV_{t-1} + T_t + r_e \times (1 - c) \times BV_{t-1}$$
$$= Db_t + [r_d \times c + r_e \times (1 - c)] \times BV_{t-1} + T_t \tag{7.6}$$

Taxable income may be expressed as the total fixed charge minus the tax deprecia-

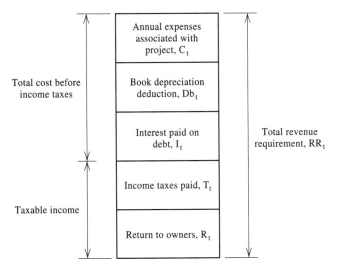

Figure 7.2 Components of total revenue requirement in year t.

tion and the interest on borrowed money. That is:

$$T_t = t' \times \text{taxable income} = t' \times (FC_t - Dt_t - r_d \times c \times BV_{t-1}) \qquad (7.7)$$

Using Equations 7.6 and 7.7, solving each for FC_t, setting both equal and rearranging terms results in:

$$T_t = [t'/(1 - t')] \times [r_e \times (1 - c) \times BV_{t-1} + Db_t - Dt_t] \qquad (7.8)$$

Then, the total revenue requirement is expressed as the fixed charge plus the annual expenses associated with the project:

$$RR_t = FC_t + C_t \qquad (7.9)$$

Now, armed with the basic equations, it is possible to determine the minimum revenue requirement each year for a specific investment.

The revenue requirements over the life of the project are then usually expressed as an after-tax present worth, annual worth, or capitalized amount. For these calculations, the after-tax cost of capital is needed. The cost of debt capital (interest on borrowed money, e.g., in the form of bonds) is deductible when calculating taxable income. Of course, the return on equity capital is not deductible, because it is paid from after-tax money. The overall after-tax cost of capital also reflects the mix of borrowed and equity money in the project (assumed to be the same mix as for the entire firm). It is expressed as:

$$r_o = (1 - t') \times c \times r_d + (1 - c) \times r_e$$

EXAMPLE 7.21

An electric utility needs to place additional transmission and distribution plant to serve the growing needs of a set of commercial customers. The utility has prepared a plan and now wants to determine the minimum revenue requirement necessary to support the plan if implemented. The installation will cost $90,000 and have a life of 20 years. Book depreciation will follow the traditional straight-line method. Tax depreciation will be accomplished using the MACRS-GDS over a recovery period of 20 years. Annual expenses are projected to be $12,000. The utility finances 40% of their investment using borrowed money at a rate of 8.5%. The remainder is equity funded and it is desired to have a 14% return on the owner's capital. There is no salvage value at the end of the project life, and the effective tax rate is 35%. Determine the after-tax revenue requirement profile, including the PW, AW, and capitalized cost.

The revenue requirement profile for the investment is given in Table 7.3. If other alternatives are to be considered, they can be evaluated in like manner, and the alternative having the lowest PW, AW, or capitalized revenue requirements can be selected. Note that the after-tax overall cost of capital is $(1 - 0.35) \times 0.4 \times 0.085 + (1 - 0.4) \times 0.14 = 0.1061$ or 10.61%. The resulting PW(10.61%) of the revenue requirements is $215,886.01. The AW(10.61%) is $25,421.70 and the capitalized revenue requirements are $2,034,740.88.

Table 7.3 Revenue Requirement Profile, Example 7.21

End of Year, A	Book Depreciation, B	Unrecovered Investment, C	Tax Depreciation, D	Return on Equity, E	Return on Debt, F	Income Taxes, G	Annual Expenses, H	Revenue Requirements I, B + E + F + G + H	PW RR @ 10.61%, J
0		$90,000.00							
1	$4,500.00	85,500.00	$3,375.00	$7,560.00	$3,060.00	$4,676.54	$12,000.00	$31,796.54	$ 28,746.53
2	4,500.00	81,000.00	6,497.10	7,182.00	2,907.00	2,791.87	12,000.00	29,380.87	24,014.63
3	4,500.00	76,500.00	6,009.30	6,804.00	2,754.00	2,850.99	12,000.00	28,908.99	21,362.39
4	4,500.00	72,000.00	5,559.30	6,426.00	2,601.00	2,889.76	12,000.00	28,416.76	18,984.41
5	4,500.00	67,500.00	5,141.70	6,048.00	2,448.00	2,911.08	12,000.00	27,907.08	16,855.53
6	4,500.00	63,000.00	4,756.50	5,670.00	2,295.00	2,914.96	12,000.00	27,379.96	14,950.87
7	4,500.00	58,500.00	4,399.20	5,292.00	2,142.00	2,903.82	12,000.00	26,837.82	13,249.10
8	4,500.00	54,000.00	4,069.80	4,914.00	1,989.00	2,877.65	12,000.00	26,280.65	11,729.54
9	4,500.00	49,500.00	4,015.80	4,536.00	1,836.00	2,703.18	12,000.00	25,575.18	10,319.75
10	4,500.00	45,000.00	4,014.90	4,158.00	1,683.00	2,500.13	12,000.00	24,841.13	9,062.07
11	4,500.00	40,500.00	4,015.80	3,780.00	1,530.00	2,296.11	12,000.00	24,106.11	7,950.40
12	4,500.00	36,000.00	4,014.90	3,402.00	1,377.00	2,093.05	12,000.00	23,372.05	6,968.90
13	4,500.00	31,500.00	4,015.80	3,024.00	1,224.00	1,889.03	12,000.00	22,637.03	6,102.28
14	4,500.00	27,000.00	4,014.90	2,646.00	1,071.00	1,685.98	12,000.00	21,902.98	5,338.04
15	4,500.00	22,500.00	4,015.80	2,268.00	918.00	1,481.95	12,000.00	21,167.95	4,664.05
16	4,500.00	18,000.00	4,014.90	1,890.00	765.00	1,278.90	12,000.00	20,433.90	4,070.44
17	4,500.00	13,500.00	4,015.80	1,512.00	612.00	1,074.88	12,000.00	19,698.88	3,547.62
18	4,500.00	9,000.00	4,014.90	1,134.00	459.00	871.82	12,000.00	18,964.82	3,087.80
19	4,500.00	4,500.00	4,015.80	756.00	306.00	667.80	12,000.00	18,229.80	2,683.42
20	4,500.00	0.00	4,014.90	378.00	153.00	464.75	12,000.00	17,496.75	2,328.33
21	0.00	0.00	2,007.90	0.00	0.00	-1,081.18	0.00	-1,081.18	-130.88
							PW	$215,886.01	
							ARR	$26,421.70	
							CAP	$2,034,740.88	

7.9 SUMMARY

Benefit-cost and cost-effectiveness analyses are accepted methods of evaluating investments in the public sector. Likewise, the minimum revenue requirement method is common and well understood in the regulated sector. A great many of today's and tomorrow's engineers will continue to be closely involved with these analysis tools.

BIBLIOGRAPHY

1. Anthony, Robert N., and Young, David W., *Management Control in Nonprofit Organizations,* Fifth Edition, Irwin, Burr Ridge, IL, 1994.
2. Arrow, Kenneth J., et al., *Benefit-Cost Analysis in Environmental, Health, and Safety Regulation: A Statement of Principles,* American Enterprise Institute, The Annapolis Center, and Resources for the Future, AEI Press, La Vergne, TN, 1996.
3. Campen, James T., *Benefit, Cost, and Beyond,* Ballenger Publishing Company, Cambridge, MA, 1986.
4. Cohn, Elchanan, *Public Expenditure Analysis,* D. C. Heath, 1972.
5. English, J. Morley (Editor), *Cost Effectiveness,* Wiley, New York, 1968.
6. Fabrycky, Wolter, J., and Blanchard, Ben S., *Life-Cycle Cost and Economic Analysis,* Prentice Hall, Englewood Cliffs, NJ, 1991.
7. Haveman, R. H., "The Opportunity Cost of Displaced Private Spending and the Social Discount Rate." *Water Resources Research, 5* (5), 1969, pp. 947–57.
8. Henderson, P. D., "The Investment Criteria for Public Enterprises," in *Public Enterprises,* edited by Ralph Turvey, Penguin Books, 1968.
9. Howe, Charles W., *Benefit-Cost Analysis for Water System Planning,* Water Resources Monograph 2, American Geophysical Union, Washington, D.C., 1971.
10. Kazanowski, A. D., "A Standardized Approach to Cost-Effectiveness Evaluations," in *Cost Effectiveness,* edited by J. Morley English, Wiley, New York, 1968.
11. Kazanowski, A. D., "Some Cost-Effectiveness Evaluation Criteria," Appendix B in *Cost Effectiveness,* edited by J. Morley English, Wiley, New York, 1968.
12. McKean, Roland N., *Efficiency in Government Through Systems Analysis,* Wiley, New York, 1958.
13. Merkhofer, Miley W., *Decision Science and Social Risk Management,* D. Reidel Publishing Company, Dordrecht, Holland, 1987.
14. Prest, A. R., and Turvey, Ralph "Cost-Benefit Analysis: A Survey," *Economic Journal, 75,* December 1965, pp. 683-735.
15. Stevens, G. T. Jr., "Revenue Requirement Analysis with Working Capital Changes: A Tutorial," *The Engineering Economist, 32,* (1), Fall 1986, pp. 1–16.
16. Wellington, A. M., *The Economic Theory of Railway Location,* Wiley, New York, 1887.
17. United States Code, 1940 edition, U.S. Government Printing Office, Washington, D.C., p. 2964.

PROBLEMS

1. Identify five projects that have been performed in your city, state, or country. For each, identify the (1) main benefit, (2) to whom the main benefit is directed, (3) the government unit(s) that funded the project, and (4) the major cost factor involved. (7.3)
2. Identify the benefits that would accrue to the public for the following projects: (a) a modern art sculpture outside City Hall, (b) a city water treatment plant, (c) a high speed limited access toll road through a busy part of a city and extending to an outlying

airport, (d) an underground tunnel linking two different areas separated by a busy street, with one of the sides having a well-lit jogging track. (7.3)

3. For each of the following projects, identify two each (1) benefits, (2) disbenefits (negative benefits), and (3) monetary costs that would impact the public for the following projects:
   a. A lakeside park with picnic shelters, green areas, and restrooms.
   b. A rural electric cooperative to distribute electrical energy they generate and/or purchase.
   c. Baseball and softball fields scattered throughout town with backstops, lights, and bleachers.
   d. An interstate highway. (7.5)

4. Three options are available for a public recreational area. Their respective annual benefits, disbenefits, costs, and savings are as follows:

	Option 1	Option 2	Option 3
Benefits	$300,000	$450,000	$600,000
Disbenefits	63,000	112,500	177,000
Costs	225,000	375,000	487,500
Savings	22,500	60,000	82,500

   a. Calculate the B/C ratios for each project. Can you tell from these calculations which should be selected?
   b. Determine which should be selected using the incremental B/C ratio.
   c. Calculate $B - C$ for each alternative. (7.5)

5. Two four-lane roads intersect and traffic is controlled by a standard green, yellow, red stoplight. From each of the four directions a left turn is permitted from the inner lane; however, this impedes the flow of traffic while the person desiring to turn left waits until a safe turn may be accomplished. The light operates on a cycle allowing 30 seconds of green–yellow, followed by 30 seconds of red light for each direction. Approximately an eighth of the 8000 vehicles using the intersection daily are held up for two full minute cycles of the light, and made to perform two extra stop-start operations, solely because of the left-turn bottleneck. These delays are only during the 250 working days/year. A stop-start costs 2.0¢/vehicle. A widening of the intersection to accommodate a left-turn lane plus a new lighting system to provide for a left-turn signal will cost $90,000. The cost of time for commercial traffic is three times that of private traffic, and private traffic accounts for 75% of the vehicles. How much must private traffic time be worth to justify the intersection changes if the interest rate is 9% and a lifetime of 10 years is expected? (7.5)

6. A dam and reservoir will be used for both flood control and electrical power generation. Costs and benefits are:
   Investment

Dam, including access roads, clearing, and foundation treatment	$39,160,000
Generation equipment and transmission apparatus	20,000,000
Land	2,750,000
Highway relocation	3,460,000
Home relocation	1,780,000
Miscellaneous	225,000

Operating and maintenance costs
   Two percent of the total investment during the first year and increasing by 5% of itself each subsequent year.
Annual benefits

Flood losses prevented	$1,250,000
Property value enhancement	200,000
Incremental value of access to power	4,000,000

For a planning horizon of 40 years and $i = 8\%$, what is the B/C ratio? (7.5)

7. A city is trying to decide between coal, fuel oil 3, low-sulphur fuel oil, and natural gas to power their electrical generators. Fuel forecasts indicate the following needs for the upcoming year, and the gradient increase during each subsequent year.

Fuel	First Year	Gradient Each Subsequent Year
Coal	500,000 tons	25,000 tons
Fuel oil 3	1,900,000 barrels	95,000 barrels
Low-sulphur fuel oil	2,000,000 barrels	100,000 barrels
Natural gas	12,000 10^6 cubic feet	600 10^6 cubic feet

The cost of the fuel, transportation, and various pollution effects have been estimated as follows:

Fuel	Cost	Transportation and Storage	Health	Crops	Uncleanliness
Coal	$29.00/ton	11.25/ton	3.50/ton	$8.75/ton	5.25/ton
Fuel oil 3	10.75/barrel	1.00/barrel	0.50/barrel	1.25/barrel	0.75/barrel
Low-sulphur fuel oil	13.75/barrel	1.00/barrel	0.15/barrel	0.375/barrel	0.225/barrel
Natural gas	0.0029/cubic foot	0.0001/cubic foot	Negligible	Negligible	Negligible

Calculate the annual equivalent benefits and costs of these fuels considering a life of 20 years and (a) $i = 6\%$ and (b) $i = 10\%$. Use an incremental B/C analysis to determine the best fuel to use. (7.5)

8. Solve problem 7 using the $B - C$ criterion. (7.5).

9. A highway is to be built connecting Coyle with Lake Carl Blackwell (Coyle is northeast of Langston). Route X follows the old road and costs $1 million initially and $52,500/year thereafter. A new route, Y, will cost $1.75 million initially and $45,000/year thereafter. Route Z is simply an enhanced version of route Y with wider lands, shoulders, and so on. It will cost $2.25 million at first, plus $60,000/year to maintain. Relevant annual user costs considering time, operation, and safety are $250,000 for X, $175,000 for Y, and $125,000 for Z. Using a *MARR* of 8%, and a 20-year study period, which should be constructed? Use a B/C analysis. Which route is preferred if $i = 0$? (7.5)

10. Solve problem 6 using the $B - C$ criterion. (7.5)

11. A city library is to be expanded to include meeting rooms, more electronic volumes,

computer facilities, and electronic check-in and check-out. The cost of the expansion will be $700,000 and the new equipment will cost another $175,000. Maintenance and renewal of the new addition and equipment will run approximately $100,000/year. The library is projected to be in operation for 20 years, with a residual value of 40% of first cost for the physical facilities. There is no salvage value for the equipment. Interest is 8%. An estimated 150,000 persons will visit the library each year, and they will receive, on average, an additional $1.50/person in benefits per visit due to the new facilities and equipment. Should the city vote to perform the expansion? Use a $B - C$ measure of worth. (7.5)

12. A metropolitan zoo is to be enlarged, and the initial cost of the enlargement for physical facilities will be $2,000,000. Animals for the addition will cost another $500,000. Maintenance, food, and animal care will run $300,000/year. The zoo is expected to be in operation for an indefinite period; however, a study period of 20 years is to be assumed, with a residual (salvage) value of 50% for all physical facilities. Interest is 10%. An estimated 400,000 persons will visit the zoo each year. How much additional benefit per person, on average, must the visitors perceive per visit to justify the new area being built? Use a $B - C$ measure of worth. (7.5)

13. For the zoo in problem 12, suppose the City Council decides to increase the price of admission by the average additional benefit per person determined. Their reasoning is that the zoo enlargement could then pay for itself due to additional income from attendees. What, if anything, is the flaw to this logic? Be specific. (7.6)

14. Reconsider problem 12 and assume that half of the 400,000 persons will derive an additional $2.50 enjoyment per person, while the other half anticipate only $1.25 in benefits. With this information available, and assuming that an incremental $1.25 entrance fee/person will be charged separately for attendance to the addition, do you recommend building the addition? (7.6)

15. A proposed expressway is under study and is found to have an unfortunately high annual cost of $2,200,000 as compared to benefits of only $1,300,000. However, with a change in design increasing the annual cost by $300,000, the highway may be eligible for incorporation into the interstate highway system, in which case 90% of the cost would be paid by the federal government. If so, it is argued that the expressway would cost only $250,000/year, yielding a handsome B/C ratio of 5.2. Is this reasoning sound? Discuss the reasoning from different points of view. (7.6)

16. Seven projects are available as summarized below. It is desired to select only the projects that government can afford on an available budget of $127,500 for first cost. Operating and maintenance costs are no worry. All have a favorable B/C ratio at the cost of capital, $i = 6\%$. Raise the interest rate until only those projects remain that continue to have a $B/C > 1$ and that government can afford. Use a 10-year planning horizon. Which projects are selected? What is the opportunity cost of those investments foregone by government? (7.6)

Project	First Cost	Operating and Maintenance	Residual Value After 10 Years	Benefits/Year
A	$51,000	$7,140	$27,200	$14,620
B	58,000	1,020	13,600	9,180
C	43,000	8,670	5,100	15,270
D	56,000	5,100	27,200	24,480
E	46,000	2,550	13,600	12,970
F	70,000	4,250	20,400	18,360
G	65,000	2,720	24,650	14,380

17. Three projects, each having a first cost of $1.4 million and annual operating costs of $140,000 are proposed. The lives of Options I, II, and III are 20, 30, and 40 years, respectively, after which the project will be over, providing no benefits and requiring no costs. The annual benefits provided over the lives of Options I, II, and III will be $375,000, $350,000, and $340,000, respectively.
   a. Select only one project to be implemented. Use a present worth basis and a $B - C$ measure of worth. Use a $MARR$ of 5%, and a study period of 40 years. Projects A and B will be void during the last 20 and 10 years of the study period, respectively.
   b. Now, use another method to select only one project to be implemented. Continually increase the interest rate, again using a study period of 40 years, until only one project remains with a favorable $B - C$ value.
   c. Do the two methods point to the same project? Why or why not? (7.6)
18. Many benefits resulting from public sector projects may be argued to be at the expense of someone else, thus counterbalancing the supposed direct benefits. For example, in Example 7.1, operating costs saved include the cost of fuel, oil, tires, wear and tear, and the like. This is money that will *not* be spent at service stations, tire stores, and new car dealers. The safety costs saved include fees that otherwise would go to lawyers, hospitals, doctors, auto repair shops, and so forth. Do you think the fact that the direct savings represent lost revenues to others should void or nullify these types of benefits in a benefit-cost analysis? Why or why not? (7.6)
19. Comment on the following analysis if you think it is incorrect. Benefits of $4 each now received by 15,000 persons and benefits are perceived to be $7 each by another 10,000 users. The annual cost of the recreational facility is $150,000. Commissioners have argued that an entrance fee of $5 should be charged to make the facility self-supporting. They argue that $5 is below the average benefit received per person. They do point out, though, that 9000 persons do not perceive but a $4 recreational value, and hence a $1 disbenefit per person should be noted. The B/C ratio is then

$$B/C = \frac{\$7(10,000) + \$4(15,000) - \$1(15,000)}{150,000 - \$5(25,000)} = \frac{115,000}{5,000} = 23.0$$

and the user fee of $5 should definitely be implemented immediately. Identify anything wrong with this analysis. Then, prepare a correct analysis using the B/C measures of worth. (7.6)
20. Reconsider problem 19. Opponents recognize that with a $5 entrance fee as in problem 19, only 10,000 persons will utilize the facility, and the benefits derived by 15,000 more persons at $4 each will be lost. They argue that the *true* benefits are those received

only by 10,000 persons at $7 each, less the $5 fee, less a disbenefit of $4(15,000) for the persons who chose not to pay $5 to enter and lose the previous recreational enjoyment from the facility. As such, the B/C ratio is as follows:

$$B/C = \frac{\$7(10,000) - \$5(10,000) - \$4(15,000)}{150,000 - \$5(10,000)} = -.38$$

Do you agree with this analysis? Why or why not? (7.6)

21. Reconsider problem 19. Prepare correct analyses using the $B - C$ measure of worth. (7.6)

22. An area near Philmont Scout Ranch in New Mexico suitable for camping, picnicking, hiking, and water sports is to be developed. It will have an annual equivalent cost of $150,000, including initial cost of construction (clearing, water, restrooms, road, etc.), operation, upkeep, and security. An average of 50 families will camp each night throughout the year. In addition, another 100 persons will be admitted for day use of the recreational facilities. The perceived benefits provided to overnight and day users will vary, due to the subjective nature of people. However, the average family camping at the area is estimated to be willing to pay $6 for the privilege, and the average day user $2. What is the B/C ratio of this recreational area? $B - C$? Should it be built? (7.5)

23. In problem 22, assume that 60% of the potential camping families perceive $6 in benefits, 25% perceive $9 in benefits, and 15% expect to receive $12 worth of recreation. If a charge of $10 is imposed on campers only, and day users are permitted free entrance, recalculate B/C and $B - C$. Should the facility be built? (7.6)

24. A government has the following estimates:

Annual benefits	$2,650,000
Annual disbenefits	1,960,000
Annual costs	1,760,000
Annual savings	1,730,000

a. Calculate the B/C ratio.
b. Calculate $B - C$.
c. Mistakenly treating disbenefits as costs and savings as benefits, calculate the B/C ratio.
d. Mistakenly treating disbenefits as costs and savings as benefits, calculate $B - C$.
e. Which do you prefer, $B - C$ or B/C? Why? (7.6)

25. Six subsystem designs have been proposed for a critical part of a communications network. Each will be "burned in" to eliminate the infant mortality or early failure problem when the subsystem is placed in service. Hence, reliability of the subsystem will have a constant hazard rate. That is, the failure density function will be exponential and reliability may be expressed as

$$R(t) = e^{-\lambda t}$$

where t is the mission time. Mission time for the subsystem will be one year, as a thorough annual preventative maintenance and recalibration procedure is standard, returning the subsystem to "new" condition. The unit will be on call 24 hours/day, seven days/week. The design data are as follows:

System	Annual Equivalent Initial Cost	Annual Maintenance Cost	λ (per hour)
1	$25,000	$5,000	0.8585165×10^{-5}
2	27,000	4,250	0.8241758×10^{-5}
3	28,000	5,250	0.8818680×10^{-5}
4	30,000	4,000	0.8025000×10^{-5}
5	30,000	4,500	0.8109375×10^{-5}
6	35,250	3,000	0.7485000×10^{-5}

Using a cost-effectiveness analysis, plot reliability versus annual cost. Which subsystems can be immediately eliminated from consideration? How might you decide among the rest? (7.7)

26. Four designs are under consideration for a space vehicle designed to carry a payload expressed in Standard Units. It is estimated that each will have equal lives; however, the first cost, operating cost, and renovation cost, as well as the salvage value, will differ, as will the size of the payload that may be carried. These characteristics are summarized below for shuttles having lives of 4 years, and fired once per year at the beginning of the year, as well as the end of the fourth year for a total of five firings.

Vehicle Design	First Cost	Operating Cost/Firing	Renovation Cost/Firing (before second and subsequent firings)	Salvage Value	Payload Capability in Standard Units
A	$ 72MM	$24.0MM	$4.2MM	$ 8.4MM	234
B	96MM	19.2MM	7.2MM	9.6MM	202
C	102MM	21.6MM	5.7MM	10.2MM	182
D	120MM	16.8MM	3.6MM	12MM	228

Based on $i = 10\%$, and using a cost-effectiveness analysis, which designs may be immediately eliminated, and which should be considered further? (7.7)

27. An electric public utility company proposes to install new meter rework area in an existing building. The rework consists of completely refurbishing used meters that have been removed from premises for inaccuracies or failure. Equipment for the new area will cost $650,000 and be required for an estimated 10 years. Depreciation for tax purposes will follow a MACRS-GDS 7-year recovery period. Depreciation for book purposes will use the straight-line method (equal depreciation deduction) over each of the 10 years. Borrowed money constitutes 45% of all capital and costs 8.5%. The remainder is equity capital and the Corporation Commission permits the utility to earn 13% on ownership (equity) capital. The effective tax rate is 35%. There is expected to be no salvage value at the end of the 10 year period. Determine:
   a. The revenue requirements during each of the 10 years.
   b. The annual worth of the revenue requirements using the after-tax weighted cost of capital as the appropriate discount factor (interest rate). (7.8)

28. A gas utility must provide a regulation and metering unit to a new subdivision. Unit A is available for $120,000 and will require operating and maintenance costs of $8,000 per year. Unit B is available for $90,000 but it will require O&M costs of $12,000 per year. The useful life of the equipment is 25 years with no salvage value after that time. The after-tax cost of capital is 11.0% and the cost of borrowed money is 8.5%. The utility uses 45% borrowed funds. Tax depreciation will follow a MACRS-GDS 20-year recovery period, but book depreciation will follow the straight-line method over the span of 25 years. The effective tax rate is 35%.

a. What is the revenue requirement in the fifth year for each alternative?

b. What is the present worth of each alternative's revenue requirements over the entire 25 years? (7.8)

Chapter 8

Cost Concepts

8.1 INTRODUCTION

Engineering economic analysis is primarily concerned with comparing alternative projects on the basis of an economic measure of effectiveness. The comparison process utilizes a variety of cost terminologies and cost concepts. This chapter provides an overview of terminology and concepts that are applicable to economic measures of effectiveness for comparing alternative projects. To implement the discussion of cost terminology, a typical production situation will now be described.

Let us assume that the business of a small manufacturing firm is job-shop machining. That is, the firm produces a variety of products and component parts according to customer order. Any given order may be for quantities of as few as five parts or as many as several hundred parts. The firm has periodically received orders to manufacture a part, which we will identify as Part No. 163H, for the B&K Corporation. The part has been manufactured in a four-stage production sequence consisting of (1) sawing bar stock to length, (2) machining on an engine lathe, (3) machining on an upright drill press, and (4) packaging. The unit cost to produce Part No. 163H by this sequence has been $25, where the unit cost consists of the major cost elements of direct labor, direct materials, and overhead (prorated costs for insurance, taxes, electric power, marketing expenses, etc.). The firm is now in the process of negotiations with the B&K Corporation to obtain a contract for producing 10,000 of these parts over a period of 4 years, or an average of 2500 units/year. A contract for this volume of parts is highly desirable but, in order to obtain the contract, the firm must lower the unit cost.

An engineer for the firm has been assigned to determine production methods to lower the unit cost. After study, the engineer recommends the purchase of a small turret lathe. With the turret lathe, the processing sequence of Part No. 163H would consist essentially of (1) machining bar stock on the turret lathe and (2) packaging. The estimated unit cost for Part No. 163H by this production method would be $15. Furthermore, the production rate by the new method would be increased over the old method because the turret lathe would replace the sawing, engine lathe, and drill press operations.

If the turret lathe is purchased, the saw, engine lathe, and drill press would not be sold but would be kept for other jobs for which the firm may receive orders. The turret lathe would be reserved for the production of Part No. 163H, but about 25% excess production capacity could be devoted to other jobs.

The incremental investment required to purchase the turret lathe and the new tooling required, as well as installing the machine, is $50,000. The physical life of the turret lathe is judged to be about 25 years, but assume that federal tax laws permit the investment capital to be recovered through annual depreciation charges in 5 years. At the end of 5 years, the firm estimates the market, or salvage, value of the turret lathe would be $25,000. If the maximum unit price that the B&K Corporation will pay for Part No. 163H is $22, should the firm accept the contract for 10,000 parts and then purchase the turret lathe in order to execute the contract?

This particular question will not be answered here or in another chapter of the book. Rather, the example situation has been cited to illustrate one type of decision with which this book is concerned. The reader can also appreciate that considerable research and investigation is required to determine or estimate the cost figures used in the turret lathe example situation. This cost information is typically obtained from a variety of sources, such as company production records, accounting records, manufacturer's catalogs, publications from the U.S. Government Printing Office, and so forth. The engineer should therefore be familiar with cost terminology, cost factors, and cost concepts as used by different specialists if effective comparisons and intelligent recommendations are to be made.

8.2   COST TERMINOLOGY

Both cost definitions and cost concepts are included in this section and will be discussed under six categories: (1) life-cycle costs; (2) past and sunk costs; (3) future and opportunity costs; (4) direct, indirect, and overhead costs; (5) fixed and variable costs; and (6) average and marginal costs.

8.2.1   Life-Cycle Costs

The *life-cycle cost* for an item is the sum of all expenditures associated with the item during its entire service life. The term *item* should be interpreted in the general sense as a machine, a unit of equipment, a product line, a project, a building, a system, and so forth. Life-cycle costs may include engineering design and development costs, fabrication and testing costs, operating and maintenance costs, and disposal costs. Life-cycle costs may also be expressed as the summation of acquisition, operation, maintenance, and disposal costs. Thus, life-cycle cost terminology may vary from author to author, but the basic meaning of the term is clear.

This textbook is predominantly concerned with the economic justification of engineering projects, the replacement of existing projects or capital assets, and the economic comparison of alternative projects. For the purpose of these types of analyses, we will define life-cycle costs to consist of (1) first cost (or initial investment), (2) operating and maintenance costs, and (3) disposal costs. There is obviously a time period involved for the life cycle and, upon disposal, the item may have a salvage value.

The *first cost* of an item is considered the total initial investment required to get the item ready for service; such costs are usually nonrecurring during the life of the item. For the purchase of a machine tool, the first cost of the machine tool may consist of the following major elements: (1) the basic machine cost, (2) costs for training personnel, (3) shipping and installation costs, (4) initial tooling costs, and (5) supporting equipment costs. The installation costs may include, for example, the costs of preparing a foundation; vibration and noise insulation; providing heat, light, and power supply; and cost of testing. Supporting equipment costs may include computer control hardware and software, and a spare parts inventory.

For some other item, a different set of first-cost elements may be appropriate. Some projects may include *working capital* for inventories, accounts receivable, and cash for wages, materials, and so forth. In any case, it is emphasized that the first cost of an item normally involves many more cost elements than just the basic purchase price. Whether the first-cost elements are aggregated or maintained separately depends on income tax considerations and whether or not a before-tax or after-tax economic analysis is desired. Certain income tax laws and depreciation methods were presented in Chapter Six, thus further discussion on this particular point is omitted here.

Operating and *maintenance* costs are recurring costs that are necessary to operate and maintain an item during its useful life. Operating costs usually consist of labor, material, and overhead items. Depending upon the accounting system used by a firm, a wide range of cost factors may be included in the major cost classification of overhead. Typical overhead items are fuel or electric power, insurance premiums, inventory charges, indirect labor (as opposed to direct labor), administrative and management expenses, and so forth. It is usually assumed that *operating and maintenance costs* are annual costs, but maintenance costs may not be on a recurring, annual basis. That is, a regular annual schedule of minor or preventive maintenance may be followed, or it could be policy that maintenance is performed only when necessary, such as when a major overhaul is required. In most cases, the maintenance policy would consist of both preventive maintenance and maintenance on an "as needed" basis. In any case, repair and upkeep result in costs that must be recognized in the economic analysis of engineering projects.

When the life cycle of an item has ended, *disposal costs* usually result. Disposal costs may include labor and material costs for removal of the item, shipping costs, or special costs; an example of special costs being the cost of disposing of hazardous materials. Although disposal costs may be incurred at the end of the life cycle, most items have some monetary value at the time of disposal. This value is the *market* or *trade-in value* (i.e., the actual dollar worth for which the item may be sold at the time of disposal). Then, after deducting the cost of disposal, the net dollar worth at the time of disposal is termed the *salvage value*.

The market value, the disposal costs, and the salvage value are usually not known with certainty and therefore must be estimated. For an item that satisfies the United States Internal Revenue Service (IRS) definition of a capital asset and that decreases in value over time through physical deterioration, the IRS has approved depreciation methods that can serve to estimate the rate of deterioration and consequent decrease in value of the asset. The value of the capital asset at the end of a given accounting period during the asset's life is termed the *book value*. *Scrap value*, on the other hand, refers only to the value of the material of which the item is made. For example, a 4-year-old automobile may have a scrap value of $500 but a market value of $5000. A distinction between these terms is not particularly important for evaluating potential investment projects—and salvage value will be used as the general term to denote the end-of-life value. For example, a trade-in value of $3000 minus disposal costs of $500 equals a net salvage value of $2500.

The life cycle obviously involves a time horizon, and the end of an item's life may be judged from either a functional or an economic point of view. The economic life of an item is generally shorter than the functional life. For example, an engine lathe may remain functionally useful for 40 years or more, but, because of periodic advancements in machine design technology, newer engine lathes have higher production rates; the economically useful life of an engine lathe may be only 10 years. The economic life of an item is usually a matter of company policy that is greatly influenced by income tax considerations.

8.2.2 Past and Sunk Costs

Past costs are historical costs that have occurred for the item under consideration. *Sunk costs* are past costs that are unrecoverable. The distinction is perhaps best made through examples. Assume that an investor purchases 100 shares of common stock in the JHP Corporation through a broker at $25/share. In addition, the investor pays $85 in brokerage fees and other charges. Just 2 months later, and before receiving any dividend payments, the purchaser resells the 100 shares of common stock through the same broker at $35/share minus $105 for selling expenses. The purchaser realizes a net profit of $810 ($3500 − $2500 − $85 − $105) on these transactions. At the time of sale, the $2500 and $85 are past costs, but because these are recovered after the sales transaction, sunk costs are not incurred. If, on the other hand, the investor were to sell the 100 shares 2 months after purchase and the market price were $20/share, with a $70 charge for selling fees, the investor would incur a capital loss of $655 ($2000 − $2500 − $85 − $70). In this instance, some of the past costs would be recovered, but the $655 capital loss would be a sunk cost. If the investor reasons that the market price will decline further or if he or she simply needs the money, the $655 sunk cost should be ignored if the shares are to be sold for $20 each. However, sunk costs are not totally irrelevant to a present decision. They may qualify as capital losses and serve to offset capital gains or other taxable income and thus reduce income taxes paid. Past costs and sunk costs provide information that can improve the accuracy of estimating future costs for similar items.

Another example of sunk costs is the purchase and sale of an item of equipment. Assume the equipment is purchased for $10,000 and the salvage value at the end of 5 years of service is estimated to be $5000. For illustrative purposes, we will further assume that the annual decrease in value for the equipment through physical deterioration, or depreciation, is $1000. The $1000 annual cost of depreciation is a cost of production that, in theory, should be allocated to the output of the equipment. After allocating this and other manufacturing costs, general and administrative costs, and marketing costs to each unit of production, the total unit cost is determined. A profit is then added to each unit of production in order to arrive at the unit selling price. Thus, when a unit is sold, a portion of each sales dollar returns a portion of the depreciation expense. In this illustration, it is assumed that sales will return, or recover, the total estimated depreciation expense of $5000 (first cost minus estimated salvage value) for the 5-year period. However, if the equipment has a market value of over $2000 at the end of 5 years, there is a $3000 ($5000 − $2000) sunk cost. The $3000 capital loss represents an error in estimating the rate of depreciation, and the owner cannot insist that the equipment is worth $5000 when the market value for the 5-year-old equipment is, in fact, only $2000. If the equipment is kept, it is argued that the true value being kept is thus only $2000.

8.2.3   Future and Opportunity Costs

All costs that may occur in the future are termed *future costs*. These future costs may be operating costs for labor and materials, maintenance costs, overhaul costs, and disposal costs. In any case, by virtue of occurring in the future, these costs are rarely known with certainty and must therefore be estimated. This is also true for future revenues or savings if these are involved in a given project. Estimates of future costs or revenues are uncertain and subject to error. Thus, the economic analysis is simplified if certainty of future costs, revenues, or savings is assumed.

The cost of forgoing the opportunity to earn interest, or a return, on investment funds is termed an *opportunity cost*. This concept is best explained by means of illustrations. For example, if a person has $1000 and stores this cash in a home safe, the person is forgoing the opportunity to earn interest on the money by establishing a savings account in a local bank that pays, for example, 5% annual compound interest. (Of course, investments other than savings accounts are possible). For a 1-year period, the person is forgoing the opportunity to earn (0.05) ($1000) = $50. The $50 amount is thus termed the opportunity cost associated with storing the $1000 in the home safe.

A similar illustration of an opportunity cost is to assume that a person has $5000 cash on hand. This amount is considered *equity capital* if the $5000 was not borrowed (i.e., there is no debt obligation involved). The person has available secure investment opportunities such as establishing a personal savings account in a commercial bank or purchasing other financial instruments. From the available investment opportunities, suppose the optimum combination of risk (security level) and interest yield on the investment results in a 10% annual interest. Thus, the investment of $5000 would yield (0.10) ($5000) = $500 each year. If the person, instead of investing the $5000, purchases an automobile for the same amount for personal use, the person will forgo the opportunity to earn $5000 interest/year. The $500 amount is again termed an annual opportunity cost associated with purchasing the automobile.

The same logic applies in defining an annual opportunity cost for investments in business and engineering projects. The purchase of an item of production machinery with $20,000 of equity funds prevents this money from being invested elsewhere with greater security and/or higher profit potential. This concept of *opportunity cost* is fundamental to the study of engineering economy and is a cost element that is included in virtually all methodologies for comparing alternative projects. In Chapter Five, the concept of opportunity costs was discussed under the heading of *minimum attractive rate of return (MARR)*.

Some individuals define *MARR* as the *cost of capital*. As used in this text, the term *cost of capital* refers to the cost of obtaining funds for financing projects through debt obligations. These funds are usually obtained from external sources by (1) borrowing money from banks or other financial organizations (e.g., insurance companies and pension funds) and (2) issuing bonds. These debt obligations are normally long-term, as opposed to short-term, obligations for the purchase of supplies and raw materials. The debt obligations result in interest payments on, say, a monthly, quarterly, semianual, or annual basis. The interest payments are thus a cost of borrowed capital. Financing projects through issuing bonds is a method of obtaining capital funds that is probably less known to the reader than borrowing money from a bank at a stated interest rate. Some elaboration of bonds is therefore appropriate.

Bonds are issued by various organizational units—partnerships, corporations (profit or nonprofit), governmental units (municipal, state, and federal), or other legal entities. The sale of bonds represents a legal debt of the issuing organization; bonds are generally secured by its assets. Examples are mortgage bonds or collateral bonds. Debenture bonds, on the other hand, are promissory notes or just a promise to pay. In any case, the purchase of a bond has legal claim to the assets of the issuing unit but has no ownership privileges in the issuing unit. Purchasers of the common or preferred stock of an organizational unit do hold ownership status but may or may not have voting privileges, depending on the stipulations of the particular stock issue. In the sense that bonds are debt obligations and not ownership shares, bonds are considered a more secure investment than either common or preferred stock. This statement should not be taken as a universal truth, however, since the security level for a bond or a stock depends on many factors, economic and otherwise; the principal factor is the financial soundness of the issuing unit. Further details on interest payments on bank loans and interest payments on issued bonds was presented in Chapter Three.

Another method of financing engineering projects is through the use of *equity funds*. Equity funds are generally obtained from one or both of the following sources: (1) common or preferred stock authorized by the company and sold through brokers to investors, or (2) earnings accumulated from prior years and retain by the company. Both of these sources of funds incur an opportunity cost. Chapter Five presented details on calculating these costs. Additional information related a company's equity funding is presented later in this chapter under the heading "General Accounting Principles."

8.2.4   Direct, Indirect, and Overhead Costs

It will be helpful to provide definitions of direct, indirect, and overhead costs in the context of a manufacturing environment. A typical cost structure for manufacturing, adapted from Ostwald, [17] is provided in Figure 8.1.

The *cost of goods sold*, as shown in Figure 8.1, is the total cost of manufacturing

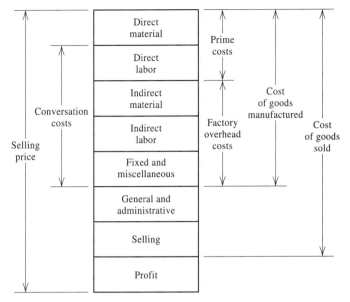

Figure 8.1 A cost structure for manufacturing.

a product. An amount of profit is then added to this total cost to arrive at a selling price. Such a cost structure is helpful in arriving at a unit cost, which is a primary objective of cost accounting. The term *cost of goods sold*, as used here, has a different meaning from the term *cost of goods sold* used in general accounting practice, particularly for retail businesses. As will be mentioned later in the section on accounting principles, general accounting defines the cost of goods sold to be beginning-of-the-year inventory plus purchases minus end-of-the-year inventory. Different meanings for the same terminology are unfortunate, but they do occur in the literature on accounting, and the reader is cautioned on this point. To simplify the treatment of the total cost of goods sold (as defined by Figure 8.1), the major cost elements can be defined as direct material, direct labor, and overhead costs.

Direct material and labor costs are the costs of material and labor that are easily measured and can be conveniently allocated to a specific operation, product, or project.

Indirect costs for both labor and material, on the other hand, are either very difficult or impossible to assign directly to a specific operation, product, or project. The expense of directly assigning such costs is prohibitive, and actual direct costs are therefore considered to be indirect for accounting purposes.

As an example of these different cost elements, suppose the raw material for a given part is a rectangular gray iron casting. The casting is milled on five sides, the unmachined surface is painted and air-dried, then four through holes are drilled and tapped. The finished parts are stacked in wooden boxes, 30/box, and are delivered to a customer.

In this example, the direct labor required per part to machine, paint, and package is probably readily determined. The labor required to receive the raw materials, handle parts between work stations, load boxes onto a truck, and deliver material to the customer is less easily identified and assigned to each part. This labor would be classified as indirect labor, especially if the labor in receiving, handling, shipping, and delivery is responsible for dealing with many different parts during the normal workday. The unit purchase price of the gray iron casting is an identifiable direct material cost. The cost of paint used per part may or may not be easily determined; if it is not, it is an example of indirect material cost. Also, any lubricating oils used during the machining processes would be an indirect material cost, not readily assigned on a cost per part basis.

Overhead costs consist of all costs of manufacturing other than direct material and direct labor. A given firm may identify different overhead categories such as factory overhead, general and administrative overhead, and marketing expenses. Furthermore, overhead amounts may be allocated to a total plant, departments within a plant, or even to a given item of equipment. Typical specific items of cost included in the general category of overhead are: indirect materials, indirect labor, taxes, insurance premiums, rent, maintenance and repairs, supervisory and administrative (technical, sales, and management) personnel, and utilities (water, electric power, etc.). Depreciation expenses are also usually included in the general overhead, but may occasionally be considered a part of direct costs. It is the task of cost accounting to assign a proportionate amount of these costs to various products manufactured or to services provided by a business organization.

8.2.5   Fixed and Variable Costs

Fixed costs do not vary in proportion to the quantity of output. General administrative expenses, taxes and insurance, rent, building and equipment depreciation, and utilities are examples of cost items that are usually invariant with production volume and hence are termed *fixed costs*. Such costs may be fixed only over a given range of production; they may then change and be fixed for another range of production. *Variable* costs vary in proportion to quantity of output. These costs are usually for direct material and direct labor.

Many cost items have both fixed and variable components. For example, a plant maintenance department may have a constant number of maintenance personnel at fixed salaries over a wide range of production output. However, the amount of maintenance work done and replacement parts required on equipment may vary in proportion to production output. Thus, total annual maintenance costs for a plant over several years would consist of both fixed and variable components. Indirect labor, equipment depreciation, and electrical power are other cost items that may consist of fixed and variable components. Determining the fixed and variable portion of such a cost item may not be possible; if it is possible, the expense of establishing detailed measurement techniques and accounting records may be prohibitive. A comprehensive discussion on this issue is outside the scope of this book, and the reader is referred to books on general cost accounting for further reading.

Certain total costs (TC) can be expressed as the sum of fixed (FC) and variable costs (VC). As an example, the total annual cost for operating a personal automobile for a given year might be expressed as

$$TC(x) = FC + VC(x) \qquad \textbf{(8.1)}$$

where x = miles per year. Costs for insurance, license tags, depreciation, certain maintenance, and interest on borrowed money if the automobile were financed are essentially fixed costs, independent of the miles traveled per year. Expenses for gasoline, oil, tire replacements, and certain maintenance are proportional to, or functional with, the mileage per year. One could argue, however, that depreciation expenses are comprised of both fixed and variable components. Arbitrarily assigning numerical values to the total cost function, assume that

$$TC(x) = \$950 + \$0.15x$$

is a valid relationship for a given year in question (the expression is restricted to a given year, since actual depreciation expenses, and hence the fixed expenses, may vary from year to year). This relationship is linear in terms of x, however, the variable cost component is often a nonlinear function. Figure 8.2 graphically illustrates the total cost function.

Now let us consider Figure 8.2 as a total cost function for a production line in a manufacturing firm where the output from the line is a single product. Furthermore, let it be assumed that each unit of production can be sold for $\$R$ and that the total revenue (TR) is a linear function of the production quantity:

$$TR(x) = \$Rx \qquad \textbf{(8.2)}$$

Adding this functional relationship to Figure 8.2 and modifying the terminology for this example yields Figure 8.3.

It is noted from Figure 8.3 that the total annual revenue equals the total annual cost at point A or at an annual production volume of x^* units. Thus, at x^*,

$$
\begin{aligned}
TR(x^*) &= TC(x^*) \qquad \textbf{(8.3)}\\
&= FC + VC(x^*)
\end{aligned}
$$

and x^* is termed the annual production volume required in order to *break even*. Certain important observations can now be made. If the production volume is less than x^*, an annual net loss will occur, the amount of which is equal to $TC(x) - TR(x)$, evaluated for a particular value of x. By the same token, if the production volume is greater than x^*, then an annual net revenue or profit will

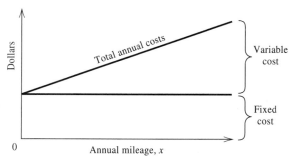

Figure 8.2 Total annual costs as a function of annual mileage.

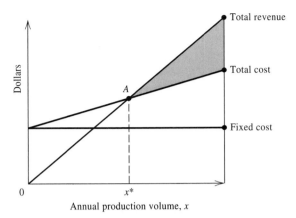

Figure 8.3 Relationship between total revenue and total cost as a function of annual production volume.

result (the shaded region of Figure 8.3). The amount of annual profit is equal to $TR(x) - TC(x)$, evaluated for a particular value of x.

It is generally desirable to have a "low" break-even value. For the general example of Figure 8.3, this can be accomplished in three independent ways: (1) increasing the slope of the total revenue line, (2) decreasing the slope of the variable cost line, and (3) decreasing the magnitude of the fixed cost line. Increasing the slope of the total revenue line means increasing the selling price of the product, which may be a poor marketing strategy in a competitive market environment where sales would be lost. Fixed costs, although not literally "fixed" in all cases, are difficult to reduce. Thus, reducing variable costs for direct material and labor usually offers the greatest opportunity to the engineer or analyst for profit improvement.

The concept of break-even analysis is general. Assuming that a breakeven point exists, then, for two relationships $y = g(\cdot)$ and $y = h(\cdot)$ that are functions of a single variable x, the value of x for break-even, say x^*, may be determined from equating $g(x) = h(x)$ and solving for x^*. The concept can be extended to more than two functions of a single variable, say $y = h(x)$, $y = g(x)$, and $y = t(x)$. If these are all linear functions, then Figure 8.4 depicts two of the possible results.

In Figure 8.4b, there is no unique break-even value of x involving all three functional relationships. The linear equations $y = h(x)$ and $y = t(x)$ intersect at point B or $x = x_1^*$, which is then the break-even value for these two relationships. Point C, or $x = x_2^*$, is the break-even value for $y = h(x)$ and $y = g(x)$. Point D, or $x = x_3^*$, is the break-even value for $y = g(x)$ and $y = t(x)$.

The concept of break-even analysis also extends to nonlinear functions, with one or more break-even values, and functions of more than a single variable, which may be of linear or nonlinear form. However, examples and problems dealing only with functions of a single variable will be presented in this chapter.

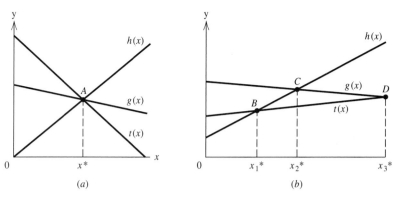

Figure 8.4 Graphical representation of break-even point.
(*a*) A single break-even value. (*b*) Three break-even values.

EXAMPLE 8.1

The cost of tooling and direct labor required to set up for a machining job on a turret lathe is $300. Once set up, the variable cost to produce one finished unit consists of $2.50 for material and $1 for labor to operate the lathe. For simplicity, it is assumed these are the only relevant fixed and variable costs. If each finished unit can be sold for $5, determine the production quantity required to break even and the net profit (or loss) if the lot size is 1000 units. Letting x = the production volume in units, then

$$TR(x) = TC(x) \; FC + VC(x)$$

and

$$\$5x = \$300 + (\$2.50 + \$1.00)x$$

Solving yields $x = x^*$ (the break-even value), or

$$x^* = \$300/\$1.50 = 200 \text{ units}$$

For a production output of 1000 units, the net profit, P, is calculated to be

$$P = \$5(1000 \text{ units}) - (\$2.50 + \$1.00)(1000 \text{ units}) - \$300$$
$$= \$1200$$

EXAMPLE 8.2

This example concerns the decision of selecting between two alternative methods of processing crude oil in a producing oil field, where the basis for the decision is the number of barrels of crude oil processed per year. The two methods of processing the crude oil are (1) a manually operated tank battery or (2) an automated tank battery. The tank batteries consist of heaters, treaters, storage tanks, and so forth, that remove salt water and sediment from crude oil prior to its entrance to pipelines for transport to an oil refinery.[1]

For each alternative, fixed costs and variable costs are involved. Fixed costs include items such as pumper labor, maintenance (fixed over the production quantity of interest), taxes, certain energy costs (power to operate control panels and motors in continuous operation), and, for manual tank batteries, a cost for oil "shrinkage." Variable costs for chemical additives, heating, and noncontinuous operating motors are proportionate with the volume of oil being processed. This relationship is assumed to be linear over the production quantity of interest, since the data given in Tables 8.1 and 8.2 are based on a production quantity of 500 barrels/day.

In addition to the fixed and variable costs given in Table 8.1 for the automatic tank battery operation, other annual fixed costs are

$$D_1 = \text{annual cost of depreciation and interest}$$
$$= \$3082$$

$$M_1 = \text{annual cost of maintenance, taxes, and labor}$$
$$= \$5485$$

Letting $x = $ the number of barrels of oil processed per year, the total annual cost, $TC_1(x)$, for the automatic tank battery operations is given by

$$TC_1(x) = FC_1 + VC_1(x)$$
$$= (\$982 + \$3082 + \$5485)$$
$$+ \$0.01136x$$
$$= \$9549 + \$0.01136x$$

In addition to the fixed and variable costs given in Table 8.2 for the manual tank battery operations, other annual fixed costs are

$$D_2 = \text{annual cost of depreciation and interest}$$
$$= \$2017$$

$$M_2 = \text{annual cost of maintenance, taxes, and labor}$$
$$= \$7921$$

Then, the total annual cost, $TC_2(x)$, for the manual tank battery operation is given by

$$TC_2(x) = FC_2 + VC_2(x)$$
$$= (\$358 + \$2017 + \$7921)$$
$$+ \$0.00810x$$
$$= \$10,296 + \$0.00810x$$

By equating the two total cost functions, the break-even production volume can be determined as

$$TC_1(x) = TC_2(x)$$
$$\$9549 + \$0.01136x = \$10,296 + \$0.00810x$$
$$x^* = 229,141 \text{ barrels/year}$$

---

[1] The example is taken, with slight modification, from Ferguson and Shamblin [10] by permission of the publisher.

Table 8.1 Cost Data for Automatic Tank Battery Operations

Fixed cost	
Control panel power	$0.15/day
Circulating pump power (3 horsepower)	0.82/day
Maintenance	1.00/day
Meter calibration	0.40/day
Chemical pump power (1/4 horsepower)	0.32/day
Total	$2.69/day
or $982/year ($2.69/day × 365 days)	
Variable cost	
Pipeline pump (5 horsepower @ 50% utilization)	$0.63/day
Chemical additives (7.5 quarts/day)	3.75/day
Inhibitor (2 quarts/day)	1.00/day
Gas (10.8 MCF/day × $0.0275/MCF)	0.30/day
Total	$5.68/day
or $0.01136/barrel based on 500 barrels/day	

The interpretation of the break-even point in this example is that $x^* = 229{,}141$ barrels/year is the *point of indifference* between the choice of the two alternatives. If production volume is less than x^*, then the first alternative, or the automatic tank battery operation, would be preferred. For instance, if $x = 200{,}000$ barrels/year, then $TC_1(x) = \$11{,}821$ and $TC_2(x) = \$11{,}916$. Similarly, if production volume is greater than x^*, the manual tank battery operation would be preferred.

8.2.6   Average and Marginal Cost

The *average cost* of one unit of output (unit cost) is the ratio of total cost and quantity of output (miles traveled, production volume, etc.). That is,

$$AC(x) = \frac{TC(x)}{x} \qquad\qquad (8.4)$$

Table 8.2 Cost Data for Manual Tank Battery Operation

Fixed cost	
Chemical pump power	$0.16/day
Circulating pump power	0.82/day
Total	$0.98/day
or $358/year ($0.98/day × 365 days)	
Variable cost	
Chemical additives (7.5 quarts/day)	$375/day
Gas	0.30/day
Total	$4.05/day
or $0.00810/barrel based on 500 barrels/day	

where $AC(x)$ = average cost per unit of x
$TC(x)$ = total cost for x units of output
x = output quantity

The average cost is usually a variable function of the output quantity, and normally decreases with an increasing quantity of output. Using the automobile example from Section 8.2.5, which had a total cost function of $\$(950 + 0.15x)$, the average cost, in dollars per mile, is given by

$$AC(x) = \frac{950 + 0.15x}{x}$$

$$= \frac{950}{x} + 0.15$$

If the automobile travels 10,000 miles/year, then the average operating cost is $\$(950/10000 + 0.15) = \$0.245$/mile. For a total annual travel distance of 20,000 miles, the unit operating cost decreases to $\$0.1975$/mile.

It should be noted from the average cost function of this example, $\$(950/x + 0.15)$, that as the output quantity x increases, the proportion of the fixed cost allocated to each unit of output decreases. This relationship is a fundamental economic principle often referred to as the *economies of scale,* a principle underlying the economic benefits of mass production. Such a relationship assumes, however, that the variable cost coefficient remains constant over the range of the output variable x. In a production environment, it is very likely that the variable cost coefficient will increase as the production volume increases (due to increased maintenance expenses, defective product, etc.).

For a total cost function that is continuous in the output (or independent) variable x, marginal cost is defined as the derivative of the total cost function (dependent variable) with respect to x, or $dTC(x)/dx$. This is true for continuous functions that are linear or nonlinear in the output variable x. In the special case of a continuous total cost function that is linear in x, such as $TC(x) = (\$950 + 0.15x)$, then $dTC(x)/dx = \$0.15$. In this case, marginal cost is the constant value $\$0.15$ and is the cost required to increase the output quantity x by one unit.

If the total cost function is discontinuous and defined only for discrete values of x (for example, $x = 1, 2, 3, \ldots$), then difference equations must be used to determine marginal costs. For example, $TC(6) - TC(5)$ is the marginal cost of increasing the output quantity from $x = 5$ to $x = 6$. Thus, in the discrete case, marginal cost is always the cost required to increase the output quantity x by one unit at a specified level of output. This is true for a discrete total cost function which has either a linear or nonlinear trend.

EXAMPLE 8.3

For a certain production process fixed costs are $\$60,000$. Variable costs are $\$30$ per unit of production. Therefore, the total cost function is given by $TC(x) = 60000 + 30x$. What is the marginal cost at $x = 10$? $x = 20$?

There are two ways to solve this problem: (1) difference equations, and (2) differentiation. In general, since the total cost function is continuous, we would use the differentiation approach. However, for purposes of illustration, both will be demonstrated here.

Difference Equations
Marginal cost at 10 = total cost at 11 − total cost at 10
$MC(10) = TC(11) − TC(10)$
$MC(10) = [60000 + 30(11)] − [60000 + 30(10)]$
$MC(10) = 60330 − 60300$
$MC(10) = 30.$
Similarly,
$MC(20) = TC(21) − TC(20)$
$MC(20) = [60000 + 30(21)] − [60000 + 30(20)]$
$MC(20) = 60630 − 60600$
$MC(20) = 30.$

Differentiation
Marginal cost at 10 = first derivative of the total cost function evaluated at 10
$MC(x) = d/dx\ TC(x)$
$MC(x) = d/dx\ [60000 + 30x]$
$MC(x) = 30$
therefore
$MC(10) = 30$ and $MC(20) = 30.$

The concept of marginalism is general and applies to other mathematical functions as well. For example, marginal revenues can be determined from total revenue functions, marginal profit values can be determined from total profit functions, and so forth. If these functions are defined for discrete value of x or are continuous functions that are linear in x, then marginal revenue (profit) is the additional revenue (profit) received from selling one more unit of the output quantity x, at a specified level of output.

Marginal and average values corresponding to a specified output quantity are generally different. If the marginal cost is smaller than the average cost, an increase in output will result in a reduction of unit cost. This can be seen by recalling the familiar automobile problem, when $TC(x) = \$950 + \$0.15x$. The average cost is $AC(x) = \$950/x + \$0.15$ and the marginal cost is $MC(x) = \$0.15$. Thus, for all nonnegative finite values of x, marginal cost is always smaller than the average cost, and the unit cost will continue to decrease as x is increased. Such a relationship is not true in general for nonlinear total cost functions. Tables 8.3 and 8.4 summarize the relationships between marginal cost and total cost

Table 8.3 The Relationship Between
Marginal Cost and Total Cost

If $MC(x) > 0$ then $TC(x + 1) > TC(x)$
If $MC(x) = 0$ then $TC(x + 1) = TC(x)$
If $MC(x) < 0$ then $TC(x + 1) < TC(x)$

Table 8.4 The Relationship Between
Marginal Cost and Average Cost (AC)

If $MC(x) < AC(x)$ then $AC(x + 1) < AC(x)$
If $MC(x) = AC(x)$ then $AC(x + 1) = AC(x)$
If $MC(x) > AC(x)$ then $AC(x + 1) > AC(x)$

(Table 8.3) and marginal cost and average cost (Table 8.4). The following example
will illustrate some of the cost, revenue, and profit relationships.

EXAMPLE 8.4

A small firm blends and bags chemicals, primarily for home gardening purposes.
The market area for the firm is local, and all sales are to wholesale distributors.
For one pesticide dust product, sales and production cost records over the past
10 seasons have been reviewed and analyzed. The following equations *approximate* the relationships among selling price, sales volume, production costs, and
profit before income taxes. (The functional form for selling price implicitly reflects
a fundamental relationship between price and demand. Namely, as the selling
price is decreased, demand for the item increases. Alternatively, in order to
increase the demand, the selling price must be reduced).

Let t = number of tons per season

$SP(t)$ = selling price in order to sell t tons
$= \$(800 - 0.8t)$

$TR(t)$ = total revenue when t tons are sold at a particular selling price
$=$ selling price $\times$ demand
$= \$(800 - 0.8t)t$
$= \$(800t - 0.8t^2)$

$MR(t)$ = the marginal revenue at a sales volume of t tons
$= \dfrac{dTR(t)}{dt} = \$(800 - 1.6t)$

$TC(t)$ = the total production cost for t tons
$= \$(10,000 + 400t)$

$TP(t)$ = total profit when t tons are sold
$= TR(t) - TC(t)$
$= \$(800t - 0.8t^2) - \$(10,000 + 400t)$
$= \$(-0.8t^2 + 400t - 10,000)$

$AP(t)$ = average profit per ton when t tons are sold
$= TP(t)/t$
$= \$(-0.8t + 400 - 10,000/t)$

The equations apply for the range, $0 \le t < 1000$.

We first note that the total revenue will be maximized when 500 tons are produced and sold per season. That is, by calculus,

$$\frac{dTR(t)}{dt} = 800 - 2(0.8)t = 0 \qquad \text{and} \qquad t = 500 \text{ tons}$$

The total revenue with a sales volume of 500 tons is

$$TR(500) = \$800(500) - \$0.8(500)^2$$
$$= \$200,000$$

The marginal revenue at the output level of 500 tons is

$$MR(500) = \$800 - \$1.6(500) = 0$$

Thus, the rate of change in the $TR(t)$ function with respect to t is zero when $TR(t)$ is evaluated at $t = 500$. The $TR(t)$ function is a strictly concave function with a unique maximum value at $TR(500)$. For sales from $t = 1$ to 500, the total revenue function is increasing at a decreasing rate. For sales volumes from $t = 500$ to 1000, total revenues are decreasing at an increasing rate.

Maximizing total revenues is not the issue in this example, however. We wish to maximize profits. Again, by calculus,

$$\frac{dTP(t)}{dt} = 2(-0.8)t + 400 = 0 \qquad \text{and} \qquad t = 250 \text{ tons}$$

Thus, total profit will be maximized for a sales volume of 250 tons, and the maximum profit per season would be

$$TP(250) = -\$0.8(250)^2 + \$400(250) - \$10,000$$
$$= \$40,000$$

The average profit per ton when 250 tons are sold is

$$AP(250) = -0.8(250) + 400 - 10,000/250$$
$$= \$160/\text{ton}$$

Finally, we note that there are two break-even points in this example. By equating $TR(t) = TC(t)$, or

$$800t - 0.8t^2 = 10,000 + 400t$$

we obtain

$$-0.8t^2 + 400t - 10,000 = 0$$

Solving for the positive roots of this quadratic equation yields

$$t = 26.39, \ 473.61$$

For a sales volume in the range, $26.39 \leq t \leq 473.61$, the firm will make a profit. Sales volumes outside this range will result in total costs exceeding total revenues and a net loss to this firm.

8.3 ESTIMATION

The estimation of future events or the outcomes of present actions is obviously a fact of life for every individual, group, and organization. Family budgeting, weather forecasts, market forecasts of demand for consumer products, and predicting the annual national revenues from income taxation are only a few examples of the almost limitless number and variety of estimates that are made in our personal and business lives. In this textbook, we are concerned with estimation in the specific context of factors relevant to comparing alternative engineering/ investment projects and making a selection from these projects. The annual revenues or savings, the initial and annual recurring costs, the life of a project, and the future salvage value of capital assets such as buildings and equipment that may be associated with a given project are rarely, if ever, known with certainty.

The National Estimating Society defines cost estimating as "the art of approximating the probable worth or cost of an activity based on information available at the time."[2] Webster's New Collegiate Dictionary goes on to say that "estimate, the comprehensive term, implies personal judgment the significance of which can only be made clear by the context." These definitions make it clear that estimating, in particular cost estimating, is not an exact science. Rather, it is an approximation that involves the availability and relevancy of appropriate historical data, personal judgments based on the experience of the estimator, and the time frame available for completing the estimating activity.

Cost estimates can be developed using two fundamentally different approaches: the top-down or parametric approach and the bottom-up or detailed approach. The parametric approach utilizes historical data from prior work and determines a new cost estimate based on the differences between the new product or project and the prior one (e.g., differences in speed, power, or size). The detailed approach requires estimates of costs for each element and subelement of a product or project and then applies appropriate pricing parameters (i.e., allowances for overhead and profit) to accumulate a total cost estimate. It is generally confirmed in practice that the detailed approach provides more accurate estimates but requires more resources and more time.

Much debate exists within the cost estimating community as to which approach is best. In actuality, both have an appropriate and preferred place in the life cycle of cost estimating. During the early stages of a decision-making process, attention typically focuses on narrowing a broad set of choices to a smaller set of attractive and viable choices which will be subjected to further study. At this stage, parametric estimating is typically employed since development of detailed estimates for a large set of choices would require considerable time and resources. Further, at this stage, detailed accuracy of the estimates is generally not needed or warranted. In the final stages of decision making, where the choices have typically been narrowed to two or three top contenders, attention typically is more detail focused. Decision makers are concerned with element and subelement definitions and costs. At this stage the detailed approach is appropriate and preferred.

---

[2] National Estimating Society By-Laws; March 1978.

The two estimating approaches highlighted above should be considered end points along a cost-estimating spectrum. As the decision-making process proceeds from initial to final, the estimating process proceeds from parametric to detailed. A simple way to implement this progressive method of estimating is to repeatedly decompose an initial parametric cost estimate into finer and finer levels of detail as the needs and conditions of the situation dictate.

Many different terms pertain to the general subject of estimation. No attempt will be made in this chapter to enumerate and explain all the terms exhaustively; selected terminology will be given as needed to explain the topics. Furthermore, an in-depth study of estimation procedures and the accuracy of estimated values is the study of mathematical statistics and probability theory, about which a vast literature exists.

It is difficult to state precisely in quantitative terms the relationship between the accuracy of an estimate and the cost of making the estimate. Intuitively, as more detailed information is obtained to provide the basis for an estimate and as more mathematical preciseness is exercised in calculating the estimate, the more accurate the estimate should be. However, as the level of detail increases, the greater the cost involved in making the estimate. Ostwald [15] has conceptualized this notion by the function

$$C_T = C(M) + C(E) \qquad\qquad (8.5)$$

where C_T = the total cost of making the estimate, dollars
  $C(M)$ = the functional cost of making the estimate, dollars
  $C(E)$ = the functional cost of errors in the estimate, dollars

As depicted in Figure 8.5, the total cost of making an estimate reaches a minimum value when the amount of detail reaches a value D_1. Quantitatively defining the amount of detail is at best difficult and may be a practical impossibility. However, this concept of the total cost of an estimate varying with the amount of detail involved in making the estimate is realistic and is important to the general subject of estimation. In the abbreviated discussion on cost estimation techniques that follows, it will be noted that the individual techniques are based

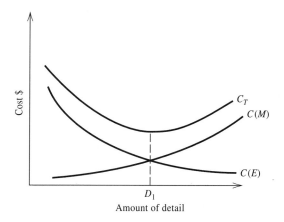

Figure 8.5 Cost of increasing detail.

on varying amounts of detail with implied differences in the cost of making the estimate.

The texts by Ostwald [14, 15] are primary references on the subject of cost estimating. He defines four categories of estimated items: operations, products, projects, and systems. The categories are based on the scope of activity involved, but the distinction among them is not sharp and unequivocal. An *operation* is considered a basic or lowest-level activity. Machining, assembly, and painting are examples of operations for cost estimation purposes. Therefore, operations are subelements of manufacturing a product or providing a service. Producing a given product or providing a given service may be a subelement of a project, and a system may consist of several projects. A system may, for example, be defined as a total manufacturing plant (or several plants), a regional health care program of facilities and personnel, or a total program of national defense.

Generally, the larger the scope of the categorical item being estimated, the more difficult it is to make accurate estimates of the costs and revenues involved. Ostwald presents a comprehensive discussion of estimating techniques for all categories. Vinton [23] provides procedural detail for estimating the cost of certain manufacturing operations and manufactured products. Stewart [18] and Stewart and Wyskida [19] provide well-founded general methodologies for cost estimating.

8.3.1   Project Estimation

For all categories of estimated items previously mentioned, three principal classes of estimates, based on accuracy and degree of detail, may be defined: (1) order-of-magnitude estimates, (2) preliminary estimates, and (3) detailed estimates. *Order-of-magnitude estimates* are usually gross estimates based on experience and judgment and made without formal examination of the details involved.

Preliminary estimates are also gross estimates, but more consideration is given to detail in making the estimate than for order-of-magnitude estimates. Certain subelements of the overall task are individually estimated, engineering specifications are considered, and so forth. Estimating the cost for manufacturing new components or products before designs and production plans are complete is an example of preliminary cost estimation. This type of estimate enables process engineering and product engineering groups to compare alternative designs or manufacturing methods and assess the economic impact of them.

Detailed estimates are expected to result in the most accurate estimate of actual cost. In preparing the estimate, each subelement of the overall task is considered, and an attempt is made to assign a realistic cost to the subelement. Pricing a product or contract bidding usually involves detailed estimates of the costs involved.

In prior chapters, examples illustrating the principles of engineering economic analysis have primarily involved the following estimated items: first cost, operating costs, maintenance costs, recurring revenues or savings, useful life, salvage value, income taxes, and cost of capital. These cost items have been discussed previously in this chapter under the section on cost terminology.

Estimates of the functionally useful physical life of an item of equipment may be obtained from manufacturers and suppliers. Alternatively, if a company repeatedly buys a particular item of equipment and keeps accurate maintenance records, these records may be used to obtain an estimate of the functional life of the item. For example, suppose records reveal that 100% of the items survive the first 3 years of service. Then, 10% of the items fail in the fourth year, 20% of the items fail in the fifth year, 50% in the sixth year, 15% in the seventh year, and the remaining 5% fail in the eighth year. A weighted-average time of failure for any equipment item is then $[(0,10)(4) + (0.20)(5) + (0.50)(6) + (0.15)(7) + (0.05)(8)]$ or 5.85 years. This is an estimate of the functional life of this particular equipment item.

In making an economic comparison of alternative equipment items, machines, or projects, the analyst may be interested only in a particular period of time or planning horizon. This planning horizon may be (and often is) different from the functionally useful life of one or more of the items, machines, or projects under consideration. For example, a job contract may cover a 5-year period. To execute the contract, machines and equipment items must be purchased. There are alternative machine and equipment items which can accomplish the job. These may have functional lives of 10, 15, or 20 years. An economic comparison of the alternative machines and equipment items would probably adopt a 5-year planning horizon (the length of the contract), and only the relevant cost and revenue figures for the 5-year period would be used in the comparison. Further discussion on this point was presented in Chapter Five.

8.3.2   General Sources of Data

There are many sources for providing data to make the various estimates required in comparing alternative investment projects. Sources may be either internal or external to the firm. Examples of sources within a firm are sales records, production control records, inventory records, quality control records, purchasing department records, work measurement and other industrial engineering studies, maintenance records, and personnel records. Some, all, or additional records are input into the accounting function of the firm, which then compiles various financial reports for management.

The accounting system can and usually does serve as an important, if not primary, internal source of detailed estimates on operating costs, maintenance costs, and material costs, among others.

Sources of data external to the firm may be grouped into two general classes: published information that is generally available and information (published or otherwise) available on request. Available published information includes the vast literature of trade journals, professional society journals, U.S. government publications, reference handbooks, other books, and technical directories. Information not generally available except by request includes many sources listed in the previous category. For instance, many professional societies and trade associations publish handbooks, other books, special reports, and research bulletins that are available on request. Manufacturers of equipment and distributors of equipment are excellent sources of technical data, and most will readily supply this information without charge. Additionally, various government agencies, commercial banks (particularly holding companies involved in leasing buildings and equipment), and research organizations (commercial, governmental, industrial, and educational) may be sources of data to aid the estimating process.

8.4   GENERAL ACCOUNTING PRINCIPLES

As already mentioned, the engineer should have some understanding of basic accounting practice and cost accounting techniques in order to obtain data from the firm's accounting system. If accounting is classified into general accounting and cost accounting, then cost accounting is judged more important to the engineer as a source of data for making cost estimates pertinent to engineering projects. Cost accounting will therefore receive the greater emphasis in this text; in either case, the treatment of accounting is general and high level and is directed toward fundamental accounting concepts instead of comprehensive accounting detail.

Accounting is the language of business. Without an understanding of this language it is virtually impossible for an engineer to acquire and correctly interpret the data needed for economic analysis or to communicate the results of an analysis in meaningful and significant terms to managerial colleagues. Without an understanding of key terms and concepts, it is like trying to communicate in a foreign language. Learning this language is also crucial or an engineer whose aspirations are to progress through a technical career track to higher levels of authority and responsibility within a company's management structure.

The American Institute of Certified Public Accountants defines accounting as "the art of recording, classifying and summarizing in a significant manner and in terms of money, transactions and events which are, in part at least, of a financial character, and interpreting the results thereof." This definition embodies the four key elements of accounting. An accounting system is concerned with the four primary functions of *recording, classifying, summarizing,* and *interpreting* the financial data of an organziation, whether nonprofit or a business organization. The discussion to follow assumes a business organization, and, in virtually all such organizations, general accounting information is summarized in two basic financial reports; the balance sheet and the income statement.

The *balance sheet* provides a statement of the financial condition of a firm *at a point* in time by providing a summarized listing of the values of the assets, liabilities, and net worth of the firm. The *income statement* provides a statement of the revenues and expenses incurred by a firm *during a period of time,* usually a month, quarter, or year. The two statements are closely related. The income statement summarizes the financial activities that occur between two balance sheets. The balance sheet reflects the financial condition as a result of the activities reported in the income statement. This relationship is illustrated in Figure 8.6.

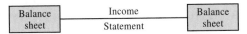

Figure 8.6 The Relationship between balance sheets and income statements.

8.4.1 Balance Sheet

The items listed on a balance sheet are usually classified into three main groups: assets, liabilities, and net worth items. Subgroups may also be identified, such as current and fixed assets or liabilities. *Assets* are properties owned by the firm, and *liabilities* are debts owed by the firm against these assets. The dollar difference between assets and liabilities is the *net worth* of the business, which measures the investment made by the owners or stockholders of the business plus any accumulated profits left in the business by the owners or stockholdes. Another term for net worth is *owners' equity*. A fundamental accounting equation is thus defined:

$$\text{Assets} - \text{Liabilities} = \text{Net Worth}$$

Rewriting, we have

$$\text{Assets} = \text{Liabilities} + \text{Net Worth} \qquad \textbf{(8.6)}$$

and the usual format of a balance sheet follows the equation in this form.

Current assets include cash and other assets that can be readily converted into cash; an arbitrary period of 1 year is usually assumed as a criterion for conversion. Similarly, *current liabilities* are the debts that are due and payable within 1 year from the date of the balance sheet in question. *Fixed assets,* then, are the properties owned by the firm that are not readily converted into cash within a 1-year period, and *fixed liabilties* are long-term debts due and payable after 1 year from the date of the balance sheet. Typical current-asset items are cash, accounts receivable, notes receivable, raw material inventory, work in process, finished goods inventory, and prepaid expenses. Fixed-asset items are land, buildings, equipment, furniture, and fixtures. Items that are typically listed under current liabilities are accounts payable, notes payable, interest payable, taxes payable, prepaid income, and dividends payable. Fixed liabilities may also be notes payable, bonds payable, mortgages payable, and so forth. Net worth items appearing on a balance sheet are less standard and, to a degree, depend on whether the business is a sole proprietorship, a partnership, or a corporation. The size of the corporation is also an influencing factor on item designation. However, items such as capital stock, retained earnings, capital surplus, or earned surplus appear under the net worth group. An example of a balance sheet for a hypothesized firm is exhibited in Table 8.5.

The sample balance sheet in Table 8.5 is "balanced" since assets = liabilities + net worth. It gives a statement of the financial condition of the organization as of a specific date—the close of an accounting period. Note the fixed asset portion of the balance sheet. The building originally cost $200,000, and depreciation expenses have been charged annually, so that the total depreciation charges (as of the date of the balance sheet) have been $50,000, the amount entered as *depreciation reserve* for the building. In theory, the depreciation reserve is an accumulated amount of funds held to repurchase the asset when its functionally useful life terminates. The first cost of the depreciable asset (building, in this case) minus the amount in the depreciation reserve equals the *book value*. If the book value were a true estimate of the salvage, or market, value, then the sum of the amount in the depreciation reserve plus the book value provides the

Table 8.5 Sample Balance Sheet

Jax Tool and Engineering Company, Inc.
Balance Sheet
December 31, 19___

Assets

Current assets			
Cash		$ 25,000	
Accounts receivable		115,000	
Raw materials		8,500	
Work in process		7,000	
Finished goods inventory		3,000	
Small tool inventory		12,500	
Total current assets			$171,000
Fixed assets			
Land		$ 30,000	
Building	$200,000		
Less: Depreciation Reserve	50,000	150,000	
Equipment	$750,000		
Less: Depreciation Reserve	150,000	600,000	
Office equipment		10,000	
Total fixed assets			$790,000
Total assets			$961,000

Liabilities and Net Worth

Current liabilities			
Accounts payable		$ 32,000	
Taxes payable		15,000	
Total current liabilities			$ 47,000
Fixed liabilities			
Mortgage loan payable		$130,000	
Equipment loan payable		350,000	
Total fixed liabilities			$480,000
Total liabilities			$527,000
Common stock		$325,000	
Retained earnings		80,000	
Earned surplus (for current year)		29,000	
Total equity			$434,000
Total liabilities and equity			$961,000

firm with an amount of funds equal to the original purchase price. This sum can be applied toward the purchase of a replacement asset. It is the usual case, however, that the market value of the asset when sold differs from the book value. The difference results in either a capital gain or loss and affects the firm's income taxes.

A similar explanation applies to the fixed asset of equipment. In this particular balance sheet, the equipment account is an aggregate for all the equipment owned by the company instead of an individual listing of each equipment item. There could be separate equipment accounts, grouped according to equipment class. In any case, a company normally keeps individual records on equipment items, which are then summarized on the balance sheet.

The Net Worth section of the balance sheet in Table 8.5 contains three entries. These entries summarize the ownership (or equity) accounts of the company. The common stock account reflects the ownership position of stockholders, frequently referred to as *contributed* capital. The reporting of contributed capital accounts on a balance sheet is subject to a number of legal complications depending upon the nature of the contributed capital (e.g., common stock, preferred stock, par value, no par value). Discussion of these issues is beyond the scope of this text. The next two entries in the net worth section of Table 8.5 summarize the equity generated through the operation of the business rather than the contributions of investors. These accounts are frequently referred to as *earned* capital. The retained earnings account summarizes the surplus (or deficit) of earnings over expenses which have been held by the company (i.e., retained) for accounting periods prior to the one reflected in the current income statement. The earned surplus account highlights this same information for the current period and should match the net profit (or loss) reflected on the current income statement (Table 8.6 or 8.7).

8.4.2   Income Statement

The second basic financial report compiled by the accounting system is the *income statement* or *profit and loss* statement. For the current accounting period, the income statement provides management with (1) a summary of the revenues received, (2) a summary of the expenses incurred to obtain the revenues, and (3) the profit or loss resulting from business operations. The format of the income statement varies, and the revenue and expense items depend on the type of business involved.

The income statement usually begins with revenues. Revenues are generated from the sales of the products or services marketed by the firm. Next, the income statements reflects the direct costs associated with generating the sales revenue. This section is generally referred to as the cost of goods. For a manufacturing company (Table 8.6) the cost of goods section includes the costs incurred in producing the product. For a retail company (Table 8.7) this section includes the costs incurred in acquiring the retail goods to be sold. In either case, the result of subtracting the cost of goods from the sales is gross profit.

After gross profit is determined, other expenses associated with operating the business are incorporated. This section includes all expenses which are not

Table 8.6 Sample Income Statement

Jax Tool and Engineering Co., Inc.
Income Statement
Year Ended December 31, 19___

Sales		$1,200,000
Less cost of goods manufactured:		
Direct labor	$420,000	
Direct materials	302,000	
Indirect labor	112,000	
Depreciation	98,000	
Repairs and maintenance	41,500	
Utilities	11,500	985,000
Gross profit		$ 215,000
Less other expenses:		
Administration	$ 76,000	
Marketing	49,000	
Interest payments	35,000	160,000
Net profit before income taxes		$ 55,000
Less income taxes		26,000
Net profit (posted to earned surplus)		$ 29,000

Table 8.7 Sample Income Statement

Jax Tool and Engineering Co., Inc.
Income Statement
Year Ended December 31, 19___

Sales		$1,200,000
Less cost of goods sold:		
Inventory, January 1, 19___	$ 26,000	
Plus purchases	432,000	
	$458,000	
Less inventory, December 31, 19___	44,000	414,000
Gross profit		$ 786,000
Less expenses:		
Direct labor	$420,000	
Depreciation—building	10,000	
Depreciation—equipment	30,000	
Repairs and maintenance	41,500	
Indirect labor	218,000	
Utilities	9,800	
Supplies, tooling	1,700	$ 731,000
Net profit before income taxes		$ 55,000
Less income taxes		26,000
Net profit (posted to earned surplus)		$ 29,000

included in the cost of goods. Generally this would include general and administrative expenses as well as the marketing expenses. The subtraction of these costs from gross profit results in net profit before taxes. Subtracting taxes results in the net profit (or loss) for the period. This final value is transferred to the earned surplus line on the balance sheet.

The format for the income statement in Table 8.6 is an oversimplification of such a statement for even a small manufacturing firm. For example, the *cost of goods sold* entry may be considerably more detailed to reflect multiple product lines, the depreciation items may be detailed to a greater extent to reflect multiple classes of assets, and several common expense items—such as employee benefits contributions, insurance premiums, and advertising—are not included.

Similarly the Income Statement in Table 8.7 is highly simplified. This format is typical for a retail business, and the major difference in form concerns the method of determining the cost of goods sold item.

8.4.3   Ratio Analysis

An important topic related to the interpretation of balance sheets and income statements is ratio analysis. Ratio analysis is not critical to an engineer focused solely on the accounting function as a source of data for economic analysis problems. However, to interpret a company's accounting statements to determine the attractiveness of a company's stock as a potential investment or to determine the economic health of a company as a potential employer, ratio analysis is of fundamental importance. If the student's interest is focused solely on data sourcing, this section can be skipped with no loss of continuity for the remainder of the chapter.

To meaningfully interpret the information contained in balance sheets and income statements (i.e., to convert data to information), a common practice is to examine relationships between the values found on these statements. This type of analysis is referred to as *ratio analysis*. Before proceeding to a presentation of several popular ratio calculations, it is important for the student to bear in mind that the values of the ratios themselves are neither good nor bad. They can only be interpreted in comparison to the ratios of peer group companies or to an individual's personal expectations of a company's performance. Ratios are only rough guides to the interpretation of financial statements, not mathematical conclusions.

Generally the analysis of financial statements focus on three primary areas: (1) the earning power of the company; (2) the short-term liability obligations; and (3) the long term liability obligations. We will use the financial statements of the Okie Manufacturing Company to illustrate the calculation of several common ratios in each of these areas. Table 8.8 contains comparative balance sheets (two balance sheets displayed side by side) for the years ended 19×5 and 19×6. Table 8.9 contains comparative income statements for the years 19×5 and 19×6.

The viability of maintaining the financial health of a company depends upon its *earning power*. Firms must earn and sustain profit over the long term to survive. The following ratios are commonly used to assess earning power.

Table 8.8 Okie Manufacturing Company Comparative
Balance Sheets

Okie Manufacturing
Comparative Balance Sheets
as of December 31, 19×5 and 19×6

ASSETS	19 × 6	19 × 5
Current Assets		
Cash	$ 61,750	$ 83,520
Accounts Receivable (net)	195,000	130,500
Inventory	65,000	50,000
Prepaid Expenses	22,750	31,900
Total Current Assets	$344,500	$295,920
Fixed Assets		
Machinery	$208,000	$187,830
Furniture	74,750	72,500
Other	22,750	23,750
Total Assets	$650,000	$580,000
LIABILITIES AND CAPITAL		
Current Liabilities		
Notes Payable	$ 92,950	$ 87,000
Accounts Payable	147,212	109,653
Taxes Payable	69,438	64,920
Total Current Liabilities	$309,600	$261,573
Fixed Liabilities		
Loans	$100,000	$ 90,000
Total Liabilities	$409,600	$351,573
Capital		
Stock	$100,000	$100,000
Retained Earnings	88,427	77,397
Earned Surplus	51,973	51,030
Total Capital	$240,400	$228,427
Total Liabilities and Capital	$650,000	$580,000

$$\text{return on assets employed} = \frac{\text{net income}}{\text{total average assets}} \qquad (8.7)$$

$$\text{return on owner's equity} = \frac{\text{net income}}{\text{average owner's equity}} \qquad (8.8)$$

The calculation of earning power ratios for Okie Manufacturing for 19×6 are
shown below.

Return on Assets Employed

$$\text{return on assets employed} = \frac{\text{net income}}{\text{total average assets}}$$

Table 8.9 Okie Manufacturing Company Comparative
Income Statements

Okie Manufacturing
Comparative Income Statements
For Years Ended December 31, 19×5 and 19×6

	19×6	19×5
Net Sales	$1,625,450	$1,450,000
Cost of Goods Sold		
Beginning Inventory	$ 50,000	$ 40,000
Direct Materials	406,000	350,000
Direct Labor	801,500	700,000
Factory Overhead	94,603	90,000
Total	$1,352,103	$1,180,000
Less: Ending Inventory	65,000	50,000
Cost of Goods Sold	$1,287,103	$1,130,000
Gross Profit	$ 338,347	$ 320,000
Other Operating Expenses		
Selling Expenses	$ 43,980	$ 37,200
General and Administrative	180,606	174,371
Total Other Operating Expenses	$ 224,586	$ 211,570
Net Operating Income	$ 113,761	$ 108,430
Less: Interest Expenses	$ 21,600	$ 18,000
Less: Income Taxes	40,188	39,400
Net Income	$ 51,973	$ 51,030

$$= \frac{\$51,973}{0.5(\$580,000 + \$650,000)} = 0.0845 \text{ or } 8.45\%$$

Return on Owner's Equity

$$\text{return on owner's equity} = \frac{\text{net income}}{\text{average owner's equity}}$$

$$= \frac{\$51,973}{0.5(\$228,427 + \$240,400)} = 0.2217 \text{ or } 22.17\%$$

The second major area of focus in ratio analysis is short term liability obligations. This is referred to as *liquidity*. Liquidity ratios measure a company's ability to meet its current obligations. The following ratios are commonly used to measure liquidity.

$$\text{current ratio} = \frac{\text{current assets}}{\text{current liabilities}} \qquad \textbf{(8.9)}$$

$$\text{acid-test ratio} = \frac{\text{cash + receivables + short-term marketable securities}}{\text{current liabilities}} \quad \textbf{(8.10)}$$

$$\text{accounts receivable turnover} = \frac{\text{net sales}}{\text{average accounts receivable}} \quad \textbf{(8.11)}$$

$$\text{inventory turnover} = \frac{\text{cost of goods sold}}{\text{average inventory}} \quad \textbf{(8.12)}$$

The calculation of liquidity ratios for Okie Manufacturing for 19×6 is shown below.

Current Ratio

$$\text{current ratio} = \frac{\text{current assets}}{\text{current liabilities}}$$
$$= \frac{\$344,500}{\$309,600} = 1.112$$

Acid-Test Ratio

$$\text{acid-text ratio} = \frac{\text{cash + receivables + short-term marketable securities}}{\text{current liabilities}}$$
$$= \frac{\$61,750 + \$195,000 + \$0}{\$309,600} = 0.829$$

Accounts Receivable Turnover

$$\text{accounts receivable turnover} = \frac{\text{net sales}}{\text{average accounts receivable}}$$
$$= \frac{\$1,625,450}{0.5\,(\$130,500 + \$195,000)} = 9.98$$

Inventory Turnover

$$\text{inventory turnover} = \frac{\text{cost of goods sold}}{\text{average inventory}}$$
$$= \frac{\$1,287,103}{0.5\,(\$50,000 + \$65,000)} = 22.38$$

The final major area of focus in ratio analysis is long-term liability obligations. This is referred to as *solvency*. Solvency ratios measure a company's ability to meet its long term obligations based on its current debt structure. The following ratios are commonly used to measure solvency.

$$\text{debt to equity ratio} = \frac{\text{total liabialities}}{\text{total capital worth}} \quad \textbf{(8.13)}$$

$$\text{times interest earned ratio} = \frac{\text{net income before income taxes and interest}}{\text{interest charges}}$$

$$\textbf{(8.14)}$$

$$\text{operating income to total assets ratio} = \frac{\text{net operating income}}{\text{total assets}} \qquad \textbf{(8.15)}$$

The calculation of liquidity ratios for Okie Manufacturing for 19×6 is shown below.

Debt to Equity Ratio

$$\text{debt to equity ratio} = \frac{\text{total liabilities}}{\text{total capital worth}}$$

$$= \frac{\$409,600}{\$240,400} = 1.70$$

Times Interest Earned Ratio

$$\text{times interest earned ratio} = \frac{\text{net income before income taxes and interest}}{\text{interest charges}}$$

$$= \frac{\$113,761}{\$21,600} = 5.27$$

Operating Income to Total Assets

$$\text{operating income to total assets ratio} = \frac{\text{net operating income}}{\text{total assets}}$$

$$= \frac{\$113,761}{\$650,000} = 0.1750 \text{ or } 17.50\%$$

As stated earlier, ratios provide a useful and powerful means to express the relationships found in balance sheets and income statements. These ratios are generally only useful when compared to a meaningful set of standards or expectations. The standards or expectations generally take the form of previous year's results, peer group comparisons, or industry averages. Many variations to the names and calculation formulas presented above can be found in financial reports. The student is cautioned to ensure that ratios are calculated in consistent ways before making any direct comparisons.

8.5   COST ACCOUNTING PRINCIPLES

8.5.1   Cost Allocation Methods

The balance sheet and the income statement in Section 8.4 are considerably removed both in time and in detail from decisions at the usual engineering project level. More important to the engineer as a source of cost information is the cost accounting system within a particular firm. The firm may be involved in manufacturing or providing services, and if it is involved in manufacturing, production may be on a job-shop or process basis. There are some fundamental differences in cost accounting procedures for determining manufacturing costs versus determining the cost of providing a service; also, there are differences in accounting procedures if manufacturing is on a job-shop or process basis. In order to concentrate on basic principles instead of details, the cost accounting system assumed will be that of a job-shop manufacturing firm. Thus, the emphasis will be on determining the per-order costs for a job order.

The total cost of producing any job order consists of direct material, direct labor, and overhead costs. An additional item of cost could be special tooling or equipment purchases strictly for the job order in question. This definition of total cost does not detail the overhead cost into factory overhead, general overhead, and marketing expenses in order to simplify the presentation. Materials for a given job order may include purchased parts and in-house fabricated parts, and the cost for direct materials is determined primarily from purchase invoices. Questions of scrap allowances and averaging material costs, which fluctuate over time, present problems in obtaining accurate direct material costs, but determining such costs is reasonably straightforward. Direct labor time spent on job order is normally recorded by operators on labor time cards, and the direct labor cost is determined by applying the appropriate labor cost rates. The labor rates, as determined by the accounting system, will normally include the cost of employee fringe benefits in addition to the basic hourly rate. Although accurately determining the direct labor cost for a given job is a major accounting problem, it is more readily determined than the overhead cost.

Overhead costs typically cannot be allocated as direct charges to any single job order and must, therefore, be prorated among all the job orders on some rational basis. Three popular methods of allocating overhead costs to manufacturing jobs are in wide use today. They are:

- Allocation based on direct labor hours
- Allocation based on direct labor dollars
- Allocation based on direct labor dollars plus direct material dollars (prime costs)

These methods can be applied at any desired manufacturing unit level (i.e., an entire plant, specific departments, workcenters, or machines). Step-by-step procedures for using each of these methods to (1) determine an appropriate overhead rate and (2) use this rate to estimate overhead on a specific job are outlined below. Example 8.5 illustrates the application of these procedures.

Allocate Overhead Based on Direct Labor Hours

1. Determine (or estimate) values for previous period direct labor hours and overhead cost for the manufacturing unit
2. Calculate the rate per direct labor hour:

$$\text{rate} = \frac{\text{overhead cost}}{\text{direct labor hours}} \qquad \textbf{(8.16)}$$

3. Determine (or estimate) the number of direct labor hours required by the particular job for which overhead cost is being estimated.
4. Calculate the overhead cost for the job:

$$\text{estimated overhead} = \text{rate} \times \text{estimated direct labor hours}$$

Allocate Overhead Based on Direct Labor Dollars

1. Determine (or estimate) values for previous period direct labor dollars and overhead cost for the manufacturing unit
2. Calculate the percentage ratio of overhead cost to direct labor dollars:

$$\text{ratio} = \frac{\text{overhead cost}}{\text{direct labor dollars}} \times 100\% \qquad \textbf{(8.17)}$$

3. Determine (or estimate) the direct labor dollars required by the particular job for which overhead cost is being estimated.
4. Calculate the overhead cost for the job:

$$\text{estimated overhead} = \text{ratio} \times \text{estimated direct labor dollars}$$

Allocate Overhead Based on Direct Labor Dollars and Direct Material Dollars

1. Determine (or estimate) values for previous period direct labor dollars, direct material dollars, and overhead cost for the manufacturing unit
2. Calculate the percentage ratio of overhead cost to direct labor dollars plus direct material dollars:

$$\text{ratio} = \frac{\text{overhead cost}}{\text{direct labor dollars} + \text{direct material dollars}} \times 100\% \qquad \textbf{(8.18)}$$

3. Determine (or estimate) the direct labor dollars and direct material dollars required by the particular job for which overhead cost is being estimated.
4. Calculate the overhead cost for the job:

$$\text{estimated overhead} = \text{ratio} \times (\text{estimated direct labor dollars} + \text{estimated direct material dollars})$$

EXAMPLE 8.5

The overhead allocation for a job is to be estimated. Assume the direct labor hours for the job are estimated to be 40 hours at a rate of $12.50 per hour. Direct material costs are estimated at $850. The overhead calculations are to be based on the following previous period cost totals:

Total direct material dollars	48,000
Total direct labor dollars	$480,000
Total direct material dollars	$600,000
Total overhead cost	$360,000

Using the procedures above, calculate the overhead allocation based on (a) direct labor hours, (b) direct labor dollars, and (c) direct labor dollars plus direct material dollars.

(a) Direct labor hours

Step 1: previous period direct labor hours = $48,000
previous period overhead cost = $360,000

Step 2: rate per direct labor hour = $360,000/48,000 = $7.50/hour

Step 3: estimated direct labor hours for job = 40

Step 4: estimate overhead = $7.50/hour × 40 hours = $300

(b) Direct labor dollars

Step 1: previous period direct labor dollars = $480,000
previous period overhead cost = $360,000

Step 2: percentage ratio per direct labor dollar = ($360,000/$480,000) × 100% = 75%

Step 3: estimated direct labor dollars for job = 40 hours × $12.50/hour = $500

Step 4: estimate overhead = 75% × $500 = $375

(c) Direct labor dollars + direct material dollars

Step 1: previous period direct labor dollars = $480,000
previous period direct material dollars = $60,000
previous period overhead cost = $360,000

Step 2: percentage ratio per (direct labor dollar + direct material dollar) = [$360,000/($480,000 + $600,000)] × 100% = 33.33%

Step 3: estimated direct labor dollars + direct material dollars for job = (40 hours × $12.50/hour) + $850 = $1350

Step 4: estimate overhead = 33.33% × $1350 = $450

Determining the overhead cost for a job order by the *rate per direct labor hour* method will yield the same result as the *percentage of direct labor cost* method, provided that the rate per direct labor hour used on the job in question is equal to the average factory labor rate. The "percentage of prime cost" method will necessarily yield a different assignment of overhead to a job order than the other two methods. The choice among these three methods is arbitary; indeed, other methods are used by cost accountants in distributing overhead costs to a given job order. The *rate per direct labor hour* method is perhaps most commonly used.

Whatever method is chosen from the above for distributing overhead costs to job orders in a current year, the rates or percentages are based on the previous year's cost figures. Thus, overhead rates may change from year to year within a particular firm.

Since an average overhead rate for the entire factory may very well be too gross an estimate when actual overhead costs differ among departments within the factory, cost accounting may determine individual overhead rates for departments or cost centers. Furthermore, the hourly rates for direct labor may vary among these cost centers. A further refinement is to determine overhead rates for individual machines within cost centers. Then, as particular job orders progress through cost centers (departments and/or machines), the direct labor time (or machine time) spent on the job order in the various cost centers is recorded, the appropriate labor or machine rates and overhead rates are applied, and the total cost for the job is calculated.

The following example illustrates the variety of methods used to distribute overhead to a given cost center; the total overhead for the cost center will then be distributed to particular products by yet another method.

EXAMPLE 8.6

The following information has been accumulated for the Deetco Company's two departments during the past year. (see Table 8.10)

The Deetco Company distributes depreciation overhead based on (1) the first cost of equipment in each department, (2) a zero salvage value of the equipment in 10 years, and (3) a constant annual (or straight-line) rate of depreciation. All overhead other than depreciation is first distributed to each department according to the number of employees in each department, and then an overhead rate per direct labor hour is computed for each department.

What selling price should the company quote on Job Order D if raw material costs are estimated as $900, estimated direct labor hours required in Departments A and B are 30 hours and 100 hours respectively, and profit is to be calculated as 25% of selling price?

For Department A, the total overhead allocation is determined as follows:

Annual depreciation	=	$ 25,000
Other factory overhead = (14/23)($150,000)	=	91,304
General overhead = (14/23)($350,000)	=	213,043
TOTAL OVERHEAD COSTS	=	$329,347

Table 8.10 Data for Example 2.4

	Department A	Department B	Total
Direct materials cost	$720,000	$240,000	$960,000
Direct labor cost	$260,000	$140,000	$400,000
Direct labor hours	25,200	16,200	41,400
Number of employees	14	9	23
First cost of equipment	$250,000	$200,000	$450,000
Annual depreciation	$ 25,000	$ 20,000	$ 45,000
Other factory overhead			$150,000
General overhead			$350,000

Thus, the overhead rate for Department A per direct labor hour is

$$\text{Rate} = \frac{\$329,347}{25,200 \text{ hours}}$$
$$= \$13.07/\text{direct labor hour}$$

For Department B, the total overhead allocation is calculated as follows:

Annual depreciation		= $ 20,000
Other factory overhead= (9/23)($150,000)	=	58,696
General overhead = (9/23)($350,000)	=	136,957
TOTAL OVERHEAD COSTS		= $215,653

Thus, the overhead rate for Department B per direct labor hour is

$$\text{Rate} = \frac{\$215,653}{16,200 \text{ hours}}$$
$$= \$13.31/\text{direct labor hour}$$

The estimated total cost for Job Order D is then computed as

Direct material cost		= $ 900.00
Direct labor cost for Dept. A = ($260,000/25,200)(30 hours)	=	309.52
Overhead cost for Dept. A = ($13.07)(30 hours)	=	392.10
Direct labor cost for Dept. B = ($140,000/16,200)(100 hours)	=	864.20
Overhead cost for Dept. B = ($13.31)(100 hours)	=	1331.00
Total cost		$3796.82

If x = the selling price of Job Order D, then

$$x = \text{total cost} + \text{profit}$$
$$= \$3796.82 + (0.25)x$$

and

$$x = \frac{3796.82}{0.75} = \$5062.43$$

8.5.2   Activity-Based Costing

A relatively new entrant in the field of cost accounting is *activity-based costing.*
Activity-based costing (ABC) has emerged due to the dramatic changes that
have occurred, and continue to occur, in the nature and characteristics of manufacturing costs. Historically, direct labor and direct material constituted the most
significant elements of the cost of goods. Overhead was the smallest element
and hence was allocated based on direct labor or prime costs (see Section 8.5.1).
In many cases today, it is no longer accurate to assume that direct labor is the
largest element in the cost pool and overhead the smallest. With the introduction
and implementation of computer controlled and automated manufacturing sys-

tems, it is not unusual for overhead costs to dominate the cost of producing items. Frequently in fact, direct labor is the least significant in the cost pool and overhead the largest.

Activity-based costing is designed to meet the challenge of a changing cost mix by associating manufacturing costs with activities which drive them. First, costs must be identified by categories. Categories need not (and probably should not) be associated with products or organizational units, rather, they are associated with clearly defined elements or cost categories. Typical examples might include material handling costs, energy costs, tooling costs, or maintenance costs.

Next the activities which drive the significant cost categories must be identified. These are the activities which are to be monitored and controlled under activity-based costing. This task is difficult and complex. Many companies have never considered their processes from the cost driver and/or value-adding viewpoint. The newness of this approach as well as the implicit challenge of considering activities from a new perspective make this task a significant undertaking. Examples of cost-driving activities include machine hours for energy costs, material moves or truck hours for material handling costs, and machine hours or production volume for tooling and maintenance costs.

Next, the expected (or actual) rate of activity for each of the cost drivers is used to predict (or monitor) the costs associated with each cost category. Such activity-based accounting of costs can be used as a basis to eliminate high cost activities, particularly if they generate low value added. Similarly, ABC analysis can be used to focus the attention of process improvement activities toward those activities that drive high costs. Better still, product and process redesign can focus on changes which ultimately eliminate the activation of high-cost driver activities.

Many companies are employing ABC to make activity-based decisions which result from a more realistic allocation of costs than was previously possible. In many cases, ABC-generated process and product redesigns have been impressive when measured in terms of cost reduction and profit improvement. ABC does, however, require information sharing and a cross-functional perspective that is new to most companies. Additional information regarding activity-based costing can be found in [2, 3, 6, 9].

8.5.3 Standard Costs

Although the first task of cost accounting is to determine per-item or per-order costs, another major purpose of cost accounting is to interpret financial data so that management can (1) measure changes in production efficiency and (2) judge the adequacy of production performance. Establishing cost standards can be of great assistance in achieving these objectives. A standard-cost system involves, in advance of manufacture, (1) the preparation of standard rates for material, labor, and overhead; and (2) the application of these rates to the standard quantities of material and labor required for a job order, or for each production operation required to complete the job order.

Since a process-type manufacturing firm such as an oil refinery outputs the same product (or a few products) over a long time period, cost standards are more readily determined for process firms than for job-shop firms where the variety of output is large and varies with customer order. However, the number and type of production operations required to complete various job orders are finite for a given manufacturing firm. Each job order is, of course, made up of single units. Thus, a standard amount of material can be determined for each unit, and standard labor times and machine times can be determined for each unit. It is usually the responsibility of the work measurement function within the firm to determine these standard quantities. Then, by applying standard unit material costs and standard labor rates, standard unit costs for material, labor, and overhead can be determined. The standard costs then serve as a basis for measuring production efficiency and performance over time. Deviations from standard costs may be caused by several factors, especially (1) raw material price variations and (2) actual quantities of material and labor used versus the standard amounts of these items. This latter factor is the one of primary concern in determining production efficiency and performance, measures of which provide information to management to aid in cost control.

8.6   ECONOMIC VALUE ADDED[3]

Since the mid-1980s, a management tool called *economic value added* has been used by an impressive set of firms to make investment decisions. It focuses management's attention on an important objective: adding value for the shareholders. In fact, it has been the principal tool used by upper management within the Coca-Cola Company. EVA is used to facilitate decisions regarding major capital investment, as well as acquisitions and divestitures. It has also been used to analyze products and operating units to determine the economic dogs and cash cows within the firm's portfolio of products and businesses.

Basically, EVA is a management tool that examines the difference between the net operating profit after taxes and the cost of capital, which includes the cost of both debt and equity capital. Hence, the interest charges and bond rates that contribute to the cost of debit capital are combined with the cost to the shareholders of providing the firm with equity capital (by purchasing its stock) [24].

Many firms that adopted EVA found that few of their managers knew how much capital was tied up in their business units. Moreover, the managers did not have the foggiest notion about the true cost of capital. EVA seeks to remedy this by focusing attention on adding value for the shareholder through more effective use of capital. The result of an increased emphasis on effective use of capital will result in lower inventories, fewer warehouses, etc. [24].

Stern Stewart argues that there are four ways to create value for the shareholder:

1. Increase profitability without using additional capital, e.g., increase profit margins and increase sales without using additional capital.
2. Invest in projects that earn more than the cost of capital.

---

[3] Stern Stewart & Co., a corporate financial advisory services firm located at 450 Park Avenue, New York, 10022, is generally credited with the development of EVA. Joel M. Stern and G. Bennett Stewart, the principals in the firm, are generally credited with being its architects. According to Stern Stewart & Co., it owns the trademark on the use of EVA [21], [22].

3. Free-up capital that earns less than the cost of capital.
4. Use debt to reduce the cost of capital [21].

Real estate, equipment, facilities, working capital, and inventories are examples of capital being used within the firm. Other, not so obvious examples of capital are investments in training and in research and development. The investments in training result in increased value of the firm's human capital. While it is not easy to quantify the value of capital in R&D and in training, they should not be completely overlooked in the quest for value.

Having touched on why the EVA concept of economic value added has merit, we will use several examples to illustrate how it is calculated.

EXAMPLE 8.7

Two firms (A and B) are being considered for acquisition. The assets of the firms are $100 million and $200 million, respectively. Both are debt free; hence, the equity equals the assets. The annual operating profits for the firms are $40 million and $70 million, respectively. Taxes equal 40% of operating profit. Consequently, the annual net incomes for the firms are $24 million and $42 million, respectively. Dividing the net income by equity yields return on equity of 24% and 21%, respectively. The cost of capital is 12%. Which firm is best from a shareholder value point of view? Some would choose A because it has the greatest ROE (return on equity) or ROA (return on assets). However, B maximizes value for shareholders.

To determine the EVA, subtract the cost of capital from net operating profit after taxes. For A, this yields: $24 million − 0.12($100 million) = $12 million. For B, the EVA equals $42 million − 0.12($200 million) = $18 million. (As was the case with the IRR method, one cannot choose the alternative that has the greatest return on equity; instead, incremental net operating profits should be compared with the cost of capital.)

It has long been known that maximizing return on investment is not the right objective. In 1924, Donaldson Brown, Chief Financial Officer, General Motors Corporation, noted, "The object of management is not necessarily the highest rate of return on capital, but . . . to assure profit with each increment of volume that will at least equal the economic cost of additional capital required [21]."

EXAMPLE 8.8

Consider two firms, C and D, each with a 12% cost of capital. The financial data for the two firms (in $M's) are:

	FIRM C	FIRM D
Equity	$100	$200
Annual Operating Profit	$ 25	$ 35
Taxes (40%)	$ 10	$ 14
Net Income	$ 15	$ 21

On the basis of net income generated, one might conclude that D is the better performing company. However, for C there is a capital cost of $12 and an EVA of $3; whereas, for D there is a capital cost of $24 and an EVA of −$3. Therefore, firm C is superior to firm D in adding value for the shareholder.

EXAMPLE 8.9

There are several ways to compute EVA. The method used in the previous examples was:

	FIRM C	FIRM D
Capital Invested	$100	$200
Operating Profits	$25	$35
Taxes (40%)	$10	$14
Net Operating Profit After Taxes (NOPAT)	$15	$21
Cost of Capital (12%)	$12	$24
EVA	+$3	−$3

Alternatively, EVA can be computed as follows:

Return on Assets (ROA)	15%	10.5%
Cost of Capital	12%	12%
Difference	+3%	−1.5%
EVA = Difference × Assets	+$3	−$3

8.7 SUMMARY

In this chapter we provided an introduction to the language of accountants, financial analysts, and managers. To be successful in selling engineering designs, one must learn to communicate; to communicate effectively, all parties must speak the same language. Additionally, the data sources for economic analyses are often to be found in the accounting systems; hence, it is essential that the engineer be familiar with accounting principles. To complete the coverage of cost concepts, a brief introduction to cost estimation, ratio analysis, activity based costing, and economic value added was presented.

BIBLIOGRAPHY

1. Apple, James M., *Material Handling Systems Design,* Ronald Press, 1972, Chapters 13 to 15.
2. Barnes, Frank, C., *IEs Can Improve Management Decisions Using Activity-Based Costing,* Industrial Engineering, September 1991, pp. 44–50.
3. Beaujon, George J., and Vinod R. Singhal, "Understanding the Activity Costs in an Activity-Based Cost System," *Journal of Cost Management, 4,* (1), Sp 90, pp. 51–72.

4. Blank, Leland T., and Anthony J. Tarquin, *Engineering Economy,* Third Edition, McGraw-Hill, 1989.

5. Clark, Forest D., and A. B. Lorenzoni, *Applied Cost Engineering,* Second Edition, Revised and Expanded, Marcel Dekker, 1985.

6. Cooper, Robin, and Robert S. Kaplan, *Profit Priorities from Activity-Based Costing,* Harvard Business Review, May–June 1991, pp. 130–135.

7. De La Mare, R. F., *Manufacturing Systems Economics,* Holt, Rinehart & Winston, 1982.

8. DeGarmo, E. Paul, William G. Sullivan, and J. A. Bontadein, *Engineering Economy,* 9 ed., Macmillan, 1993.

9. Drucker, Peter F., *The Information Executives Really Need,* Harvard Business Review, January–February 1995, pp. 54–62.

10. Ferguson, Earl J., and James E. Shamblin, "Break-Even Analysis," *The Journal of Industrial Engineering, 18* (8), August 1967.

11. Hartley, Ronald V., *Cost and Managerial Accounting,* Allyn & Bacon, 1983.

12. Kepner, Charles H., and Benjamin B. Tregoe, *The Rational Manager,* McGraw-Hill, 1965.

13. Niswonger, C. Rollin, and Philip E. Fess, *Accounting Principles,* 12 ed., South-Western, 1977.

14. Ostwald, Phillips F., *Cost Estimating for Engineering and Management,* Prentice-Hall, 1974.

15. Ostwald, Phillip F., *Cost Estimating,* Third Edition, Prentice-Hall, 1992.

16. Polimeni, Ralph S., Frank J. Fabozzi, and Arthur H. Adelberg, *Cost Accounting: Concepts and Applications for Managerial Decision Making,* 2 ed., McGraw-Hill, 1986.

17. Riggs, James L., D. S. Bedworth, and S. U. Randhawa, *Engineering Economics,* 4th Edition, McGraw-Hill, 1996.

18. Stewart, R. D., *Cost Estimating,* John Wiley & Sons, New York, 1982.

19. Stewart, R. D. and R. M. Wyskida, *Cost Estimator's Reference Manual,* John Wiley & Sons, New York, 1987.

20. Thuesen, G. J. and W. J. Fabrycky, *Engineering Economy,* 7 ed., Prentice-Hall, Inc., 1989.

21. Viewgraphs used in a seminar conducted by Stern Steward & Co., *EVA*™, Stern Steward & Co., 450 Park Avenue, New York, 1986.

22. Viewgraphs used in a seminar conducted by Stern Steward & Co., *The Quest for Value: How Value is Created and How to Create Value,* Stern Stewart & Co., 450 Park Avenue, New York, 1986.

23. Vinton, Ivan R. (Editor), *Realistic Cost Estimating for Manufacturing,* Society of Manufacturing Engineers, Dearborn, MI., 1968.

24. White, J. A., "Three Ways to Build Economic Value," *Modern Materials Handling, 49* (7), June 1994, p. 31.

PROBLEMS

1. A small manufacturing firm is converting from a manual inventory system to a computer system. The incremental investment required would be $10,000. Due to the rapid increase in technology, it is assumed that the salvage value will be zero any time after the system is purchased. It is estimated that a net annual savings of $1,500 in operating costs will result when the new computer if purchased. Assume that the time value of money is zero. (8.2.5)

    a. How many years of savings are required to recover the incremental investment?

    b. What annual savings are required to recover the incremental investment in 4 years?

2. The Rossler Metal Machining Company produces widgets according to customer order. The company has determined that widgets can be produced on three different machine tools: M1, M2, or M3. An analysis of widget production cost reveals the following data:

Machine Tool	Fixed Cost/Order	Variable Cost/Unit
M1	$300	$9
M2	$750	$3
M3	$500	$5

a. By either the graphical or analytical approach, determine the most economical machine tool to use for *all* order sizes between 1 and 200 units. Hint: Determine the subranges within the overall range of 1 to 200 for which each machine tool is preferred.

b. For an order of size 75, which machine tool should be used to produce the order and what is the total production cost?

c. For an order of size 160 assume that the preferred (most economical) machine is unavailable. What penalty (expressed in dollars of *additional* production cost) must be paid if the second most economical machine is used? the third? (8.2.5)

3. An engine lathe or a turret lathe can be used to produce a job order of Part 173. If the job order is produced on the engine lathe, Machinist B performs all required operations (i.e., the tooling setup and the operation of the lathe during the machining cycle). On the engine lathe, a tooling setup time of 15 minutes and machining time of 30 minutes are required to produce each part. Machinist B earns $11/hour, and the overhead rate for each machining hour of engine lathe operation is $14.00. The tooling costs per lot for production on the engine lathe are estimated to be $500. If the job order is produced on the turret lathe, Machinist A must do the initial tooling setup for the job, and 4 hours are required for this. Once the setup is done, no further tooling setup per part is required, and Machinist B will operate the turret lathe for the machining cycles. Machinist A earns $12/hour. The machining time for Part No. 173 is 10 minutes on the turret lathe, and the overhead rate for each machining hour is $25. The tooling costs per lot for production on the turret lathe is $1,000. For what job-order size would the turret lathe be economically preferred to the engine lathe? (8.2.5)

4. In a stable economic environment, the A. B. Jax Specialty Foundry Co. can produce a maximum of 1500 gray iron railroad car wheels per month and sell these for $300 each. Now, after prolonged labor problems in the coal industry, the foundry can only sell an average of 500 wheels/month at $250 each. The coal industry strike is temporary, but the future over the next 6 months to a year appears very uncertain.

The foundry has a depreciable investment in buildings and equipment of $300,000, and the value decreases at a rate of $6\frac{2}{3}\%$ each year.

At a production rate of 500 wheels/month, direct labor costs will increase from $55/wheel at maximum production rate to $75/wheel. Direct material costs per wheel will remain at $50. Other annual overhead costs theoretically vary in linear fashion from $60,000 at zero output to $105,000 at maximum production capacity. Would you recommend that the A. B. Jax Specialty Foundary Co. shut down temporarily or continue to operate at this reduced capacity? (8.2.5)

5. A commercial machine shop regularly produces a stainless steel component for a major electronics manufacturer. The machine shop purchases the component from a nearby specialty steel company in semifinished condition, performs drilling and milling operations on the part, and ships it to the electronics firm.

The machining operations are those that can readily be performed on a tape-controlled drill press with a turret head. Thus, the management of the machine shop feels the purchase of such a machine to produce only this part is economically justified. An engineer is then assigned the task of determining the production quantity for break-even, assuming a time value of money (minimum attractive rate of return) equal to zero.

The engineer compiles the following information and cost estimates. The tape-controlled machine will have an installed first cost of $60,000, which includes the necessary electronic software, cutting tools and holders, and work-holding devices. Training of the operator for the machine is included in the purchase price. The economic life of the machine is assumed to be 10 years with a salvage value of $20,000 at that time. The decrease in asset value is estimated at $4000/year and judged to be an annual fixed cost. Other fixed costs are $1500/year. The steel parts are sold to the electronics firm for $17.30/unit. The variable unit costs are estimated as $1.70 for direct labor, $7.00 for direct material, and $3.50 for overhead (excluding depreciation of the machine—the $4000/year fixed cost mentioned previously).

What annual sales volume (number of parts) is required in order to break even on the machine purchase if linearity is assumed? (8.2.5)

6. A subsidiary plant of a major furniture company manufactures wooden pallets primarily for the parent company but also sells pallets to other industrial customers. Although the plant produces multiple pallet styles and sizes, approximatey 95% of sales is due to a single pallet style and size. The subsidiary plant can therefore be considered a single-product plant. The plant has the capacity to produce 300,000 pallets per year. Presently, the plant is operating at 70% of capacity. The selling price of the pallet is $18.25 and the variable cost per pallet is $15.75. (The selling price and variable cost per pallet are linearly related to production quantity.) At zero output, the subsidiary plant's annual fixed costs are about $550,000 and are approximately constant up to the maximum production quantity per year. (8.2.5)

   a. With the present 70% of capacity production, what is the expected annual profit or loss for the subsidiary plant?
   b. What annual volume of sales (units) is required in order for the plant to break even?
   c. What would be the annual profit or loss if the plant were operating at 90% of capacity?
   d. If fixed costs could be reduced by 40%, what would be the effect on the break-even sales volume?

7. A first production run of a new product is made with the following data compiled:

$$
\begin{aligned}
\text{Direct material cost} &= \$\ 5.00/\text{unit} \\
\text{Direct labor cost} &= \$\ 2.90/\text{unit} \\
\text{Overhead cost} &= \$\ 4.10/\text{unit} \\
\text{Selling price} &= \$20.00/\text{unit}
\end{aligned}
$$

The above costs are based on all good units produced and no rejects or scrap. Parts are inspected only once, after all the manufacturing operations are performed but prior to shipping. During the production run, a scrap rate of 30% has occurred. What is the maximum scrap rate permissible in order to break even (i.e., total costs equal to total revenues)? (8.2.5)

8. A typical gasoline-powered farm tractor with PTO rating of 35 horsepower has an operating cost of about $3.50/hour of use (for fuel, lubricants, oil filters, repairs, and maintenance), excluding labor cost, and an annual fixed cost of about $1200 (for depreciation, insurance and taxes, housing, and opportunity cost).

Similar data for a diesel-powered tractor of the same size are $2.75/hour of use for operating cost and $1425 annual fixed cost[4] (8.2.5)

a. What is the number of operating hours per year for break-even between the two tractors?

b. If the estimated number of operating hours per year is 600, what annual savings are estimated if the diesel-powered tractor is purchased instead of the gasoline-powered tractor?

c. For the diesel-powered tractor operated at 600 hours per year, what revenues/hour must be generated in order to break-even?

9. A new engineering building at State University is to contain 10,000,000 square feet. The total cost of the building (TC) is given by

$$TC = (200 + 80X + 2X^2)A$$
where X = number of floors
A = floor area in ft^2/per floor

a. Create a table which shows the total building cost, average cost per floor, and marginal cost per floor (using the difference equation approach) for configurations ranging from 1 floor to 12 floors, inclusive. Hint: You may want to create the table using your favorite spreadsheet program.

b. Based on your table, what is the optimal number of floors for the building? Justify your answer based on the "total cost" column.

c. Justify your answer in part (b) based on the "marginal cost per floor" column.

d. Demonstrate, using differential calculus, that your answer in parts (b) and (c) is correct. Note: For this part assume that X is a continuous variable. (8.2.6)

10. An oil refinery produces one base type of crude oil for the southeastern U.S. market. The two equations below give the relationships that approximate the total cost and the total profit per week in dollars. (*Note:* This problem is for illustration purposes only; the total cost function and the cost coefficients are not necessarily realistic.)

$$\text{Total cost, } TC(x) = 50,000 + 20.2x + 0.0001x^2, \text{ for } x \geq 0$$
$$= 0 \text{ otherwise}$$
$$\text{Total profit, } TP(x) = Sx - TC(x)$$

where x = amount of crude oil produced, barrels/week
S = sales prive of crude oil, dollars/barrel
= $35/barrel.

a. At what level of production in barrels/week is the cost/barrel minimum? What is the minimum cost/barrel?

---

[4] Example problem based on Publication 510 (revised February 1974) entitled "Farm Machinery Performance and Costs," Extension Division. Virginia Polytechnic Institute and State University, Blacksburg, Virginia.

b. What is the maximum weekly profit that the company can make? At what level of production is the maximum weekly profit attainable?

c. Over what range of production is profit possible?

d. Suppose the refinery capacity is limited to 50,000 barrels/week. What is the range of production over which profit is possible? (8.2.6)

11. Production of a particular type of annual crop is a function of fertilizer used and is given by the following relationship:

$$P(t) = \text{crop production, barrels/acre}$$
$$= 0.417t - 0.00125t^2$$

where t = amount of fertilizer used, pounds/acre.

The crop can be sold at a price of $15/barrel. Given that one barrel of crop weighs about 120 pounts, the total cost of crop production is

$$TC(t) = \$(220 + 2t) \text{ per acre}$$

a. How much fertilizer should the farmer use for this crop per acre of land to maximize his annual profit?

b. What is the maximum annual profit per acre of land for this crop?

c. Between what range of fertilizer application is a profit possible? (8.2.6)

12. Assume the total annual inventory cost for a particular item carried in inventory is

$$TC(Q) = \text{annual cost of ordering} + \text{annual cost}$$
$$\text{of carrying inventory, dollars/year}$$

$$= P_c\left(\frac{A}{Q}\right) + (0.60\,Q + 150),$$

where P_c = cost of preparing a purchase order (a constant value, regardless of the quantity ordered)

= $15 per purchase order.

Q = the number of items ordered each time a purchase order is placed.

A = total annual demand for this item

= 2000 units.

A/Q = the number of orders placed each year.

Note that the total annual inventory cost $TC(Q)$ is a function of the order size Q. (8.2.6)

a. Determine the value of Q that will minimize the total annual inventory cost (i.e., determine the economic order quantity Q^*).

b. What will be the corresponding total annual inventory cost, $TC(Q^*)$?

c. For the economic order quantity Q^*, determine the number of orders that will be placed each year.

d. If the cost of placing a purchase order is decreased by 50%, what is the effect on the economic order quantity?

13. A worker in a company is on an incentive system. The worker earns $1.50 for each part produced. In *addition,* the worker earns a guaranteed wage of $5/hour regardless of production output. The product is sold at a price of $(5 - 0.4t)$, dollars/unit, where t is the number of units produced per hour. (8.2.6)

a. At what rate of production will the company earn the most profit from this worker's output? (Assume all units produced can be sold.)

b. What is the amount of maximum profit per hour?

c. At what rate of production will the revenue be maximized? What is the corresponding profit per hour?

d. Within what range of production can the company make a profit?

14. When electrical current flows in a conductor there is a power loss due to electrical resistance. Resistance is inversely proportional to the size of the conductor and thus, the cost of power loss will decrease with increased conductor size. However, as the size of the conductor increases, the investment in the conductor also increases. Define A to be the cross-sectional area (square inches) of a conductor and assume the total cost per year (TC) of transmitting a given quantity of current over a fixed distance by a copper conductor is given by:

$$TC = \$20 + \$30A + \frac{\$270}{A}$$

a. Determine the cross-sectional area of the copper conductor which will result in minimum total cost/year.

b. Using the cross-sectional area determined in (a), what is the minimum total cost per year to operate the conductor system?

c. For 2-inch and 4-inch diameter copper conductors, determine the total cost per year. (8.2.6)

15. A large coal mining company owns and operates a fleet of four-wheel drive vehicles. Maintenance records over a period of five years reveal that about 20% of the time, tires are replaced after 10,000 miles of use. Fifteen percent of the tires are replaced after 15,000 miles, 25% are replaced after 20,000 miles, 30% are replaced after 25,000 miles, and 10% are replaced after 30,000 miles of use. Determine a weighted-average estimate of tire life. (8.3.1)

16. On January 3, 19__, a manufacturing firm purchased a drill press for $15,000, a fork-lift truck for $25,000, and a microcomputer system for $10,000. For depreciation purposes, the estimated lives of these items are 10, 7, and 5 years, respectively. Annual depreciation expenses for the items are $1300, $3000, and $2000, respectively. Assume it is now three years later and December 31, 19__. How would these items appear under the Fixed Assets portion of the Balance Sheet? What is the total Book Value and the total amount in the Depreciation Reserve for the items? (8.4.1)

17. K. Z. Moley purchased a small retail restaurant business and opened on July 1, 19__. At the date of opening, he had invested $20,000 of equity funds with the following breakdown: $10,000 in equipment, $7000 in inventory items, and $3000 in operating cash.

a. Prepare a balance sheet for the business as of July 1, 19__. (2.4.1)

The data below summarize the gross sales and expenses for the restaurant during the first 3-month period.

Gross sales	$32,000
Purchases	12,500
Salaries	6,000
Advertising expense	1,000
Rent expense	1,800
Expense for utilities	900
Expense for misc. supplies	600

For the purchases, $10,000 was paid with cash and $2500 is still owed. At the close of the 3-month period, the end-of-period inventory is worth $5500.

b. Prepare an income statement for the 3-month period covered. (2.4.2)

c. If, at the end of the 3-month period on October 1, 19__, the net worth (or ownership) account is $22,000, determine the amount of the cash account in order to balance the accounting equation as of October 1, 19__. (8.4.1)

18. A successful building contractor purchased a 150-acre farm to operate on a parttime basis and raise beef cattle. The purchase price of the farm was $150,000. The contractor paid 30% of the purchase price in cash and financed the remainder over 15 years at a 10% annual simple interest rate with a mortgage, payable to a local bank. Soon after the purchase of the farm, the contractor purchased cattle for $25,000, paid $10,000 in cash, and gave a promissory note to the seller for the balance. The note carried an 8% annual simple interest rate and was to be paid off within 5 years. During the first full year of operations, the farm resulted in the following revenues and expenses.

Calves sold (income)	$14,600
Timber sold (income)	5,000
Labor expenses	1,500
Expenses for machinery hired	2,750
Veterinarian fees	1,350
Fertilizer expenses	1,800
Property taxes	1,625
Feed purchased	275
Expenses for repairs	950
Interest expenses	11,700
Insurance expenses	1,500
Legal fees	1,350
Expenses for miscellaneous supplies	425

Prepare an income statement for the farming operation and determine the net profit (loss) before income taxes. (8.4.2)

19. An income statement for the WAC Company covering a calendar year ending December 31 is as follows.

Income Statement

Gross income from sales	$247,000	
Less: cost of goods sold	138,800	
Net income from sales	$108,200	
Operating expenses		
Rent	$ 9,700	
Salaries	35,200	
Depreciation	5,800	
Advertising	4,300	
Insurance	1,500	$56,500
Net profit before income taxes		51,700
Less: income taxes		23,973
Net profit after income taxes		$27,727

The Balance Sheet Accounts as of December 31 are:

Cash	$89,227
Accounts receivable	8,000
Notes payable	25,000
Raw material inventory	10,000
Work-in-process inventory	15,000
Accounts payable	6,000
Declared dividends	20,000
Finished goods inventory	18,500
Land	30,000
Original cost of building	80,000
Building—Reserve for depreciation	8,000
Original cost of equipment	40,000
Equipment—Reserve for depreciation	4,000

a. Determine the Earned Surplus for the year in question.
b. Prepare a Balance Sheet as of December 31. (8.4)
20. Comparative Balance Sheets and Income Statements for B&K Manufacturing for 19×5 and 19×6 are shown below. Based on these financial statements, determine the following ratios.
a. Return on Assets Employed
b. Current Ratio
c. Acid-Test Ratio
d. Accounts Receivable Turnover
e. Inventory Turnover
f. Debt to Equity Ratio
g. Operating Income to Total Assets Ratio
h. If the industry average for Return on Assets Employed is 9.9%, assess B&K's position relative to the industry on this measure.
i. If the industry average for Current Ratio is 2.15, assess B&K's position relative to the industry on this measure.
j. If the industry average for Debt to Equity Ratio is 85%, assess B&K's position relative to the industry on this measure.

B&K Manufacturing
Comparative Balance Sheets
as of December 31, 19×5 and 19×6
(in thousands)

ASSETS	19×6	19×5
Current Assets		
Cash	$120	$390
Temporary Investments	–	250
Accounts Receivable (net)	2,400	2,000
Inventory	1,900	2,100
Prepaid Expenses	65	60
Total Current Assets	$4,485	$4,800

Fixed Assets Net	$2,015	$1,200
Total Assets	$6,500	$6,000

LIABILITIES AND CAPITAL

Current Liabilities

Notes Payable	$125	$50
Accounts Payable	2,345	2,230
Taxes Payable	130	120
Total Current Liabilities	$2,600	$2,400

Fixed Liabilities

Mortgage Payable	$146	$100
Long Term Loans	$270	–
Total Liabilities	$3,016	$2,500

Capital

Stock	$1,950	$2,000
Retained Earnings	809	900
Earned Surplus	725	600
Total Capital	$3,484	$3,500
Total Liabilities and Capital	6,500	6,000

B&K Manufacturing
Comparative Income Statements
For Years Ended December 31, 19×5 and 19×6
(in thousands)

	19×6	19×5
Net Sales	$25,000	$20,000
Cost of Goods Sold	20,000	16,000
Gross Profit	5,000	4,000
Other Operating Expenses	4,025	3,185
Net Operating Income	975	815
Less: Interest Expenses	45	15
Less: Income Taxes	205	200
Net Income	$725	$600

21. An order for 5000 units of Part D142 is received by the J. T. Kling Engineering Company, a small machine shop. The raw material for this part is SAE 1020 steel rectangular bar stock. Each unit of bar stock costs $30 and yields 20 D142 parts. The basic manufacturing sequence, with standard operation times per part and machine overhead rates (dollars/hour), is given in the following table.

Manufacturing Operation	Standard Time Per Part	Machine Overhead Rate
Cutoff on power hacksaw	2 minutes	$1.20/hour
Mill two sides; deburr	5 minutes	2.00/hour
Drill three 3/8-inch diam. holes	3 minutes	1.60/hour
Surface grind one side; deburr	3 minutes	1.40/hour
Package	0.25 minutes	—

The direct labor time per part is the same as the machine time per part (packaging is a manual operation only). The tooling cost for this job order is estimated to be $500. In addition to tooling costs and machine overhead costs, factory overhead costs (for indirect labor, indirect materials, utilities, etc.) are $15 per direct labor hour. The average direct labor hour rate, including fringe benefits, is $10 per hour. (8.5.1)

a. Assuming a zero scrap rate during production, determine the total estimated costs for the job order of 7000 units.

b. Determine the *unit* selling price if profit is to be 30% of the total cost.

22. The welding department of a mining equipment manufacturing plant consists of four cost centers: manual arc welding (A), semiautomatic welding (B), furnace brazing and heat treating (C), and finishing (D). Some oxyacetylene cutting is also done in Center B. Assume that it is possible to allocate departmental overhead expenses directly to each cost center, and that the following data for the welding department were compiled last year by the accounting system.

Cost Center	Departmental Expenses	Direct Labor Hours	Cost	Direct Material
A	$21,000	10,000	$75,000	$12,000
B	10,000	4,000	18,000	12,000
C	5,400	1,500	9,000	4,500
D	4,800	2,800	11,200	3,000

Compute the overhead rate (or rates) applicable by the following methods. (8.5.2)

a. Percentage of direct labor cost for each cost center.

b. Percentage of prime cost for each cost center.

c. Rate per direct labor hour for each cost center.

23. Given the welding department and four cost centers found in Problem 20, the direct labor hours and costs for each cost center remain the same, but new and additional data are given below (assume cost data are for the previous year).

Cost Center	Square Feet Occupied	Cost of Machinery	Number of Direct Labor Employees
A	900	$10,000	5
B	400	9,000	2
C	600	11,700	1
D	500	4,800	1
Total	2,400	$35,500	9

Expenses other than for direct labor and materials chargeable to the welding department last year were

Maintenance	$ 6,400
Gas and electricity	20,000
Supervision and other indirect labor	30,000
Miscellaneous supplies	7,500
Equipment depreciation	4,600
Building depreciation	8,000

Determine an overhead rate per direct labor hour for each cost center if the welding department expenses above are first allocated to each cost center as follows. (8.5.1)

a. Maintenance expenses and equipment depreciation expenses are allocated according to the value of equipment (percent of total) in each cost center.

b. Building depreciation expenses are allocated according to the floor space occupied (percent of total) by each cost center.

c. Supervision and other indirect labor expenses are allocated according to the number of direct labor employees (percent of total) of each cost center.

d. Supplies and gas and electricity expenses are allocated according to the number of direct labor hours (percent of total) for each cost center.

Appendix A

Discrete Compounding

Section I Discrete Compound Interest Factors
Section II Geometric Series Factors

Time Value of Money Factors—Discrete Compounding   i = 0.5%

	Single Sums		Uniform Series				Gradient Series	
	To Find F Given P $(F\|P, i\%, n)$	To Find P Given F $(P\|F, i\%, n)$	To Find F Given A $(F\|A, i\%, n)$	To Find A Given F $(A\|F, i\%, n)$	To Find P Given A $(P\|A, i\%, n)$	To Find A Given P $(A\|P, i\%, n)$	To Find P Given G $(P\|G, i\%, n)$	To Find A Given G $(A\|G, i\%, n)$
n								
1	1.0050	0.9950	1.0000	1.0000	0.9950	1.0050	0.0000	0.0000
2	1.0100	0.9901	2.0050	0.4988	1.9851	0.5038	0.9901	0.4988
3	1.0151	0.9851	3.0150	0.3317	2.9702	0.3367	2.9604	0.9967
4	1.0202	0.9802	4.0301	0.2481	3.9505	0.2531	5.9011	1.4938
5	1.0253	0.9754	5.0503	0.1980	4.9259	0.2030	9.8026	1.9900
6	1.0304	0.9705	6.0755	0.1646	5.8964	0.1696	14.6552	2.4855
7	1.0355	0.9657	7.1059	0.1407	6.8621	0.1457	20.4493	2.9801
8	1.0407	0.9609	8.1414	0.1228	7.8230	0.1278	27.1755	3.4738
9	1.0459	0.9561	9.1821	0.1089	8.7791	0.1139	34.8244	3.9668
10	1.0511	0.9513	10.2280	0.0978	9.7304	0.1028	43.3865	4.4589
11	1.0564	0.9466	11.2792	0.0887	10.6770	0.0937	52.8526	4.9501
12	1.0617	0.9419	12.3356	0.0811	11.6189	0.0861	63.2136	5.4406
13	1.0670	0.9372	13.3972	0.0746	12.5562	0.0796	74.4602	5.9302
14	1.0723	0.9326	14.4642	0.0691	13.4887	0.0741	86.5835	6.4190
15	1.0777	0.9279	15.5365	0.0644	14.4166	0.0694	99.5743	6.9069
16	1.0831	0.9233	16.6142	0.0602	15.3399	0.0652	113.4238	7.3940
17	1.0885	0.9187	17.6973	0.0565	16.2586	0.0615	128.1231	7.8803
18	1.0939	0.9141	18.7858	0.0532	17.1728	0.0582	143.6634	8.3658
19	1.0994	0.9096	19.8797	0.0503	18.0824	0.0553	160.0360	8.8504
20	1.1049	0.9051	20.9791	0.0477	18.9874	0.0527	177.2322	9.3342
21	1.1104	0.9006	22.0840	0.0453	19.8880	0.0503	195.2434	9.8172
22	1.1160	0.8961	23.1944	0.0431	20.7841	0.0481	214.0611	10.2993
23	1.1216	0.8916	24.3104	0.0411	21.6757	0.0461	233.6768	10.7806
24	1.1272	0.8872	25.4320	0.0393	22.5629	0.0443	254.0820	11.2611
25	1.1328	0.8828	26.5591	0.0377	23.4456	0.0427	275.2686	11.7407
26	1.1385	0.8784	27.6919	0.0361	24.3240	0.0411	297.2281	12.2195
27	1.1442	0.8740	28.8304	0.0347	25.1980	0.0397	319.9523	12.6975
28	1.1499	0.8697	29.9745	0.0334	26.0677	0.0384	343.4332	13.1747
29	1.1556	0.8653	31.1244	0.0321	26.9330	0.0371	367.6625	13.6510
30	1.1614	0.8610	32.2800	0.0310	27.7941	0.0360	392.6324	14.1265
36	1.1967	0.8356	39.3361	0.0254	32.8710	0.0304	557.5598	16.9621
42	1.2330	0.8110	46.6065	0.0215	37.7983	0.0265	747.1886	19.7678
48	1.2705	0.7871	54.0978	0.0185	42.5803	0.0235	959.9188	22.5437
54	1.3091	0.7639	61.8167	0.0162	47.2214	0.0212	1.194E + 03	25.2899
60	1.3489	0.7414	69.7700	0.0143	51.7256	0.0193	1.449E + 03	28.0064
66	1.3898	0.7195	77.9650	0.0128	56.0970	0.0178	1.722E + 03	30.6932
72	1.4320	0.6983	86.4089	0.0116	60.3395	0.0166	2.012E + 03	33.3504
120	1.8194	0.5496	163.8793	6.102E − 03	90.0735	0.0111	4.824E + 03	53.5508
180	2.4541	0.4075	290.8187	3.439E − 03	118.5035	8.439E − 03	9.031E + 03	76.2115
360	6.0226	0.1660	1.005E + 03	9.955E − 04	166.7916	5.996E − 03	2.140E + 04	128.3236

Time Value of Money Factors—Discrete Compounding   i = 0.75%

	Single Sums		Uniform Series				Gradient Series	
	To Find F Given P (F|P, i%, n)	To Find P Given F (P|F, i%, n)	To Find F Given A (F|A, i%, n)	To Find A Given F (A|F, i%, n)	To Find P Given A (P|A, i%, n)	To Find A Given P (A|P, i%, n)	To Find P Given G (P|G, i%, n)	To Find A Given G (A|G, i%, n)
n								
1	1.0075	0.9926	1.0000	1.0000	0.9926	1.0075	0.0000	0.0000
2	1.0151	0.9852	2.0075	0.4981	1.9777	0.5056	0.9852	0.4981
3	1.0227	0.9778	3.0226	0.3308	2.9556	0.3383	2.9408	0.9950
4	1.0303	0.9706	4.0452	0.2472	3.9261	0.2547	5.8525	1.4907
5	1.0381	0.9633	5.0756	0.1970	4.8894	0.2045	9.7058	1.9851
6	1.0459	0.9562	6.1136	0.1636	5.8458	0.1711	14.4866	2.4782
7	1.0537	0.9490	7.1595	0.1397	6.7946	0.1472	20.1808	2.9701
8	1.0616	0.9420	8.2132	0.1218	7.7366	0.1293	26.7747	3.4608
9	1.0696	0.9350	9.2748	0.1078	8.6716	0.1153	34.2544	3.9502
10	1.0776	0.9280	10.3443	0.0967	9.5996	0.1042	42.6064	4.4384
11	1.0857	0.9211	11.4219	0.0876	10.5207	0.0951	51.8174	4.9253
12	1.0938	0.9142	12.5076	0.0800	11.4349	0.0875	61.8740	5.4110
13	1.1020	0.9074	13.6014	0.0735	12.3423	0.0810	72.7632	5.8954
14	1.1103	0.9007	14.7034	0.0680	13.2430	0.0755	84.4720	6.3766
15	1.1186	0.8940	15.8137	0.0632	14.1370	0.0707	96.9876	6.8606
16	1.1270	0.8873	16.9323	0.0591	15.0243	0.0666	110.2973	7.3413
17	1.1354	0.8807	18.0593	0.0554	15.9050	0.0629	124.3887	7.8207
18	1.1440	0.8742	19.1947	0.0521	16.7792	0.0596	139.2494	8.2989
19	1.1525	0.8676	20.3387	0.0492	17.6468	0.0567	154.8671	8.7759
20	1.1612	0.8612	21.4912	0.0465	18.5080	0.0540	171.2297	9.2516
21	1.1699	0.8548	22.6524	0.0441	19.3628	0.0516	188.3253	9.7261
22	1.1787	0.8484	23.8223	0.0420	20.2112	0.0495	206.1420	10.1994
23	1.1875	0.8421	25.0010	0.0400	21.0533	0.0475	224.6682	10.6714
24	1.1964	0.8358	26.1885	0.0382	21.8891	0.0457	243.8923	11.1422
25	1.2054	0.8296	27.3849	0.0365	22.7188	0.0440	263.8029	11.6117
26	1.2144	0.8234	28.5903	0.0350	23.5422	0.0425	284.3888	12.0800
27	1.2235	0.8173	29.8047	0.0336	24.3595	0.0411	305.6387	12.5470
28	1.2327	0.8112	31.0282	0.0322	25.1707	0.0397	327.5416	13.0128
29	1.2420	0.8052	32.2609	0.0310	25.9759	0.0385	350.0867	13.4774
30	1.2513	0.7992	33.5029	0.0298	26.7751	0.0373	373.2631	13.9407
36	1.3086	0.7641	41.1527	0.0243	31.4468	0.0318	524.9924	16.6946
42	1.3686	0.7306	49.1533	0.0203	35.9137	0.0278	696.8709	19.4040
48	1.4314	0.6986	57.5207	0.0174	40.1848	0.0249	886.8404	22.0691
54	1.4907	0.6680	66.2718	0.0151	44.2686	0.0226	1.093E + 03	24.6898
60	1.5657	0.6387	75.4241	0.0133	48.1734	0.0208	1.314E + 03	27.2665
66	1.6375	0.6107	84.9961	0.0118	51.9070	0.0193	1.547E + 03	29.7992
72	1.7126	0.5839	95.0070	0.0105	55.4768	0.0180	1.791E + 03	32.2882
120	2.4514	0.4079	193.5143	5.168E − 03	78.9417	0.0127	3.999E + 03	50.6521
180	3.8380	0.2605	378.4058	2.643E − 03	98.5934	0.0101	6.893E + 03	69.9094
360	14.7306	0.0679	1.831E + 03	5.462E − 04	124.2819	8.046E − 03	1.331E + 04	107.1145

414 Appendix A   Discrete Compounding

Time Value of Money Factors—Discrete Compounding   i = 1%

	Single Sums		Uniform Series				Gradient Series	
	To Find F Given P $(F\|P, i\%, n)$	To Find P Given F $(P\|F, i\%, n)$	To Find F Given A $(F\|A, i\%, n)$	To Find A Given F $(A\|F, i\%, n)$	To Find P Given A $(P\|A, i\%, n)$	To Find A Given P $(A\|P, i\%, n)$	To Find P Given G $(P\|G, i\%, n)$	To Find A Given G $(A\|G, i\%, n)$
n								
1	1.0100	0.9901	1.0000	1.0000	0.9901	1.0100	0.0000	0.0000
2	1.0201	0.9803	2.0100	0.4975	1.9704	0.5075	0.9803	0.4975
3	1.0303	0.9706	3.0301	0.3300	2.9410	0.3400	2.9215	0.9934
4	1.0406	0.9610	4.0604	0.2463	3.9020	0.2563	5.8044	1.4876
5	1.0510	0.9515	5.1010	0.1960	4.8534	0.2060	9.6103	1.9801
6	1.0615	0.9420	6.1520	0.1625	5.7955	0.1725	14.3205	2.4710
7	1.0721	0.9327	7.2135	0.1386	6.7282	0.1486	19.9168	2.9602
8	1.0829	0.9235	8.2857	0.1207	7.6517	0.1307	26.3812	3.4478
9	1.0937	0.9143	9.3685	0.1067	8.5660	0.1167	33.6959	3.9337
10	1.1046	0.9053	10.4622	0.0956	9.4713	0.1056	41.8435	4.4179
11	1.1157	0.8963	11.5668	0.0865	10.3676	0.0965	50.8067	4.9005
12	1.1268	0.8874	12.6825	0.0788	11.2551	0.0888	60.5687	5.3815
13	1.1381	0.8787	13.8093	0.0724	12.1337	0.0824	71.1126	5.8607
14	1.1495	0.8700	14.9474	0.0669	13.0037	0.0769	82.4221	6.3384
15	1.1610	0.8613	16.0969	0.0621	13.8651	0.0721	94.4810	6.8143
16	1.1726	0.8528	17.2579	0.0579	14.7179	0.0679	107.2734	7.2886
17	1.1843	0.8444	18.4304	0.0543	15.5623	0.0643	120.7834	7.7613
18	1.1961	0.8360	19.6147	0.0510	16.3983	0.0610	134.9957	8.2323
19	1.2081	0.8277	20.8109	0.0481	17.2260	0.0581	149.8950	8.7017
20	1.2202	0.8195	22.0190	0.0454	18.0456	0.0554	165.4664	9.1694
21	1.2324	0.8114	23.2392	0.0430	18.8570	0.0530	181.6950	9.6354
22	1.2447	0.8034	24.4716	0.0409	19.6604	0.0509	198.5663	10.0998
23	1.2572	0.7954	25.7163	0.0389	20.4558	0.0489	216.0660	10.5626
24	1.2697	0.7876	26.9735	0.0371	21.2434	0.0471	234.1800	11.0237
25	1.2824	0.7798	28.2432	0.0354	22.0232	0.0454	252.8945	11.4831
26	1.2953	0.7720	29.5256	0.0339	22.7952	0.0439	272.1957	11.9409
27	1.3082	0.7644	30.8209	0.0324	23.5596	0.0424	292.0702	12.3971
28	1.3213	0.7568	32.1291	0.0311	24.3164	0.0411	312.5047	12.8516
29	1.3345	0.7493	33.4504	0.0299	25.0658	0.0399	333.4863	13.3044
30	1.3478	0.7419	34.7849	0.0287	25.8077	0.0387	355.0021	13.7557
36	1.4308	0.6989	43.0769	0.0232	30.1075	0.0332	494.6207	16.4285
42	1.5188	0.6584	51.8790	0.0193	34.1581	0.0293	650.4514	19.0424
48	1.6122	0.6203	61.2226	0.0163	37.9740	0.0263	820.1460	21.5976
54	1.7114	0.5843	71.1410	0.0141	41.5687	0.0241	1.002E + 03	24.0945
60	1.8167	0.5504	81.6697	0.0122	44.9550	0.0222	1.193E + 03	26.5333
66	1.9285	0.5185	92.8460	0.0108	48.1452	0.0208	1.392E + 03	28.9146
72	2.0471	0.4885	104.7099	9.550E + 03	51.1504	0.0196	1.598E + 03	31.2386
120	3.3004	0.3030	230.0387	4.347E - 03	69.7005	0.0143	3.334E + 03	47.8349
180	5.9958	0.1668	499.5802	2.002E - 03	83.3217	0.0120	5.330E + 03	63.9697
360	35.9496	0.0278	3.495E + 03	2.861E - 04	97.2183	0.0103	8.720E + 03	89.6995

Time Value of Money Factors—Discrete Compounding   i = 1.5%

	Single Sums		Uniform Series				Gradient Series	
	To Find F Given P (F|P, i%, n)	To Find P Given F (P|F, i%, n)	To Find F Given A (F|A, i%, n)	To Find A Given F (A|F, i%, n)	To Find P Given A (P|A, i%, n)	To Find A Given P (A|P, i%, n)	To Find P Given G (P|G, i%, n)	To Find A Given G (A|G, i%, n)
n								
1	1.0150	0.9852	1.0000	1.0000	0.9852	1.0150	0.0000	0.0000
2	1.0302	0.9707	2.0150	0.4963	1.9559	0.5113	0.9707	0.4963
3	1.0457	0.9563	3.0452	0.3284	2.9122	0.3434	2.8833	0.9901
4	1.0614	0.9422	4.0909	0.2444	3.8544	0.2594	5.7098	1.4814
5	1.0773	0.9283	5.1523	0.1941	4.7826	0.2091	9.4229	1.9702
6	1.0934	0.9145	6.2296	0.1605	5.6972	0.1755	13.9956	2.4566
7	1.1098	0.9010	7.3230	0.1366	6.5982	0.1516	19.4018	2.9405
8	1.1265	0.8877	8.4328	0.1186	7.4859	0.1336	25.6157	3.4219
9	1.1434	0.8746	9.5593	0.1046	8.3605	0.1196	32.6125	3.9008
10	1.1605	0.8617	10.7027	0.0934	9.2222	0.1084	40.3675	4.3772
11	1.1779	0.8489	11.8633	0.0843	10.0711	0.0993	48.8568	4.8512
12	1.1956	0.8364	13.0412	0.0767	10.9075	0.0917	58.0571	5.3227
13	1.2136	0.8240	14.2368	0.0702	11.7315	0.0852	67.9454	5.7917
14	1.2318	0.8118	15.4504	0.0647	12.5434	0.0797	78.4994	6.2582
15	1.2502	0.7999	16.6821	0.0599	13.3432	0.0749	89.6974	6.7223
16	1.2690	0.7880	17.9324	0.0558	14.1313	0.0708	101.5178	7.1839
17	1.2880	0.7764	19.2014	0.0521	14.9076	0.0671	113.9400	7.6431
18	1.3073	0.7649	20.4894	0.0488	15.6726	0.0638	126.9435	8.0997
19	1.3270	0.7536	21.7967	0.0459	16.4262	0.0609	140.5084	8.5539
20	1.3469	0.7425	23.1237	0.0432	17.1686	0.0582	154.6154	9.0057
21	1.3671	0.7315	24.4705	0.0409	17.9001	0.0559	169.2453	9.4550
22	1.3876	0.7207	25.8376	0.0387	18.6208	0.0537	184.3798	9.9018
23	1.4084	0.7100	27.2251	0.0367	19.3309	0.0517	200.0006	10.3462
24	1.4295	0.6995	28.6335	0.0349	20.0304	0.0499	216.0901	10.7881
25	1.4509	0.6892	30.0630	0.0333	20.7196	0.0483	232.6310	11.2276
26	1.4727	0.6790	31.5140	0.0317	21.3986	0.0467	249.6065	11.6646
27	1.4948	0.6690	32.9867	0.0303	22.0676	0.0453	267.0002	12.0992
28	1.5172	0.6591	34.4815	0.0290	22.7267	0.0440	284.7958	12.5313
29	1.5400	0.6494	35.9987	0.0278	23.3761	0.0428	302.9779	12.9610
30	1.5631	0.6398	37.5387	0.0266	24.0158	0.0416	321.5310	13.3883
36	1.7091	0.5851	47.2760	0.0212	27.6607	0.0362	439.8303	15.9009
42	1.8688	0.5351	57.9231	0.0173	30.9941	0.0323	568.0201	18.3267
48	2.0435	0.4894	69.5652	0.0144	34.0426	0.0294	703.5463	20.6667
54	2.2344	0.4475	82.2952	0.0122	36.8305	0.0272	844.2184	22.9217
60	2.4432	0.4093	96.2147	0.0104	39.3803	0.0254	988.1674	25.0930
66	2.6715	0.3743	111.4348	8.974E − 03	41.7121	0.0240	1.134E + 03	27.1817
72	2.9212	0.3423	128.0772	7.808E − 03	43.8447	0.0228	1.280E + 03	29.1893
120	5.9693	0.1675	331.2882	3.019E − 03	55.4985	0.0180	2.360E + 03	42.5185
180	14.5844	0.0686	905.6245	1.104E − 03	62.0956	0.0161	3.317E + 03	53.4161
360	212.7038	0.0047	1.411E + 04	7.085E − 05	66.3532	0.0151	4.311E + 03	64.9662

Time Value of Money Factors—Discrete Compounding   i = 2%

	Single Sums		Uniform Series				Gradient Series	
	To Find F Given P	To Find P Given F	To Find F Given A	To Find A Given F	To Find P Given A	To Find A Given P	To Find P Given G	To Find A Given G
n	$(F|P, i\%, n)$	$(P|F, i\%, n)$	$(F|A, i\%, n)$	$(A|F, i\%, n)$	$(P|A, i\%, n)$	$(A|P, i\%, n)$	$(P|G, i\%, n)$	$(A|G, i\%, n)$
1	1.0200	0.9804	1.0000	1.0000	0.9804	1.0200	0.0000	0.0000
2	1.0404	0.9612	2.0200	0.4950	1.9416	0.5150	0.9612	0.4950
3	1.0612	0.9423	3.0604	0.3268	2.8839	0.3468	2.8458	0.9868
4	1.0824	0.9238	4.1216	0.2426	3.8077	0.2626	5.6173	1.4752
5	1.1041	0.9057	5.2040	0.1922	4.7135	0.2122	9.2403	1.9604
6	1.1262	0.8880	6.3081	0.1585	5.6014	0.1785	13.6801	2.4423
7	1.1487	0.8706	7.4343	0.1345	6.4720	0.1545	18.9035	2.9208
8	1.1717	0.8535	8.5830	0.1165	7.3255	0.1365	24.8779	3.3961
9	1.1951	0.8368	9.7546	0.1025	8.1622	0.1225	31.5720	3.8681
10	1.2190	0.8203	10.9497	0.0913	8.9826	0.1113	38.9551	4.3367
11	1.2434	0.8043	12.1687	0.0822	9.7868	0.1022	46.9977	4.8021
12	1.2682	0.7885	13.4121	0.0746	10.5753	0.0946	55.6712	5.2642
13	1.2936	0.7730	14.6803	0.0681	11.3484	0.0881	64.9475	5.7231
14	1.3195	0.7579	15.9739	0.0626	12.1062	0.0826	74.7999	6.1786
15	1.3459	0.7430	17.2934	0.0578	12.8493	0.0778	85.2021	6.6309
16	1.3728	0.7284	18.6393	0.0537	13.5777	0.0737	96.1288	7.0799
17	1.4002	0.7142	20.0121	0.0500	14.2919	0.0700	107.5554	7.5256
18	1.4282	0.7002	21.4123	0.0467	14.9920	0.0667	119.4581	7.9681
19	1.4568	0.6864	22.8406	0.0438	15.6785	0.0638	131.8139	8.4073
20	1.4859	0.6730	24.2974	0.0412	16.3514	0.0612	144.6003	8.8433
21	1.5157	0.6598	25.7833	0.0388	17.0112	0.0588	157.7959	9.2760
22	1.5460	0.6468	27.2990	0.0366	17.6580	0.0566	171.3795	9.7055
23	1.5769	0.6342	28.8450	0.0347	18.2922	0.0547	185.3309	10.1317
24	1.6084	0.6217	30.4219	0.0329	18.9139	0.0529	199.6305	10.5547
25	1.6406	0.6095	32.0303	0.0312	19.5235	0.0512	214.2592	10.9745
26	1.6734	0.5976	33.6709	0.0297	20.1210	0.0497	229.1987	11.3910
27	1.7069	0.5859	35.3443	0.0283	20.7069	0.0483	244.4311	11.8043
28	1.7410	0.5744	37.0512	0.0270	21.2813	0.0470	259.9392	12.2145
29	1.7758	0.5631	38.7922	0.0258	21.8444	0.0458	275.7064	12.6214
30	1.8114	0.5521	40.5681	0.0246	22.3965	0.0446	291.7164	13.0251
36	2.0399	0.4902	51.9944	0.0192	25.4888	0.0392	392.0405	15.3809
42	2.2972	0.4353	64.8622	0.0154	28.2348	0.0354	497.6010	17.6237
48	2.5871	0.3865	79.3535	0.0126	30.6731	0.0326	605.9657	19.7556
54	2.9135	0.3432	95.6731	0.0105	32.8383	0.0305	715.1815	21.7789
60	3.2810	0.3048	114.0515	8.768E − 03	34.7609	0.0288	823.6975	23.6961
66	3.6950	0.2706	134.7487	7.421E − 03	36.4681	0.0274	930.3000	25.5100
72	4.1611	0.2403	158.0570	6.327E − 03	37.9841	0.0263	1.034E + 03	27.2234
120	10.7652	0.0929	488.2582	2.048E − 03	45.3554	0.0220	1.710E + 03	37.7114
180	35.3208	0.0283	1.716E + 03	5.827E − 04	48.5844	0.0206	2.174E + 03	44.7554
360	1.248E + 03	8.016E − 04	6.233E + 04	1.604E − 05	49.9599	0.0200	2.484E + 03	49.7112

Time Value of Money Factors—Discrete Compounding   i = 3%

	Single Sums		Uniform Series				Gradient Series	
	To Find F Given P	To Find P Given F	To Find F Given A	To Find A Given F	To Find P Given A	To Find A Given P	To Find P Given G	To Find A Given G
n	$(F\vert P, i\%, n)$	$(P\vert F, i\%, n)$	$(F\vert A, i\%, n)$	$(A\vert F, i\%, n)$	$(P\vert A, i\%, n)$	$(A\vert P, i\%, n)$	$(P\vert G, i\%, n)$	$(A\vert G, i\%, n)$
1	1.0300	0.9709	1.0000	1.0000	0.9709	1.0300	0.0000	0.0000
2	1.0609	0.9426	2.0300	0.4926	1.9135	0.5226	0.9426	0.4926
3	1.0927	0.9151	3.0909	0.3235	2.8286	0.3535	2.7729	0.9803
4	1.1255	0.8885	4.1836	0.2390	3.7171	0.2690	5.4383	1.4631
5	1.1593	0.8626	5.3091	0.1884	4.5797	0.2184	8.8888	1.9409
6	1.1941	0.8375	6.4684	0.1546	5.4172	0.1846	13.0762	2.4138
7	1.2299	0.8131	7.6625	0.1305	6.2303	0.1605	17.9547	2.8819
8	1.2668	0.7894	8.8923	0.1125	7.0197	0.1425	23.4806	3.3450
9	1.3048	0.7664	10.1591	0.0984	7.7861	0.1284	29.6119	3.8032
10	1.3439	0.7441	11.4639	0.0872	8.5302	0.1172	36.3088	4.2565
11	1.3842	0.7224	12.8078	0.0781	9.2526	0.1081	43.5330	4.7049
12	1.4258	0.7014	14.1920	0.0705	9.9540	0.1005	51.2482	5.1485
13	1.4685	0.6810	15.6178	0.0640	10.6350	0.0940	59.4196	5.5872
14	1.5126	0.6611	17.0863	0.0585	11.2961	0.0885	68.0141	6.0210
15	1.5580	0.6419	18.5989	0.0538	11.9379	0.0838	77.0002	6.4500
16	1.6047	0.6232	20.1569	0.0496	12.5611	0.0796	86.3477	6.8742
17	1.6528	0.6050	21.7616	0.0460	13.1661	0.0760	96.0280	7.2936
18	1.7024	0.5874	23.4144	0.0427	13.7535	0.0727	106.0137	7.7081
19	1.7535	0.5703	25.1169	0.0398	14.3238	0.0698	116.2788	8.1179
20	1.8061	0.5537	26.8704	0.0372	14.8775	0.0672	126.7987	8.5229
21	1.8603	0.5375	28.6765	0.0349	15.4150	0.0649	137.5496	8.9231
22	1.9161	0.5219	30.5368	0.0327	15.9369	0.0627	148.5094	9.3186
23	1.9736	0.5067	32.4529	0.0308	16.4436	0.0608	159.6566	9.7093
24	2.0328	0.4919	34.4265	0.0290	16.9355	0.0590	170.9711	10.0954
25	2.0938	0.4776	36.4593	0.0274	17.4131	0.0574	182.4336	10.4768
26	2.1566	0.4637	38.5530	0.0259	17.8768	0.0559	194.0260	10.8535
27	2.2213	0.4502	40.7096	0.0246	18.3270	0.0546	205.7309	11.2255
28	2.2879	0.4371	42.9309	0.0233	18.7641	0.0533	217.5320	11.5930
29	2.3566	0.4243	45.2189	0.0221	19.1885	0.0521	229.4137	11.9558
30	2.4273	0.4120	47.5754	0.0210	19.6004	0.0510	241.3613	12.3141
36	2.8983	0.3450	63.2759	0.0158	21.8323	0.0458	313.7028	14.3688
42	3.4607	0.2890	82.0232	0.0122	23.7014	0.0422	385.5024	16.2650
48	4.1323	0.2420	104.4084	9.578E − 03	25.2667	0.0396	455.0255	18.0089
54	4.9341	0.2027	131.1375	7.626E − 03	26.5777	0.0376	521.1157	19.6073
60	5.8916	0.1697	163.0534	6.133E − 03	27.6756	0.0361	583.0526	21.0674
66	7.0349	0.1421	201.1627	4.971E − 03	28.5950	0.0350	640.4407	22.3969
72	8.4000	0.1190	246.6672	4.054E − 03	29.3651	0.0341	693.1226	23.6036
120	34.7110	0.0288	1.124E + 03	8.899E − 03	32.3730	0.0309	963.8635	29.7737
180	204.5034	4.890E − 03	6.783E + 03	1.474E − 04	33.1703	0.0301	1.076E + 03	32.4488
360	4.182E + 04	2.391E − 05	1.394E + 06	7.173E − 07	33.3325	0.0300	1.111E + 03	33.3247

Time Value of Money Factors—Discrete Compounding   i = 4%

	Single Sums		Uniform Series				Gradient Series	
n	To Find F Given P $(F\|P, i\%, n)$	To Find P Given F $(P\|F, i\%, n)$	To Find F Given A $(F\|A, i\%, n)$	To Find A Given F $(A\|F, i\%, n)$	To Find P Given A $(P\|A, i\%, n)$	To Find A Given P $(A\|P, i\%, n)$	To Find P Given G $(P\|G, i\%, n)$	To Find A Given G $(A\|G, i\%, n)$
1	1.0400	0.9615	1.0000	1.0000	0.9615	1.0400	0.0000	0.0000
2	1.0818	0.9246	2.0400	0.4902	1.8861	0.5302	0.9246	0.4902
3	1.1249	0.8890	3.1216	0.3203	2.7751	0.3603	2.7025	0.9739
4	1.1699	0.8548	4.2465	0.2355	3.6299	0.2755	5.2670	1.4510
5	1.2167	0.8219	5.4163	0.1846	4.4518	0.2246	8.5547	1.9216
6	1.2653	0.7903	6.6330	0.1508	5.2421	0.1908	12.5062	2.3857
7	1.3159	0.7599	7.8983	0.1266	6.0021	0.1666	17.0657	2.8433
8	1.3688	0.7307	9.2142	0.1085	6.7327	0.1485	22.1806	3.2944
9	1.4233	0.7026	10.5828	0.0945	7.4353	0.1345	27.8013	3.7391
10	1.4802	0.6756	12.0061	0.0833	8.1109	0.1233	33.8814	4.1773
11	1.5395	0.6496	13.4864	0.0741	8.7605	0.1141	40.3772	4.6090
12	1.6010	0.6246	15.0258	0.0666	9.3851	0.1066	47.2477	5.0343
13	1.6651	0.6006	16.6268	0.0601	9.9856	0.1001	54.4546	5.4533
14	1.7317	0.5775	18.2919	0.0547	10.5631	0.0947	61.9618	5.8659
15	1.8009	0.5553	20.0236	0.0499	11.1184	0.0899	69.7355	6.2721
16	1.8730	0.5339	21.8245	0.0458	11.6523	0.0858	77.7441	6.6720
17	1.9479	0.5134	23.6975	0.0422	12.1657	0.0822	85.9581	7.0656
18	2.0258	0.4936	25.6454	0.0390	12.6593	0.0790	94.3498	7.4530
19	2.1068	0.4746	27.6712	0.0361	13.1339	0.0761	102.8933	7.8342
20	2.1911	0.4564	29.7781	0.0336	13.5903	0.0736	111.5647	8.2091
21	2.2788	0.4388	31.9692	0.0313	14.0292	0.0713	120.3414	8.5779
22	2.3699	0.4220	34.2480	0.0292	14.4511	0.0692	129.2024	8.9407
23	2.4647	0.4057	36.6179	0.0273	14.8568	0.0673	138.1284	9.2973
24	2.5633	0.3901	39.0826	0.0256	15.2470	0.0656	147.1012	9.6479
25	2.6658	0.3751	41.6459	0.0240	15.6221	0.0640	156.1040	9.9925
26	2.7725	0.3607	44.3117	0.0226	15.9828	0.0626	165.1212	10.3312
27	2.8834	0.3468	47.0842	0.0212	16.3296	0.0612	174.1385	10.6640
28	2.9987	0.3335	49.9676	0.0200	16.6631	0.0600	183.1424	10.9909
29	3.1187	0.3207	52.9663	0.0189	16.9837	0.0589	192.1206	11.3120
30	3.2434	0.3083	56.0849	0.0178	17.2920	0.0578	201.0618	11.6274
36	4.1039	0.2437	77.5983	0.0129	18.9083	0.0529	253.4052	13.4018
42	5.1928	0.1926	104.8196	9.540E − 03	20.1856	0.0495	302.4370	14.9828
48	6.5705	0.1522	139.2632	7.181E − 03	21.1951	0.0472	347.2446	16.3832
54	8.3138	0.1203	182.8454	5.469E − 03	21.9930	0.0455	387.4436	17.6167
60	10.5196	0.0951	237.9907	4.202E − 03	22.6235	0.0442	422.9966	18.6972
66	13.3107	0.0751	307.7671	3.249E − 03	23.1218	0.0432	454.0847	19.6388
72	16.8423	0.0594	396.0566	2.525E − 03	23.5156	0.0425	481.0170	20.4552
120	110.6626	9.036E − 03	2.742E + 03	3.648E − 04	24.7741	0.0404	592.2428	23.9057
180	1.164E + 03	8.590E − 04	2.908E + 04	3.439E − 05	24.9785	0.0400	620.5976	24.8452
360	1.355E + 06	7.379E − 07	3.388E + 07	2.952E − 08	25.0000	0.0400	624.9929	24.9997

Time Value of Money Factors—Discrete Compounding   i = 5%

	Single Sums		Uniform Series				Gradient Series	
	To Find F Given P	To Find P Given F	To Find F Given A	To Find A Given F	To Find P Given A	To Find A Given P	To Find P Given G	To Find A Given G
n	$(F|P, i\%, n)$	$(P|F, i\%, n)$	$(F|A, i\%, n)$	$(A|F, i\%, n)$	$(P|A, i\%, n)$	$(A|P, i\%, n)$	$(P|G, i\%, n)$	$(A|G, i\%, n)$
1	1.0500	0.9524	1.0000	1.0000	0.9524	1.0500	0.0000	0.0000
2	1.1025	0.9070	2.0500	0.4878	1.8594	0.5378	0.9070	0.4878
3	1.1576	0.8638	3.1525	0.3172	2.7232	0.3672	2.6347	0.9675
4	1.2155	0.8227	4.3101	0.2320	3.5460	0.2820	5.1028	1.4391
5	1.2763	0.7835	5.5256	0.1810	4.3295	0.2310	8.2369	1.9025
6	1.3401	0.7462	6.8019	0.1470	5.0757	0.1970	11.9680	2.3579
7	1.4071	0.7107	8.1420	0.1228	5.7864	0.1728	16.2321	2.8052
8	1.4775	0.6768	9.5491	0.1047	6.4632	0.1547	20.9700	3.2445
9	1.5513	0.6446	11.0266	0.0907	7.1078	0.1407	26.1268	3.6758
10	1.6289	0.6139	12.5779	0.0795	7.7217	0.1295	31.6520	4.0991
11	1.7103	0.5847	14.2068	0.0704	8.3064	0.1204	37.4988	4.5144
12	1.7959	0.5568	15.9171	0.0628	8.8633	0.1128	46.6241	4.9219
13	1.8856	0.5303	17.7130	0.0565	9.3936	0.1065	49.9879	5.3215
14	1.9799	0.5051	19.5986	0.0510	9.8986	0.1010	56.5538	5.7133
15	2.0789	0.4810	21.5786	0.0463	10.3797	0.0963	63.2880	6.0973
16	2.1829	0.4581	23.6575	0.0423	10.8378	0.0923	70.1597	6.4736
17	2.2920	0.4363	25.8404	0.0387	11.2741	0.0887	77.1405	6.8423
18	2.4066	0.4155	28.1324	0.0355	11.6896	0.0855	84.2043	7.2034
19	2.5270	0.3957	30.5390	0.0327	12.0853	0.0827	91.3275	7.5569
20	2.6533	0.3769	33.0660	0.0302	12.4622	0.0802	98.4884	7.9030
21	2.7860	0.3589	35.7193	0.0280	12.8212	0.0780	105.6673	8.2416
22	2.9253	0.3418	38.5052	0.0260	13.1630	0.0760	112.8461	8.5730
23	3.0715	0.3256	41.4305	0.0241	13.4886	0.0741	120.0087	8.8971
24	3.2251	0.3101	44.5020	0.0225	13.7986	0.0725	127.1402	9.2140
25	3.3864	0.2953	47.7271	0.0210	14.0939	0.0710	134.2275	9.5238
26	3.5557	0.2812	51.1135	0.0196	14.3752	0.0696	141.2585	9.8266
27	3.7335	0.2678	54.6691	0.0183	14.6430	0.0683	148.2226	10.1224
28	3.9201	0.2551	58.4026	0.0171	14.8981	0.0671	155.1101	10.4114
29	4.1161	0.2429	62.3227	0.0160	15.1411	0.0660	161.9126	10.6936
30	4.3219	0.2314	66.4388	0.0151	15.3725	0.0651	168.6226	10.9691
36	5.7918	0.1727	95.8363	0.0104	16.5469	0.0604	206.6237	12.4872
42	7.7616	0.1288	135.2318	7.395E − 03	17.4232	0.0574	240.2389	13.7884
48	10.4013	0.0961	188.0254	5.318E − 03	18.0772	0.0553	269.2467	14.8943
54	13.9387	0.0717	258.7739	3.864E − 03	18.5651	0.0539	293.8208	15.8265
60	18.6792	0.0535	353.5837	2.828E − 03	18.9293	0.0528	314.3432	16.6062
66	25.0319	0.0399	480.6379	2.081E − 03	19.2010	0.0521	331.2877	17.2536
72	33.5451	0.0298	650.9027	1.536E − 03	19.4038	0.0515	345.1485	17.7877
120	348.9120	2.866E − 03	6.958E + 03	1.437E − 04	19.9427	0.0501	391.9751	19.6551
180	6.517E + 03	1.534E − 04	1.303E + 04	7.673E − 05	19.9969	0.0500	399.3863	19.9724
360	4.248E + 07	2.354E − 07	8.495E + 07	1.177E − 08	20.0000	0.0500	399.9998	20.0000

Time Value of Money Factors—Discrete Compounding   i = 6%

	Single Sums		Uniform Series				Gradient Series	
	To Find F Given P	To Find P Given F	To Find F Given A	To Find A Given F	To Find P Given A	To Find A Given P	To Find P Given G	To Find A Given G
n	$(F\|P, i\%, n)$	$(P\|F, i\%, n)$	$(F\|A, i\%, n)$	$(A\|F, i\%, n)$	$(P\|A, i\%, n)$	$(A\|P, i\%, n)$	$(P\|G, i\%, n)$	$(A\|G, i\%, n)$
1	1.0600	0.9434	1.0000	1.0000	0.9434	1.0600	0.0000	0.0000
2	1.1236	0.8900	2.0600	0.4854	1.8334	0.5454	0.8900	0.4854
3	1.1910	0.8396	3.1836	0.3141	2.6730	0.3741	2.5692	0.9612
4	1.2625	0.7921	4.3746	0.2286	3.4651	0.2886	4.9455	1.4272
5	1.3382	0.7473	5.6371	0.1774	4.2124	0.2374	7.9345	1.8836
6	1.4185	0.7050	6.9753	0.1434	4.9173	0.2034	11.4594	2.3304
7	1.5036	0.6651	8.3938	0.1191	5.5824	0.1791	15.4497	2.7676
8	1.5938	0.6274	9.8975	0.1010	6.2098	0.1610	19.8416	3.1952
9	1.6895	0.5919	11.4913	0.0870	6.8017	0.1470	24.5768	3.6133
10	1.7908	0.5584	13.1808	0.0759	7.3601	0.1359	29.6023	4.0220
11	1.8983	0.5268	14.9716	0.0668	7.8869	0.1268	34.8702	4.4213
12	2.0122	0.4970	16.8699	0.0593	8.3838	0.1193	40.3369	4.8113
13	2.1329	0.4688	18.8821	0.0530	8.8527	0.1130	45.9629	5.1920
14	2.2609	0.4423	21.0151	0.0476	9.2950	0.1076	51.7128	5.5635
15	2.3966	0.4173	23.2760	0.0430	9.7122	0.1030	57.5546	5.9260
16	2.5404	0.3936	25.6725	0.0390	10.1059	0.0990	63.4592	6.2794
17	2.6928	0.3714	28.2129	0.0354	10.4773	0.0954	69.4011	6.6240
18	2.8543	0.3503	30.9057	0.0324	10.8276	0.0924	75.3569	6.9597
19	3.0256	0.3305	33.7600	0.0296	11.1581	0.0896	81.3062	7.2867
20	3.2071	0.3118	36.7856	0.0272	11.4699	0.0872	87.2304	7.6051
21	3.3996	0.2942	39.9927	0.0250	11.7641	0.0850	93.1136	7.9151
22	3.6035	0.2775	43.3923	0.0230	12.0416	0.0830	98.9412	8.2166
23	3.8197	0.2618	46.9958	0.0213	12.3034	0.0813	104.7007	8.5099
24	4.0489	0.2470	50.8156	0.0197	12.5504	0.0797	110.3812	8.7951
25	4.2919	0.2330	54.8645	0.0182	12.7834	0.0782	115.9732	9.0722
26	4.5494	0.2198	59.1564	0.0169	13.0032	0.0769	121.4684	9.3414
27	4.8223	0.2074	63.7058	0.0157	13.2105	0.0757	126.8600	9.6029
28	5.1117	0.1956	68.5281	0.0146	13.4062	0.0746	132.1420	9.8568
29	5.4184	0.1846	73.6398	0.0136	13.5907	0.0736	137.3096	10.1032
30	5.7435	0.1741	79.0582	0.0126	13.7648	0.0726	142.3588	10.3402
36	8.1473	0.1227	119.1209	8.395E − 03	14.6210	0.0684	170.0387	11.6298
42	11.5570	0.0865	175.9505	5.683E − 03	15.2245	0.0657	193.1732	12.6883
48	16.3939	0.0610	256.5645	3.898E − 03	15.6500	0.0639	212.0351	13.5485
54	23.2550	0.0430	370.9170	2.696E − 03	15.9500	0.0627	227.1316	14.2402
60	32.9877	0.0303	533.1282	1.876E − 03	16.1614	0.0619	239.0428	14.7909
66	46.7937	0.0214	763.2278	1.310E − 03	16.3105	0.0613	248.3341	15.2254
72	66.3777	0.0151	1.090E + 03	9.177E − 04	16.4156	0.0609	255.5146	15.5654
120	1.088E + 03	9.190E − 04	1.812E + 04	5.519E − 05	16.6514	0.0601	275.6846	16.5563
180	3.590E + 04	2.786E − 05	5.983E + 05	1.672E − 06	16.6662	0.0600	277.6865	16.6617
360	1.289E + 09	7.760E − 10	2.148E + 10	4.656E − 11	16.6667	0.0600	277.7778	16.6667

Time Value of Money Factors—Discrete Compounding $\qquad\qquad\qquad\qquad\qquad\qquad$ i = 7%

	Single Sums		Uniform Series				Gradient Series	
	To Find F Given P	To Find P Given F	To Find F Given A	To Find A Given F	To Find P Given A	To Find A Given P	To Find P Given G	To Find A Given G
n	(F|P, i%, n)	(P|F, i%, n)	(F|A, i%, n)	(A|F, i%, n)	(P|A, i%, n)	(A|P, i%, n)	(P|G, i%, n)	(A|G, i%, n)
1	1.0700	0.9346	1.0000	1.0000	0.9346	1.0700	0.0000	0.0000
2	1.1449	0.8734	2.0700	0.4831	1.8080	0.5531	0.8734	0.4831
3	1.2250	0.8163	3.2149	0.3111	2.6243	0.3811	2.5060	0.9549
4	1.3108	0.7629	4.4399	0.2252	3.3872	0.2952	4.7947	1.4155
5	1.4026	0.7130	5.7507	0.1739	4.1002	0.2439	7.6467	1.8650
6	1.5007	0.6663	7.1533	0.1398	4.7665	0.2098	10.9784	2.3032
7	1.6058	0.6227	8.6540	0.1156	5.3893	0.1856	14.7149	2.7304
8	1.7182	0.5820	10.2598	0.0975	5.9713	0.1675	18.7889	3.1465
9	1.8385	0.5439	11.9780	0.0835	6.5152	0.1535	23.1404	3.5517
10	1.9672	0.5083	13.8164	0.0724	7.0236	0.1424	27.7156	3.9461
11	2.1049	0.4751	15.7836	0.0634	7.4987	0.1334	32.4665	4.3296
12	2.2522	0.4440	17.8885	0.0559	7.9427	0.1259	37.3506	4.7025
13	2.4098	0.4150	20.1406	0.0497	8.3577	0.1197	42.3302	5.0658
14	2.5785	0.3878	22.5505	0.0443	8.7455	0.1143	47.3718	5.4167
15	2.7590	0.3624	25.1290	0.0398	9.1079	0.1098	52.4461	5.7583
16	2.9522	0.3387	27.8881	0.0359	9.4466	0.1059	57.5271	6.0897
17	3.1588	0.3166	30.8402	0.0324	9.7632	0.1024	62.5923	6.4110
18	3.3799	0.2959	33.9990	0.0294	10.0591	0.0994	67.6219	6.7225
19	3.6165	0.2765	37.3790	0.0268	10.3356	0.0968	72.5991	7.0242
20	3.8697	0.2584	40.9955	0.0244	10.5940	0.0944	77.5091	7.3163
21	4.1406	0.2415	44.8652	0.0223	10.8355	0.0923	82.3393	7.5990
22	4.4304	0.2257	49.0057	0.0204	11.0612	0.0904	87.0793	7.8725
23	4.7405	0.2109	53.4361	0.0187	11.2722	0.0887	91.7201	8.1369
24	5.0724	0.1971	58.1767	0.0172	11.4693	0.0872	96.2545	8.3923
25	5.4274	0.1842	63.2490	0.0158	11.6536	0.0858	100.6765	8.6391
26	5.8074	0.1722	68.6765	0.0146	11.8258	0.0846	104.9814	8.8773
27	6.2139	0.1609	74.4838	0.0134	11.9867	0.0834	109.1656	9.1072
28	6.6488	0.1504	80.6977	0.0124	12.1371	0.0824	113.2264	9.3289
29	7.1143	0.1406	87.3465	0.0114	12.2777	0.0814	117.1622	9.5427
30	7.6123	0.1314	94.4608	0.0106	12.4090	0.0806	120.9718	9.7487
36	11.4239	0.0875	148.9135	6.715E − 03	13.0352	0.0767	141.1990	10.8321
42	17.1443	0.0583	230.6322	4.336E − 03	13.4524	0.0743	157.1807	11.6842
48	25.7289	0.0389	353.2701	2.831E − 03	13.7305	0.0728	169.4981	12.3447
54	38.6122	0.0259	537.3164	1.861E − 03	13.9157	0.0719	178.8173	12.8500
60	57.9464	0.0173	813.5204	1.229E − 03	14.0392	0.0712	185.7677	13.2321
66	86.9620	0.0115	1.228E + 03	8.143E − 04	14.1214	0.0708	190.8927	13.5179
72	130.5065	7.662E − 03	1.850E + 03	5.405E − 04	14.1763	0.0705	194.6365	13.7298
120	3.358E + 03	2.978E − 04	4.795E + 04	2.085E − 05	14.2815	0.0700	203.5103	14.2500
180	1.946E + 05	5.139E − 06	2.780E + 06	3.598E − 07	14.2856	0.0700	204.0674	14.2848
360	3.786E + 10	2.641E − 11	5.408E + 11	1.849E − 12	14.2857	0.0700	204.0816	14.2857

Time Value of Money Factors—Discrete Compounding   i = 8%

	Single Sums		Uniform Series				Gradient Series	
n	To Find F Given P $(F\|P, i\%, n)$	To Find P Given F $(P\|F, i\%, n)$	To Find F Given A $(F\|A, i\%, n)$	To Find A Given F $(A\|F, i\%, n)$	To Find P Given A $(P\|A, i\%, n)$	To Find A Given P $(A\|P, i\%, n)$	To Find P Given G $(P\|G, i\%, n)$	To Find A Given G $(A\|G, i\%, n)$
1	1.0800	0.9259	1.0000	1.0000	0.9259	1.0800	0.0000	0.0000
2	1.1664	0.8573	2.0800	0.4808	1.7833	0.5608	0.8573	0.4808
3	1.2597	0.7938	3.2464	0.3080	2.5771	0.3880	2.4450	0.9487
4	1.3605	0.7350	4.5061	0.2219	3.3121	0.3019	4.6501	1.4040
5	1.4693	0.6806	5.8666	0.1705	3.9927	0.2505	7.3724	1.8465
6	1.5869	0.6302	7.3359	0.1363	4.6229	0.2163	10.5233	2.2763
7	1.7138	0.5835	8.9228	0.1121	5.2064	0.1921	14.0242	2.6937
8	1.8509	0.5403	10.6366	0.0940	5.7466	0.1740	17.8061	3.0985
9	1.9990	0.5002	12.4876	0.0801	6.2469	0.1601	21.8081	3.4910
10	2.1589	0.4632	14.4866	0.0690	6.7101	0.1490	25.9768	3.8713
11	2.3316	0.4289	16.6455	0.0601	7.1390	0.1401	30.2657	4.2395
12	2.5182	0.3971	18.9771	0.0527	7.5361	0.1327	34.6339	4.5957
13	2.7196	0.3677	21.4953	0.0465	7.9038	0.1265	39.0463	4.9402
14	2.9372	0.3405	24.2149	0.0413	8.2442	0.1213	43.4723	5.2731
15	3.1722	0.3152	27.1521	0.0368	8.5595	0.1168	47.8857	5.5945
16	3.4259	0.2919	30.3243	0.0330	8.8514	0.1130	52.2640	5.9046
17	3.7000	0.2703	33.7502	0.0296	9.1216	0.1096	56.5883	6.2037
18	3.9960	0.2502	37.4502	0.0267	9.3719	0.1067	60.8426	6.4920
19	4.3157	0.2317	41.4463	0.0241	9.6036	0.1041	65.0134	6.7697
20	4.6610	0.2145	45.7620	0.0219	9.8181	0.1019	69.0898	7.0369
21	5.0338	0.1987	50.4229	0.0198	10.0168	0.0998	73.0629	7.2940
22	5.4365	0.1839	55.4568	0.0180	10.2007	0.0980	76.9257	7.5412
23	5.8715	0.1703	60.8933	0.0164	10.3711	0.0964	80.6726	7.7786
24	6.3412	0.1577	66.7648	0.0150	10.5288	0.0950	84.2997	8.0066
25	6.8485	0.1460	73.1059	0.0137	10.6748	0.0937	87.8041	8.2254
26	7.3964	0.1352	79.9544	0.0125	10.8100	0.0925	91.1842	8.4352
27	7.9881	0.1252	87.3508	0.0114	10.9352	0.0914	94.4390	8.6363
28	8.6271	0.1159	95.3388	0.0105	11.0511	0.0905	97.5687	8.8289
29	9.3173	0.1073	103.9659	9.619E − 03	11.1584	0.0896	100.5738	9.0133
30	10.0627	0.0994	113.2832	8.827E − 03	11.2578	0.0888	103.4558	9.1897
36	15.9682	0.0626	187.1021	5.345E − 03	11.7172	0.0853	118.2839	10.0949
42	25.3395	0.0395	304.2435	3.287E − 03	12.0067	0.0833	129.3651	10.7744
48	40.2106	0.0249	490.1322	2.040E − 03	12.1891	0.0820	137.4428	11.2758
54	63.8091	0.0157	785.1141	1.274E − 03	12.3041	0.0813	143.2229	11.6403
60	101.2571	9.876E − 03	1.253E + 03	7.979E − 04	12.3766	0.0808	147.3000	11.9015
66	160.6822	6.223E − 03	1.996E + 03	5.010E − 04	12.4222	0.0805	150.1432	12.0867
72	254.9825	3.922E − 03	3.175E + 03	3.150E − 04	12.4510	0.0803	152.1076	12.2165
120	1.025E + 04	9.753E − 05	1.281E + 05	7.803E − 06	12.4988	0.0800	156.0885	12.4883
180	1.038E + 06	9.632E − 07	1.298E + 07	7.706E − 08	12.5000	0.0800	156.2477	12.4998
360	1.078E + 12	9.278E − 13	1.347E + 13	7.422E − 14	12.5000	0.0800	156.2500	12.5000

Time Value of Money Factors—Discrete Compounding $\qquad$ i = 9%

	Single Sums		Uniform Series				Gradient Series	
	To Find F Given P	To Find P Given F	To Find F Given A	To Find A Given F	To Find P Given A	To Find A Given P	To Find P Given G	To Find A Given G
n	$(F\|P, i\%, n)$	$(P\|F, i\%, n)$	$(F\|A, i\%, n)$	$(A\|F, i\%, n)$	$(P\|A, i\%, n)$	$(A\|P, i\%, n)$	$(P\|G, i\%, n)$	$(A\|G, i\%, n)$
1	1.0900	0.9174	1.0000	1.0000	0.9174	1.0900	0.0000	0.0000
2	1.1881	0.8417	2.0900	0.4785	1.7591	0.5685	0.8417	0.4785
3	1.2950	0.7722	3.2781	0.3051	2.5313	0.3951	2.3860	0.9426
4	1.4116	0.7084	4.5731	0.2187	3.2397	0.3087	4.5113	1.3925
5	1.5386	0.6499	5.9847	0.1671	3.8897	0.2571	7.1110	1.8282
6	1.6771	0.5963	7.5233	0.1329	4.4859	0.2229	10.0924	2.2498
7	1.8280	0.5470	9.2004	0.1087	5.0330	0.1987	13.3746	2.6574
8	1.9926	0.5019	11.0285	0.0907	5.5348	0.1807	16.8877	3.0512
9	2.1719	0.4604	13.0210	0.0768	5.9952	0.1668	20.5711	3.4312
10	2.3674	0.4224	15.1929	0.0658	6.4177	0.1558	24.3728	3.7978
11	2.5804	0.3875	17.5603	0.0569	6.8052	0.1469	28.2481	4.1510
12	2.8127	0.3555	20.1407	0.0497	7.1607	0.1397	32.1590	4.4910
13	3.0658	0.3262	22.9534	0.0436	7.4869	0.1336	36.0731	4.8182
14	3.3417	0.2992	26.0192	0.0384	7.7862	0.1284	39.9633	5.1326
15	3.6425	0.2745	29.3609	0.0341	8.0607	0.1241	43.8069	5.4346
16	3.9703	0.2519	33.0034	0.0303	8.3126	0.1203	47.5849	5.7245
17	4.3276	0.2311	36.9737	0.0270	8.5436	0.1170	51.2821	6.0024
18	4.7171	0.2120	41.3013	0.0242	8.7556	0.1142	54.8860	6.2687
19	5.1417	0.1945	46.0185	0.0217	8.9501	0.1117	58.3868	6.5236
20	5.6044	0.1784	51.1601	0.0195	9.1285	0.1095	61.7770	6.7674
21	6.1088	0.1637	56.7645	0.0176	9.2922	0.1076	65.0509	7.0006
22	6.6586	0.1502	62.8733	0.0159	9.4424	0.1059	68.2048	7.2232
23	7.2579	0.1378	69.5319	0.0144	9.5802	0.1044	71.2359	7.4357
24	7.9111	0.1264	76.7898	0.0130	9.7066	0.1030	74.1433	7.6384
25	8.6231	0.1160	84.7009	0.0118	9.8226	0.1018	76.9265	7.8316
26	9.3992	0.1064	93.3240	0.0107	9.9290	0.1007	79.5863	8.0156
27	10.2451	0.0976	102.7231	9.735E − 03	10.0266	0.0997	82.1241	8.1906
28	11.1671	0.0895	112.9682	8.852E − 03	10.1161	0.0989	84.5419	8.3571
29	12.1722	0.0822	124.1354	8.056E − 03	10.1983	0.0981	86.8422	8.5154
30	13.2677	0.0754	136.3075	7.336E − 03	10.2737	0.0973	89.0280	8.6657
36	22.2512	0.0449	236.1247	4.235E − 03	10.6118	0.0942	99.9319	9.4171
42	37.3175	0.0268	403.5281	2.478E − 03	10.8134	0.0925	107.6432	9.9546
48	62.5852	0.0160	684.2804	1.461E − 03	10.9336	0.0915	112.9625	10.3317
54	104.9617	9.527E − 03	1.155E + 03	8.657E − 04	11.0053	0.0909	116.5642	10.5917
60	176.0313	5.681E − 03	1.945E + 03	5.142E − 04	11.0480	0.0905	118.9683	10.7683
66	295.2221	3.387E − 03	3.269E + 03	3.059E − 04	11.0735	0.0903	120.5546	10.8868
72	495.1170	2.020E − 03	5.490E + 03	1.821E − 04	11.0887	0.0902	121.5917	10.9654
120	3.099E + 04	3.227E − 05	3.443E + 05	2.905E − 06	11.1108	0.0900	123.4098	11.1072
180	5.455E + 06	1.833E − 07	6.061E + 07	1.650E − 08	11.1111	0.0900	123.4564	11.1111
360	2.975E + 13	3.361E − 14	3.306E + 14	3.025E − 15	11.1111	0.0900	123.4568	11.1111

Time Value of Money Factors—Discrete Compounding   i = 10%

	Single Sums		Uniform Series				Gradient Series	
	To Find F Given P ($F\|P$, i%, n)	To Find P Given F ($P\|F$, i%, n)	To Find F Given A ($F\|A$, i%, n)	To Find A Given F ($A\|F$, i%, n)	To Find P Given A ($P\|A$, i%, n)	To Find A Given P ($A\|P$, i%, n)	To Find P Given G ($P\|G$, i%, n)	To Find A Given G ($A\|G$, i%, n)
n								
1	1.1000	0.9091	1.0000	1.0000	0.9091	1.1000	0.0000	0.0000
2	1.2100	0.8264	2.1000	0.4762	1.7355	0.5762	0.8264	0.4762
3	1.3310	0.7513	3.3100	0.3021	2.4869	0.4021	2.3291	0.9366
4	1.4641	0.6830	4.6410	0.2155	3.1699	0.3155	4.3781	1.3812
5	1.6105	0.6209	6.1051	0.1638	3.7908	0.2638	6.8618	1.8101
6	1.7716	0.5645	7.7156	0.1296	4.3553	0.2296	9.6842	2.2236
7	1.9487	0.5132	9.4872	0.1054	4.8684	0.2054	12.7631	2.6216
8	2.1436	0.4665	11.4359	0.0874	5.3349	0.1874	16.0287	3.0045
9	2.3579	0.4241	13.5795	0.0736	5.7590	0.1736	19.4215	3.3742
10	2.5937	0.3855	15.9374	0.0627	6.1446	0.1627	22.8913	3.7255
11	2.8531	0.3505	18.5312	0.0540	6.4951	0.1540	26.3963	4.0641
12	3.1384	0.3186	21.3843	0.0468	6.8137	0.1468	29.9012	4.3884
13	3.4523	0.2897	24.5227	0.0408	7.1034	0.1408	33.3772	4.6988
14	3.7975	0.2633	27.9750	0.0357	7.3667	0.1357	36.8005	4.9955
15	4.1772	0.2394	31.7725	0.0315	7.6061	0.1315	40.1520	5.2789
16	4.5950	0.2176	35.9497	0.0278	7.8237	0.1278	43.4164	5.5493
17	5.0545	0.1978	40.5447	0.0247	8.0216	0.1247	46.5819	5.8071
18	5.5599	0.1799	45.5992	0.0219	8.2014	0.1219	49.6395	6.0526
19	6.1159	0.1635	51.1591	0.0195	8.3649	0.1195	52.5827	6.2861
20	6.7275	0.1486	57.2750	0.0175	8.5136	0.1175	55.4069	6.5081
21	7.4002	0.1351	64.0025	0.0156	8.6487	0.1156	58.1095	6.7189
22	8.1403	0.1228	71.4027	0.0140	8.7715	0.1140	60.6893	6.9189
23	8.9543	0.1117	79.5430	0.0126	8.8832	0.1126	63.1462	7.1085
24	9.8497	0.1015	88.4973	0.0113	8.9847	0.1113	65.4813	7.2881
25	10.8347	0.0923	98.3471	0.0102	9.0770	0.1102	67.6964	7.4580
26	11.9182	0.0839	109.1818	9.159E − 03	9.1609	0.1092	69.7940	7.6186
27	13.1100	0.0763	121.0999	8.258E − 03	9.2372	0.1083	71.7773	7.7704
28	14.4210	0.0693	134.2099	7.451E − 03	9.3066	0.1075	73.6495	7.9137
29	15.8631	0.0630	148.6309	6.728E − 03	9.3696	0.1067	75.4146	8.0489
30	17.4494	0.0573	164.4940	6.079E − 03	9.4269	0.1061	77.0766	8.1762
36	30.9127	0.0323	299.1268	3.343E − 03	9.6765	0.1033	85.1194	8.7965
42	54.7637	0.0183	537.6370	1.860E − 03	9.8174	0.1019	90.5047	9.2188
48	97.0172	0.0103	960.1723	1.041E − 03	9.8969	0.1010	94.0217	9.5001
54	171.8719	5.818E − 03	1.709E + 03	5.852E − 04	9.9418	0.1006	96.2763	9.6840
60	304.4816	3.284E − 03	3.035E + 03	3.295E − 04	9.9672	0.1003	97.7010	9.8023
66	539.4078	1.854E − 03	5.384E + 03	1.857E − 04	9.9815	0.1002	98.5910	9.8774
72	955.5938	1.046E − 03	9.546E + 03	1.048E − 04	9.9895	0.1001	99.1419	9.9246
120	9.271E + 04	1.079E − 05	9.271E + 05	1.079E − 06	9.9999	0.1000	99.9860	9.9987
180	2.823E + 07	3.543E − 08	2.823E + 08	3.543E − 09	10.0000	0.1000	99.9999	10.0000
360	7.968E + 14	1.255E − 15	7.968E + 15	1.255E − 16	10.0000	0.1000	100.0008	10.0000

Time Value of Money Factors—Discrete Compounding   i = 12%

	Single Sums		Uniform Series				Gradient Series	
	To Find F Given P $(F\|P, i\%, n)$	To Find P Given F $(P\|F, i\%, n)$	To Find F Given A $(F\|A, i\%, n)$	To Find A Given F $(A\|F, i\%, n)$	To Find P Given A $(P\|A, i\%, n)$	To Find A Given P $(A\|P, i\%, n)$	To Find P Given G $(P\|G, i\%, n)$	To Find A Given G $(A\|G, i\%, n)$
n								
1	1.1200	0.8929	1.0000	1.0000	0.8929	1.1200	0.0000	0.0000
2	1.2544	0.7972	2.1200	0.4717	1.6901	0.5917	0.7972	0.4717
3	1.4049	0.7118	3.3744	0.2963	2.4018	0.4163	2.2208	0.9246
4	1.5735	0.6355	4.7793	0.2092	3.0373	0.3292	4.1273	1.3589
5	1.7623	0.5674	6.3528	0.1574	3.6048	0.2774	6.3970	1.7746
6	1.9738	0.5066	8.1152	0.1232	4.1114	0.2432	8.9302	2.1720
7	2.2107	0.4523	10.0890	0.0991	4.5638	0.2191	11.6443	2.5515
8	2.4760	0.4039	12.2997	0.0813	4.9676	0.2013	14.4714	2.9131
9	2.7731	0.3606	14.7757	0.0677	5.3282	0.1877	17.3563	3.2574
10	3.1058	0.3220	17.5487	0.0570	5.6502	0.1770	20.2541	3.5847
11	3.4785	0.2875	20.6546	0.0484	5.9377	0.1684	23.1288	3.8953
12	3.8960	0.2567	24.1331	0.0414	6.1944	0.1614	25.9523	4.1897
13	4.3635	0.2292	28.0291	0.0357	6.4354	0.1557	28.7024	4.4683
14	4.8871	0.2046	32.3926	0.0309	6.6282	0.1509	31.3624	4.7317
15	5.4736	0.1827	37.2797	0.0268	6.8109	0.1468	33.9202	4.9803
16	6.1304	0.1631	42.7533	0.0234	6.9740	0.1434	36.3670	5.2147
17	6.8660	0.1456	48.8837	0.0205	7.1196	0.1405	38.6973	5.4353
18	7.6900	0.1300	55.7497	0.0179	7.2497	0.1379	40.9080	5.6427
19	8.6128	0.1161	63.4397	0.0158	7.3658	0.1358	42.9979	5.8375
20	9.6463	0.1037	72.0524	0.0139	7.4694	0.1339	44.9676	6.0202
21	10.8038	0.0926	81.6987	0.0122	7.5620	0.1322	46.8188	6.1913
22	12.1003	0.0826	92.5026	0.0108	7.6446	0.1308	48.5543	6.3514
23	13.5523	0.0738	104.6029	9.560E − 03	7.7184	0.1296	50.1776	6.5010
24	15.1786	0.0659	118.1552	8.463E − 03	7.7843	0.1285	51.6929	6.6406
25	17.0001	0.0588	133.3339	7.500E − 03	7.8431	0.1275	53.1046	6.7708
26	19.0401	0.0525	150.3339	6.652E − 03	7.8957	0.1267	54.4177	6.8921
27	21.3249	0.0469	169.3740	5.904E − 03	7.9426	0.1259	55.6369	7.0049
28	23.8839	0.0419	190.6989	5.244E − 03	7.9844	0.1252	56.7674	7.1098
29	26.7499	0.0374	214.5828	4.660E − 03	8.0218	0.1247	57.8141	7.2071
30	29.9599	0.0334	241.3327	4.144E − 03	8.0552	0.1241	58.7821	7.2974
36	59.1356	0.0169	484.4631	2.064E − 03	8.1924	0.1221	63.1970	7.7141
42	116.7231	8.567E − 03	964.3595	1.037E − 03	8.2619	0.1210	65.8509	7.9704
48	230.3908	4.340E − 03	1.912E + 03	5.231E − 04	8.2972	0.1205	67.4068	8.1241
54	454.7505	2.199E − 03	3.781E + 03	2.645E − 04	8.3150	0.1203	68.3022	8.2143
60	897.5969	1.114E − 03	7.472E + 03	1.338E − 04	8.3240	0.1201	68.8100	8.2664
66	1.772E + 03	5.644E − 04	1.476E + 04	6.777E − 05	8.3286	0.1201	69.0948	8.2961
72	3.497E + 03	2.860E − 04	2.913E + 04	3.432E − 05	8.3310	0.1200	69.2530	8.3127
120	8.057E + 05	1.241E − 06	6.714E + 06	1.489E − 07	8.3333	0.1200	69.4431	8.3332
180	7.232E + 08	1.383E − 09	6.026E + 09	1.659E − 10	8.3333	0.1200	69.4444	8.3333
360	5.230E + 17	1.912E − 18	4.358E + 18	2.295E − 19	8.3333	0.1200	69.4444	8.3333

Time Value of Money Factors—Discrete Compounding   i = 15%

	Single Sums		Uniform Series				Gradient Series	
n	To Find F Given P $(F\|P, i\%, n)$	To Find P Given F $(P\|F, i\%, n)$	To Find F Given A $(F\|A, i\%, n)$	To Find A Given F $(A\|F, i\%, n)$	To Find P Given A $(P\|A, i\%, n)$	To Find A Given P $(A\|P, i\%, n)$	To Find P Given G $(P\|G, i\%, n)$	To Find A Given G $(A\|G, i\%, n)$
1	1.1500	0.8696	1.0000	1.0000	0.8696	1.1500	0.0000	0.0000
2	1.3225	0.7561	2.1500	0.4651	1.6257	0.6151	0.7561	0.4651
3	1.5209	0.6575	3.4725	0.2880	2.2832	0.4380	2.0712	0.9071
4	1.7490	0.5718	4.9934	0.2003	2.8550	0.3503	3.7864	1.3263
5	2.0114	0.4972	6.7424	0.1483	3.3522	0.2983	5.7751	1.7228
6	2.3131	0.4323	8.7537	0.1142	3.7845	0.2642	7.9368	2.0972
7	2.6600	0.3759	11.0668	0.0904	4.1604	0.2404	10.1924	2.4498
8	3.0590	0.3269	13.7268	0.0729	4.4873	0.2229	12.4807	2.7813
9	3.5179	0.2843	16.7858	0.0596	4.7716	0.2096	14.7548	3.0922
10	4.0456	0.2472	20.3037	0.0493	5.0188	0.1993	16.9795	3.3832
11	4.6524	0.2149	24.3493	0.0411	5.2337	0.1911	19.1289	3.6549
12	5.3503	0.1869	29.0017	0.0345	5.4206	0.1845	21.1849	3.9082
13	6.1528	0.1625	34.3519	0.0291	5.5831	0.1791	23.1352	4.1438
14	7.0757	0.1413	40.5047	0.0247	5.7245	0.1747	24.9725	4.3624
15	8.1371	0.1229	47.5804	0.0210	5.8474	0.1710	26.6930	4.5650
16	9.3576	0.1069	55.7175	0.0179	5.9542	0.1679	28.2960	4.7522
17	10.7613	0.0929	65.0751	0.0154	6.0472	0.1654	29.7828	4.9251
18	12.3755	0.0808	75.8364	0.0132	6.1280	0.1632	31.1565	5.0843
19	14.2318	0.0703	88.2118	0.0113	6.1982	0.1613	32.4213	5.2307
20	16.3665	0.0611	102.4436	9.761E − 03	6.2593	0.1598	33.5822	5.3651
21	18.8215	0.0531	118.8101	8.417E − 03	6.3125	0.1584	34.6448	5.4883
22	21.6447	0.0462	137.6316	7.266E − 03	6.3587	0.1573	35.6150	5.6010
23	24.8915	0.0402	159.2764	6.278E − 03	6.3988	0.1563	36.4988	5.7040
24	28.6252	0.0349	184.1678	5.430E − 03	6.4338	0.1554	37.3023	5.7979
25	32.9190	0.0304	212.7930	4.699E − 03	6.4641	0.1547	38.0314	5.8834
26	37.8568	0.0264	245.7120	4.070E − 03	6.4906	0.1541	38.6918	5.9612
27	43.5353	0.0230	283.5688	3.526E − 03	6.5135	0.1535	39.2890	6.0319
28	50.0656	0.0200	327.1041	3.057E − 03	6.5335	0.1531	39.8283	6.0960
29	57.5755	0.0174	377.1697	2.651E − 03	6.5509	0.1527	40.3146	6.1541
30	66.2118	0.0151	434.7451	2.300E − 03	6.5660	0.1523	40.7526	6.2066
36	153.1519	6.529E − 03	1.014E + 03	9.859E − 04	6.6231	0.1510	42.5872	6.4301
42	354.2495	2.823E − 03	2.355E + 03	4.246E − 04	6.6478	0.1504	43.5286	6.5478
48	819.4007	1.220E − 03	5.456E + 03	1.833E − 04	6.6585	0.1502	43.9997	6.6080
54	1.895E + 03	5.276E − 04	1.263E + 04	7.918E − 05	6.6631	0.1501	44.2311	6.6382
60	4.384E + 03	2.281E − 04	2.922E + 04	3.422E − 05	6.6651	0.1500	44.3431	6.6530
66	1.014E + 04	9.861E − 05	6.760E + 04	1.479E − 05	6.6660	0.1500	44.3967	6.6602
72	2.346E + 04	4.263E − 05	1.564E + 05	6.395E − 06	6.6664	0.1500	44.4221	6.6636
120	1.922E + 07	5.203E − 08	1.281E + 08	7.805E − 09	6.6667	0.1500	44.4444	6.6667
180	8.426E + 10	1.187E − 11	5.617E + 11	1.780E − 12	6.6667	0.1500	44.4444	6.6667
360	7.099E + 21	1.409E − 22	4.733E + 22	2.113E − 23	6.6667	0.1500	44.4444	6.6667

Time Value of Money Factors—Discrete Compounding   i = 18%

	Single Sums		Uniform Series				Gradient Series	
	To Find F Given P	To Find P Given F	To Find F Given A	To Find A Given F	To Find P Given A	To Find A Given P	To Find P Given G	To Find A Given G
n	$(F\|P, i\%, n)$	$(P\|F, i\%, n)$	$(F\|A, i\%, n)$	$(A\|F, i\%, n)$	$(P\|A, i\%, n)$	$(A\|P, i\%, n)$	$(P\|G, i\%, n)$	$(A\|G, i\%, n)$
1	1.1800	0.8475	1.0000	1.0000	0.8475	1.1800	0.0000	0.0000
2	1.3924	0.7182	2.1800	0.4587	1.5656	0.6387	0.7182	0.4587
3	1.6430	0.6086	3.5724	0.2799	2.1743	0.4599	1.9354	0.8902
4	1.9388	0.5158	5.2154	0.1917	2.6901	0.3717	3.4828	1.2947
5	2.2878	0.4371	7.1542	0.1398	3.1272	0.3198	5.2312	1.6728
6	2.6996	0.3704	9.4420	0.1059	3.4976	0.2859	7.0834	2.0252
7	3.1855	0.3139	12.1415	0.0824	3.8115	0.2624	8.9670	2.3526
8	3.7589	0.2660	15.3270	0.0652	4.0776	0.2452	10.8292	2.6558
9	4.4355	0.2255	19.0859	0.0524	4.3030	0.2324	12.6329	2.9358
10	5.2338	0.1911	23.5213	0.0425	4.4941	0.2225	14.3525	3.1936
11	6.1759	0.1619	28.7551	0.0348	4.6560	0.2148	15.9716	3.4303
12	7.2876	0.1372	34.9311	0.0286	4.7932	0.2086	17.4811	3.6470
13	8.5994	0.1163	42.2187	0.0237	4.9095	0.2037	18.8765	3.8449
14	10.1472	0.0985	50.8180	0.0197	5.0081	0.1997	20.1576	4.0250
15	11.9737	0.0835	60.9653	0.0164	5.0916	0.1964	21.3269	4.1887
16	14.1290	0.0708	72.9390	0.0137	5.1624	0.1937	22.3885	4.3369
17	16.6722	0.0600	87.0680	0.0115	5.2223	0.1915	23.3482	4.4708
18	19.6733	0.0508	103.7403	9.639E − 03	5.2732	0.1896	24.2123	4.5916
19	23.2144	0.0431	123.4135	8.103E − 03	5.3162	0.1881	24.9877	4.7003
20	27.3930	0.0365	146.6280	6.820E − 03	5.3527	0.1868	25.6813	4.7978
21	32.3238	0.0309	174.0210	5.746E − 03	5.3837	0.1857	26.3000	4.8851
22	38.1421	0.0262	206.3448	4.846E − 03	5.4099	0.1848	26.8506	4.9632
23	45.0076	0.0222	244.4868	4.090E − 03	5.4321	0.1841	27.3394	5.0329
24	53.1090	0.0188	289.4945	3.454E − 03	5.4509	0.1835	27.7725	5.0950
25	62.6686	0.0160	342.6035	2.919E − 03	5.4669	0.1829	28.1555	5.1502
26	73.9490	0.0135	405.2721	2.467E − 03	5.4804	0.1825	28.4935	5.1991
27	87.2598	0.0115	479.2211	2.087E − 03	5.4919	0.1821	28.7915	5.2425
28	102.0666	9.712E − 03	566.4809	1.765E − 03	5.5016	0.1818	29.0537	5.2810
29	121.5005	8.230E − 03	669.4475	1.494E − 03	5.5098	0.1815	29.2842	5.3149
30	143.3706	6.975E − 03	790.9480	1.264E − 03	5.5168	0.1813	29.4864	5.3448
36	387.0368	2.584E − 03	2.145E + 03	4.663E − 04	5.5412	0.1805	30.2677	5.4623
42	1.045E + 03	9.571E − 04	5.799E + 03	1.724E − 04	5.5502	0.1802	30.6113	5.5153
48	2.821E + 03	3.545E − 04	1.566E + 04	6.384E − 05	5.5536	0.1801	30.7587	5.5385
54	7.614E + 03	1.313E − 04	4.230E + 04	2.364E − 05	5.5548	0.1800	30.8207	5.5485
60	2.056E + 04	4.865E − 05	1.142E + 05	8.757E − 06	5.5553	0.1800	30.8465	5.5526
66	5.549E + 04	1.802E − 05	3.083E + 05	3.244E − 06	5.5555	0.1800	30.8570	5.5544
72	1.498E + 05	6.676E − 06	8.322E + 05	1.202E − 06	5.5555	0.1800	30.8613	5.5551
120	4.225E + 08	2.367E − 09	2.347E + 09	4.260E − 10	5.5556	0.1800	30.8642	5.5556
180	8.685E + 12	1.151E − 13	4.825E + 13	2.073E − 14	5.5556	0.1800	30.8642	5.5556
360	7.543E + 25	1.326E − 26	4.190E + 26	2.386E − 27	5.5556	0.1800	30.8642	5.5556

Time Value of Money Factors—Discrete Compounding   i = 20%

	Single Sums		Uniform Series				Gradient Series	
	To Find F Given P $(F\|P, i\%, n)$	To Find P Given F $(P\|F, i\%, n)$	To Find F Given A $(F\|A, i\%, n)$	To Find A Given F $(A\|F, i\%, n)$	To Find P Given A $(P\|A, i\%, n)$	To Find A Given P $(A\|P, i\%, n)$	To Find P Given G $(P\|G, i\%, n)$	To Find A Given G $(A\|G, i\%, n)$
n								
1	1.2000	0.8333	1.0000	1.0000	0.8333	1.2000	0.0000	0.0000
2	1.4400	0.6944	2.2000	0.4545	1.5278	0.6545	0.6944	0.4545
3	1.7280	0.5787	3.6400	0.2747	2.1065	0.4747	1.8519	0.8791
4	2.0736	0.4823	5.3680	0.1863	2.5887	0.3863	3.2986	1.2742
5	2.4883	0.4019	7.4416	0.1344	2.9906	0.3344	4.9061	1.6405
6	2.9860	0.3349	9.9299	0.1007	3.3255	0.3007	6.5806	1.9788
7	3.5832	0.2791	12.9159	0.0774	3.6046	0.2774	8.2551	2.2902
8	4.2998	0.2326	16.4991	0.0606	3.8372	0.2606	9.8831	2.5756
9	5.1598	0.1938	20.7989	0.0481	4.0310	0.2481	11.4335	2.8364
10	6.1917	0.1615	25.9587	0.0385	4.1925	0.2385	12.8871	3.0739
11	7.4301	0.1346	32.1504	0.0311	4.3271	0.2311	14.2330	3.2893
12	8.9161	0.1122	39.5805	0.0253	4.4392	0.2253	15.4667	3.4841
13	10.6993	0.0935	48.4966	0.0206	4.5327	0.2206	16.5883	3.6597
14	12.8392	0.0779	59.1959	0.0169	4.6106	0.2169	17.6008	3.8175
15	15.4070	0.0649	72.0351	0.0139	4.6755	0.2139	18.5095	3.9588
16	18.4884	0.0541	87.4421	0.0114	4.7296	0.2114	19.3208	4.0851
17	22.1861	0.0451	105.9306	9.440E − 03	4.7746	0.2094	20.0419	4.1976
18	26.6233	0.0376	128.1167	7.805E − 03	4.8122	0.2078	20.6805	4.2975
19	31.9480	0.0313	154.7400	6.462E − 03	4.8435	0.2065	21.2439	4.3861
20	38.3376	0.0261	186.6880	5.357E − 03	4.8696	0.2054	21.7395	4.4643
21	46.0051	0.0217	225.0256	4.444E − 03	4.8913	0.2044	22.1742	4.5334
22	55.2061	0.0181	271.0307	3.690E − 03	4.9094	0.2037	22.5546	4.5941
23	66.2474	0.0151	326.2369	3.065E − 03	4.9245	0.2031	22.8867	4.6475
24	79.4968	0.0126	392.4842	2.548E − 03	4.9371	0.2025	23.1760	4.6943
25	95.3962	0.0105	471.9811	2.119E − 03	4.9476	0.2021	23.4276	4.7352
26	114.4855	8.735E − 03	567.3773	1.762E − 03	4.9563	0.2018	23.6460	4.7709
27	137.3706	7.280E − 03	681.8528	1.467E − 03	4.9636	0.2015	23.8353	4.8020
28	164.8447	6.066E − 03	819.2233	1.221E − 03	4.9697	0.2012	23.9991	4.8291
29	197.8136	5.055E − 03	984.0680	1.016E − 03	4.9747	0.2010	24.1406	4.8527
30	237.3763	4.213E − 03	1.182E + 03	8.461E − 04	4.9789	0.2008	24.2628	4.8731
36	708.8019	1.411E − 03	3.539E + 03	2.826E − 04	4.9920	0.2003	24.7108	4.9491
42	2.116E + 03	4.725E − 04	1.058E + 04	9.454E − 05	4.9976	0.2001	24.8890	4.9801
48	6.320E + 03	1.582E − 04	3.159E + 04	3.165E − 05	4.9992	0.2000	24.9581	4.9924
54	1.887E + 04	5.299E − 05	9.435E + 04	1.060E − 05	4.9997	0.2000	24.9844	4.9971
60	5.635E + 04	1.775E − 05	2.817E + 05	3.549E − 06	4.9999	0.2000	24.9942	4.9989
66	1.683E + 05	5.943E − 06	8.413E + 05	1.189E − 06	5.0000	0.2000	24.9979	4.9996
72	5.024E + 05	1.990E − 06	2.512E + 06	3.981E − 07	5.0000	0.2000	24.9992	4.9999
120	3.175E + 09	3.150E − 10	1.588E + 10	6.299E − 11	5.0000	0.2000	25.0000	5.0000
180	1.789E + 14	5.590E − 15	8.945E + 14	1.118E − 15	5.0000	0.2000	25.0000	5.0000
360	3.201E + 28	3.124E − 29	1.600E + 29	6.249E − 30	5.0000	0.2000	25.0000	5.0000

Time Value of Money Factors—Discrete Compounding

i = 25%

	Single Sums		Uniform Series				Gradient Series	
	To Find F Given P $(F\|P, i\%, n)$	To Find P Given F $(P\|F, i\%, n)$	To Find F Given A $(F\|A, i\%, n)$	To Find A Given F $(A\|F, i\%, n)$	To Find P Given A $(P\|A, i\%, n)$	To Find A Given P $(A\|P, i\%, n)$	To Find P Given G $(P\|G, i\%, n)$	To Find A Given G $(A\|G, i\%, n)$
n								
1	1.2500	0.8000	1.0000	1.0000	0.8000	1.2500	0.0000	0.0000
2	1.5625	0.6400	2.2500	0.4444	1.4400	0.6944	0.6400	0.4444
3	1.9531	0.5120	3.8125	0.2623	1.9520	0.5123	1.6640	0.8525
4	2.4414	0.4096	5.7656	0.1734	2.3616	0.4234	2.8928	1.2249
5	3.0518	0.3277	8.2070	0.1218	2.6893	0.3718	4.2035	1.5631
6	3.8147	0.2621	11.2588	0.0888	2.9514	0.3388	5.5142	1.8683
7	4.7684	0.2097	15.0735	0.0663	3.1611	0.3163	6.7725	2.1424
8	5.9605	0.1678	19.8419	0.0504	3.3289	0.3004	7.9469	2.3872
9	7.4506	0.1342	25.8023	0.0388	3.4631	0.2888	9.0207	2.6048
10	9.3132	0.1074	33.2529	0.0301	3.5705	0.2801	9.9870	2.7971
11	11.6415	0.0859	42.5661	0.0235	3.6564	0.2735	10.8460	2.9663
12	14.5519	0.0687	54.2077	0.0184	3.7251	0.2684	11.6020	3.1145
13	18.1899	0.0550	68.7596	0.0145	3.7801	0.2645	12.2617	3.2437
14	22.7374	0.0440	86.9495	0.0115	3.8241	0.2615	12.8334	3.3559
15	28.4217	0.0352	109.6868	9.117E − 03	3.8593	0.2591	13.3260	3.4530
16	35.5271	0.0281	138.1085	7.241E − 03	3.8874	0.2572	13.7482	3.5366
17	44.4089	0.0225	176.6357	5.759E − 03	3.9099	0.2558	14.1085	3.6084
18	55.5112	0.0180	218.0446	4.586E − 03	3.9279	0.2546	14.4147	3.6698
19	69.3889	0.0144	273.5558	3.656E − 03	3.9424	0.2537	14.6741	3.7222
20	86.7362	0.0115	342.9447	2.916E − 03	3.9539	0.2529	14.8932	3.7667
21	108.4202	9.223E − 03	429.6809	2.327E − 03	3.9631	0.2523	15.0777	3.8045
22	135.5253	7.379E − 03	538.1011	1.858E − 03	3.9705	0.2519	15.2326	3.8365
23	169.4066	5.903E − 03	673.6264	1.485E − 03	3.9764	0.2515	15.3625	3.8634
24	211.7582	4.722E − 03	843.0329	1.186E − 03	3.9811	0.2512	15.4711	3.8861
25	264.6978	3.778E − 03	1055E + 03	9.481E − 04	3.9849	0.2509	15.5618	3.9052
26	330.8722	3.022E − 03	1.319E + 03	7.579E − 03	3.9879	0.2508	15.6373	3.9212
27	413.5903	2.418E − 03	1.650E + 03	6.059E − 04	3.9903	0.2506	15.7002	3.9346
28	516.9879	1.934E − 03	2.064E + 03	4.845E − 04	3.9923	0.2505	15.7524	3.9457
29	646.2349	1.547E − 03	2.581E + 03	3.875E − 04	3.9938	0.2504	15.7957	3.9551
30	807.7936	1.238E − 03	3.227E + 03	3.099E − 04	3.9950	0.2503	15.8316	3.9628
36	3.081E + 03	3.245E − 04	1.232E + 04	8.116E − 05	3.9987	0.2501	15.9481	3.9883
42	1.175E + 04	8.507E − 05	4.702E + 04	2.127E − 05	3.9997	0.2500	15.9843	3.9964
48	4.484E + 04	2.230E − 05	1.794E + 05	5.575E − 06	3.9999	0.2500	15.9954	3.9989
54	1.711E + 05	5.846E − 06	6.842E + 05	1.462E − 06	4.0000	0.2500	15.9986	3.9997
60	6.525E + 05	1.532E − 06	2.610E + 06	3.831E − 07	4.0000	0.2500	15.9996	3.9999
66	2.489E + 06	4.017E − 07	9.957E + 06	1.004E − 07	4.0000	0.2500	15.9999	4.0000
72	9.496E + 06	1.053E − 07	3.798E + 07	2.633E − 08	4.0000	0.2500	16.0000	4.0000
120	4.258E + 11	2.349E − 12	1.703E + 12	5.871E − 13	4.0000	0.2500	16.0000	4.0000
180	2.778E + 17	3.599E − 18	1.111E + 18	8.998E − 19	4.0000	0.2500	16.0000	4.0000
360	7.720E + 34	1.295E − 35	3.088E + 35	3.238E − 36	4.0000	0.2500	16.0000	4.0000

Time Value of Money Factors—Discrete Compounding   i = 30%

	Single Sums		Uniform Series					Gradient Series	
n	To Find F Given P $(F\|P, i\%, n)$	To Find P Given F $(P\|F, i\%, n)$	To Find F Given A $(F\|A, i\%, n)$	To Find A Given F $(A\|F, i\%, n)$	To Find P Given A $(P\|A, i\%, n)$	To Find A Given P $(A\|P, i\%, n)$	To Find P Given G $(P\|G, i\%, n)$	To Find A Given G $(A\|G, i\%, n)$
1	1.3000	0.7692	1.0000	1.0000	0.7692	1.3000	0.0000	0.0000
2	1.6900	0.5917	2.3000	0.4348	1.3609	0.7348	0.5917	0.4348
3	2.1970	0.4552	3.9900	0.2506	1.8161	0.5506	1.5020	0.8271
4	2.8561	0.3501	6.1870	0.1616	2.1662	0.4616	2.5524	1.1783
5	3.7129	0.2693	9.0431	0.1106	2.4356	0.4106	3.6297	1.4903
6	4.8268	0.2072	12.7560	0.0784	2.6427	0.3784	4.6656	1.7654
7	6.2749	0.1594	17.5828	0.0569	2.8021	0.3569	5.6218	2.0063
8	8.1573	0.1226	23.8577	0.0419	2.9247	0.3419	6.4800	2.2156
9	10.6045	0.0943	32.0150	0.0312	3.0190	0.3312	7.2343	2.3963
10	13.7858	0.0725	42.6195	0.0235	3.0915	0.3235	7.8872	2.5512
11	17.9216	0.0558	56.4053	0.0177	3.1473	0.3177	8.4452	2.6833
12	23.2981	0.0429	73.3270	0.0135	3.1903	0.3135	8.9173	2.7952
13	30.2875	0.0330	97.6250	0.0102	3.2233	0.3102	9.3135	2.8895
14	39.3738	0.0254	127.9125	7.818E − 03	3.2487	0.3078	9.6437	2.9685
15	51.1859	0.0195	167.2863	5.978E − 03	3.2682	0.3060	9.9172	3.0344
16	66.5417	0.0150	218.4722	4.577E − 03	3.2832	0.3046	10.1426	3.0892
17	86.5042	0.0116	285.0139	3.509E − 03	3.2948	0.3035	10.3276	3.1345
18	112.4554	8.892E − 03	371.5180	2.692E − 03	3.3037	0.3027	10.4788	3.1718
19	146.1920	6.840E − 03	483.9734	2.066E − 03	3.3105	0.3021	10.6019	3.2025
20	190.0496	5.262E − 03	630.1655	1.587E − 03	3.3158	0.3016	10.7019	3.2275
21	247.0645	4.048E − 03	820.2151	1.219E − 03	3.3198	0.3012	10.7828	3.2480
22	321.1839	3.113E − 03	1.067E + 03	9.370E − 04	3.3230	0.3009	10.8482	3.2646
23	417.5391	2.395E − 03	1.388E + 03	7.202E − 04	3.3254	0.3007	10.9009	3.2781
24	542.8008	1.842E − 03	1.806E + 03	5.537E − 04	3.3272	0.3006	10.9433	3.2890
25	705.6410	1.417E − 03	2.349E + 03	4.257E − 04	3.3286	0.3004	10.9773	3.2979
26	917.3333	1.090E − 03	3.274E + 03	3.274E − 04	3.3297	0.3003	11.0045	3.3050
27	1.193E + 03	8.386E − 04	3.972E + 03	2.518E − 04	3.3305	0.3003	11.0263	3.3107
28	1.550E + 03	6.450E − 04	5.164E + 03	1.936E − 04	3.3312	0.3002	11.0437	3.3153
29	2.105E + 03	4.962E − 04	6.715E + 03	1.489E − 04	3.3317	0.3001	11.0576	3.3189
30	2.620E + 03	3.817E − 04	8.730E + 03	1.145E − 04	3.3321	0.3001	11.0687	3.3219
36	1.265E + 04	7.908E − 05	4.215E + 04	2.372E − 05	3.3331	0.3000	11.1007	3.3305
42	6.104E + 04	1.638E − 05	2.035E + 05	4.915E − 06	3.3333	0.3000	11.1086	3.3326
48	2.946E + 05	3.394E − 06	9.821E + 05	1.018E − 06	3.3333	0.3000	11.1105	3.3332
54	1.422E + 06	7.032E − 07	4.740E + 06	2.110E − 07	3.3333	0.3000	11.1110	3.3333
60	6.864E + 06	1.457E − 07	2.288E + 07	4.370E − 08	3.3333	0.3000	11.1111	3.3333
66	3.313E + 07	3.018E − 08	1.104E + 08	9.054E − 09	3.3333	0.3000	11.1111	3.3333
72	1.599E + 08	6.253E − 09	5.331E + 08	1.876E − 09	3.3333	0.3000	11.1111	3.3333
120	4.712E + 13	2.122E − 14	1.571E + 14	6.367E − 15	3.3333	0.3000	11.1111	3.3333
180	3.234E + 04	3.092E − 21	1.078E + 21	9.275E − 22	3.3333	0.3000	11.1111	3.3333
360	1.046E + 41	9.559E − 42	3.487E + 41	2.868E − 42	3.3333	0.3000	11.1111	3.3333

Time Value of Money Factors—Discrete Compounding $\qquad$ i = 40%

	Single Sums		Uniform Series				Gradient Series	
n	To Find F Given P $(F\|P, i\%, n)$	To Find P Given F $(P\|F, i\%, n)$	To Find F Given A $(F\|A, i\%, n)$	To Find A Given F $(A\|F, i\%, n)$	To Find P Given A $(P\|A, i\%, n)$	To Find A Given P $(A\|P, i\%, n)$	To Find P Given G $(P\|G, i\%, n)$	To Find A Given G $(A\|G, i\%, n)$
1	1.4000	0.7143	1.0000	1.0000	0.7143	1.4000	0.0000	0.0000
2	1.9600	0.5102	2.4000	0.4167	1.2245	0.8167	0.5102	0.4167
3	2.7440	0.3644	4.3600	0.2294	1.5889	0.6294	1.2391	0.7798
4	3.8416	0.2603	7.1040	0.1408	1.8492	0.5408	2.0200	1.0923
5	5.3782	0.1859	10.9456	0.0914	2.0352	0.4914	2.7637	1.3580
6	7.5295	0.1328	16.3238	0.0613	2.1680	0.4613	3.4278	1.5811
7	10.5414	0.0949	28.8534	0.0419	2.2628	0.4419	3.9970	1.7664
8	14.7579	0.0678	34.3947	0.0291	2.3306	0.4219	4.4713	1.9185
9	20.6610	0.0484	49.1526	0.0203	2.3790	0.4203	4.8585	2.0422
10	28.9255	0.0346	69.8137	0.0143	2.4136	0.4143	5.1696	2.1419
11	40.4957	0.0247	98.7391	0.0101	2.4383	0.4101	5.4166	2.2215
12	56.6939	0.0176	139.2348	7.182E − 03	2.4559	0.4072	5.6106	2.2845
13	79.3715	0.0126	195.9287	5.104E − 03	2.4685	0.4051	5.7618	2.3341
14	111.1201	8.999E − 03	275.3002	3.632E − 03	2.4775	0.4036	5.8788	2.3729
15	155.5681	6.428E − 03	386.4202	2.588E − 03	2.4839	0.4026	5.9688	2.4030
16	217.7953	4.591E − 03	541.9883	1.845E − 03	2.4885	0.4018	6.0376	2.4262
17	304.9135	3.280E − 03	759.7837	1.316E − 03	2.4918	0.4013	6.0901	2.4441
18	426.8789	2.343E − 03	1.065E + 03	9.392E − 04	2.4941	0.4009	6.1299	2.4577
19	597.6304	1.673E − 03	1.492E + 03	6.704E − 04	2.4958	0.4007	6.1601	2.4682
20	836.6826	1.195E − 03	2.089E + 03	4.787E − 04	2.4970	0.4005	6.1828	2.4761
21	1.171E + 03	8.537E − 04	2.926E + 03	3.418E − 04	2.4979	0.4003	6.1998	2.4821
22	1.640E + 03	6.098E − 04	4.097E + 03	2.441E − 04	2.4985	0.4002	6.2127	2.4866
23	2.296E + 03	4.356E − 04	5.737E + 03	1.743E − 04	2.4989	0.4002	6.2222	2.4900
24	3.214E + 03	3.111E − 04	8.033E + 04	1.245E − 04	2.4992	0.4001	6.2294	2.4925
25	4.500E + 03	2.222E − 04	1.125E + 04	8.891E − 05	2.4994	0.4001	6.2347	2.4944
26	6.300E + 03	1.587E − 04	1.575E + 04	6.350E − 05	2.4996	0.4001	6.2387	2.4959
27	8.820E + 04	1.134E − 04	2.205E + 04	4.536E − 05	2.4997	0.4000	6.2416	2.4969
28	1.235E + 04	8.099E − 05	3.087E + 04	3.240E − 05	2.4998	0.4000	6.2438	2.4977
29	1.729E + 04	5.785E − 05	4.321E + 04	2.314E − 05	2.4999	0.4000	6.2454	2.4983
30	2.420E + 04	4.132E − 05	6.050E + 04	1.653E − 05	2.4999	0.4000	6.2466	3.4988
36	1.822E + 05	5.488E − 06	4.556E + 05	2.195E − 06	2.5000	0.4000	6.2495	2.4998
42	1.372E + 06	7.288E − 07	3.430E + 06	2.915E − 07	2.5000	0.4000	6.2499	2.5000
48	1.033E + 07	9.680E − 08	2.583E + 07	3.872E − 08	2.5000	0.4000	6.2500	2.5000
54	7.779E + 07	1.286E − 08	1.945E + 08	5.142E − 09	2.5000	0.4000	6.2500	2.5000
60	5.857E + 08	1.707E − 09	1.464E + 09	6.829E − 10	2.5000	0.4000	6.2500	2.5000
66	4.410E + 09	2.268E − 10	1.103E + 10	9.070E − 11	2.5000	0.4000	6.2500	2.5000
72	3.321E + 10	3.011E − 11	8.302E + 10	1.205E − 11	2.5000	0.4000	6.2500	2.5000
120	3.431E + 17	2.915E − 18	8.576E + 17	1.166E − 18	2.5000	0.4000	6.2500	2.5000
180	2.009E + 26	4.977E − 27	5.023E + 26	1.991E − 27	2.5000	0.4000	6.2500	2.5000
360	4.037E + 52	2.477E − 53	1.009E + 53	9.908E − 54	2.5000	0.4000	6.2500	2.5000

Discrete Compounding $i = 5\%$

	Geometric series present worth factor,				$(P\|A_1\ i, i, n)$
n	$j = 4\%$	$j = 6\%$	$j = 8\%$	$j = 10\%$	$j = 15\%$
1	0.9524	0.9524	0.9524	0.9524	0.9524
2	1.8957	1.9138	1.9320	1.9501	1.9955
3	2.8300	2.8844	2.9396	2.9954	3.1379
4	3.7554	3.8643	3.9759	4.0904	4.3891
5	4.6721	4.8535	5.0419	5.2375	5.7595
6	5.5799	5.8521	6.1383	6.4393	7.2604
7	6.4792	6.8602	7.2661	7.6983	8.9043
8	7.3699	7.8779	8.4261	9.0173	10.7047
9	8.2521	8.9053	9.6192	10.3991	12.6765
10	9.1258	9.9425	10.8464	11.8467	14.8362
11	9.9913	10.9896	12.1087	13.3632	17.2016
12	10.8485	12.0466	13.4070	14.9519	19.7922
13	11.6976	13.1137	14.7425	16.6163	22.6295
14	12.5386	14.1910	16.1161	18.3599	25.7371
15	13.3715	15.2785	17.5289	20.1866	29.1407
16	14.1966	16.3764	18.9821	22.1002	32.8683
17	15.0137	17.4848	20.4769	24.1050	36.9510
18	15.8231	18.6037	22.0143	26.2052	41.4226
19	16.6248	19.7332	23.5956	28.4055	46.3200
20	17.4189	20.8736	25.2222	30.7105	51.6838
21	18.2054	22.0247	26.8952	33.1253	57.5584
22	18.9844	23.1869	28.6160	35.6550	63.9925
23	19.7559	24.3601	30.3860	38.3053	71.0394
24	20.5202	25.5445	32.2066	41.0817	78.7575
25	21.2771	26.7401	34.0791	43.9904	87.2106
26	22.0269	27.9472	36.0052	47.0375	96.4487
27	22.7695	29.7695	37.9863	50.2298	106.6086
28	23.5050	30.3959	40.0240	53.5741	117.7142
29	24.2335	31.6377	42.1199	57.0776	129.8774
30	24.9551	32.8914	44.2757	60.7480	143.1991
31	25.6698	34.1571	46.4931	64.5931	157.7895
32	26.3777	35.4348	48.7739	68.6213	173.7695
33	27.0789	36.7246	51.1198	72.8414	191.2713
34	27.7734	38.0267	53.5328	77.2624	210.4400
35	28.4612	39.3413	56.0146	81.8940	231.4343
36	29.1426	40.6683	58.5674	86.7461	254.4280
37	29.8174	42.0030	61.1932	91.8292	279.6116
38	30.4858	43.3605	63.8939	97.1544	307.1937
39	31.1478	44.7258	66.6719	102.7332	337.4026
40	31.8036	46.1042	69.5291	108.5776	370.4886
48	36.8296	57.6141	95.5310	166.5488	777.7891
50	38.0271	60.6306	102.9998	184.7384	934.9897
54	40.3544	66.8378	119.2613	226.6112	1349.7568
60	43.6826	76.6013	147.3617	306.0117	2236.9967
65	46.3138	85.1726	174.6930	391.0117	3688.7517
70	48.8221	94.1599	206.1585	499.1191	5819.0515
80	53.4925	113.4643	284.0874	806.6104	14467.1576
90	57.7367	134.6830	387.3737	1296.2390	35945.7801
100	61.5936	158.0219	524.2687	2075.8906	89290.5489
120	68.2835	211.8803	946.1871	5294.1808	550829.3461

Discrete Compounding $i = 5\%$

	Geometric series future worth factor,				$(F \mid A_1 \, i, i, n)$
n	$j = 4\%$	$j = 6\%$	$j = 8\%$	$j = 10\%$	$j = 15\%$
1	1.0000	1.0000	1.0000	1.0000	1.0000
2	2.0900	2.1100	2.1300	2.1500	2.2000
3	3.2761	3.3391	3.4029	3.4675	3.6325
4	4.5648	4.6971	4.8328	4.9719	5.3350
5	5.9629	6.1944	6.4349	6.6846	7.3508
6	7.4777	7.8423	8.2260	8.6293	9.7297
7	9.1169	9.6530	10.2241	10.8323	12.5292
8	10.8886	11.6393	12.4492	13.3227	15.8157
9	12.8016	13.8151	14.9225	16.1324	19.6655
10	14.8650	16.1953	17.6677	19.2970	24.1666
11	17.0885	18.7959	20.7100	22.8556	29.4205
12	19.4824	21.6340	24.0771	26.8515	35.5439
13	22.0576	24.7279	27.7992	31.3325	42.6714
14	24.8255	28.0972	31.9087	36.3514	50.9577
15	27.7985	31.7630	36.4414	41.9664	60.5813
16	30.9893	35.7477	41.4356	48.2420	71.7475
17	34.4118	40.0754	46.9333	55.2491	84.6925
18	38.0803	44.7720	52.9800	63.0660	99.6883
19	42.0101	49.8649	59.6250	71.7792	117.0482
20	46.2175	55.3838	66.9220	81.4841	137.1324
21	50.7195	61.3601	74.9290	92.2858	160.3556
22	55.5342	67.8277	83.7093	104.3004	187.1948
23	60.6808	74.8226	93.3313	117.6557	218.1993
24	66.1796	82.3835	103.8694	132.4928	254.0008
25	72.0519	90.5516	115.4040	148.9672	295.3260
26	78.3203	69.3710	128.0227	167.2503	343.0112
27	85.0088	108.8890	141.8202	187.5310	398.0186
28	92.1426	119.1558	156.8992	210.0176	461.4548
29	99.7484	130.2252	173.3713	234.9395	534.5932
30	107.8545	142.1549	191.3572	262.5496	618.8983
31	116.4906	155.0061	210.9877	293.1266	716.0550
32	125.6883	168.8445	232.4047	326.9773	828.0013
33	135.4807	183.7401	255.7620	364.4400	956.9664
34	145.9032	199.7677	281.2262	405.8872	1105.5145
35	156.9926	217.0071	308.9776	451.7293	1276.5951
36	168.7884	235.5436	339.2119	502.4183	1473.6003
37	181.3317	255.4680	372.1406	558.4520	1700.4322
38	194.6664	276.8775	407.9933	620.3787	1961.5784
39	208.8385	299.8756	447.0182	688.8021	2262.2007
40	223.8968	324.5729	489.4844	764.3871	2608.2355
48	383.0741	599.2602	993.6434	1732.3243	8089.9942
50	436.0716	695.2754	1181.1404	2118.4755	10721.9001
54	562.4882	931.6324	1662.3477	3158.6756	18813.8487
60	815.9558	1430.8505	2752.5959	5716.0709	43653.1937
65	1104.1165	2030.5071	4164.6649	9330.6558	87939.4708
70	1485.4807	2864.9505	6272.6660	15186.4808	177052.9271
80	2651.1642	5623.4552	14079.7798	39976.9935	717013.1369
90	4661.1032	10873.4146	31272.8241	104646.5051	2901915.7503
100	8099.6310	20780.0826	68941.9999	272984.1812	11741818.5907
120	23824.9425	73927.5762	330136.0319	1847219.6697	192190942.7662

Discrete Compounding $i = 8\%$

	Geometric series present worth factor,				$(P\|A_1\ i,\ i,\ n)$
n	$j = 4\%$	$j = 6\%$	$j = 8\%$	$j = 10\%$	$j = 15\%$
1	0.9259	0.9259	0.9259	0.9259	0.9259
2	1.8176	1.8347	1.8519	1.8690	1.9119
3	2.6762	2.7267	2.7778	2.8295	2.9617
4	3.5030	3.6021	3.7037	3.8079	4.0796
5	4.2992	4.4613	4.6296	4.8043	5.2699
6	5.0659	5.3046	5.5556	5.8192	6.5374
7	5.8042	6.1323	6.4815	6.8529	7.8871
8	6.5151	6.9447	7.4074	7.9057	9.3242
9	7.1997	7.7420	8.3333	8.9780	10.8545
10	7.8590	8.5246	9.2593	10.0702	12.4839
11	8.4939	9.2926	10.1852	11.1826	14.2190
12	9.1052	10.0465	11.1111	12.3157	16.0665
13	9.6939	10.7863	12.0370	13.4696	18.0338
14	10.2608	11.5125	12.9630	14.6450	20.1286
15	10.8067	12.2252	13.8889	15.8421	22.3592
16	11.3324	12.9248	14.8148	17.0614	24.7343
17	11.8386	13.6114	15.7407	18.3033	27.2634
18	12.3260	14.2852	16.6667	19.5682	29.9564
19	12.7954	14.9466	17.5926	20.8565	32.8239
20	13.2475	15.5957	18.5185	22.1687	35.8773
21	13.6827	16.2329	19.4444	23.5051	39.1286
22	14.1019	16.8582	20.3704	24.8663	42.5906
23	14.5055	17.4719	21.2963	26.2527	46.2771
24	14.8942	18.0743	22.2222	27.6648	50.2024
25	15.2685	18.6655	23.1482	29.1031	54.3822
26	15.6289	19.2458	24.0741	30.5679	58.8329
27	15.9760	19.8153	25.0000	32.0599	63.5721
28	16.3102	20.3743	25.9259	33.5796	68.6184
29	16.6321	20.9229	26.8519	35.1273	73.9919
30	16.9420	21.4614	27.7778	36.7038	79.7136
31	17.2404	21.9899	28.7037	38.3094	85.8061
32	17.5278	22.5086	29.6296	39.9447	92.2935
33	17.8046	23.0177	30.5556	41.6104	99.2015
34	18.0711	23.5173	31.4815	43.3069	106.5571
35	18.3277	24.0078	32.4074	45.0348	114.3895
36	18.5749	24.4891	33.3333	46.7947	122.7296
37	18.8126	24.9615	34.2593	48.5872	131.6102
38	19.0419	25.4252	35.1852	50.4129	141.0664
39	19.2626	25.8803	36.1111	52.2724	151.1355
40	19.4751	26.3269	37.0370	54.1663	161.8573
48	20.9149	29.6150	44.4444	70.6365	276.8249
50	21.2119	30.3630	46.2963	75.1459	315.7844
54	21.7427	31.7777	50.0000	84.6766	410.0431
60	22.4027	33.7109	55.5556	100.3508	604.2248
65	22.8494	35.1643	60.1852	114.7974	832.3907
70	23.2192	36.4881	64.8148	130.6322	1144.7262
80	23.7790	38.7917	74.0741	167.0123	2157.5589
90	24.1629	40.7026	83.3333	210.7196	4055.4818
100	24.4260	42.2878	92.5926	263.2297	7611.9539
120	24.7302	44.6933	111.1111	402.1073	26764.5779

Discrete Compounding $i = 8\%$

	Geometric series present worth factor,				$(F\|A_1\ i,\ i,\ n)$
n	$j = 4\%$	$j = 6\%$	$j = 8\%$	$j = 10\%$	$j = 15\%$
1	1.0000	1.0000	1.0000	1.0000	1.0000
2	2.1200	2.1400	2.1000	2.1000	2.2300
3	3.3712	3.4348	3.4902	3.5644	3.7300
4	4.7658	4.9006	5.0388	5.1806	5.5502
5	6.3169	6.5551	6.8024	7.0591	7.7433
6	8.0389	8.4176	6.8160	9.2343	10.3741
7	9.9473	10.5097	11.1061	11.7446	13.5171
8	12.0590	12.8541	13.7106	14.6329	17.2585
9	14.3923	15.4763	16.6584	17.9472	21.6982
10	16.9670	18.4039	19.9900	21.7409	26.9519
11	19.8046	21.6670	23.7482	26.0739	33.1536
12	22.9284	25.2987	27.9797	31.0129	40.4583
13	26.3638	29.3348	32.7362	36.6324	49.0452
14	30.1379	33.8145	38.0747	43.0153	59.1216
15	34.2806	38.7805	44.0579	50.2540	70.9270
16	38.8240	44.2795	50.7547	58.4516	84.7383
17	43.8029	50.3623	58.2410	67.7227	100.8749
18	49.2551	57.0840	66.6003	78.1950	119.7062
19	55.2213	64.5051	75.9243	90.0105	141.6581
20	61.7459	72.6911	86.3139	103.3272	167.2226
21	68.8766	81.7135	97.8800	118.3209	196.9669
22	76.6655	91.6501	110.7442	135.1869	231.5458
23	85.1687	102.5857	125.0403	154.1421	271.7142
24	94.4469	114.6123	140.9150	175.4276	318.3428
25	104.5660	127.8302	158.5293	199.3116	372.4354
26	115.5971	142.3485	178.0601	226.0914	435.1492
27	127.6173	158.2858	199.7013	256.0970	507.8179
28	140.7101	175.7710	223.6654	289.6947	591.9786
29	154.9656	194.9443	250.1858	327.2914	689.4025
30	170.4815	215.9583	279.5179	369.3378	802.1302
31	187.3634	238.9784	311.9419	416.3343	932.5124
32	205.7256	264.1848	347.7649	468.8354	1083.2569
33	225.6917	291.7730	387.3232	527.4561	1257.4825
34	247.3954	321.9554	430.9850	592.8778	1458.7810
35	270.9814	354.9629	479.1539	665.8558	1691.2882
36	296.6060	391.0460	532.2715	747.2268	1959.7668
37	324.4384	430.4769	590.8214	835.9177	2269.7000
38	354.6616	473.5512	655.3327	938.9552	2627.4006
39	387.4733	520.5895	726.3845	1051.4760	3040.1360
40	423.0875	571.9402	804.6105	1176.7390	3516.2717
48	841.0011	1190.8351	1787.1327	2840.3400	1131.2874
50	994.8732	1424.0729	2171.3660	3524.4711	14810.7971
54	1387.3838	2027.7053	3190.4485	5403.1563	26164.4921
60	2268.4359	3413.4646	5625.3771	10161.2610	61182.0216
65	3399.5278	5231.7437	8954.3161	17079.6034	123842.9595
70	5075.8697	7976.5238	14168.8885	28557.1360	250244.4686
80	11222.6259	18307.9420	34959.4895	78822.6198	1018270.2942
90	24619.8939	41472.5289	84909.2410	214706.4761	4132191.3076
100	53731.4077	93022.9586	203680.6644	579045.9229	16744490.0582
120	253558.2595	458240.2597	1139215.1626	4122833.7485	274417003.9699

Discrete Compounding $i = 10\%$

	Geometric series present worth factor,				$(P\|A_1\ i, i, n)$
n	$j = 4\%$	$j = 6\%$	$j = 8\%$	$j = 10\%$	$j = 15\%$
1	0.9091	0.9091	0.9091	0.9091	0.9091
2	1.7686	1.7851	1.8017	1.8182	1.8596
3	2.5812	2.6293	2.6780	2.7273	2.8531
4	3.3495	3.4428	3.5384	3.6364	3.8919
5	4.0759	4.2267	4.3831	4.5455	4.9779
6	4.7627	4.9821	5.2125	5.4545	6.1123
7	5.4120	5.7100	6.0269	6.3636	7.3002
8	6.0259	6.4115	6.8264	7.2727	8.5411
9	6.6063	7.0874	7.6113	8.1818	9.8385
10	7.1550	7.7388	8.3820	9.0909	11.1948
11	7.6738	8.3664	9.1387	10.0000	12.6127
12	8.1644	8.9713	9.8817	10.9091	14.0951
13	8.6281	9.5542	10.6111	11.8182	15.6449
14	9.0666	10.1158	11.3273	12.7273	17.2651
15	9.4811	10.6571	12.0304	13.6364	18.9590
16	9.8731	11.1786	12.7208	14.5455	20.7296
17	10.2436	11.6812	13.3986	15.4545	22.5812
18	10.5940	12.1656	14.0640	16.3636	24.5167
19	10.9252	12.6323	14.7174	17.2727	26.5402
20	11.2384	13.0820	15.3589	18.1818	28.6556
21	11.5345	13.5154	15.9888	19.0909	30.8672
22	11.8144	13.9330	16.6071	20.0000	33.1794
23	12.0791	14.3354	17.2143	20.9091	35.5966
24	12.3293	14.7232	17.8104	21.8182	38.1238
25	12.5659	15.0969	18.3957	22.7273	40.7658
26	12.7896	15.4570	18.9703	23.6364	43.5278
27	13.0011	15.8041	19.5345	24.5455	46.4155
28	13.2010	16.1385	20.0884	25.4545	49.4343
30	13.5688	16.7712	21.1662	27.2727	55.8900
31	13.7377	17.0704	21.6904	28.1818	59.3396
32	13.8975	17.3588	22.2052	29.0909	62.9459
33	14.0485	17.6367	22.7105	30.0000	64.7162
34	14.1913	17.9044	23.2067	30.9091	70.6578
35	14.3264	18.1624	23.6938	31.8182	74.7786
36	14.4540	18.4111	24.1721	32.7273	79.0867
37	14.5747	18.6507	24.6417	33.6364	83.5907
38	14.6888	18.8816	25.1028	34.5455	88.2994
39	14.7967	19.1040	25.5555	35.4545	93.2221
40	14.8987	19.3184	25.9999	36.3636	98.3685
48	15.5379	20.7755	29.2766	43.6364	148.9196
50	15.6577	21.0772	30.0233	45.4545	164.6238
54	15.8605	21.6174	31.4370	49.0909	200.5507
60	16.0908	22.2915	33.3722	54.5455	267.9647
65	16.2317	22.7494	34.8299	50.0909	339.6376
70	16.3380	23.1290	36.1597	63.6364	429.1494
80	16.4791	23.7088	38.4799	72.7273	680.5553
90	16.5596	24.1085	40.4112	81.8182	1072.6824
100	16.6056	24.3845	42.0186	90.9091	1684.2979
120	16.6468	24.7066	44.4703	109.0909	4126.1845

Discrete Compounding $i = 10\%$

	Geometric series present worth factor,				$(F|A_1\ i, i, n)$
n	$j = 4\%$	$j = 6\%$	$j = 8\%$	$j = 10\%$	$j = 15\%$
1	1.0000	1.0000	1.0000	1.0000	1.0000
2	2.1400	2.1600	2.1800	2.2000	2.2500
3	3.4356	3.4996	3.5644	3.6300	3.7975
4	4.9040	5.0406	5.1806	5.3240	5.6961
5	6.5643	6.8071	7.0591	7.3205	8.0169
6	8.4374	8.8260	9.2343	9.6631	10.8300
7	10.5464	11.1272	11.7446	12.4009	14.2261
8	12.9179	13.7435	14.6329	15.5897	18.3067
9	15.5773	16.7117	17.9471	19.2923	23.1986
10	18.5583	20.0724	21.7409	23.5795	29.0343
11	21.8944	23.8705	26.0739	28.5312	35.9856
12	25.6233	28.1558	31.0129	34.2374	44.2364
13	29.7886	32.9836	36.6324	40.7996	54.0103
14	34.4304	38.4149	43.0152	48.3318	65.5641
15	39.6051	44.5172	50.2539	54.9625	79.1963
16	45.3665	51.3655	58.4515	66.8360	95.2530
17	51.7762	59.0424	67.7226	76.1146	114.1359
18	58.9017	67.6395	78.1948	90.9805	136.3107
19	66.8177	77.2577	90.0103	105.6385	162.3173
20	75.6063	88.0091	103.3271	122.3183	192.7807
21	85.3580	100.0172	118.3207	141.2776	228.4254
22	96.1726	113.4184	135.1866	162.8057	270.0894
23	108.1598	128.3638	154.1418	187.2265	318.7431
24	121.4405	145.0200	175.4274	214.9035	375.5089
25	136.1478	163.5709	199.3113	246.2436	441.6849
26	152.4284	184.2198	226.0909	281.7027	518.7724
27	170.4438	207.1912	256.0964	321.7911	608.5064
28	190.3715	232.7327	289.6940	367.0803	712.8924
29	212.4074	261.1176	327.2905	418.2093	834.2472
30	236.7667	292.6478	369.3368	475.8934	975.2474
31	263.6868	327.6560	416.3331	540.9322	1138.9839
32	293.4286	366.5098	468.8341	614.2198	1329.0258
33	326.2796	409.6141	527.4545	696.7556	1549.4935
34	362.5559	457.4161	592.8760	789.6564	1805.1426
35	402.6058	510.4088	665.8537	894.1698	2101.4617
36	446.8125	569.1357	747.2244	1011.6893	2444.7834
37	495.5976	634.1965	837.9149	1143.7710	2842.4136
38	549.4255	706.2523	938.9520	1292.1522	3302.7795
39	608.8069	786.0318	1051.4724	1458.7719	3835.6008
40	657.3039	874.3384	1176.7348	1645.7940	4452.0857
48	1507.4451	2015.5841	2840.3278	4233.4883	14447.8602
50	1838.0695	2474.2675	3524.4553	5335.9507	19325.3312
54	2725.9689	3715.4232	5403.1300	8437.3705	34469.0327
60	4899.3669	6787.3487	10161.2061	16608.1339	81590.3302
65	7959.5332	11155.6438	17079.5033	28976.5362	166548.3266
70	12902.9223	18266.7757	28556.9556	50256.7821	338919.4468
80	33755.8403	48565.1055	78822.0497	148975.0958	1394049.5518
90	87981.7213	128088.9525	214704.7272	434703.6079	5699185.7115
100	228835.1232	336032.7564	579040.6774	1252788.5754	23210655.1891
120	1543306.7710	2290522.0267	4122788.8750	10113771.2878	382534696.2027

Discrete Compounding $i = 15\%$

	Geometric series present worth factor,				$(P\|A_1\ i, i, n)$
n	$j = 4\%$	$j = 6\%$	$j = 8\%$	$j = 10\%$	$j = 15\%$
1	0.8696	0.8696	0.8696	0.8696	0.8696
2	1.6560	1.6711	1.6862	1.7013	1.7391
3	2.3671	2.4099	2.4531	2.4969	2.6087
4	3.0103	3.0908	3.1734	3.2579	3.4783
5	3.5919	3.7185	3.8498	3.9858	4.3478
6	4.1179	4.2971	4.4850	4.6821	5.2174
7	4.5936	4.8303	5.0616	5.3481	6.0870
8	5.0237	5.3219	5.6418	5.9851	6.9565
9	5.4128	5.7749	6.1680	6.5945	7.8261
10	5.7646	6.1926	6.6621	7.1773	8.6957
11	6.0828	6.5775	7.1261	7.7348	9.5652
12	6.3705	6.9323	7.5619	8.2681	10.4348
13	6.6307	7.2593	7.9712	8.7782	11.3043
14	6.8660	7.5608	8.3556	9.2661	12.1739
15	7.0789	7.8386	8.7165	9.7328	13.0435
16	7.2713	8.0947	9.0555	10.1792	13.9130
17	7.4454	8.3308	9.3739	10.6062	14.7826
18	7.6028	8.5484	9.6729	11.0146	15.6522
19	7.7451	8.7489	9.9537	11.4053	16.5217
20	7.8738	8.9338	10.2173	11.7790	17.3913
21	7.9903	9.1042	10.4650	12.1364	18.2609
22	8.0955	9.2613	10.6976	12.4783	19.1304
23	8.1907	9.4060	10.9160	12.8053	20.0000
24	8.2768	9.5395	11.1211	13.1181	20.8696
25	8.3547	9.6625	11.3137	13.4173	21.7391
26	8.4251	9.7759	11.4946	13.7035	22.6087
27	8.4888	9.8803	11.6645	13.9773	23.4783
28	8.5464	9.9767	11.8241	14.2392	24.3478
29	8.5985	10.0655	11.9739	14.4896	25.2174
30	8.6456	10.1473	12.1146	14.7292	26.0870
31	8.6882	10.2227	12.2468	14.9584	26.9565
32	8.7267	10.2922	12.3709	15.1776	27.8261
33	8.7615	10.3563	12.4874	15.3873	28.6956
34	8.7930	10.4154	12.5969	15.5878	29.5652
35	8.8215	10.4698	12.6997	15.7796	30.4348
36	8.8473	10.5200	12.7962	15.9631	31.3043
37	8.8706	10.5663	12.8869	16.1386	32.1739
38	8.8917	10.6089	12.9720	16.3065	33.0435
39	8.9107	10.6482	13.0520	16.4671	33.9130
40	8.9280	10.6845	13.1271	16.6207	34.7826
48	9.0180	10.8888	13.5847	17.6320	41.7391
50	9.0313	10.9222	13.6674	17.8334	43.4783
54	9.0510	10.9748	13.8048	18.1964	46.9565
60	9.0691	11.0275	13.9558	18.6109	52.1739
65	9.0777	11.0555	14.0447	18.8878	56.5217
70	9.0829	11.0741	14.1096	19.1094	60.8696
80	9.0880	11.0947	14.1917	19.4290	69.5652
90	9.0898	11.1039	14.2356	19.6339	78.2609
100	9.0905	11.1079	14.2590	19.7653	86.9565
120	9.0909	11.1105	14.2781	19.9035	104.3478

Discrete Compounding $\qquad\qquad i = 15\%$

	Geometric series present worth factor,				$(F|A_1\,i,\,i,\,n)$
n	$j = 4\%$	$j = 6\%$	$j = 8\%$	$j = 10\%$	$j = 15\%$
1	1.0000	1.0000	1.0000	1.0000	1.0000
2	2.1900	2.2100	2.2300	2.5000	2.3000
3	3.6001	3.6651	3.7300	3.7975	3.9675
4	5.2650	5.4059	5.5502	5.6981	6.0625
5	7.2246	7.4792	7.7433	8.0169	8.7450
6	9.5249	9.9394	10.3741	10.8300	12.0681
7	12.2190	12.8488	13.5171	14.2261	16.1914
8	15.3678	16.2797	17.2585	18.3087	21.2802
9	19.0415	20.3155	21.6982	23.1986	27.5312
10	23.3210	25.0523	26.9519	29.0363	36.1788
11	28.2994	30.6010	33.1536	35.9855	44.5011
12	34.0838	37.0895	40.4583	44.2365	55.8287
13	40.7974	44.6651	49.0452	54.0104	69.5532
14	48.5821	53.4978	59.1216	65.5642	96.1390
15	57.6011	63.7834	70.9270	79.1963	106.1356
16	68.0422	75.7474	84.7382	95.2530	130.1930
17	80.1215	89.6499	100.8749	114.1359	159.0795
18	94.0876	105.7902	119.7061	136.3108	193.7027
19	110.2266	124.5130	141.6581	162.3174	235.1336
20	128.8674	146.2156	167.2225	192.7809	284.6354
21	150.3886	171.3550	196.9668	228.4255	343.6972
22	175.2257	200.4579	231.5456	270.0896	414.0733
23	203.8795	234.1301	271.7140	318.7434	497.8291
24	236.9261	273.0694	318.3425	375.5092	597.3949
25	275.0283	318.0787	372.4351	441.6853	715.6293
26	318.9884	370.0824	435.1488	518.7729	855.8927
27	369.5631	430.1441	507.8175	608.5070	1022.1333
28	427.8810	499.4881	591.9781	712.8931	1218.9886
29	495.0618	579.5230	689.4019	834.2481	1451.9025
30	572.4398	671.8698	802.1295	975.2484	1727.2634
31	661.5491	778.3937	932.5115	1138.9851	2052.5646
32	764.1546	901.2409	1083.2559	1329.0273	2436.5929
33	882.2859	1042.8804	1257.4813	1549.4852	2889.6568
34	1018.2772	1206.1531	1458.7795	1805.1447	3423.7937
35	1174.8130	1394.3271	1691.2865	2101.4641	4053.1675
36	1354.9811	1611.1622	1959.7648	2444.7863	4794.3182
37	1562.3322	1860.9838	2269.6977	2842.4170	5666.6177
38	1800.9501	2142.7675	2627.3979	3302.7836	6692.7350
39	2075.5314	2480.2368	3040.1368	3835.6057	7899.1885
40	2391.4775	2861.9759	3516.2679	4452.0914	9316.9916
48	7389.3653	8922.2982	11131.2743	14447.6900	34201.0678
50	9786.8251	11835.9690	14810.7795	19325.3598	47115.5338
54	17154.6348	20800.7624	26164.4602	34469.0859	88897.7921
60	38758.9011	48344.5673	61181.9447	81590.4721	228730.3333
65	80045.2514	97484.9157	123842.8007	166548.6034	498396.5994
70	161092.2584	196407.1548	250244.1432	338920.0390	1079565.3949
80	652071.1782	796056.4823	1018268.9462	1394052.0739	4991364.6825
90	2638529.1443	3223142.8966	4132185.7913	5699196.5809	22716960.7653
100	10475117.6887	13044157.2069	16744457.6481	23210700.6830	102114195.2907
120	174721221.2670	213537297.9353	274416638.3034	382535463.5055	2005506937.5163

Appendix **B**

Continuous Compounding

Time Value of Money Factors—Continuous Compounding $r = 0.5\%$

	Single Sums		Uniform Series				Gradient Series	
	To Find F Given P $(F\|P, i\%, n)_\infty$	To Find P Given F $(P\|F, i\%, n)_\infty$	To Find F Given A $(F\|A, i\%, n)_\infty$	To Find A Given F $(A\|F, i\%, n)_\infty$	To Find P Given A $(P\|A, i\%, n)_\infty$	To Find A Given P $(A\|P, i\%, n)_\infty$	To Find P Given G $(P\|G, i\%, n)_\infty$	To Find A Given G $(A\|G, i\%, n)_\infty$
n								
1	1.0050	0.9950	1.0000	1.0000	0.9950	1.0050	0.0000	0.0000
2	1.0101	0.9900	2.0050	0.4988	1.9851	0.5038	0.9900	0.4988
3	1.0151	0.9851	3.0151	0.3317	2.9702	0.3367	2.9603	0.9967
4	1.0202	0.9802	4.0302	0.2481	3.9504	0.2531	5.9009	1.4938
5	1.0253	0.9753	5.0504	0.1980	4.9257	0.2030	9.8021	1.9900
6	1.0305	0.9704	6.0757	0.1646	5.8961	0.1696	14.6543	2.4854
7	1.0356	0.9656	7.1061	0.1407	6.8617	0.1457	20.4480	2.9800
8	1.0408	0.9608	8.1418	0.1228	7.8225	0.1278	27.1735	3.4738
9	1.0460	0.9560	9.1826	0.1089	8.7785	0.1139	34.8215	3.9667
10	1.0513	0.9512	10.2286	0.0978	9.7298	0.1028	43.3825	4.4588
11	1.0565	0.9465	11.2799	0.0887	10.6762	0.0937	52.8474	4.9500
12	1.0618	0.9418	12.3364	0.0811	11.6180	0.0861	63.2068	5.4404
13	1.0672	0.9371	13.3983	0.0746	12.5551	0.0796	74.4516	5.9300
14	1.0725	0.9324	14.4654	0.0691	13.4875	0.0741	86.5727	6.4188
15	1.0779	0.9277	15.5379	0.0644	14.4152	0.0694	99.5611	6.9067
16	1.0833	0.9231	16.6158	0.0602	15.3383	0.0652	113.4079	7.3938
17	1.0887	0.9185	17.6991	0.0565	16.2568	0.0615	128.1041	7.8800
18	1.0942	0.9139	18.7878	0.0532	17.1708	0.0582	143.6409	8.3654
19	1.0997	0.9094	19.8820	0.0503	18.0801	0.0553	160.0096	8.8500
20	1.1052	0.9048	20.9816	0.0477	18.9850	0.0527	177.2015	9.3338
21	1.1107	0.9003	22.0868	0.0453	19.8853	0.0503	195.2080	9.8167
22	1.1163	0.8958	23.1975	0.0431	20.7811	0.0481	214.0205	10.2988
23	1.1219	0.8914	24.3138	0.0411	21.6725	0.0461	233.6306	10.7800
24	1.1275	0.8869	25.4357	0.0393	22.5594	0.0443	254.0298	11.2605
25	1.1331	0.8825	26.5632	0.0376	23.4419	0.0427	275.2097	11.7401
26	1.1388	0.8781	27.6963	0.0361	24.3200	0.0411	297.1621	12.2188
27	1.1445	0.8737	28.8351	0.0347	25.1937	0.0397	319.8787	12.6968
28	1.1503	0.8694	29.9797	0.0334	26.0631	0.0384	343.3514	13.1739
29	1.1560	0.8650	31.1300	0.0321	26.9281	0.0371	367.5720	13.6501
30	1.1618	0.8607	32.2860	0.0310	27.7888	0.0360	392.5325	14.1256
36	1.1972	0.8353	39.3449	0.0254	32.8637	0.0304	557.3910	16.9607
42	1.2337	0.8106	46.6189	0.0215	37.7885	0.0265	746.9263	19.7660
48	1.2712	0.7866	54.1143	0.0185	42.5678	0.0235	959.5358	22.5413
54	1.3100	0.7634	61.8380	0.0162	47.2059	0.0212	1193.6902	25.2869
60	1.3499	0.7408	69.7970	0.0143	51.7069	0.0193	1447.9301	28.0027
66	1.3910	0.7189	77.9983	0.0128	56.0748	0.0178	1720.8638	30.6887
72	1.4333	0.6977	86.4494	0.0116	60.3137	0.0166	2011.1644	33.3451
120	1.8221	0.5488	164.0130	$6.097E - 03$	90.0123	0.0111	4818.9067	53.5361
180	2.4596	0.4066	291.1914	$3.434E - 03$	118.3896	0.0084	9018.8276	76.1792
360	6.0496	0.1653	$1.007E + 03$	$9.926E - 04$	166.5232	0.0060	21349.6607	128.2083

Time Value of Money Factors—Continuous Compounding $r = 1\%$

	Single Sums		Uniform Series				Gradient Series	
	To Find F Given P	To Find P Given F	To Find F Given A	To Find A Given F	To Find P Given A	To Find A Given P	To Find P Given G	To Find A Given G
n	$(F\|P, i\%, n)_\infty$	$(P\|F, i\%, n)_\infty$	$(F\|A, i\%, n)_\infty$	$(A\|F, i\%, n)_\infty$	$(P\|A, i\%, n)_\infty$	$(A\|P, i\%, n)_\infty$	$(P\|G, i\%, n)_\infty$	$(A\|G, i\%, n)_\infty$
1	1.0101	0.9900	1.0000	1.0000	0.9900	1.0101	0.0000	0.0000
2	1.0202	0.9802	2.0101	0.4975	1.9702	0.5076	0.9802	0.4975
3	1.0305	0.9704	3.0303	0.3300	2.9407	0.3401	2.9211	0.9933
4	1.0408	0.9608	4.0607	0.2463	3.9015	0.2563	5.8035	1.4875
5	1.0513	0.9512	5.1015	0.1960	4.8527	0.2061	9.6084	1.9800
6	1.0618	0.9418	6.1528	0.1625	5.7945	0.1726	14.3172	2.4708
7	1.0725	0.9324	7.2146	0.1386	6.7269	0.1487	19.9116	2.9600
8	1.0833	0.9231	8.2871	0.1207	7.6500	0.1307	26.3734	3.4475
9	1.0942	0.9139	9.3704	0.1067	8.5639	0.1168	33.6848	3.9333
10	1.1052	0.9048	10.4646	0.0956	9.4688	0.1056	41.8284	4.4175
11	1.1163	0.8958	11.5698	0.0864	10.3646	0.0965	50.7867	4.9000
12	1.1275	0.8869	12.6860	0.0788	11.2515	0.0889	60.5428	5.3809
13	1.1388	0.8781	13.8135	0.0724	12.1296	0.0824	71.0800	5.8600
14	1.1503	0.8694	14.9524	0.0669	12.9990	0.0769	82.3816	6.3376
15	1.1618	0.8607	16.1026	0.0621	13.8597	0.0722	94.4315	6.8134
16	1.1735	0.8521	17.2645	0.0579	14.7118	0.0680	107.2137	7.2876
17	1.1853	0.8437	18.4380	0.0542	15.5555	0.0643	120.7123	7.7601
18	1.1972	0.8353	19.6233	0.0510	16.3908	0.0610	134.9119	8.2310
19	1.2092	0.8270	20.8205	0.0480	17.2177	0.0581	149.7972	8.7002
20	1.2214	0.8187	22.0298	0.0454	18.0364	0.0554	165.3531	9.1677
21	1.2337	0.8106	23.2512	0.0430	18.8470	0.0531	181.5648	9.6336
22	1.2461	0.8025	24.4848	0.0408	19.6495	0.0509	198.4177	10.0978
23	1.2586	0.7945	25.7309	0.0389	20.4441	0.0489	215.8974	10.5604
24	1.2712	0.7866	26.9895	0.0371	21.2307	0.0471	233.9898	11.0213
25	1.2840	0.7788	28.2608	0.0354	22.0095	0.0454	252.6811	11.4805
26	1.2969	0.7711	29.5448	0.0338	22.7806	0.0439	271.9573	11.9381
27	1.3100	0.7634	30.8417	0.0324	23.5439	0.0425	291.8052	12.3941
28	1.3231	0.7558	32.1517	0.0311	24.2997	0.0412	312.2114	12.8484
29	1.3364	0.7483	33.4748	0.0299	25.0480	0.0399	333.1628	13.3010
30	1.3499	0.7408	34.8112	0.0287	25.7888	0.0388	354.6465	13.7520
36	1.4333	0.6977	43.1166	0.0232	30.0815	0.0332	494.0326	16.4232
42	1.5220	0.6570	51.9356	0.0193	34.1241	0.0293	649.5574	19.0351
48	1.6161	0.6188	61.2999	0.0163	37.9314	0.0264	818.8687	21.5882
54	1.7160	0.5827	71.2433	0.0140	41.5169	0.0241	999.8332	24.0826
60	1.8221	0.5488	81.8015	0.0122	44.8936	0.0223	1190.5195	26.5187
66	1.9348	0.5169	93.0126	0.0108	48.0737	0.0208	1389.1816	28.8969
72	2.0544	0.4868	104.9170	9.531E − 03	51.0686	0.0196	1594.2437	31.2177
120	3.3201	0.3012	230.8536	4.332E − 03	69.5318	0.0144	3322.1789	47.7793
180	6.0496	0.1653	502.4441	1.990E − 03	83.0535	0.0120	5303.3602	63.8548
360	36.5982	0.0273	3.542E + 03	2.823E − 04	96.7821	0.0103	8651.1557	89.3880

Time Value of Money Factors—Continuous Compounding $r = 2\%$

	Single Sums		Uniform Series				Gradient Series	
	To Find F Given P	To Find P Given F	To Find F Given A	To Find A Given F	To Find P Given A	To Find A Given P	To Find P Given G	To Find A Given G
n	$(F\|P, i\%, n)_\infty$	$(P\|F, i\%, n)_\infty$	$(F\|A, i\%, n)_\infty$	$(A\|F, i\%, n)_\infty$	$(P\|A, i\%, n)_\infty$	$(A\|P, i\%, n)_\infty$	$(P\|G, i\%, n)_\infty$	$(A\|G, i\%, n)_\infty$
1	1.0202	0.9802	1.0000	1.0000	0.9802	1.0202	0.0000	0.0000
2	1.0408	0.9608	2.0202	0.4950	1.9410	0.5152	0.9608	0.4950
3	1.0618	0.9418	3.0610	0.3267	2.8828	0.3469	2.8443	0.9867
4	1.0833	0.9231	4.1228	0.2426	3.8059	0.2628	5.6137	1.4750
5	1.1052	0.9048	5.2061	0.1921	4.7107	0.2123	9.2330	1.9600
6	1.1275	0.8869	6.3113	0.1584	5.5976	0.1786	13.6676	2.4417
7	1.1503	0.8694	7.4388	0.1344	6.4670	0.1546	18.8838	2.9200
8	1.1735	0.8521	8.5891	0.1164	7.3191	0.1366	24.8488	3.3950
9	1.1972	0.8353	9.7626	0.1024	8.1544	0.1226	31.5309	3.8667
10	1.2214	0.8187	10.9598	0.0912	8.9731	0.1114	38.8995	4.3351
11	1.2461	0.8025	12.1812	0.0821	9.7756	0.1023	46.9247	4.8002
12	1.2712	0.7866	13.4273	0.0745	10.5623	0.0947	55.5776	5.2619
13	1.2969	0.7711	14.6985	0.0680	11.3333	0.0882	64.8302	5.7203
14	1.3231	0.7558	15.9955	0.0625	12.0891	0.0827	74.6554	6.1754
15	1.3499	0.7408	17.3186	0.0577	12.8299	0.0779	85.0269	6.6272
16	1.3771	0.7261	18.6685	0.0536	13.5561	0.0738	95.9191	7.0757
17	1.4049	0.7118	20.0456	0.0499	14.2678	0.0701	107.3074	7.5209
18	1.4333	0.6977	21.4505	0.0466	14.9655	0.0668	119.1679	7.9628
19	1.4623	0.6839	22.8839	0.0437	15.6494	0.0639	131.4774	8.4014
20	1.4918	0.6703	24.3461	0.0411	16.3197	0.0613	144.2135	8.8368
21	1.5220	0.6570	25.8380	0.0387	16.9768	0.0589	157.3545	9.2688
22	1.5527	0.6440	27.3599	0.0365	17.6208	0.0568	170.8792	9.6976
23	1.5841	0.6313	28.9126	0.0346	18.2521	0.0548	184.7675	10.1231
24	1.6161	0.6188	30.4967	0.0328	18.8709	0.0530	198.9995	10.5453
25	1.6487	0.6065	32.1128	0.0311	19.4774	0.0513	213.5562	10.9643
26	1.6820	0.5945	33.7615	0.0296	20.0719	0.0498	228.4192	11.3800
27	1.7160	0.5827	35.4435	0.0282	20.6547	0.0484	243.5707	11.7925
28	1.7507	0.5712	37.1595	0.0269	21.2259	0.0471	258.9933	12.2018
29	1.7860	0.5599	38.9102	0.0257	21.7858	0.0459	274.6705	12.6078
30	1.8221	0.5488	40.6963	0.0246	22.3346	0.0448	290.5860	13.0106
36	2.0544	0.4868	52.1962	0.0192	25.4066	0.0394	390.2482	15.3601
42	2.3164	0.4317	65.1624	0.0153	28.1313	0.0355	494.9887	17.5957
48	2.6117	0.3829	79.7817	0.0125	30.5478	0.0327	602.3844	19.7194
54	2.9447	0.3396	96.2649	0.0104	32.6911	0.0306	710.4956	21.7336
60	3.3201	0.3012	114.8497	8.707E − 03	34.5921	0.0289	817.7873	23.6409
66	3.7434	0.2671	135.8039	7.364E − 03	36.2780	0.0276	923.0622	25.4441
72	4.2207	0.2369	159.4298	6.272E − 03	37.7733	0.0265	1025.4047	27.1462
120	11.0232	0.0907	496.1639	2.015E − 03	45.0110	0.0222	1689.2356	37.5294
180	36.5982	0.0273	1.762E + 03	5.675E − 04	48.1491	0.0208	2139.9980	44.4452
360	1.339E + 03	7.466E − 04	6.625E + 04	1.509E − 05	49.4647	0.0202	2435.2809	49.2327

Time Value of Money Factors—Continuous Compounding $r = 3\%$

	Single Sums		Uniform Series				Gradient Series	
	To Find F Given P $(F\|P, i\%, n)_\infty$	To Find P Given F $(P\|F, i\%, n)_\infty$	To Find F Given A $(F\|A, i\%, n)_\infty$	To Find A Given F $(A\|F, i\%, n)_\infty$	To Find P Given A $(P\|A, i\%, n)_\infty$	To Find A Given P $(A\|P, i\%, n)_\infty$	To Find P Given G $(P\|G, i\%, n)_\infty$	To Find A Given G $(A\|G, i\%, n)_\infty$
n								
1	1.0305	0.9704	1.0000	1.0000	0.9704	1.0305	0.0000	0.0000
2	1.0618	0.9418	2.0305	0.4925	1.9122	0.5230	0.9418	0.4925
3	1.0942	0.9139	3.0923	0.3234	2.8261	0.3538	2.7696	0.9800
4	1.1275	0.8869	4.1865	0.2389	3.7131	0.2693	5.4304	1.4625
5	1.1618	0.8607	5.3140	0.1882	4.5738	0.2186	8.8732	1.9400
6	1.1972	0.8353	6.4758	0.1544	5.4090	0.1849	13.0496	2.4125
7	1.2337	0.8106	7.6730	0.1303	6.2196	0.1608	17.9131	2.8801
8	1.2712	0.7866	8.9067	0.1123	7.0063	0.1427	23.4195	3.3427
9	1.3100	0.7634	10.1779	0.0983	7.7696	0.1287	29.5265	3.8002
10	1.3499	0.7408	11.4879	0.0870	8.5104	0.1175	36.1939	4.2529
11	1.3910	0.7189	12.8378	0.0779	9.2294	0.1083	43.3831	4.7005
12	1.4333	0.6977	14.2287	0.0703	9.9270	0.1007	51.0575	5.1433
13	1.4770	0.6771	15.6621	0.0638	10.6041	0.0943	59.1822	5.5811
14	1.5220	0.6570	17.1390	0.0583	11.2612	0.0888	67.7238	6.0139
15	1.5683	0.6376	18.6610	0.0536	11.8988	0.0840	76.6506	6.4419
16	1.6161	0.6188	20.2293	0.0494	12.5176	0.0799	85.9324	6.8649
17	1.6653	0.6005	21.8454	0.0458	13.1181	0.0762	95.5403	7.2831
18	1.7160	0.5827	23.5107	0.0425	13.7008	0.0730	105.4470	7.6964
19	1.7683	0.5655	25.2267	0.0396	14.2663	0.0701	115.6265	8.1048
20	1.8221	0.5488	26.9950	0.0370	14.8151	0.0675	126.0539	8.5084
21	1.8776	0.5326	28.8171	0.0347	15.3477	0.0652	136.7057	8.9072
22	1.9348	0.5169	30.6947	0.0326	15.8646	0.0630	147.5596	9.3012
23	1.9937	0.5016	32.6295	0.0306	16.3662	0.0611	158.5943	9.6904
24	2.0544	0.4868	34.6232	0.0289	16.8529	0.0593	169.7896	10.0748
25	2.1170	0.4724	36.6776	0.0273	17.3253	0.0577	181.1264	10.4545
26	2.1815	0.4584	38.7946	0.0258	17.7837	0.0562	192.5866	10.8294
27	2.2479	0.4449	40.9761	0.0244	18.2285	0.0549	204.1529	11.1996
28	2.3164	0.4317	43.2240	0.0231	18.6603	0.0536	215.8090	11.5652
29	2.3869	0.4190	45.5404	0.0220	19.0792	0.0524	227.5397	11.9261
30	2.4596	0.4066	47.9273	0.0209	19.4858	0.0513	239.3302	12.2823
36	2.9447	0.3396	63.8552	0.0157	21.6849	0.0461	310.6103	14.3238
42	3.5254	0.2837	82.9243	0.0121	23.5218	0.0425	381.1698	16.2049
48	4.2207	0.2369	105.7542	9.456E − 03	25.0561	0.0399	449.3118	17.9322
54	5.0531	0.1979	133.0866	7.514E − 03	26.3377	0.0380	513.9181	19.5127
60	6.0496	0.1653	165.8094	6.031E − 03	27.4081	0.0365	574.3044	20.9538
66	7.2427	0.1381	204.9857	4.878E − 03	28.3022	0.0353	630.1080	22.2636
72	8.6711	0.1153	251.8882	3.970E − 03	29.0490	0.0344	681.2000	23.4500
120	36.5982	0.0273	1.169E + 03	8.555E − 04	31.9386	0.0313	941.0681	29.4649
180	2.214E + 02	4.517E − 03	7.237E + 03	1.382E − 04	32.6875	0.0306	1046.6272	32.0192
360	4.902E + 04	2.040E − 05	1.610E + 06	6.213E − 07	32.8352	0.0305	1077.9288	32.8285

Time Value of Money Factors—Continuous Compounding $r = 4\%$

	Single Sums		Uniform Series				Gradient Series	
	To Find F Given P $(F\|P, i\%, n)_\infty$	To Find P Given F $(P\|F, i\%, n)_\infty$	To Find F Given A $(F\|A, i\%, n)_\infty$	To Find A Given F $(A\|F, i\%, n)_\infty$	To Find P Given A $(P\|A, i\%, n)_\infty$	To Find A Given P $(A\|P, i\%, n)_\infty$	To Find P Given G $(P\|G, i\%, n)_\infty$	To Find A Given G $(A\|G, i\%, n)_\infty$
n								
1	1.0408	0.9608	1.0000	1.0000	0.9608	1.0408	0.0000	0.0000
2	1.0833	0.9231	2.0408	0.4900	1.8839	0.5308	0.9231	0.4900
3	1.1275	0.8869	3.1241	0.3201	2.7708	0.3609	2.6970	0.9733
4	1.1735	0.8521	4.2516	0.2352	3.6230	0.2760	5.2534	1.4500
5	1.2214	0.8187	5.4251	0.1843	4.4417	0.2251	8.5283	1.9201
6	1.2712	0.7866	6.6465	0.1505	5.2283	0.1913	12.4615	2.3834
7	1.3231	0.7558	7.9178	0.1263	5.9841	0.1671	16.9962	2.8402
8	1.3771	0.7261	9.2409	0.1082	6.7103	0.1490	22.0792	3.2904
9	1.4333	0.6977	10.6180	0.0942	7.4079	0.1350	27.6606	3.7339
10	1.4918	0.6703	12.0513	0.0830	8.0783	0.1238	33.6935	4.1709
11	1.5527	0.6440	13.5432	0.0738	8.7223	0.1146	40.1339	4.6013
12	1.6161	0.6188	15.0959	0.0662	9.3411	0.1071	46.9405	5.0252
13	1.6820	0.5945	16.7120	0.0598	9.9356	0.1006	54.0747	5.4425
14	1.7507	0.5712	18.3940	0.0544	10.5068	0.0952	61.5004	5.8534
15	1.8221	0.5488	20.1447	0.0496	11.0556	0.0905	69.1838	6.2578
16	1.8965	0.5273	21.9668	0.0455	11.5829	0.0863	77.0932	6.6558
17	1.9739	0.5066	23.8633	0.0419	12.0895	0.0827	85.1991	7.0473
18	2.0544	0.4868	25.8371	0.0387	12.5763	0.0795	93.4738	7.4326
19	2.1383	0.4677	27.8916	0.0359	13.0439	0.0767	101.8918	7.8114
20	2.2255	0.4493	30.0298	0.0333	13.4933	0.0741	110.4291	8.1840
21	2.3164	0.4317	32.2554	0.0310	13.9250	0.0718	119.0633	8.5503
22	2.4109	0.4148	34.5717	0.0289	14.3398	0.0697	127.7737	8.9104
23	2.5093	0.3985	36.9826	0.0270	14.7383	0.0679	136.5412	9.2644
24	2.6117	0.3829	39.4919	0.0253	15.1212	0.0661	145.3477	9.6122
25	2.7183	0.3679	42.1036	0.0238	15.4891	0.0646	154.1768	9.9539
26	2.8292	0.3535	44.8219	0.0223	15.8425	0.0631	163.0132	10.2896
27	2.9447	0.3396	47.6511	0.0210	16.1821	0.0618	171.8427	10.6193
28	3.0649	0.3263	50.5958	0.0198	16.5084	0.0606	180.6522	10.9431
29	3.1899	0.3135	53.6607	0.0186	16.8219	0.0594	189.4298	11.2609
30	3.3201	0.3012	56.8506	0.0176	17.1231	0.0584	198.1645	11.5730
36	4.2207	0.2369	78.9178	0.0127	18.6978	0.0535	249.1600	13.3256
42	5.3656	0.1864	106.9707	9.348E − 03	19.9365	0.0502	296.7070	14.8826
48	6.8210	0.1466	142.6329	7.011E − 03	20.9110	0.0478	339.9553	16.2573
54	8.6711	0.1153	187.9684	5.320E − 03	21.6775	0.0461	378.5747	17.4640
60	11.0232	0.0907	245.6012	4.072E − 03	22.2804	0.0449	412.5715	18.5172
66	14.0132	0.0714	318.8669	3.136E − 03	22.7547	0.0439	442.1602	19.4316
72	17.8143	0.0561	412.0057	2.427E − 03	23.1278	0.0432	467.6741	20.2213
120	121.5104	8.230E − 03	2.953E + 03	3.386E − 04	24.3017	0.0411	571.2733	23.5076
180	1.339E + 03	7.466E − 04	3.280E + 04	3.049E − 05	24.4850	0.0408	596.6722	24.3688
360	1.794E + 06	5.574E − 07	4.396E + 07	2.275E − 08	24.5033	0.0408	600.4081	24.5031

Time Value of Money Factors—Continuous Compounding $r = 5\%$

	Single Sums		Uniform Series				Gradient Series	
n	To Find F Given P $(F\|P, i\%, n)_\infty$	To Find P Given F $(P\|F, i\%, n)_\infty$	To Find F Given A $(F\|A, i\%, n)_\infty$	To Find A Given F $(A\|F, i\%, n)_\infty$	To Find P Given A $(P\|A, i\%, n)_\infty$	To Find A Given P $(A\|P, i\%, n)_\infty$	To Find P Given G $(P\|G, i\%, n)_\infty$	To Find A Given G $(A\|G, i\%, n)_\infty$
1	1.0513	0.9512	1.0000	1.0000	0.9512	1.0513	0.0000	0.0000
2	1.1052	0.9048	2.0513	0.4875	1.8561	0.5388	0.9048	0.4875
3	1.1618	0.8607	3.1564	0.3168	2.7168	0.3681	2.6263	0.9667
4	1.2214	0.8187	4.3183	0.2316	3.5355	0.2828	5.0824	1.4375
5	1.2840	0.7788	5.5397	0.1805	4.3143	0.2318	8.1976	1.9001
6	1.3499	0.7408	6.8237	0.1465	5.0551	0.1978	11.9017	2.3544
7	1.4191	0.7047	8.1736	0.1223	5.7598	0.1736	16.1299	2.8004
8	1.4918	0.6703	9.5926	0.1042	6.4301	0.1555	20.8221	3.2382
9	1.5683	0.6376	11.0845	0.0902	7.0678	0.1415	25.9231	3.6678
10	1.6487	0.6065	12.6528	0.0790	7.6743	0.1303	31.3819	4.0892
11	1.7333	0.5769	14.3015	0.0699	8.2512	0.1212	37.1514	4.5025
12	1.8221	0.5488	16.0347	0.0624	8.8001	0.1136	43.1883	4.9077
13	1.9155	0.5220	17.8569	0.0560	9.3221	0.1073	49.4529	5.3049
14	2.0138	0.4966	19.7724	0.0506	9.8187	0.1018	55.9085	5.6941
15	2.1170	0.4724	21.7862	0.0459	10.2911	0.0972	62.5216	6.0753
16	2.2255	0.4493	23.9032	0.0418	10.7404	0.0931	69.2616	6.4487
17	2.3396	0.4274	26.1287	0.0383	11.1678	0.0895	76.1002	6.8143
18	2.4596	0.4066	28.4683	0.0351	11.5744	0.0864	83.0119	7.1720
19	2.5857	0.3867	30.9279	0.0323	11.9611	0.0836	89.9732	7.5221
20	1.7183	0.3679	33.5137	0.0298	12.3290	0.0811	96.9629	7.8646
21	2.8577	0.3499	36.2319	0.0276	12.6789	0.0789	103.9617	8.1996
22	3.0042	0.3329	39.0896	0.0256	13.0118	0.0769	110.9520	8.5270
23	3.1582	0.3166	42.0938	0.0238	13.3284	0.0750	117.9180	8.8471
24	3.3201	0.3012	45.2519	0.0221	13.6296	0.0734	124.8455	9.1599
25	3.4903	0.2865	48.5721	0.0206	13.9161	0.0719	131.7216	9.4654
26	3.6693	0.2725	52.0624	0.0192	14.1887	0.0705	138.5349	9.7638
27	3.8574	0.2592	55.7317	0.0179	14.4479	0.0692	145.2751	10.0551
28	4.0552	0.2466	59.5891	0.0168	14.6945	0.0681	151.9332	10.3395
29	4.2631	0.2346	63.6443	0.0157	14.9291	0.0670	158.5012	10.6170
30	4.4817	0.2231	67.9074	0.0147	15.1522	0.0660	164.9720	10.8877
36	6.0496	0.1653	98.4892	0.0102	16.2801	0.0614	201.4661	12.3750
42	8.1662	0.1225	139.7702	7.155E − 03	17.1158	0.0584	233.5153	13.6433
48	11.0232	0.0907	195.4937	5.115E − 03	17.7348	0.0564	260.9721	14.7153
54	14.8797	0.0672	270.7126	3.694E − 03	18.1934	0.0550	284.0642	15.6136
60	20.0855	0.0498	372.2475	2.686E − 03	18.5331	0.0540	303.2096	16.3604
66	27.1126	0.0369	509.3053	1.963E − 03	18.7848	0.0532	318.9029	16.9767
72	36.5982	0.0273	694.3139	1.440E − 03	18.9712	0.0527	331.6475	17.4816
120	403.4288	2.479E − 03	7.849E + 03	1.274E − 04	19.4558	0.0514	373.6680	19.2060
180	8.103E + 03	1.234E − 04	1.580E + 05	6.328E − 06	19.5018	0.0513	379.9323	19.4819
360	6.566E + 07	1.523E − 08	1.281E + 09	7.809E − 10	19.5042	0.0513	380.4124	19.5042

Time Value of Money Factors—Continuous Compounding $r = 8\%$

	Single Sums		Uniform Series					Gradient Series	
n	To Find F Given P $(F\|P, i\%, n)_\infty$	To Find P Given F $(P\|F, i\%, n)_\infty$	To Find F Given A $(F\|A, i\%, n)_\infty$	To Find A Given F $(A\|F, i\%, n)_\infty$	To Find P Given A $(P\|A, i\%, n)_\infty$	To Find A Given P $(A\|P, i\%, n)_\infty$	To Find P Given G $(P\|G, i\%, n)_\infty$	To Find A Given G $(A\|G, i\%, n)_\infty$
1	1.0833	0.9231	1.0000	1.0000	0.9231	1.0833	0.0000	0.0000
2	1.1735	0.8521	2.0833	0.4800	1.7753	0.5633	0.8521	0.4800
3	1.2712	0.7866	3.2568	0.3071	2.5619	0.3903	2.4254	0.9467
4	1.3771	0.7261	4.5280	0.2208	3.2880	0.3041	4.6038	1.4002
5	1.4918	0.6703	5.9052	0.1693	3.9584	0.2526	7.2851	1.8404
6	1.6161	0.6188	7.3970	0.1352	4.5771	0.2185	10.3790	2.2676
7	1.7507	0.5712	9.0131	0.1109	5.1483	0.1942	13.8063	2.6817
8	1.8965	0.5273	10.7637	0.0929	5.6756	0.1762	17.4973	3.0829
9	2.0544	0.4868	12.6602	0.0790	6.1624	0.1623	21.3914	3.4713
10	2.2255	0.4493	14.7147	0.0680	6.6117	0.1512	25.4353	3.8470
11	2.4109	0.4148	16.9402	0.0590	7.0265	0.1423	29.5832	4.2102
12	2.6117	0.3829	19.3511	0.0517	7.4094	0.1350	33.7950	4.5611
13	2.8292	0.3535	21.9628	0.0455	7.7629	0.1288	38.0364	4.8998
14	3.0649	0.3263	24.7920	0.0403	8.0891	0.1236	42.2781	5.2265
15	3.3201	0.3012	27.8569	0.0359	8.3903	0.1192	46.4948	5.5415
16	3.5966	0.2780	31.1770	0.0321	8.6684	0.1154	50.6653	5.8449
17	3.8962	0.2567	34.7736	0.0288	8.9250	0.1120	54.7719	6.1369
18	4.2207	0.2369	38.6698	0.0259	9.1620	0.1091	58.7997	6.4178
19	4.5722	0.2187	42.8905	0.0233	9.3807	0.1066	62.7365	6.6879
20	4.9530	0.2019	47.4627	0.0211	9.5826	0.1044	66.5725	6.9473
21	5.3656	0.1864	52.4158	0.0191	9.7689	0.1024	70.3000	7.1963
22	5.8124	0.1720	57.7813	0.0173	9.9410	0.1006	73.9130	7.4352
23	6.2965	0.1588	63.5938	0.0157	10.0998	0.0990	77.4069	7.6642
24	6.8210	0.1466	69.8903	0.0143	10.2464	0.0976	80.7789	7.8836
25	7.3891	0.1353	76.7113	0.0130	10.3817	0.0963	84.0270	8.0937
26	8.0045	0.1249	84.1003	0.0119	10.5067	0.0952	87.1502	8.2947
27	8.6711	0.1153	92.1048	0.0109	10.6220	0.0941	90.1487	8.4870
28	9.3933	0.1065	100.7759	9.923E − 03	10.7285	0.0932	93.0230	8.6707
29	10.1757	0.0983	110.1693	9.077E − 03	10.8267	0.0924	95.7747	8.8461
30	11.0232	0.0907	120.3449	8.309E − 03	10.9174	0.0916	98.4055	9.0136
36	17.8143	0.0561	201.8834	4.953E − 03	11.3327	0.0882	111.8039	9.8656
42	28.7892	0.0347	333.6555	2.997E − 03	11.5896	0.0863	121.6363	10.4953
48	46.5255	0.0215	546.6092	1.829E − 03	11.7486	0.0851	128.6743	10.9523
54	75.1886	0.0133	890.7581	1.123E − 03	11.8470	0.0844	133.6196	11.2788
60	121.5104	8.230E − 03	1.447E + 03	6.911E − 04	11.9079	0.0840	137.0449	11.5088
66	196.3699	5.092E − 03	2.346E + 03	4.263E − 04	11.9455	0.0837	139.3905	11.6688
72	317.3483	3.151E − 03	3.798E + 03	2.633E − 04	11.9688	0.0836	140.9817	11.7791
120	1.476E + 04	6.773E − 05	1.773E + 05	5.641E − 06	12.0059	0.0833	144.0527	11.9985
180	1.794E + 06	5.574E − 07	2.154E + 07	4.642E − 08	12.0067	0.0833	144.1587	12.0066
360	3.219E + 12	3.107E − 13	3.865E + 13	2.588E − 14	12.0067	0.0833	144.1600	12.0067

Time Value of Money Factors—Continuous Compounding $r = 10\%$

	Single Sums		Uniform Series				Gradient Series	
	To Find F Given P $(F\vert P, i\%, n)_\infty$	To Find P Given F $(P\vert F, i\%, n)_\infty$	To Find F Given A $(F\vert A, i\%, n)_\infty$	To Find A Given F $(A\vert F, i\%, n)_\infty$	To Find P Given A $(P\vert A, i\%, n)_\infty$	To Find A Given P $(A\vert P, i\%, n)_\infty$	To Find P Given G $(P\vert G, i\%, n)_\infty$	To Find A Given G $(A\vert G, i\%, n)_\infty$
n								
1	1.1052	0.9048	1.0000	1.0000	0.9048	1.1052	0.0000	0.0000
2	1.2214	0.8187	2.1052	0.4750	1.7236	0.5802	0.8187	0.4750
3	1.3499	0.7408	3.3266	0.3006	2.4644	0.4058	2.3004	0.9334
4	1.4918	0.6703	4.6764	0.2138	3.1347	0.3190	4.3113	1.3754
5	1.6487	0.6065	6.1683	0.1621	3.7412	0.2673	6.7374	1.8009
6	1.8221	0.5488	7.8170	0.1279	4.2900	0.2331	9.4815	2.2101
7	2.0138	0.4966	9.6391	0.1037	4.7866	0.2089	12.4610	2.6033
8	2.2255	0.4493	11.6528	0.0858	5.2360	0.1910	15.6063	2.9806
9	2.4596	0.4066	13.8784	0.0721	5.6425	0.1772	18.8589	3.3423
10	2.7183	0.3679	16.3380	0.0612	6.0104	0.1664	22.1698	3.6886
11	3.0042	0.3329	19.0563	0.0525	6.3433	0.1576	25.4985	4.0198
12	3.3201	0.3012	22.0604	0.0453	6.6445	0.1505	28.8116	4.3362
13	3.6693	0.2725	25.3806	0.0394	6.9170	0.1446	32.0820	4.6381
14	4.0552	0.2466	29.0499	0.0344	7.1636	0.1396	35.2878	4.9260
15	4.4817	0.2231	33.1051	0.0302	7.3867	0.1354	38.4116	5.2001
16	4.9530	0.2019	37.5867	0.0266	7.5886	0.1318	41.4401	5.4608
17	5.4739	0.1827	42.5398	0.0235	7.7713	0.1287	44.3630	5.7086
18	6.0496	0.1653	48.0137	0.0208	7.9366	0.1260	47.1731	5.9437
19	6.6859	0.1496	54.0634	0.0185	8.0862	0.1237	49.8653	6.1667
20	7.3891	0.1353	60.7493	0.0165	8.2215	0.1216	52.4367	6.3780
21	8.1662	0.1225	68.1383	0.0147	8.3440	0.1198	54.8858	6.5779
22	9.0250	0.1108	76.3045	0.0131	8.4548	0.1183	57.2127	6.7669
23	9.9742	0.1003	85.3295	0.0117	8.5550	0.1169	59.4184	6.9454
24	11.0232	0.0907	95.3037	0.0105	8.6458	0.1157	61.5049	7.1139
25	12.1825	0.0821	106.3269	9.405E − 03	8.7278	0.1146	63.4749	7.2727
26	13.4637	0.0743	118.5094	8.438E − 03	8.8021	0.1136	65.3318	7.4223
27	14.8797	0.0672	131.9731	7.577E − 03	8.8693	0.1127	67.0791	7.5630
28	16.4446	0.0608	146.8528	6.810E − 03	8.9301	0.1120	68.7210	7.6954
29	18.1741	0.0550	163.2975	6.124E − 03	8.9852	0.1113	70.2616	7.8197
30	20.0855	0.0498	181.4716	5.511E − 03	9.0349	0.1107	71.7054	7.9365
36	36.5982	0.0273	338.4798	2.954E − 03	9.2485	0.1081	78.5852	8.4970
42	66.6863	0.0150	624.5674	1.601E − 03	9.3657	0.1068	83.0642	8.8689
48	121.5104	8.230E − 03	1.146E + 03	8.727E − 04	9.4301	0.1060	85.9083	9.1100
54	221.4064	4.517E − 03	2.096E + 03	4.772E − 04	9.4654	0.1056	87.6810	9.2633
60	403.4288	2.479E − 03	3.826E + 03	2.613E − 04	9.4848	0.1054	88.7701	9.3592
66	735.0952	1.360E − 03	6.980E + 03	1.433E − 04	9.4954	0.1053	89.4317	9.4184
72	1.339E + 03	7.466E − 04	1.273E + 04	7.858E − 05	9.5012	0.1052	89.8298	9.4545
120	1.628E + 05	6.144E − 06	1.548E + 06	6.462E − 07	9.5083	0.1052	90.4008	9.5076
180	6.566E + 07	1.523E − 08	6.243E + 08	1.602E − 09	9.5083	0.1052	90.4083	9.5083
360	4.311E + 15	2.320E − 16	4.099E + 16	2.439E − 17	9.5083	0.1052	90.4084	9.5083

Time Value of Money Factors—Continuous Compounding $r = 12\%$

	Single Sums		Uniform Series				Gradient Series	
	To Find F Given P	To Find P Given F	To Find F Given A	To Find A Given F	To Find P Given A	To Find A Given P	To Find P Given G	To Find A Given G
n	$(F|P, i\%, n)_\infty$	$(P|F, i\%, n)_\infty$	$(F|A, i\%, n)_\infty$	$(A|F, i\%, n)_\infty$	$(P|A, i\%, n)_\infty$	$(A|P, i\%, n)_\infty$	$(P|G, i\%, n)_\infty$	$(A|G, i\%, n)_\infty$
1	1.1275	0.8869	1.0000	1.0000	0.8869	1.1275	0.0000	0.0000
2	1.2712	0.7866	2.1275	0.4700	1.6735	0.5975	0.7866	0.4700
3	1.4333	0.6977	3.3987	0.2942	2.3712	0.4217	2.1820	0.9202
4	1.6161	0.6188	4.8321	0.2070	2.9900	0.3344	4.0383	1.3506
5	1.8221	0.5488	6.4481	0.1551	3.5388	0.2826	6.2336	1.7615
6	2.0544	0.4868	8.2703	0.1209	4.0256	0.2484	8.6673	2.1531
7	2.3164	0.4317	10.3247	0.0969	4.4573	0.2244	11.2576	2.5257
8	2.6117	0.3829	12.6411	0.0791	4.8402	0.2066	13.9379	2.8796
9	2.9447	0.3396	15.2528	0.0656	5.1798	0.1931	16.6546	3.2153
10	3.3201	0.3012	18.1974	0.0550	5.4810	0.1824	19.3654	3.5332
11	3.7434	0.2671	21.5176	0.0465	5.7481	0.1740	22.0367	3.8337
12	4.2207	0.2369	25.2610	0.0396	5.9850	0.1671	24.6429	4.1174
13	4.7588	0.2101	29.4817	0.0339	6.1952	0.1614	27.1646	4.3848
14	5.3656	0.1864	34.2405	0.0292	6.3815	0.1567	29.5874	4.6364
15	6.0496	0.1653	39.6061	0.0252	6.5468	0.1527	31.9016	4.8728
16	6.8210	0.1466	45.6557	0.0219	6.6934	0.1494	34.1007	5.0946
17	7.6906	0.1300	52.4767	0.0191	6.8235	0.1466	36.1812	5.3025
18	8.6711	0.1153	60.1673	0.0166	6.9388	0.1441	38.1417	5.4969
19	9.7767	0.1023	68.8384	0.0145	7.0411	0.1420	39.9828	5.6785
20	11.0232	0.0907	78.6151	0.0127	7.1318	0.1402	41.7064	5.8480
21	12.4286	0.0805	89.6383	0.0112	7.2123	0.1387	43.3156	6.0058
22	14.0132	0.0714	102.0669	9.797E − 03	7.2836	0.1373	44.8142	6.1527
23	15.7998	0.0633	116.0801	8.615E − 03	7.3469	0.1361	46.2066	6.2893
24	17.8143	0.0561	131.8799	7.583E − 03	7.4030	0.1351	47.4977	6.4160
25	20.0855	0.0498	149.6942	6.680E − 03	7.4528	0.1342	48.6926	6.5334
26	22.6464	0.0442	169.7797	5.890E − 03	7.4970	0.1334	49.7966	6.6422
27	25.5337	0.0392	192.4261	5.197E − 03	7.5362	0.1327	50.8148	6.7428
28	28.7892	0.0347	217.9598	4.588E − 03	7.5709	0.1321	51.7527	6.8357
29	32.4597	0.0308	246.7490	4.053E − 03	7.6017	0.1315	52.6153	6.9215
30	36.5982	0.0273	279.2087	3.582E − 03	7.6290	0.1311	53.4077	7.0006
36	75.1886	0.0133	581.8860	1.719E − 03	7.7390	0.1292	56.9443	7.3581
42	154.4700	6.474E − 03	1.204E + 03	8.308E − 04	7.7926	0.1283	58.9870	7.5697
48	317.3483	3.151E − 03	2.481E + 03	4.030E − 04	7.8186	0.1279	60.1377	7.6916
54	651.9709	1.534E − 03	5.106E + 03	1.959E − 04	7.8313	0.1277	60.7739	7.7604
60	1.339E + 03	7.466E − 04	1.050E + 04	9.526E − 05	7.8375	0.1276	61.1206	7.7985
66	2.752E + 03	3.634E − 04	2.158E + 04	4.635E − 05	7.8405	0.1275	61.3074	7.8193
72	5.653E + 03	1.769E − 04	4.433E + 04	2.256E − 05	7.8419	0.1275	61.4071	7.8306
120	1.794E + 06	5.574E − 07	1.407E + 07	7.107E − 08	7.8433	0.1275	61.5173	7.8433
180	2.403E + 09	4.161E − 10	1.885E + 10	5.306E − 11	7.8433	0.1275	61.5178	7.8433
360	5.775E + 18	1.732E − 19	4.529E + 19	2.208E − 20	7.8433	0.1275	61.5178	7.8433

Time Value of Money Factors—Continuous Compounding $r = 15\%$

	Single Sums		Uniform Series				Gradient Series	
n	To Find F Given P $(F\|P, i\%, n)_\infty$	To Find P Given F $(P\|F, i\%, n)_\infty$	To Find F Given A $(F\|A, i\%, n)_\infty$	To Find A Given F $(A\|F, i\%, n)_\infty$	To Find P Given A $(P\|A, i\%, n)_\infty$	To Find A Given P $(A\|P, i\%, n)_\infty$	To Find P Given G $(P\|G, i\%, n)_\infty$	To Find A Given G $(A\|G, i\%, n)_\infty$
1	1.1618	0.8807	1.0000	1.0000	0.8607	1.1618	0.0000	0.0000
2	1.3499	0.7408	2.1618	0.4628	1.6015	0.6244	0.7408	0.4626
3	1.5683	0.6376	3.5117	0.2848	2.2392	0.4466	2.0161	0.9004
4	1.8221	0.5488	5.0800	0.1969	2.7880	0.3587	3.6625	1.3137
5	2.1170	0.4724	6.9021	0.1449	3.2603	0.3067	5.5520	1.7029
6	2.4596	0.4066	9.0191	0.1109	3.6669	0.2727	7.5848	2.0685
7	2.8577	0.3499	11.4787	0.0871	4.0168	0.2490	9.6845	2.4110
8	3.3201	0.3012	14.3364	0.0698	4.3180	0.2316	11.7928	2.7311
9	3.8574	0.2592	17.6565	0.0566	4.5773	0.2185	13.8667	3.0295
10	4.4817	0.2231	21.5139	0.0465	4.8004	0.2083	15.8749	3.3070
11	5.2070	0.1920	25.9956	0.0385	4.9925	0.2003	17.7954	3.5645
12	6.0496	0.1653	31.2026	0.0320	5.1578	0.1939	19.6137	3.8028
13	7.0287	0.1423	37.2522	0.0268	5.3000	0.1887	21.3210	4.0228
14	8.1662	0.1225	44.2809	0.0226	5.4225	0.1844	22.9129	4.2255
15	9.4877	0.1054	52.4471	0.0191	5.5279	0.1809	24.3885	4.4119
16	11.0232	0.0907	61.9348	0.0161	5.6186	0.1780	25.7493	4.5829
17	12.8071	0.0781	72.9580	0.0137	5.6967	0.1755	26.9986	4.7394
18	14.8797	0.0672	85.7651	0.0117	5.7639	0.1735	28.1411	4.8823
19	17.2878	0.0578	100.6448	9.936E − 03	5.8217	0.1718	29.1823	5.0126
20	20.0855	0.0498	117.9326	8.479E − 03	5.8715	0.1703	30.1282	5.1312
21	23.3361	0.0429	138.0182	7.245E − 03	5.9144	0.1691	30.9853	5.2390
22	27.1126	0.0369	161.3542	6.198E − 03	5.9513	0.1680	31.7598	5.3367
23	31.5004	0.0317	188.4669	5.306E − 03	5.9830	0.1671	32.4582	5.4251
24	36.5982	0.0273	219.9673	4.546E − 03	6.0103	0.1664	33.0867	5.5050
25	42.5211	0.0235	256.5655	3.898E − 03	6.0338	0.1657	33.6511	5.5771
26	49.4024	0.0202	299.0866	3.344E − 03	6.0541	0.1652	34.1571	5.6420
27	57.3975	0.0174	348.4890	2.870E − 03	6.0715	0.1647	34.6101	5.7004
28	66.6863	0.0150	405.8865	2.464E − 03	6.0865	0.1643	35.0150	5.7529
29	77.4785	0.0129	472.5728	2.116E − 03	6.0994	0.1640	35.3764	5.8000
30	90.0171	0.0111	550.0513	1.818E − 03	6.1105	0.1637	35.6985	5.8421
36	221.4064	4.517E − 03	1.362E + 03	7.343E − 04	6.1513	0.1626	37.0049	6.0158
42	544.5719	1.836E − 03	3.359E + 03	2.977E − 04	6.1678	0.1621	37.6354	6.1019
48	1.339E + 03	7.466E − 04	8.270E + 03	1.209E − 04	6.1745	0.1620	37.9321	6.1433
54	3.294E + 03	3.035E − 04	2.035E + 04	4.914E − 05	6.1773	0.1619	38.0692	6.1628
60	8.103E + 03	1.234E − 04	5.006E + 04	1.997E − 05	6.1784	0.1619	38.1316	6.1718
66	1.993E + 04	5.017E − 05	1.231E + 05	8.120E − 06	6.1789	0.1618	38.1597	6.1759
72	4.902E + 04	2.040E − 05	3.029E + 05	3.301E − 06	6.1790	0.1618	38.1722	6.1777
120	6.566E + 07	1.523E − 08	5.057E + 08	2.465E − 09	6.1792	0.1618	38.1820	6.1792
180	5.320E + 11	1.880E − 12	3.288E + 12	3.042E − 13	6.1792	0.1618	38.1820	6.1792
360	2.831E + 23	3.533E − 24	1.749E + 24	5.717E − 25	6.1792	0.1618	38.1820	6.1792

Time Value of Money Factors—Continuous Compounding $r = 20\%$

	Single Sums		Uniform Series				Gradient Series	
	To Find F Given P	To Find P Given F	To Find F Given A	To Find A Given F	To Find P Given A	To Find A Given P	To Find P Given G	To Find A Given G
n	$(F\|P, i\%, n)_\infty$	$(P\|F, i\%, n)_\infty$	$(F\|A, i\%, n)_\infty$	$(A\|F, i\%, n)_\infty$	$(P\|A, i\%, n)_\infty$	$(A\|P, i\%, n)_\infty$	$(P\|G, i\%, n)_\infty$	$(A\|G, i\%, n)_\infty$
1	1.2214	0.8187	1.0000	1.0000	0.8187	1.2214	0.0000	0.0000
2	1.4918	0.6703	2.2214	0.4502	1.4891	0.6716	0.6703	0.4502
3	1.8221	0.5488	3.7132	0.2693	2.0379	0.4907	1.7679	0.8675
4	2.2255	0.4493	5.5353	0.1807	2.4872	0.4021	3.1159	1.2528
5	2.7183	0.3679	7.7609	0.1289	2.8551	0.3503	4.5874	1.6068
6	3.3201	0.3012	10.4792	0.0954	3.1563	0.3168	6.0934	1.9306
7	4.0552	0.2466	13.7993	0.0725	3.4029	0.2939	7.5730	2.2255
8	4.9530	0.2019	17.8545	0.0560	3.6048	0.2774	8.9863	2.4929
9	6.0496	0.1653	22.8075	0.0438	3.7701	0.2652	10.3087	2.7344
10	7.3891	0.1353	28.8572	0.0347	3.9054	0.2561	11.5267	2.9515
11	9.0250	0.1108	36.2462	0.0276	4.0162	0.2490	12.6347	3.1459
12	11.0232	0.0907	45.2712	0.0221	4.1069	0.2435	13.6326	3.3194
13	13.4637	0.0743	56.2944	0.0178	4.1812	0.2392	14.5239	3.4736
14	16.4446	0.0608	69.7581	0.0143	4.2420	0.2357	15.3144	3.6102
15	20.0855	0.0498	86.2028	0.0116	4.2918	0.2330	16.0114	3.7307
16	24.5325	0.0408	106.2883	9.408E − 03	4.3325	0.2308	16.6229	3.8367
17	29.9641	0.0334	130.8209	7.644E − 03	4.3659	0.2290	17.1569	3.9297
18	36.5982	0.0273	160.7850	6.219E − 03	4.3932	0.2276	17.6214	4.0110
19	44.7012	0.0224	197.3832	5.066E − 03	4.4156	0.2265	18.0240	4.0819
20	54.5982	0.0183	242.0844	4.131E − 03	4.4339	0.2255	18.3720	4.1435
21	66.6863	0.0150	296.6825	3.371E − 03	4.4489	0.2248	18.6719	4.1970
22	81.4509	0.0123	363.3689	2.752E − 03	4.4612	0.2242	18.9298	4.2432
23	99.4843	0.0101	444.8197	2.248E − 03	4.4713	0.2237	19.1509	4.2831
24	121.5104	8.230E − 03	544.3040	1.837E − 03	4.4795	0.2232	19.3402	4.3175
25	148.4132	6.738E − 03	665.8145	1.502E − 03	4.4862	0.2229	19.5019	4.3471
26	181.2722	5.517E − 03	814.2276	1.228E − 03	4.4917	0.2226	19.6398	4.3724
27	221.4064	4.517E − 03	995.4999	1.005E − 03	4.4963	0.2224	19.7572	4.3942
28	270.4264	3.698E − 03	1.217E + 03	8.218E − 04	4.5000	0.2222	19.8571	4.4127
29	330.2996	3.028E − 03	1.487E + 03	6.723E − 04	4.5030	0.2221	19.9419	4.4286
30	403.4288	2.479E − 03	1.818E + 03	5.502E − 04	4.5055	0.2220	20.0137	4.4421
36	1.339E + 03	7.466E − 04	6.045E + 03	1.654E − 04	4.5133	0.2216	20.2636	4.4898
42	4.447E + 03	2.249E − 04	2.008E + 04	4.980E − 05	4.5156	0.2215	20.3529	4.5072
48	1.476E + 04	6.773E − 05	6.668E + 04	1.500E − 05	4.5163	0.2214	20.3841	4.5134
54	4.902E + 04	2.040E − 05	2.214E + 05	4.517E − 06	4.5166	0.2214	20.3948	4.5156
60	1.628E + 05	6.144E − 06	7.351E + 05	1.360E − 06	4.5166	0.2214	20.3984	4.5163
66	5.404E + 05	1.851E − 06	2.441E + 06	4.097E − 07	4.5166	0.2214	20.3996	4.5165
72	1.794E + 06	5.574E − 07	8.103E + 06	1.234E − 07	4.5167	0.2214	20.4000	4.5166
120	2.649E + 10	3.775E − 11	1.196E + 11	8.358E − 12	4.5167	0.2214	20.4002	4.5167
180	4.311E + 15	2.320E − 16	1.947E + 16	5.135E − 17	4.5167	0.2214	20.4002	4.5167
360	1.859E + 31	5.380E − 32	8.395E + 31	1.191E − 32	4.5167	0.2214	20.4002	4.5167

Time Value of Money Factors—Continuous Compounding $r = 30\%$

	Single Sums		Uniform Series				Gradient Series	
n	To Find F Given P $(F\|P, i\%, n)_\infty$	To Find P Given F $(P\|F, i\%, n)_\infty$	To Find F Given A $(F\|A, i\%, n)_\infty$	To Find A Given F $(A\|F, i\%, n)_\infty$	To Find P Given A $(P\|A, i\%, n)_\infty$	To Find A Given P $(A\|P, i\%, n)_\infty$	To Find P Given G $(P\|G, i\%, n)_\infty$	To Find A Given G $(A\|G, i\%, n)_\infty$
1	1.3499	0.7408	1.0000	1.0000	0.7408	1.3499	0.0000	0.0000
2	1.8221	0.5488	2.3499	0.4256	1.2896	0.7754	0.5488	0.4256
3	2.4596	0.4066	4.1720	0.2397	1.6962	0.5896	1.3620	0.8029
4	3.3201	0.3012	6.6316	0.1508	1.9974	0.5007	2.2655	1.1342
5	4.4817	0.2231	9.9517	0.1005	2.2205	0.4503	3.1581	1.4222
6	6.0496	0.1653	14.4334	0.0693	2.3858	0.4191	3.9845	1.6701
7	8.1662	0.1225	20.4830	0.0488	2.5083	0.3987	4.7193	1.8815
8	11.0232	0.0907	28.6492	0.0349	2.5990	0.3848	5.3543	2.0601
9	14.8797	0.0672	39.6724	0.0252	2.6662	0.3751	5.8920	2.2099
10	20.0855	0.0498	54.5521	0.0183	2.7160	0.3682	6.3400	2.3343
11	27.1126	0.0369	74.6376	0.0134	2.7529	0.3633	6.7089	2.4370
12	36.5982	0.0273	101.7503	9.828E − 03	2.7802	0.3597	7.0094	2.5212
13	49.4024	0.0202	138.3485	7.228E − 03	2.8004	0.3571	7.2523	2.5897
14	66.6863	0.0150	187.7510	5.326E − 03	2.8154	0.3552	7.4473	2.6452
15	90.0171	0.0111	254.4373	3.930E − 03	2.8265	0.3538	7.6028	2.6898
16	121.5104	8.230E − 03	344.4544	2.903E − 03	2.8348	0.3528	7.7263	2.7255
17	164.0219	6.097E − 03	465.9649	2.146E − 03	2.8409	0.3520	7.8238	2.7540
18	221.4064	4.517E − 03	629.9868	1.587E − 03	2.8454	0.3514	7.9006	2.7766
19	298.8674	3.346E − 03	851.3932	1.175E − 03	2.8487	0.3510	7.9608	2.7945
20	403.4288	2.479E − 03	1.150E + 03	8.694E − 04	2.8512	0.3507	8.0079	2.8086
21	544.5719	1.836E − 03	1.554E + 03	6.436E − 04	2.8530	0.3505	8.0446	2.8197
22	735.0952	1.360E − 03	2.098E + 03	4.766E − 04	2.8544	0.3503	8.0732	2.8283
23	992.2747	1.008E − 03	2.833E + 03	3.529E − 04	2.8554	0.3502	8.0954	2.8351
24	1.339E + 03	7.466E − 04	3.826E + 03	2.614E − 04	2.8562	0.3501	8.1125	2.8404
25	1.808E + 03	5.531E − 04	5.165E + 03	1.936E − 04	2.8567	0.3501	8.1258	2.8445
26	2.441E + 03	4.097E − 04	6.973E + 03	1.434E − 04	2.8571	0.3500	8.1361	2.8476
27	3.294E + 03	3.035E − 04	9.414E + 03	1.062E − 04	2.8574	0.3500	8.1440	2.8501
28	4.447E + 03	2.249E − 04	1.271E + 04	7.869E − 05	2.8577	0.3499	8.1500	2.8520
29	6.003E + 03	1.666E − 04	1.716E + 04	5.829E − 05	2.8578	0.3499	8.1547	2.8535
30	8.103E + 03	1.234E − 04	2.316E + 04	4.318E − 05	2.8579	0.3499	8.1583	2.8546
36	4.902E + 04	2.040E − 05	1.401E + 05	7.137E − 06	2.8582	0.3499	8.1676	2.8576
42	2.966E + 05	3.372E − 06	8.476E + 05	1.180E − 06	2.8583	0.3499	8.1694	2.8582
48	1.794E + 06	5.574E − 07	5.128E + 06	1.950E − 07	2.8583	0.3499	8.1698	2.8583
54	1.085E + 07	9.214E − 08	3.102E + 07	3.223E − 08	2.8583	0.3499	8.1698	2.8583
60	6.566E + 07	1.523E − 08	1.877E + 08	5.328E − 09	2.8583	0.3499	8.1699	2.8583
66	3.972E + 08	2.517E − 09	1.135E + 09	8.808E − 10	2.8583	0.3499	8.1699	2.8583
72	2.403E + 09	4.161E − 10	6.869E + 09	1.456E − 10	2.8583	0.3499	8.1699	2.8583
120	4.311E + 15	2.320E − 16	1.232E + 16	8.115E − 17	2.8583	0.3499	8.1699	2.8583
180	2.831E + 23	3.533E − 24	8.091E + 23	1.236E − 24	2.8583	0.3499	8.1699	2.8583
360	8.013E + 46	1.248E − 47	2.290E + 47	4.366E − 48	2.8583	0.3499	8.1699	2.8583

	Continuous Compounding			$r = 4\%$

	Continuous flow, Uniform series			
	Present worth factor	Capital recovery factor	Compound amount factor	Sinking fund factor
n	To find P given $\overline{A}$ $P\|\overline{A}\,r,n$	To find $\overline{A}$ given P $\overline{A}\|P\,r,n$	To find F given $\overline{A}$ $F\|\overline{A}\,r,n$	To find $\overline{A}$ given F $\overline{A}\|F\,r,n$
1	0.3003	1.0201	1.0203	0.9801
2	1.9221	0.5203	2.0822	0.4803
3	2.8270	0.3537	3.1874	0.3137
4	3.6944	0.2705	4.3378	0.2305
5	4.5317	0.2207	5.5351	0.1807
6	5.3343	0.1875	6.7812	0.1475
7	6.1054	0.1638	8.0782	0.1238
8	6.8463	0.1481	9.4282	0.1061
9	7.5581	0.1323	10.8332	0.0923
10	8.2420	0.1213	12.2956	0.0813
11	8.8991	0.1124	13.8177	0.0724
12	9.5304	0.1049	15.4019	0.0649
13	10.1370	0.0986	17.0507	0.0586
14	10.7198	0.0933	18.7668	0.0533
15	11.2797	0.0887	20.5530	0.0487
16	11.8177	0.0846	22.4120	0.0446
17	12.3346	0.0811	24.3469	0.0411
18	12.8312	0.0779	25.3808	0.0379
19	13.3083	0.0751	28.4569	0.0351
20	13.7668	0.0726	30.6385	0.0326
21	14.2072	0.0704	32.9092	0.0304
22	14.6304	0.0684	35.2725	0.0284
23	15.0370	0.0665	37.7323	0.0265
24	15.4277	0.0648	40.2924	0.0248
25	15.8030	0.0633	42.9570	0.0233
26	16.1636	0.0619	45.7304	0.0219
27	16.5101	0.0406	48.6170	0.0206
28	16.8430	0.0594	51.6214	0.0194
29	17.1628	0.0583	54.7483	0.0183
30	17.4701	0.0572	58.0029	0.0172
31	17.7654	0.0583	61.3803	0.0163
32	18.0491	0.0554	64.9160	0.0154
33	18.3216	0.0546	68.5855	0.0146
34	18.5835	0.0538	72.4048	0.0138
35	18.8351	0.0531	76.3800	0.0131
36	19.0768	0.0524	80.5174	0.0124
37	19.3091	0.0518	84.8236	0.0118
38	19.5322	0.0512	89.3056	0.0112
39	19.7466	0.0506	93.9705	0.0106
40	19.9526	0.0501	98.8258	0.0101
45	20.8475	0.0479	126.2412	0.0079
48	21.3348	0.0469	145.5240	0.0069
50	21.6166	0.0463	150.7264	0.0063
54	22.1169	0.0452	191.7784	0.0052
60	22.7321	0.0440	250.5794	0.0040
65	23.1432	0.0432	311.5935	0.0032
70	23.4797	0.0426	386.1162	0.0026
75	23.7553	0.0421	477.1384	0.0021
80	23.9809	0.0417	588.3133	0.0017
90	24.3169	0.0411	889.9559	0.0011
100	24.5421	0.0407	1339.9538	0.0007
120	24.7943	0.0403	3012.7604	0.0003

Continuous Compounding $r = 5\%$

	Continuous flow, Uniform series			
	Present worth factor	Capital recovery factor	Compound amount factor	Sinking fund factor
n	To find P given $\overline{A}$ $P\|\overline{A}\,r,n$	To find $\overline{A}$ given P $\overline{A}\|P\,r,n$	To find F given $\overline{A}$ $F\|\overline{A}\,r,n$	To find $\overline{A}$ given F $\overline{A}\|F\,r,n$
1	0.9754	1.0252	1.0254	0.9752
2	1.9033	0.5254	2.1034	0.4754
3	2.7858	0.3590	3.2347	0.3090
4	3.6254	0.2758	4.4281	0.2258
5	4.4240	0.2260	5.6805	0.1760
6	5.1836	0.1929	6.9972	0.1429
7	5.9062	0.1693	8.3814	0.1183
8	6.5936	0.1517	9.8365	0.1017
9	7.2474	0.1380	11.3662	0.0880
10	7.8694	0.1271	12.9744	0.0771
11	8.4610	0.1182	14.6651	0.0482
12	9.0238	0.1108	16.4424	0.0608
13	9.5591	0.1046	18.3108	0.0546
14	10.0683	0.0993	20.2751	0.0493
15	10.5527	0.0948	22.3400	0.0448
16	11.0134	0.0908	24.5108	0.0408
17	11.4517	0.0873	26.7929	0.0373
18	11.8686	0.0843	29.1921	0.0343
19	12.2652	0.0815	31.7142	0.0315
20	12.6424	0.0791	34.3656	0.0291
21	13.0012	0.0769	37.1530	0.0269
22	13.3426	0.0749	40.0833	0.0249
23	13.6673	0.0732	43.1639	0.0232
24	13.9761	0.0716	46.4023	0.0216
25	14.2699	0.0701	49.8069	0.0201
26	14.5494	0.0687	53.3859	0.0187
27	14.8152	0.0675	57.1485	0.0175
28	15.0681	0.0664	61.1040	0.0164
29	15.3086	0.0653	65.2623	0.0153
30	15.5374	0.0644	69.6338	0.0144
31	15.7550	0.0635	74.2294	0.0135
32	15.9621	0.0626	79.0606	0.0126
33	16.1590	0.0619	84.1396	0.0119
34	16.3463	0.0612	89.4789	0.0112
35	16.5245	0.0605	95.0921	0.0105
36	16.6940	0.0599	100.9929	0.0099
37	16.8553	0.0593	107.1944	0.0093
38	17.0096	0.0588	113.7179	0.0088
39	17.1545	0.0583	120.5738	0.0083
40	17.2833	0.0578	127.7811	0.0078
45	17.8920	0.0559	169.7547	0.0059
48	18.1856	0.0550	200.4635	0.0050
50	18.3583	0.0545	223.6490	0.0045
54	18.6559	0.0536	277.5946	0.0036
60	19.0043	0.0526	381.7107	0.0026
65	19.2245	0.0520	495.8068	0.0020
70	19.3961	0.0516	642.3099	0.0016
75	19.5296	0.0512	830.4216	0.0012
80	19.6337	0.0509	1071.8630	0.0009
90	19.7778	0.0506	1780.3426	0.0006
100	19.8652	0.0502	2948.2632	0.0003
120	19.9504	0.0501	3048.5759	0.0001

Continuous Compounding $r = 6\%$

	Continuous flow, Uniform series			
	Present worth factor	Capital recovery factor	Compound amount factor	Sinking fund factor
n	To find P given $\overline{A}$ $P\|\overline{A}\,r,n$	To find $\overline{A}$ given P $\overline{A}\|P\,r,n$	To find F given $\overline{A}$ $F\|\overline{A}\,r,n$	To find $\overline{A}$ given F $\overline{A}\|F\,r,n$
1	0.9706	1.0283	1.0306	0.9703
2	1.8847	0.5296	2.1249	0.4706
3	2.7455	0.3642	3.2870	0.3042
4	3.5582	0.2812	4.5208	0.2212
5	4.3197	0.2315	5.8310	0.1715
6	5.0387	0.1985	7.2222	0.1385
7	5.7159	0.1750	8.6994	0.1150
8	6.3536	0.1574	10.2679	0.0974
9	6.9542	0.1438	11.9334	0.0838
10	7.5198	0.1330	13.7020	0.0730
11	8.0525	0.1242	15.5799	0.0642
12	8.5541	0.1169	17.5739	0.0569
13	9.0266	0.1104	19.6912	0.0508
14	9.4715	0.1054	21.9394	0.0456
15	9.8905	0.1011	24.3267	0.0411
16	10.2851	0.0972	26.8616	0.0372
17	10.6568	0.0938	29.5532	0.0338
18	11.0067	0.0909	32.4113	0.0309
19	11.3363	0.0882	35.4461	0.0282
20	11.6468	0.0859	38.6686	0.0259
21	11.9391	0.0838	42.0904	0.0238
22	12.2144	0.0819	45.7237	0.0219
23	12.4737	0.0802	49.5817	0.0202
24	12.7179	0.0786	53.6783	0.0186
25	12.9478	0.0772	58.0282	0.0172
26	13.1644	0.0760	42.6470	0.0160
27	13.3684	0.0748	67.5515	0.0148
28	13.5604	0.0737	72.7583	0.0137
29	13.7413	0.0728	78.2891	0.0128
30	13.9117	0.0719	84.1608	0.0119
31	14.0721	0.0711	90.3856	0.0111
32	14.2232	0.0703	97.0160	0.0103
33	14.3655	0.0696	104.0457	0.0096
34	14.4995	0.0690	111.5102	0.0090
35	14.6257	0.0684	119.4362	0.0084
36	14.7446	0.0678	127.8523	0.0078
37	14.8565	0.0673	136.7888	0.0073
38	14.9619	0.0668	146.2780	0.0088
39	15.0612	0.0664	156.3539	0.0064
40	15.1547	0.0640	167.0529	0.0060
45	15.5466	0.0643	231.3289	0.0043
48	15.7311	0.0636	280.2379	0.0036
50	15.8369	0.0631	318.0923	0.0031
54	16.0139	0.0624	408.8954	0.0024
60	16.2113	0.0617	593.3039	0.0017
65	16.3283	0.0612	806.7075	0.0012
70	16.4167	0.0609	1094.7722	0.0009
75	16.4815	0.0607	1483.6189	0.0007
80	16.5295	0.0605	2008.5070	0.0005
90	16.5914	0.0603	3673.4403	0.0003
100	16.6254	0.0601	6707.1466	0.0001
120	16.6542	0.0600	22307.1794	0.0000

Continuous Compounding $r = 8\%$

n	Present worth factor — To find P given $\overline{A}$ $P\|\overline{A}\,r,n$	Capital recovery factor — To find $\overline{A}$ given P $\overline{A}\|P\,r,n$	Compound amount factor — To find F given $\overline{A}$ $F\|\overline{A}\,r,n$	Sinking fund factor — To find $\overline{A}$ given F $\overline{A}\|F\,r,n$
		Continuous flow, Uniform series		
1	0.9610	1.0405	1.0411	0.9605
2	1.8482	0.5411	2.1689	0.4611
3	2.6672	0.3749	3.3906	0.2949
4	3.4231	0.2921	4.7141	0.2121
5	4.1210	0.2427	6.1478	0.1627
6	4.7652	0.2099	7.7009	0.1299
7	5.3599	0.1864	9.3834	0.1066
8	5.9088	0.1692	11.2060	0.0892
9	6.4156	0.1550	13.1804	0.0750
10	6.8834	0.1453	15.3193	0.0653
11	7.3152	0.1367	17.6362	0.0547
12	7.7138	0.1296	20.1462	0.0496
13	8.0818	0.1237	22.8652	0.0437
14	8.4215	0.1187	25.8107	0.0387
15	8.7351	0.1145	29.0015	0.0345
16	9.0245	0.1108	32.4580	0.0306
17	9.2917	0.1076	34.2024	0.0276
18	9.5384	0.1048	40.2587	0.0248
19	9.7661	0.1024	44.6528	0.0224
20	9.9763	0.1002	49.4129	0.0202
21	10.1703	0.0983	54.5694	0.0183
22	10.3494	0.0966	60.1555	0.0166
23	10.5148	0.0951	66.2067	0.0151
24	10.6874	0.0937	72.7620	0.0137
25	10.8083	0.0925	79.8632	0.0125
26	10.9384	0.0914	87.5559	0.0114
27	11.0584	0.0904	95.8892	0.0104
28	11.1693	0.0895	104.9166	0.0095
29	11.2716	0.0887	114.6959	0.0087
30	11.3660	0.0880	125.2897	0.0080
31	11.4532	0.0873	136.7658	0.0073
32	11.5337	0.0867	149.1977	0.0067
33	11.6080	0.0861	162.6450	0.0061
34	11.6766	0.0856	177.2540	0.0056
35	11.7399	0.0852	193.0581	0.0052
36	11.7983	0.0848	210.1784	0.0048
37	11.8523	0.0844	228.7246	0.0044
38	11.9021	0.0840	248.8155	0.0040
39	11.9480	0.0837	270.5797	0.0037
40	11.9905	0.0834	294.1566	0.0034
45	12.1585	0.0822	444.9779	0.0022
48	12.2313	0.0818	569.0684	0.0018
50	12.2711	0.0815	669.9769	0.0015
54	12.3338	0.0811	927.3579	0.0011
60	12.3971	0.0807	1506.3802	0.0007
65	12.4310	0.0804	2253.4030	0.0004
70	12.4538	0.0803	3367.8301	0.0003
75	12.4690	0.0802	5030.3598	0.0002
80	12.4792	0.0801	7510.5430	0.0001
90	12.4907	0.0801	16730.3846	0.0001
100	12.4958	0.0800	37249.4748	0.0000
120	12.4992	0.0800	184547.2696	0.0000

Continuous Compounding $r = 10\%$

n	Present worth factor To find P given $\overline{A}$ $P\|\overline{A}\ r,n$	Capital recovery factor To find $\overline{A}$ given P $\overline{A}\|P\ r,n$	Compound amount factor To find F given $\overline{A}$ $F\|\overline{A}\ r,n$	Sinking fund factor To find $\overline{A}$ given F $\overline{A}\|F\ r,n$
		Continuous flow, Uniform series		
1	0.9516	1.0508	1.0517	0.9508
2	1.8127	0.5517	2.2140	0.4517
3	2.5918	0.3858	3.4986	0.2858
4	3.2968	0.3033	4.9182	0.2033
5	3.9347	0.2541	6.4872	0.1541
6	4.5119	0.2216	8.2212	0.1216
7	5.0341	0.1956	10.1375	0.0984
8	5.5067	0.1816	12.2554	0.0816
9	5.9343	0.1685	14.5960	0.0485
10	6.3212	0.1582	17.1828	0.0582
11	6.6713	0.1499	20.0417	0.0499
12	6.9881	0.1431	23.2012	0.0431
13	7.2747	0.1375	26.6930	0.0375
14	7.5340	0.1327	30.5520	0.0327
15	7.7687	0.1287	34.8169	0.0287
16	7.9810	0.1253	39.5303	0.0253
17	8.1732	0.1224	44.7395	0.0224
18	8.3470	0.1198	50.4965	0.0196
19	8.5043	0.1176	56.8589	0.0176
20	8.6466	0.1157	63.8906	0.0157
21	8.7754	0.1140	71.6617	0.0140
22	8.8920	0.1125	80.2501	0.0125
23	8.9974	0.1111	89.7418	0.0111
24	9.0928	0.1100	100.2318	0.0100
25	9.1792	0.1089	111.8249	0.0089
26	9.2573	0.1080	124.6374	0.0080
27	9.3279	0.1072	138.7973	0.0072
28	9.3919	0.1065	154.4465	0.0065
29	9.4498	0.1058	171.7415	0.0058
30	9.5021	0.1052	190.8554	0.0052
31	9.5495	0.1047	211.9795	0.0047
32	9.5924	0.1042	235.3253	0.0042
33	9.6312	0.1038	261.1264	0.0038
34	9.6663	0.1035	289.6410	0.0035
35	9.6980	0.1031	321.1545	0.0031
36	9.7268	0.1028	355.9823	0.0028
37	9.7528	0.1025	394.4730	0.0025
38	9.7763	0.1023	437.0118	0.0023
39	9.7976	0.1021	484.0245	0.0021
40	9.8168	0.1019	535.9815	0.0019
45	9.8889	0.1011	890.1713	0.0011
48	9.9177	0.1008	1205.1042	0.0006
50	9.9326	0.1007	1474.1316	0.0007
54	9.9548	0.1005	2204.0642	0.0005
60	9.9752	0.1002	4024.2879	0.0002
65	9.9850	0.1002	6641.4163	0.0002
70	9.9909	0.1001	10954.3316	0.0001
75	9.9945	0.1001	18070.4241	0.0001
80	9.9966	0.1000	29799.5790	0.0000
90	9.9988	0.1000	81020.8393	0.0000
100	9.9995	0.1000	220254.6579	0.0000
120	9.9999	0.1000	1627537.9142	0.0000

Continuous Compounding $r = 15\%$

	Continuous flow, Uniform series			
	Present worth factor	Capital recovery factor	Compound amount factor	Sinking fund factor
n	To find P given $\overline{A}$ $P\|\overline{A}\,r,n$	To find $\overline{A}$ given P $\overline{A}\|P\,r,n$	To find F given $\overline{A}$ $F\|\overline{A}\,r,n$	To find $\overline{A}$ given F $\overline{A}\|F\,r,n$
1	0.9286	1.0769	1.0789	0.9269
2	1.7279	0.5787	2.3324	0.4287
3	2.4158	0.4130	3.7887	0.2639
4	3.0079	0.3325	5.4908	0.1825
5	3.5176	0.2843	7.4487	0.1343
6	3.9562	0.2528	9.7307	0.1028
7	4.3337	0.2307	12.3843	0.0807
8	4.6587	0.2147	15.4674	0.0647
9	4.9384	0.2025	19.0495	0.0525
10	5.1791	0.1931	23.2113	0.0431
11	5.3863	0.1857	28.0465	0.0357
12	5.5647	0.1797	33.0643	0.0297
13	5.7182	0.1749	40.1913	0.0249
14	5.8503	0.1709	47.7745	0.0209
15	5.9640	0.1677	54.5849	0.0177
16	6.0619	0.1650	66.8212	0.0150
17	6.1461	0.1627	78.7140	0.0127
18	6.2186	0.1608	92.5315	0.0108
19	6.2810	0.1592	108.5852	0.0092
20	6.3348	0.1579	127.2369	0.0079
21	6.3810	0.1567	148.9071	0.0047
22	6.4208	0.1557	174.0843	0.0057
23	6.4550	0.1549	203.3359	0.0049
24	6.4845	0.1542	237.3216	0.0042
25	6.5099	0.1536	276.8072	0.0036
26	6.5317	0.1531	322.6830	0.0031
27	6.5505	0.1527	375.9830	0.0027
28	6.5687	0.1523	437.9089	0.0023
29	6.5806	0.1520	508.8544	0.0020
30	6.5926	0.1517	593.4475	0.0017
31	6.6029	0.1514	690.5666	0.0014
32	6.6118	0.1512	803.4028	0.0012
33	6.6194	0.1511	934.4998	0.0011
34	6.6260	0.1509	1086.8127	0.0009
35	6.6317	0.1508	1263.7751	0.0008
36	6.6366	0.1507	1469.3761	0.0007
37	6.6408	0.1506	1708.2504	0.0006
38	6.6444	0.1505	1985.7827	0.0005
39	6.6475	0.1504	2308.2292	0.0004
40	6.6501	0.1504	2682.8586	0.0004
45	6.6589	0.1502	5687.0584	0.0002
48	6.6817	0.1501	8922.8718	0.0001
50	6.6630	0.1501	12046.9494	0.0001
54	6.6646	0.1500	21956.4538	0.0000
60	6.6658	0.1500	54013.8929	0.0000
65	6.6663	0.1500	114354.8587	0.0000
70	6.6665	0.1500	242096.6845	0.0000
75	6.6666	0.1500	512526.1318	0.0000
80	6.6666	0.1500	1085025.2761	0.0000
90	6.6667	0.1500	4862769.1323	0.0000
100	6.6667	0.1500	21793442.4831	0.0000
120	6.6667	0.1500	437733120.9155	0.0000

Continuous Compounding $r = 20\%$

	Continuous flow, Uniform series			
	Present worth factor	Capital recovery factor	Compound amount factor	Sinking fund factor
n	To find P given $\overline{A}$ $P\|\overline{A}\,r,n$	To find $\overline{A}$ given P $\overline{A}\|P\,r,n$	To find F given $\overline{A}$ $F\|\overline{A}\,r,n$	To find $\overline{A}$ given F $\overline{A}\|F\,r,n$
1	0.9063	1.1033	1.1070	0.9033
2	1.6494	0.6066	2.4591	0.4066
3	2.2559	0.4433	4.1106	0.2433
4	2.7534	0.3432	6.1277	0.1632
5	3.1606	0.3164	8.5914	0.1164
6	3.4940	0.2862	11.6006	0.0662
7	3.7670	0.2655	15.2760	0.0655
8	3.9805	0.2508	19.7652	0.0506
9	4.1735	0.2396	25.2482	0.0396
10	4.3233	0.2313	31.9453	0.0313
11	4.4460	0.2249	40.1251	0.0249
12	4.5464	0.2200	50.1159	0.0200
13	4.6286	0.2160	62.3187	0.0160
14	4.6950	0.2129	77.2232	0.0129
15	4.7511	0.2105	95.4277	0.0105
16	4.7962	0.2085	117.6627	0.0085
17	4.8331	0.2069	144.8205	0.0049
18	4.8634	0.2058	177.9912	0.0056
19	4.8881	0.2046	218.5059	0.0046
20	4.9084	0.2037	267.9908	0.0037
21	4.9250	0.2030	328.4317	0.0030
22	4.9386	0.2025	402.2543	0.0025
23	4.9497	0.2020	492.4216	0.0020
24	4.9589	0.2017	602.5521	0.0017
25	4.9663	0.2014	737.0658	0.0014
26	4.9724	0.2011	901.3612	0.0011
27	4.9774	0.2009	1102.0321	0.0009
28	4.9815	0.2007	1347.1320	0.0007
29	4.9849	0.2006	1646.4978	0.0006
30	4.9876	0.2005	2012.1440	0.0005
31	4.9899	0.2004	2458.7452	0.0004
32	4.9017	0.2003	3004.2252	0.0003
33	4.9032	0.2003	3670.4759	0.0003
34	4.9944	0.2002	4484.2365	0.0002
35	4.9954	0.2002	5478.1658	0.0002
36	4.9965	0.2001	6692.1528	0.0001
37	4.9969	0.2001	8174.9221	0.0001
38	4.9975	0.2001	9985.9795	0.0001
39	4.9980	0.2001	12198.0099	0.0001
40	4.9983	0.2001	14899.7899	0.0001
45	4.9994	0.2000	40510.4196	0.0000
48	4.9997	0.2000	73818.9078	0.0000
50	4.9998	0.2000	110127.3290	0.0000
54	4.9999	0.2000	245099.0057	0.0000
60	5.0000	0.2000	813768.9571	0.0000
65	5.0000	0.2000	2212061.9600	0.0000
70	5.0000	0.2000	6013016.4208	0.0000
75	5.0000	0.2000	16345081.8624	0.0000
80	5.0000	0.2000	44430547.6025	0.0000
90	5.0000	0.2000	328299840.6887	0.0000

Continuous Compounding $r = 25\%$

	Continuous flow, Uniform series			
	Present worth factor	Capital recovery factor	Compound amount factor	Sinking fund factor
n	To find P given $\overline{A}$ $P\|\overline{A}\,r,n$	To find $\overline{A}$ given P $\overline{A}\|P\,r,n$	To find F given $\overline{A}$ $F\|\overline{A}\,r,n$	To find $\overline{A}$ given F $\overline{A}\|F\,r,n$
1	0.8848	1.1302	1.1361	0.8802
2	1.5739	0.6354	2.5949	0.3854
3	2.1105	0.4738	4.4680	0.2238
4	2.5285	0.3955	6.8731	0.1455
5	2.8540	0.3504	9.9614	0.1004
6	3.1075	0.3218	13.9268	0.0718
7	3.3049	0.3026	19.0184	0.0526
8	3.4587	0.2891	25.5542	0.0391
9	3.5784	0.2795	33.9509	0.0295
10	3.6717	0.2724	44.7300	0.0224
11	3.7443	0.2671	58.5705	0.0171
12	3.8009	0.2631	76.3421	0.0131
13	3.8449	0.2601	99.1614	0.0101
14	3.8792	0.2578	128.4618	0.0078
15	3.9058	0.2560	166.0843	0.0060
16	3.9267	0.2547	214.3926	0.0047
17	3.9429	0.2536	276.4216	0.0038
18	3.9556	0.2528	354.0685	0.0028
19	3.9654	0.2522	458.3371	0.0022
20	3.9730	0.2517	589.6526	0.0017
21	3.9790	0.2513	758.2651	0.0013
22	3.9837	0.2510	974.7677	0.0010
23	3.9873	0.2508	1252.7626	0.0008
24	3.9901	0.2506	1609.7152	0.0006
25	3.9923	0.2505	2068.0513	0.0005
26	3.9940	0.2504	2656.5665	0.0004
27	3.9953	0.2503	3412.2351	0.0003
28	3.9964	0.2502	4382.5326	0.0002
29	3.9972	0.2502	5628.4194	0.0002
30	3.9978	0.2501	7228.1697	0.0001
31	3.9983	0.2501	9282.2897	0.0001
32	3.9987	0.2501	11919.8319	0.0001
33	3.9990	0.2501	15306.5033	0.0001
34	3.9992	0.2501	19455.0754	0.0001
35	3.9994	0.2500	25238.7524	0.0000
36	3.9995	0.2500	32408.3357	0.0000
37	3.9996	0.2500	41614.2629	0.0000
38	3.9997	0.2500	53434.9073	0.0000
39	3.9998	0.2500	68612.9152	0.0000
40	3.9998	0.2500	88101.8632	0.0000
45	3.9999	0.2500	307515.6791	0.0000
48	4.0000	0.2500	651015.1657	0.0000
50	4.0000	0.2500	1073345.1461	0.0000
54	4.0000	0.2500	2917661.4704	0.0000
60	4.0000	0.2500	13076065.4899	0.0000

Continuous Compounding: $r = 5\%$

	Geometric series present worth factor,				$(P\|A_1\ r,c,n)_\infty$
n	$c = 4\%$	$c = 6\%$	$c = 8\%$	$c = 10\%$	$c = 15\%$
1	0.9512	0.9512	0.9512	0.9512	0.9512
2	1.8938	1.9120	1.9313	1.9512	2.0025
3	2.8254	2.8825	2.9412	3.0025	3.1643
4	3.7485	3.8427	3.9817	4.1077	4.4484
5	4.6424	4.8527	5.0538	5.2695	5.8474
6	5.5673	5.8527	6.1584	6.4909	7.4357
7	6.4631	6.8428	7.2945	7.7749	9.1690
8	7.3500	7.8830	8.4692	9.1248	11.0845
9	8.2281	8.9134	9.6775	10.5439	13.2015
10	9.0975	9.9542	10.9224	12.0357	15.5412
11	9.9582	11.0055	12.2052	13.6040	18.1269
12	10.8103	12.0473	13.5269	15.2527	20.9845
13	11.6540	13.1398	14.8887	16.9840	24.1427
14	12.4883	14.2231	16.2918	18.8081	27.6331
15	13.3162	15.3173	17.7375	20.7234	31.4905
16	14.1350	16.4225	19.2271	22.7374	35.7534
17	14.9455	17.5388	20.7619	24.8544	40.4651
18	15.7481	18.6663	22.3433	27.0800	45.6721
19	16.5426	19.8051	23.9726	29.4196	51.4267
20	17.3292	20.9554	25.6515	31.8792	57.7865
21	18.1080	22.1172	27.3813	34.4649	64.8152
22	18.8791	23.2907	29.1636	37.1832	72.5831
23	19.6425	24.4760	30.9999	40.0409	81.1679
24	20.3982	25.6732	32.8921	43.0450	90.6557
25	21.1465	26.8825	34.8416	46.2032	101.1412
26	21.8873	28.1039	36.8503	49.5234	112.7296
27	22.6204	29.3376	38.9200	53.0137	125.5367
28	23.3469	30.5834	41.0525	56.6830	139.6907
29	24.0458	31.8422	43.2498	60.5405	155.3334
30	24.7776	33.1135	45.5137	64.5957	172.6211
31	25.4823	34.3975	47.8463	68.8588	191.7271
32	26.1800	35.6944	50.2496	73.3405	212.8424
33	26.8707	37.0044	52.7262	78.0520	234.1785
34	27.5546	38.3275	55.2777	83.0050	261.9688
35	28.2316	39.6640	57.9067	88.2120	290.4716
36	28.9010	41.0138	60.6155	93.6880	321.9720
37	29.5454	42.3772	63.4065	99.4406	356.7853
38	30.2226	43.7544	66.2822	105.4903	395.2590
39	30.8732	45.1453	69.2452	111.8501	437.7810
40	31.5172	46.5503	72.2981	118.5361	484.7741
48	36.4441	58.3103	100.3046	185.9400	1089.9691
50	37.6154	61.4002	108.4172	207.4688	1333.2938
54	39.8890	67.7687	126.1728	257.5108	1993.4889
60	43.1333	77.8120	157.1240	354.0943	3439.8093
65	45.6921	86.8542	187.5146	459.9358	6006.8985
70	48.1261	95.9498	222.8057	595.8390	9909.5887
80	52.6438	115.9951	311.3781	994.4092	26952.5414
90	56.7316	138.1487	430.8188	1651.5409	73280.1440
100	60.4303	162.6321	591.8858	2734.9690	199211.6230
120	66.8054	219.5947	1101.9828	7466.3106	1472043.6146

Continuous Compounding: $r = 5\%$

	Geometric series future worth factor,				$(F|A_1\ r,c,n)_\infty$
n	$c = 4\%$	$c = 6\%$	$c = 8\%$	$c = 10\%$	$c = 15\%$
1	1.0000	1.0000	1.0000	1.0000	1.0000
2	2.0921	2.1131	2.1346	2.1584	2.2131
3	3.2826	3.3489	3.4175	3.4884	3.6764
4	4.5784	4.7179	4.8640	5.0171	5.4332
5	5.9867	6.2310	6.4905	6.7682	7.5330
6	7.5150	7.9003	8.3151	8.7618	10.0372
7	9.1716	9.7387	10.3575	11.0332	13.0114
8	10.9450	11.7600	12.6392	13.6126	16.5342
9	12.9043	13.9790	15.1837	16.5341	20.7041
10	14.9992	16.4118	18.0164	19.8435	25.6231
11	17.2601	19.0753	21.1659	23.5792	31.4185
12	19.6977	21.9881	24.6620	27.7923	38.2343
13	22.3237	25.1699	28.5381	32.5373	46.2464
14	25.1503	28.6419	32.8305	37.8749	55.6462
15	28.1905	32.4267	37.5786	43.8720	66.6854
16	31.4579	36.5489	42.8254	50.6030	79.5711
17	34.9673	41.0345	48.6178	58.1505	94.6740
18	38.7340	45.9116	55.0046	66.6059	112.3352
19	42.7743	51.2102	62.0476	76.0705	132.9744
20	47.1057	54.9626	69.8010	86.6547	157.0800
21	51.7464	63.2032	78.3328	98.4887	185.2192
22	56.7159	69.9491	87.7146	111.7045	218.0516
23	62.0347	77.2999	98.0242	126.4548	254.3440
24	67.7245	85.2381	109.3466	142.9146	300.9874
25	73.8085	93.8290	121.7738	161.2652	353.0176
26	80.3111	103.1215	135.4064	181.7159	413.6383
27	87.2579	113.1674	150.3532	204.4965	484.2484
28	94.6764	124.0227	166.7331	229.8410	566.4738
29	102.5954	135.7471	184.6750	258.0909	642.2039
30	111.0455	148.4043	204.3192	289.4977	773.6343
31	120.0591	162.0428	225.8180	324.4262	903.3165
32	129.6703	176.7957	249.3371	343.2579	1054.2155
33	139.9152	192.6812	275.0547	404.4151	1229.7767
34	150.8323	209.8029	303.1723	454.3652	1434.0037
35	162.4618	228.2503	333.8944	507.6252	1671.5485
36	174.8466	248.1191	367.4604	566.7673	1947.8169
37	188.0319	269.5116	404.1147	632.4244	2269.0900
38	202.0454	292.5371	444.1320	705.2970	2642.6663
39	216.9977	317.3125	487.8083	786.1597	3077.0261
40	232.8823	343.9427	535.4851	875.8696	3582.0229
48	401.7293	642.7645	1108.8901	2049.8694	12014.9216
50	458.2495	748.0082	1324.8240	2527.4875	16242.8432
54	593.5376	1008.3801	1883.7045	3831.6909	29662.5802
60	866.3558	1562.8958	3167.9343	7112.1736	73107.5250
65	1178.4151	2234.8418	4856.3452	11861.9002	154919.9539
70	1593.7181	3177.4206	7412.2338	19731.4790	328159.8458
80	2874.2537	6333.1204	17092.8416	54292.9010	1471558.8994
90	5106.8118	12435.7489	39024.4542	148666.9722	6596468.3423
100	8968.6539	24136.7505	88472.2371	405905.3880	29585826.2924
120	26951.2135	88590.8273	448564.3760	3012124.6728	593864779.4096

Continuous Compounding: $r = 8\%$

	Geometric series present worth factor,				$(P\|A_1\ r,c,n)_\infty$
n	$c = 4\%$	$c = 6\%$	$c = 8\%$	$c = 10\%$	$c = 15\%$
1	0.9231	0.9231	0.9231	0.9231	0.9231
2	1.8100	1.8280	1.8461	1.8649	1.9132
3	2.6622	2.7149	2.7691	2.8257	2.9750
4	3.4809	3.5842	3.6919	3.8059	4.1138
5	4.2675	4.4364	4.6147	4.8059	5.3352
6	5.0233	5.2716	5.5373	5.8261	6.6452
7	5.7495	6.0904	6.4599	6.8669	8.0501
8	6.4471	6.8929	7.3823	7.9287	9.5570
9	7.1175	7.6795	8.3047	9.0120	11.1730
10	7.7615	8.4506	9.2270	10.1172	12.9063
11	8.3803	9.2064	10.1482	11.2447	14.7652
12	8.9748	9.9472	11.0713	12.3950	16.7589
13	9.5460	10.6733	11.9933	13.5685	18.8972
14	10.0948	11.3851	12.9152	14.7657	21.1905
15	10.6221	12.0828	13.8371	15.9871	23.6501
16	11.1287	12.7666	14.7588	17.2332	26.2881
17	11.6155	13.4370	15.6804	18.5044	29.1173
18	12.0832	14.0940	16.6020	19.8014	32.1517
19	12.5325	14.7380	17.5234	21.1245	35.4040
20	12.9642	15.3693	18.4448	22.4743	38.8964
21	13.3790	15.0881	19.3461	23.8515	42.6398
22	13.7775	16.5946	20.2873	25.2564	46.6546
23	14.1604	17.1892	21.2083	26.6898	50.9606
24	14.5283	17.7719	22.1293	28.1520	55.5788
25	14.8817	18.3431	23.0502	29.6439	60.5318
26	15.2213	18.9030	23.9710	31.1658	65.8440
27	15.5476	19.4518	24.8918	32.7186	71.5413
28	15.8611	19.9898	25.8124	34.3026	77.6518
29	16.1623	20.5171	26.7329	35.9187	84.2053
30	16.4517	21.0339	27.6534	37.5674	91.2340
31	16.7297	21.5405	28.5737	39.2495	98.7723
32	16.9968	22.0371	29.4940	40.9655	106.8572
33	17.2535	22.5239	30.4142	42.7162	115.5283
34	17.5001	23.0010	31.3342	44.5022	124.8282
35	17.7370	23.4686	32.2542	46.3243	134.8024
34	17.9647	23.9271	33.1741	48.1833	145.4998
37	18.1834	24.3764	34.0939	50.0797	154.9728
38	18.3935	24.8168	35.0134	52.0145	169.2777
39	18.5954	25.2485	35.9332	53.9884	182.4749
40	18.7894	25.6717	36.8527	54.0022	194.6289
48	20.0910	28.7689	44.2054	73.6480	353.7898
50	20.3564	29.4688	46.0429	78.5185	408.8683
54	20.8275	30.7873	49.7164	88.8640	545.0997
60	21.4068	32.5776	55.2239	106.0200	836.2660
65	21.7940	33.9138	59.8110	121.9761	1192.0531
70	22.1109	35.1228	64.3857	139.6104	1694.9391
80	22.5829	37.2067	73.5584	180.6379	3430.1220
90	22.8993	38.9129	82.7119	230.7490	6920.3236
100	23.1113	40.3097	91.8562	291.9549	13948.7265
120	23.3488	42.3897	110.1175	458.0208	54603.7706

Continuous Compounding: $r = 8\%$

n	Geometric series future worth factor,				$(F\|A_1\ r,c,n)_\infty$
	$c = 4\%$	$c = 6\%$	$c = 8\%$	$c = 10\%$	$c = 15\%$
1	1.0000	1.0000	1.0000	1.0000	1.0000
2	2.1241	2.1461	2.1046	2.1885	2.2451
3	3.3843	3.4613	3.5206	3.5021	3.7820
4	4.7937	4.9250	5.0850	5.2412	5.0453
5	6.3664	6.6183	6.8856	7.1695	7.9582
6	8.1181	8.5184	8.9500	9.4154	10.7391
7	10.0454	10.6623	11.3125	12.0217	14.0022
8	12.2269	13.0722	14.0054	15.0347	18.1246
9	14.6224	15.7771	17.0483	18.5146	22.9643
10	17.2735	18.1071	20.5443	22.5162	28.7236
11	20.2040	22.1956	24.4809	27.1098	35.5978
12	23.4395	25.9790	28.9308	32.3719	43.7693
13	27.0078	30.1972	33.9520	38.3881	53.4643
14	30.9392	34.1837	29.6090	45.2547	64.9450
15	35.2667	40.1162	45.9728	53.0790	78.5212
16	40.0261	45.9176	53.1218	61.9815	94.5487
17	45.2542	52.3530	61.1428	72.0948	113.4468
18	50.9993	59.4865	70.1314	83.5755	135.7023
19	57.3014	67.2854	80.1931	96.5859	161.8843
20	64.2121	76.1247	91.4444	111.3162	182.6550
21	71.7857	85.7851	104.0136	127.9764	228.7962
22	80.0809	96.4553	118.0421	146.8014	271.1772
23	89.1614	108.2322	133.6859	168.0531	320.8754
24	98.0967	121.2214	151.1167	192.0239	379.1005
25	109.9619	135.5383	170.5238	219.0403	447.2729
26	121.8386	151.2086	192.1152	249.9640	527.0461
27	134.8154	168.1494	216.1204	283.7071	620.2446
28	148.9884	187.7705	242.7915	322.2160	729.4088
29	164.4621	208.7749	272.4062	345.4971	856.8454
30	181.3496	231.1605	305.2696	414.1125	1005.6880
31	199.7738	257.2211	341.7180	468.6883	1179.4680
32	219.6480	285.0681	382.1199	529.9220	1382.2852
33	241.7768	315.6315	426.8813	598.5902	1618.9221
34	265.6571	349.1423	476.4481	675.5578	1894.9323
35	291.6791	385.9336	531.3104	761.7872	2216.7776
36	320.0274	426.2430	592.0042	858.3498	2501.9728
37	350.9022	470.4147	859.1269	946.4376	3029.2570
38	384.5208	518.3015	733.3216	1087.3768	3538.7925
39	421.1186	571.7877	815.3029	1222.6426	4132.3955
40	460.9152	629.7914	905.8534	1373.8755	4823.8050
48	934.7453	1338.4850	2061.5196	3426.5061	164460.2384
50	1111.4222	1604.3405	2520.0161	4286.9639	22323.4534
54	1545.9909	2314.8578	3748.0141	6681.5593	40985.2972
60	2601.1507	3958.5138	6730.0753	12882.5306	101615.0248
65	3950.6391	6147.8265	10876.7642	22110.8837	216004.1379
70	5979.3767	9498.1411	17474.3923	37754.3366	358897.1434
80	13591.4050	22382.6784	44445.6638	106716.0047	2064001.8789
90	30671.9919	52121.9609	111280.0438	309072.2980	9289294.3290
100	68893.9547	120161.6118	275175.7425	870305.3412	41580587.7757
120	344739.8525	625875.2745	1635543.6143	6762576.6413	835742300.0391

Continuous Compounding: $r = 10\%$

	Geometric series present worth factor,				$(P \vert A_1\ r,c,n)_\infty$
n	$c = 4\%$	$c = 6\%$	$c = 8\%$	$c = 10\%$	$c = 15\%$
1	0.9048	0.9048	0.9048	0.9048	0.9048
2	1.7570	1.7742	1.7917	1.8097	1.8561
3	2.5595	2.6095	2.6609	2.7148	2.8561
4	3.3153	3.4120	3.5127	3.6194	3.9073
5	4.0271	4.1830	4.3477	4.5242	5.0125
6	4.6974	4.9239	5.1640	5.4290	6.1743
7	5.3287	5.6356	5.9490	6.3339	7.3957
8	5.9232	6.3195	6.7541	7.2387	8.6796
9	6.4831	6.9765	7.5245	8.1435	10.0296
10	7.0104	7.6078	8.2797	9.0484	11.4487
11	7.5070	8.2143	9.0197	9.9532	12.9406
12	7.9746	8.7971	9.7451	10.8581	14.5088
13	8.4151	9.3570	10.4560	11.7629	16.1576
14	8.8296	9.8949	11.1528	12.6677	17.8906
15	9.2205	10.4118	11.8357	13.5726	19.7129
16	9.5883	10.9084	12.5050	14.4774	21.6285
17	9.9348	11.3855	13.1610	15.3822	23.6422
18	10.2611	11.8439	13.8039	16.2871	25.7592
19	10.5684	12.2843	14.4341	17.1919	27.9848
20	10.8577	12.7075	15.0517	18.0968	30.3246
21	11.1303	13.1141	15.6570	19.0016	32.7840
22	11.3869	13.5047	16.2503	19.9064	35.3697
23	11.6286	13.8800	16.8317	20.8113	38.0880
24	11.8543	14.2406	17.4016	21.7161	40.9457
25	12.0707	14.5870	17.9402	22.6210	43.9498
26	12.2724	14.9199	18.5076	23.5258	47.1080
27	12.4427	15.2397	19.0442	24.4304	50.4281
28	12.6418	15.5470	19.5701	25.3355	53.9185
29	12.8104	15.8422	20.0855	26.2403	57.5878
30	12.9492	16.1259	20.5906	27.1452	61.4452
31	13.1188	16.3984	21.0457	28.0500	65.5004
32	13.2507	16.4403	21.5710	28.9548	69.7635
33	13.3923	16.9119	22.0445	29.8597	74.2452
34	13.5172	17.1534	22.5127	30.7645	78.9547
35	13.6349	17.3858	22.9495	31.6694	83.9097
36	13.7457	17.6089	23.4173	32.5742	89.1167
37	13.8500	17.8233	23.8581	33.4790	94.5906
38	13.9483	18.0293	24.2862	34.3839	100.3452
39	14.0409	18.2272	24.7078	35.2887	106.3949
40	14.1280	18.4173	25.1210	36.1936	112.7547
48	14.6654	19.6932	28.1437	43.4323	176,8900
50	14.7640	19.9533	28.8264	45.2420	197.3498
54	14.9291	20.4151	30.1122	48.8614	244.9509
60	15.1130	20.9829	31.8571	54.2904	336.8234
65	15.2231	21.3624	33.1588	58.8146	437.5024
70	15.3046	21.6731	34.3360	63.3388	566.7767
80	15.4097	22.1357	36.3635	72.3873	945.9054
90	15.4674	22.4458	38.0218	81.4357	1570.8831
100	13.4991	22.6537	30.3782	90.4842	2601.5619
120	15.5260	22.8865	41.3949	108.5811	7201.1027

Continuous Compounding: $r = 10\%$

	Geometric series future worth factor,				$(F\|A_1\ l,l,n)_\infty$
n	$c = 4\%$	$c = 6\%$	$c = 8\%$	$c = 10\%$	$c = 15\%$
1	1.0000	1.0000	1.0000	1.0000	1.0000
2	2.1460	2.1670	2.1885	2.2103	2.2670
3	3.4550	3.5224	3.5921	3.6642	3.8583
4	4.9458	5.0901	5.2412	5.3994	5.8291
5	6.6395	6.8967	7.1695	7.4591	8.2642
6	8.5592	8.9718	9.4154	9.8923	11.2504
7	10.7306	11.3488	12.0217	12.7548	14.8932
8	13.1823	14.0643	15.0367	16.1100	19.3172
9	15.9458	17.1595	18.5146	20.0299	24.6689
10	19.0542	20.6802	22.5162	24.5960	31.1208
11	22.5521	24.6773	27.1098	29.9011	38.8755
12	26.4767	29.2074	32.3718	36.0500	48.1710
13	30.8773	34.3338	38.3881	43.1615	59.2869
14	35.8067	40.1260	45.2546	51.3702	72.5508
15	41.3232	46.6624	53.0789	60.8280	88.3472
16	47.4914	54.0295	61.9814	71.7071	107.1265
17	54.3826	62.3236	72.0987	84.2016	129.4163
18	62.0759	71.6514	83.5753	98.5311	155.8342
19	70.6589	82.1317	96.5857	114.9434	167.1032
20	80.2285	93.8963	111.3159	133.7180	244.0688
21	90.8917	107.0916	127.9762	155.1703	267.7198
22	102.7672	121.8800	146.8011	179.6559	319.2122
23	115.9863	138.4416	168.0527	207.5755	379.8967
24	130.6939	154.9766	192.0235	239.3807	451.3512
25	147.0508	177.7086	219.0397	275.5797	535.4184
26	165.2346	200.8779	249.4654	316.7452	634.2499
27	185.4417	226.7632	283.7063	363.5214	750.3570
28	207.8894	255.6452	322.2151	416.6331	886.6702
29	232.8182	287.9193	365.4981	476.8954	1046.6085
30	260.4938	323.8974	414.1113	545.2252	1234.1597
31	291.2103	364.0116	468.6869	622.6526	1453.9746
32	325.2928	408.7188	529.9203	710.3355	1711.4754
33	363.1008	458.5251	598.5883	809.5748	2012.9832
34	405.0318	513.9913	675.5555	921.8312	2345.8655
35	451.5256	575.7389	761.7846	1048.7453	2778.7076
36	503.0682	644.4560	858.3468	1192.1584	3261.5131
37	560.1970	720.9052	966.4341	1354.1371	3825.9350
38	623.5064	805.9308	1067.3728	1537.0004	4485.5506
39	693.6533	900.4679	1222.6379	1743.3495	5258.1675
40	771.3643	1005.5522	1373.8702	1976.1018	6158.1978
48	1781.9951	2392.9306	3426.4900	5277.4767	21493.9790
50	2191.1713	2961.3357	4286.9429	6714.5054	29289.3015
54	3305.3885	4520.0308	6481.5230	10818.2164	54233.6964
60	6097.0283	8465.1178	12882.4546	21902.3127	135884.2753
65	10125.4885	14209.0254	22110.7422	39120.0527	291001.0399
70	16783.5006	23767.4343	37754.0761	694594.6980	621546.0867
80	45935.6564	63985.6707	108715.1458	2157834.3870	2819704.3620
90	125333.5559	181880.5081	309089.5473	659890.5555	12728807.8516
100	341389.3494	498981.2068	870296.7252	1993046.9108	57303214.4912

Continuous Compounding: $r = 15\%$

	Geometric series present worth factor,				$(P\|A_1\,r,c,n)_\infty$
n	$c = 4\%$	$c = 6\%$	$c = 8\%$	$c = 10\%$	$c = 15\%$
1	0.8607	0.8607	0.8607	0.8607	0.8607
2	1.6318	1.6473	1.6431	1.6794	1.7214
3	2.3225	2.3463	2.4113	2.4582	2.5821
4	2.9413	3.0233	3.1087	3.1991	3.4428
5	3.4956	3.6238	3.7590	3.9037	4.3035
6	3.9922	4.1726	4.3452	4.5741	5.1642
7	4.4370	4.6742	4.9304	5.2117	6.0250
8	4.8356	5.1326	5.4573	5.8182	6.8857
9	5.1926	5.5515	5.9486	6.3952	7.7484
10	5.5124	5.9344	6.4066	6.9440	8.6071
11	5.7989	6.2844	6.8335	7.4660	9.4678
12	6.0556	6.6042	7.2316	7.9626	10.3285
13	6.2855	6.8945	7.6028	8.4350	11.1892
14	6.4915	7.1634	7.9488	8.8843	12.0499
15	6.6760	7.4078	8.2713	9.3117	12.9106
16	6.8413	7.6309	8.5721	9.7183	13.7713
17	6.9894	7.8348	8.8525	10.1050	14.6320
18	7.1220	8.0212	9.1139	10.4729	15.4927
19	7.2409	8.1915	9.3576	10.8229	16.3535
20	7.3473	8.3472	9.5858	11.1557	17.2142
21	7.4427	8.4895	9.7946	11.4724	18.0749
22	7.5281	8.6195	9.9941	11.7736	18.9354
23	7.6046	8.7383	10.1782	12.0401	19.7943
24	7.6732	8.8470	10.3499	12.3326	20.8570
25	7.7346	8.9482	10.5099	12.5918	21.5177
26	7.7897	9.0349	10.6591	12.8384	22.3784
27	7.8389	9.1194	10.7982	13.0730	23.2391
28	7.8831	9.1954	10.9278	13.2941	24.0998
29	7.9227	9.2649	11.0487	13.5084	24.9405
30	7.9581	9.3282	11.1615	13.7103	25.8212
31	7.9898	9.3860	11.2665	13.9023	26.6819
32	8.0183	9.4389	11.3645	14.0850	27.5427
33	8.0438	9.4872	11.4889	14.2588	28.4034
34	8.0446	9.5313	11.5410	14.4241	29.2641
35	8.0870	9.5717	11.6204	14.5813	30.1248
36	8.1053	9.6086	11.6944	14.7309	30.9855
37	8.1218	9.6423	11.7634	14.8732	31.8442
38	8.1365	9.6731	11.8277	15.0085	32.7069
39	8.1496	9.7013	11.8877	15.1373	33.5476
40	8.1614	9.7270	11.9436	15.2597	34.4283
48	8.2208	9.8672	12.2742	16.0471	41.3140
50	8.2291	9.8891	12.3317	16.1995	43.0354
54	8.2411	9.9227	12.4251	16.4621	46.4782
60	8.2516	9.9551	12.5242	16.7695	51.6425
65	8.2564	9.9714	12.5802	16.9438	55.9460
70	8.2591	9.9619	12.6197	17.1152	60.2496
80	8.2616	9.9928	12.6670	17.3249	63.8546
90	8.2624	9.9972	12.6905	17.4521	77.4437
100	8.2627	9.9999	12.7022	17.5292	84.0706
120	8.2628	10.0000	12.7108	17.6044	103.2850

Continuous Compounding: $r = 15\%$

	Geometric series future worth factor,				$(F\|A_1\ r,c,n)_\infty$
n	$c = 4\%$	$c = 6\%$	$c = 8\%$	$c = 10\%$	$c = 15\%$
1	1.0000	1.0000	1.0000	1.0000	1.0000
2	2.2026	2.2237	2.2451	2.2670	2.3237
3	3.6424	3.7110	3.7820	3.8553	4.0496
4	5.3594	5.5088	5.6453	5.8291	6.2732
5	7.4002	7.6716	7.9592	8.2642	9.1106
6	9.8192	10.2630	10.7391	11.2504	12.7020
7	12.6795	13.3572	14.0832	14.8932	17.2172
8	16.0546	17.0408	18.1246	19.3172	22.8412
9	20.0300	21.4147	22.9543	24.6689	29.8811
10	24.7048	26.5963	28.7235	31.1208	38.5743
11	30.1947	32.7226	35.5974	38.8755	49.2986
12	36.6340	39.9531	43.7692	48.1711	62.4838
13	44.1787	48.4733	53.4643	59.2869	78.6454
14	53.0104	58.4994	64.9458	72.5508	98.4016
15	63.3399	70.2829	78.5212	88.3473	122.4926
16	75.4126	84.1167	94.5487	107.1266	151.8038
17	89.5134	100.3414	113.4465	129.4164	187.3940
18	105.9736	119.3533	135.7023	155.8343	230.5279
19	125.1782	141.6134	161.8842	187.1033	282.7149
20	147.5746	167.6581	192.6548	224.0689	345.7556
21	173.6828	198.1110	228.7860	267.7200	421.7943
22	204.1070	233.6976	271.1770	319.2125	513.3934
23	239.5494	275.2612	320.8751	379.8970	623.5907
24	280.8260	323.7828	379.1002	451.3516	756.0094
25	328.8850	380.4027	447.2726	535.4189	914.9559
26	384.8281	446.4466	527.0454	634.2506	1105.5482
27	449.9357	523.4557	620.3441	750.3578	1333.8462
28	525.6954	613.2219	729.4081	884.6712	1607.1289
29	613.8357	717.8277	854.8446	1048.6096	1933.9036
30	716.3653	839.6942	1005.6871	1234.1611	2324.3539
31	835.6179	981.6351	1179.4648	1453.9762	2790.5312
32	974.3051	1146.9210	1382.2839	1711.4774	3346.7197
33	1135.5776	1339.3531	1618.9205	2012.9854	4009.8439
34	1323.9064	1543.3490	1894.9304	2365.3584	4799.9489
35	1541.1149	1824.0430	2216.7753	2778.7111	5740.7670
36	1794.5752	2127.4018	2591.9701	3261.5172	6840.3858
37	2089.2197	2480.3594	3029.2538	2825.9048	8192.0378
38	2431.7199	2890.9738	3536.7887	4485.5545	9775.0276
39	2829.8277	3368.6090	4132.3910	5256.1745	11655.8289
40	3292.5495	3924.1466	4823.7997	6156.2961	13889.3755
48	11011.1685	13216.4694	16460.2193	21494.0113	55337.2208
50	14878.5469	17879.9807	22323.4269	29289.3445	77809.8292
54	27150.0807	32690.0967	40985.2475	54233.7831	153121.0486
60	66863.5666	80666.7153	101614.8986	135884.5034	418463.3444
65	141631.7453	171052.2043	216085.8459	291001.5449	959710.8181
70	299934.0414	362496.5147	458894.5402	621547.1956	2187993.0460
80	1344617.3776	1626370.2786	2064399.2283	2819709.0127	11204747.8079
90	6026763.0769	7292117.6314	9269282.3964	12729832.2906	56503304.4551
100	27010982.3646	32686892.4830	41580514.3450	57303326.9635	281346935.9595

Answers to Even-Numbered Problems

Chapter 1
2. 2
4. (a) A
    (b) $0.46/mile
6. ceramic insert
8. (a) A for 100 units
        B for 500 units
    (b) 280 units
10. 100″ Ingots
12. 32″ × 32″ × 6″
14. 450
16. 25
18. (a) 1095
    (b) 1200
20. (a) 2 machines/operator
    (b) 39%

Chapter 2
2. $1041.94
4. $3186.31
6. (a) $3863.68
    (b) $3350.37
8. (a) $3860.16
    (b) $2259.41
    (c) $1303.63
10. $315.47
12. $3384.25
14. $680.25 at t = 4
    $26,399.13 at t = 15
16. $5967.59
18. $14,129.33
20. $5945.45
22. $12,075.69
24. $292,647.50

26. $8473.15
28. $423.95
30. $3213.19
32. $11,323.90
34. $13,427.44
36. 19.56%
38. $13,903.98
40. 16.96%
42. (a) $10,817.80
    (b) $573.34
44. $2323.81
46. (a) $2755.51
    (b) $2335.92
48. (a) $18,372.64
    (b) $47,653.12
    (c) $2989.23
50. $75,854.40
52. (a) $47,155.26
    (b) $99,827.69
54. Do not pay cash
56. $3689.15
58. −$634.42
60. $1083.46
62. $3123.75
64. $1710.75
66. $4347.22
68. $2117.40
70. −$985.29
72. −$1212.34
74. $25,037.50
76. X = $908.52, Y = $768.12
78. 12.36%
80. $2111.20

Chapter 3

2. $7374.25

4. then current: $470,400; $526,848; $590,070; $660,878; $740,184; $829,006
   constant worth: $431,560; $443,437; $455,642; $468,183; $481,068; $494,309

6. 9.52%

8. (a) $66,147.65
   (b) $66,147.65

10. (a) constant worth: $200,000, $234,000, $273,780, $320,323, $374,777
    (b) then current: $200,000, $248,040, $307,619, $381,509, $473,148
    (c) $1,181,106
    (d) $2,028,651

12. $243.51

14. $1350.00

16. $2027.58

18. $8081.80

20. 11.34%

22. $2188.47

24. let $X = 1/(1 + i)$ and use the closed form of the 1st n terms of a geometric series

26. As n approaches infinity only the period interest may be withdrawn, leaving the principal intact for the next period. In the limit the amount Pi represents an endowed amount (or a perpetuity).

28. 1st element of a gradient series is zero.

Chapter 4

2. (a) PW = $44,864.36
   (b) AW = $5,269.75
   (c) FW = $301,825.00
   (d) IRR = 22.27%
   (e) ERR= 12.76%
   (f) SIR = 1.64

(g) PBP = 4 yrs
(h) DPBP = 6 yrs

4. (a) PW = −$233,943.85
   (b) AW = −$41,404.36
   (c) FW = −$726,594.07

6. (a) PW = $5,625.43
   (b) AW = $1,486.45
   (c) FW = $13,011.96
   (d) IRR = $0.24
   (e) ERR = $0.20
   (f) SIR = $1.28

8. (a) PW = $68,958.38
   (b) AW = $12,204.54
   (c) FW = $214,174.27
   (d) IRR = 34.22%
   (e) ERR = 22.14%
   (f) SIR = 2.38

10. (a) PW = $999.31
    (b) AW = $209.43
    (c) FW = $3,515.46
    (d) IRR = 17.23%
    (e) ERR = 11.84%
    (f) SIR = 1.09

12. $8,333.33

14. 12.55%
    $95,610.82

16. $3,157,676

18. IRR is undefined

20. multiple roots exist (0%, 100%, and 200%)

22. U = −$4,000
    V = $25,000
    W = $2,000
    X = 2
    Y = A|F

24. CR $35,686

26. CR $10,052

28. B10 = npv(b1,b5:b9) + b4 = $3,182.44
    B11 = pmt(b1,5,−b10) = $882.84
    B12 = fv(b1,5,−b11) = $5,608.55
    B13 = irr(b4:b9) = 18.67%

30. B10 = npv(b1,b5:b9) + b4 =
     $805.45
     B11 = fv(b1,5,−pmt
     (b1,5,−b10)) = $1,620.04
     B12 = irr(b4:b9) = 16.26%
     ERR = 15.85%

Chapter 5

2. {A}, {B}, {C}, {B,C}

8. {Do Nothing}, {C}, {B,C}

10. {Do Nothing}, {B}, {B,D},
     {B,C,D}, {A,D} {A,C,D},

12. {X}, {Z}, {W,Z}, {Y,Z}

14. (a) 0.1664
     (b) 0.1065

16. (a) 0.1255
     (b) 0.0803

18. Yes 0.07895

20. 9.34%

24. (a) {Do Nothing}, {A}, {B}
     (b) PW(A) = $772,370, PW(B)
         = $888,328
     (c) PW(B-A) = $115,957
     (d) AW(A) = $153,896,
         AW(B) = $177,001
     (e) AW(B-A) = $23,105
     (f) FW(A) = $3,124,669,
         FW(B) = $3,593,781
     (g) FW(B-A) = $469,112
     (h) IRR(A) = 43.96%,
         IRR(B-A) = 28.20%
     (i) ERR(A) = 24.92%,
         ERR(B-A) = 20.38%
     (j) SIR(A) = 2.29, SIR(B-A)
         = 1.58
     (k) PBP(A) = 3 yrs, PBP(B)
         = 3 yrs
     (l) DPBP(A) = 3 yrs,
         DPBP(B) = 4 yrs

26. (a) {Do Nothing}, {A}, {B}, {C},
         {B,D}
     (b) PW(Do Nothing) = $0,
         PW(A) = $203,011, PW(C)
         = $298,794, PW(B&D) =
         $383,131

(c) PW(Do Nothing) = $0,
     PW(A-Do Nothing) =
     $203,011, PW(C-A) =
     $95,783, PW(B&D-C) =
     $84,337
(d) AW(Do Nothing) = $0,
     AW(A) = $33,039, AW(C)
     = $48,627, AW(B&D) =
     $62,353
(e) AW(Do Nothing) = $0,
     AW(A-Do Nothing) =
     $33,039, PW(C-A) =
     $15,588, PW(B&D-C) =
     $13,725
(f) FW(Do Nothing) = $0,
     FW(A) = $526,558, FW(C)
     = $774,994, FW(B&D) =
     $993,742
(g) FW(Do Nothing) = $0,
     FW(A-Do Nothing) =
     $526,558, FW(C-A) =
     $248,536, PW(B&D-C) =
     $218,748
(h) IRR(A) = 20.45%, IRR(C-
     A) = 19.55%, IRR(B&D-
     C) = 27.32%
(i) ERR(A) = 14.61%
     ERR(C-A) = 14.39%,
     ERR(B&D-C) = 16.94%
(j) SIR(A) = 1.51, SIR(C-A)
     = 1.48, SIR(B&D-C) =
     1.84
(k) PBP(A) = 5 yrs, PBP(c) =
     5 yrs, PBP(B&D) = 5 yrs
(l) DPBP(A) = 6 yrs,
     DPBP(C) = 6 yrs,
     DPBP(B&D) = 6 yrs

28. (b) PW(1″) = $3.37, PW(2″) =
         $3.411
     (c) PW(1″) = $3.37, PW(2″ −
         1″) = $0.041
     (d) AW(1″) = $1.253, AW(2″)
         = $1.268
     (e) AW(1″) = $1.253, AW(2″
         − 1″) = $0.015
     (f) FW(1″) = $10,285, FW(2″)
         = $10.411

(g) FW(1″) = $10,285, FW(2″ − 1″) = $0.126

(h) IRR(1″) = 161.16%, IRR(2″ − 1″) = 27.22%

(i) ERR(1″) = 67.88%, ERR(2″ − 1″) = 26.13%

(j) SIR(1″) = 4.37, SIR(2″ − 1″) = 1.05

(k) PBP(1″) = 1 yr, PBP(2″) = 1 yr

(l) DPBP(1″) = 1 yr, DPBP(2″) = 1 yr

30. (a) PW(1) = $5,035,28, PW(2) = $1,165.82, PW(3) = $408.30

(b) PBP(1) = 4 yrs, PBP(2) = 1 yr, PBP(3) = 3 yrs

32. (a) PW(1) = $0, PW(3) = $888,328, PW(5) = $772,370

(b) AW(1) = $0, AW(3) = $177,001, AW(5) = $153,896

(c) CW(1) = $0, CW(3) = $1,180,007, CW(5) = $1,025,976

34. (a) CW(wood) = $125,875
(b) CW(steel) = $107,500
(c) CW(concrete) = $105,000

36. (a) 83,333.33 units/yr
(b) 115,384.62 units/yr
(c) 187,500 units/yr

38. 30 yrs

40. (a) 22.9 days/yr
(b) 26.85 days/yr
(c) 29.01 days/yr

42. $10,776

46. insensitive

48. (b) 0.55

50. (b) 0.5

52. 0.682

54. 0.9538

56. 0.8942

58. 0.0715

60. (a) E[PW(A)] = $13,074.71, SD[PW(A)] = $1,374.53, E[PW(B)] = $15,275.86, SD[PW(B)] = $1,965.60 E[AW(A)] = $3,454.82, SD[AW(A)] = $363.20, E[AW(B)] = $4,036.45, SD[AW(B)] = $519.38

(b) Pr[PW(A)>0] = 1,0000, Pr[PW(B)>0] = 1.0000

(c) E[PW(B-A)] = $2,201.15, SD[PW(B-A)] = $2,398.52

(d) Pr[PW(B-A)>0] = 0.8206

62. (a) B
(b) A
(c) B

66. Z

68. (b) PW(U) = −$23,101, PW(V) = −$23,594

(c) AW(U) = −$4,650.30, AW(V) = −$4,749.53

(d) FW(U) = −$57,197.29, FW(V) = −$58,417.77

(e) IRR(V-U) = 24.32%

(f) ERR(V-U) = 17.48%

(g) SIR(V-U) = 1.47

70. (b) PW(tank) = −$233,029.86, PW(pond) = −$188,204.32

(c) AW(tank) = −$37,229.19, AW(pond) = −$30,067.80

(d) FW(tank) = −$3,813,891.93, FW(pond) = −$3,080,253.00

(e) DPBP(tank-pond) > 20-yr planning horizon

(f) IRR(tank-pond) = 3.12%

(g) ERR(tank-pond) = 10.19%

(h) SIR(tank-pond) = 0.43

72. (b) PW(A) = −$26,499.09, PW(B) = −$27,909.72

(c) AW(A) = −$5,334.34, AW(B) = −$5,618.31

(d) FW(A) = −$65,610.77, FW(B) = −$69,103.43

(e) DPBP(A-B) = 5 yrs

(f) IRR(A-B) = 27.98%

(g) ERR(A-B) = 19.16%
(h) SIR(A-B) = 1.64

74. (b) PW(conv) =
    −$1,507,688.07, PW(truck)
    = −$1,384,376.49
    (c) AW(conv) = −$266,833.37,
    AW(truck) = −$245,012.72
    (d) FW(conv) =
    −$4,682,588.16, FW(truck)
    = −$4,299,663.23
    (e) DPBP(conv-truck) > 10-yr
    planning horizon
    (f) IRR(conv-truck) = 7.2%
    (g) ERR(conv-truck) = 9.61%
    (h) SIR(conv-truck) = 0.81

76. (b) PW(1) = −$101,657.92,
    PW(2) = −$86,241.47
    (c) AW(1) −$11,314.50,
    AW(2) = −$9,598.66
    (d) FW(1) = −$1,001,303.30,
    FW(2) = −$849,455.43
    (e) DPBP(2-1) = 6 yrs
    (f) IRR(2-1) = 23.41%
    (g) ERR(2-1) = 12.46%
    (h) SIR(2-1) = 1.53

78. (b) PW(X) = −$83,037.93,
    PW(Y) = −85,171.98
    (c) AW(X) = −$21,941.69,
    AW(Y) = −22,505.58
    (d) FW(X) = −$192,071.78,
    FW(Y) = −$197,007.97
    (e) DPBP(Y-X) = 2 yrs
    (f) IRR(Y-X) = 8.36%
    (g) ERR(Y-X) = 10.94%
    (h) SIR(Y-X) = 0.81

80. (a) AW(1) = −$60,827.85,
    AW(2) = −$55,969.45
    (b) AW(1) = −$58,529.45,
    AW(2) = −$55,969.45

82. (a) AW(build) =
    −$154,699.40, AW(lease)
    = $119,699.40
    (b) SIR(build-lease) = 0.75

84. (a) X = $17,381.40/yr
    (b) X = $17,381.40/yr

86. (a) X = $5,187.78/yr
    (b) X = $5,187.78/yr

88. (a) PW(keep) = −$25,358.92,
    PW(replace) =
    −$22,117.27
    (b) PW(keep) = −$29,358.92,
    PW(replace) =
    −$26,117.27

90. (a) AW(keep) = −$6,442.16,
    AW(replace) = −$7,065.05
    (b) AW(keep) = −$9,807.21,
    AW(replace) =
    −$10,430.09

92. (a) AW(keep) = −$10,634.84,
    AW(trade) = −$8,989.89,
    AW(sub) = −$9,613.48
    (b) AW(keep) = −$13,021.36,
    AW(trade) = −$11,376.42,
    AW(sub) = −$12,000.00

94. PW(gas) = −$12,616.35,
    PW(electric) = −$5,003.48

96. AW(keep) = −$149,529.19,
    AW(trade) = −$140,039.80

98. AW(keep) = −$7,747.47,
    AW(trade) = −$7,143.26

100. AW(do) = $46,488.81,
    AW(contract) = $40,882.09

102. replace old machine

104. 8-yr replacement interval

Chapter 6

2. (a) $10,400.00; 26.00%
   (b) $15,100.00; 37.75%
   (c) $13,600.00; 34.00%

4. (a) ATCF0 = −$30,000.00
   ATCF1 = $12,599.65
   ATCF4 = $9878.05
   ATCF5 = $9100.00 +
   $1300.00
   (b) ATCF0 = −$30,000.00
   ATCF1 = $10,412.50
   ATCF4 = $11,725.00
   ATCF5 = $10,412.50 +
   $1300.00

(c) ATCF0 = −$30,000.00
ATCF1 = $11,725.00
ATCF4 = $11,725.00
ATCF5 = $9100.00 +
$1300.00 (Originally
assumed 4 years and $0
salvage)

6. (a) ATCF0 = −$60,000.00
ATCF1 = $30,399.30
ATCF4 = $14,556.10 +
$9750.00
PW = $14,129.53

(b) ATCF0 = −$60,000.00
ATCF1 = 26,900.70
ATCF4 = $16,500.70 +
$9750.00
PW = $12,857.60

8. (a) D2 = $2700.00
D5 = $2700.00
D8 = $2700.00
B2 = $24,600.00
B5 = $16,500.00
B8 = $8400.00

(b) D2 = $4800.00
D5 = $2457.60
D8 = $1258.29
B2 = $19,200.00
B5 = $9830.40
B8 = $5033.16

(c) D2 = $4418.18
D5 = $2945.45
D8 = $1472.73
B2 = $20,672.73
B5 = $10,363.64
B8 = $4472.73

10. (a) D1 = $5083.33
D4 = $5083.33
D8 = $5083.33

(b) D1 = $13,200.00
D4 = $6758.40
D8 = $2768.24

(c) D1 = $9384.62
D4 = $7038.46
D8 = $3910.26

12. (b) Five-Year Property; 200%
DBSLH-GDS (MACRS-
GDS(5))

(d) Nonresidential Real
Property; SLM-GDS
(MACRS-GDS(39))

(f) Seven-Year Property; 200%
DBSLH-GDS (MACRS-
GDS(7))

(h) Five-Year Property; 200%
DBSLH-GDS (MACRS-
GDS(5))
Could be Seven-Year
Property if used in
production of wood,
tobacco, etc., products.
Could be Ten-Year
Property if used in
production of grain, cereal,
etc., products.

(j) Five-Year Property; 200%
DBSLH-GDS (MACRS-
GDS(5))

14. (a) D1 = $3400.00
D3 = $3264.00
D5 = $1958.40
B1 = $13,600.00
B3 = $4896.00
B5 = $979.20

(b) D1 = $1700.00
D3 = $3400.00
D5 = $3400.00
B1 = $15,300.00
B3 = $8500.00
B5 = $1700.00

16. (a) D1 = $1600.00
D3 = $1536.00
D6 = $460.80

(b) D1 = $800.00
D3 = $1600.00
D6 = $800.00

18. (a) D1 = $18,000.00
D3 = $25,920.00
D6 = $6633.00

(b) D1 = $5292.00
D3 = $10584.00
D6 = $5292.00

20. (a) D1 = $7383.00
D10 = $7692.00

(b) D1 = $7188.00
    D10 = $7500.00

22. (a) D1 = $749.00
        D3 = $3589.60
        D5 = $2243.50
        B1 = $139,251.00
        B3 = $132,071.80
        B5 = $126,238.70
    (b) D1 = $729.40
        D3 = $3500.00
        D5 = $2187.50
        B1 = $139,270.60
        B3 = $132,270.60
        B5 = $126,583.10

24. (a) ATCF1 = $13,182.93
        ATCF3 = $10,914.23
        ATCF5 = $9100.00 +
        $1137.50
    (b) ATCF1 = $10,631.25
        ATCF3 = $12,162.50
        ATCF5 = $10,631.25 +
        $1137.50
    (c) ATCF1 = $11,427.50
        ATCF3 = $11,427.50
        ATCF5 = $11,427.50 +
        $1750.00
    (d) ATCF1 = $12,979.17
        ATCF3 = $11.427.50
        ATCF5 = $9875.83 +
        $1750.00

26. (a) AW(A) = $20,792.25/yr
        AW(B) = $27,141.48
    (b) AW(A) = $22,478.03/yr
        AW(B) = $24,161.17
    (c) AW(A) = $20,792.25/yr
        AW(B) = $24,161.17

28. (a) CR1 = $7145.00
        CR5 = $4465.00
        R1 = $7500.00
        R5 = $2343.00
        CR + R1 = $14,645.00
        CR + R5 = $3384.78
        AECR + R = $9051.63
    (b) CR1 = $2275.00
        CR5 = $4545.00
        R1 = $7500.00
        R5 = $5113.50

        CR + R1 = $9775.00
        CR + R5 = $6808.00
        AECR + R = 9051.63
    (c) CR1 = $3750.00
        CR5 = $3750.00
        R1 = $7500.00
        R5 = $5250.00
        CR + R1 = $11,250.00
        CR + R5 = $9000.00
        AECR + R = 9051.63
    (d) CR1 = $6923.08
        CR5 = $4615.38
        R1 = $7500.00
        R5 = $3865.38
        CR + R1 = $14,423.08
        CR + R5 = $8480.77
        AECR + R = 9051.63

30. D1 = $6000.00
    D3 = $9000.00

32. Using MACRS-GDS/Using
    MACRS-ADS
    (a) $54,000.00 $54,000.00
    (b) $34,953.03 $23,849.37
    (c) $68,250.00 $68,250.00
    (d) $26,118.62 $30,004.90
    (e) Depreciation and taxes
        remain same regardless of
        depreciation method.
    (f) Accelerated depreciation
        increases PW of
        depreciation and reduces
        PW of taxes.

34. (a) BT&LCF1 = $30,000.00
        BT&LCF3 = $30,000.00
        BT&LCF7 = $30,000.00
    (b) LPP1 = $0.00
        LPP3 = $40,000.00
        LPP7 = $0.00
    (c) LIP1 = $4,400.00
        LIP3 = $4,400.00
        LIP7 = $0.00
    (d) D1 = $14,290.00
        D3 = $17,490.00
        D7 = $8,930.00
    (e) T11 = $11,310.00
        T13 = $8,110.00
        T17 = $21,070.00

(f) ATCF1 = $21,641.50
     ATCF3 = -$17,238.50
     ATCF7 = $22,625.50

36. ATCF0 = -$360,000.00
     ATCF3 = $71,080.00
     ATCF5 = $141,192.00

38. EUAC = $11,313.89/YR

40. (a) $145,860.00
     (b) $136,040.00

42. (a) $30,440.00
     (b) $26,690.00

44. (a) $15,000.00 treated as depreciation recapture and taxed at regular rate.
     (b) $15,000.00 is used to partially recover the asset. It misses by $15,000.00 − $44,440.00 = -$29,440.00 which is then negatively taxed at the regular rate.

46. (a) $53,030.00 depreciation recapture taxed as ordinary income, plus $20,000.00 Section 1231 gain taxed as a capital gain.
     (b) $33,030.00 depreciation recapture taxed as ordinary income.
     (c) $56,970.00 Section 1231 loss (negatively) taxed as ordinary income.

48. (a) $44,241.00 Section 1231 gain taxed as a capital gain.
     (b) $45,759.00 Section 1231 loss (negatively) taxed as ordinary income.

50. (a) 62,056.25 Section 1231 gain taxed as a capital gain.
     (b) $22,056.25 Section 1231 gain taxed as a capital gain.
     (c) $127,945.75 Section 1231 loss (negatively) taxed as ordinary income.

52. ATCF1 = $93,625.00
     ATCF3 = $67,960.00

ATCF6 = $56,788.00 plus $13,000 from salvage.

54. $13,726.31

56. (a) ATCF1 = $19,100.00
         ATCF3 = $22,860.00
         ATCF6 = $26,358.00
         PWATCF = $39,099.79
     (b) ATCF1 = $18,190.48
         ATCF3 = $19,747.33
         ATCF6 = $19,668.75
         PWATCF = $39,099.79

58. (a) PWATCF = -$31,502.81; do not invest
     (b) PWATCF = -$31,502.81; do not invest

60. (a) $200,000.00
     (b) $138,750.00

Chapter 7

2.   (c) Time saved by users; reduced traffic through busy part of city; and easier access to airport.

4.   (a) $(B/C)_1 = 1.17$
         $(B/C)_2 = 1.07$
         $(B/C)_3 = 1.04$
     (b) $\Delta(B/C)_{2-1} = 0.893$.
         $\Delta(B/C)_{3-1} = 0.919$.
         Choose Alt. 1
     (c) $(B-C)_1 = \$34,500$
         $(B-C)_2 = \$22,500$
         $(B-C)_3 = \$18,000$

6. $(B/C) = 0.642$

8.   (a) Use Fuel Oil 3.
         $(B-C)_{Fuel\ Oil\ 3} = -\$37,370,404/year$
     (b) Use Fuel Oil 3.
         $(B-C)_{Fuel\ Oil\ 3} = -\$35,885,341/year$

10. Route Y. $(B-C)_Y = -\$398,325$

12. $1.44/person

14. Build. (B-C) = $173,750/yr

16. Select projects D and E. Opportunity cost is 17%.

18. Opinion question.

20. No. Overcounting of disbenefits.

22. Build. (B/C) = 1.217.
    (B-C) = $32,500

24. (a) (B/C) = 23
    (b) (B-C) = $660,000
    (c) $(B/C)_{Incorrect}$ = 1.177
    (d) (B-C) = $660,000
    (e) (B-C) is preferred.

26. Choose A. Eliminate B, C, and D.

28. (a) Unit A: RR_5 = $28,751.99
    Unit B: RR_5 = $27,563.99
    (b) Unit A: PW(11%)
    = $225,151.23
    Unit B: PW(11%)
    = $219,393.89

Chapter 8

2. (a) 1 to 50 units use M1
    50 to 125 units use M3
    125 to 200 units use M2
    (b) M3, $875
    (c) $70, $510

4. operate since TR > TC

6. (a) −$25,000
    (b) 220,000 pallets/yr
    (c) $125,000
    (d) 132,000 pallets/yr

8. (a) 300 hrs/yr
    (b) $225
    (c) $5.13/hr

10. (a) 22,360 bbls/wk, $24.67/bbl
    (b) $497,600; 74,000 bbls
    (c) 3459 to 144,540 bbls/wk
    (d) 3459 to 50,000 bbls/wk

12. (a) 224 units/order
    (b) $418.33
    (c) 9 orders
    (d) 158 units/order

14. (a) 3.0 square inches
    (b) $200
    (c) $200.19, $418.48

16. (b) book value = $31,000
    (c) reserve = $18,900

18. net loss = $5125

20. (a) 11.6%
    (b) 1.70 to 1
    (c) 0.97 to 1
    (d) 11.4 times
    (e) 10.0 times
    (f) 87%
    (g) 15%
    (h) strong
    (i) weak
    (j) neutral

22. (a) A = 28.00%, B = 55.56%,
    C = 60.00%, D = 42.86%
    (b) A = 24.14%, B = 33.33%,
    C = 40.00%, D = 33.80%
    (c) A = $2.10/hr, B = $2.50/hr,
    C = $3.60/hr, D = $1.71/hr

Index